PROCEEDINGS OF SYMPOSIA
IN PURE MATHEMATICS
Volume 49, Part 2

Theta Functions
Bowdoin 1987

Leon Ehrenpreis and
Robert C. Gunning, Editors

AMERICAN MATHEMATICAL SOCIETY
PROVIDENCE, RHODE ISLAND

PROCEEDINGS OF THE SUMMER RESEARCH INSTITUTE
ON THETA FUNCTIONS
HELD AT BOWDOIN COLLEGE
BRUNSWICK, MAINE
JULY 6–24, 1987

with support from the National Science Foundation, Grant DMS–8611435

1980 *Mathematics Subject Classification* (1985 *Revision*). Primary 00A11.

Library of Congress Cataloging-in-Publication Data

Theta functions, Bowdoin 1987/Leon Ehrenpreis and Robert C. Gunning, editors.
  p. cm. – (Proceedings of symposia in pure mathematics, ISSN 0082-0717; v. 49)
Proceedings of the thirty fifth Summer Research Institute on Theta Functions held at Bowdoin College, Brunswick, Me., July 6–24, 1987; sponsored by the American Mathematical Society.
  Includes bibliographies.
  1. Functions, Theta–Congresses. I. Ehrenpreis, Leon. II. Gunning, R. C. (Robert Clifford), 1931-. III. Summer Research Institute on Theta Functions (35th: 1987: Brunswick, Me.) IV. American Mathematical Society. V. Series.

QA345.T47   1989   515$'$.984–dc20   89-6723
ISBN 0-8218-1483-4 (part 1)   CIP
ISBN 0-8218-1484-2 (part 2)
ISBN 0-8218-1485-0 (set) (alk. paper)

CB-Stk
SIMON

Contents

iii

Kac-Moody Algebras

Lattice Models

Physics

Jacobian Varieties

Prym Varieties

Algebraic Geometry

PART 2

Modular Forms

Number Theory

Combinatorics

Preface

Theta functions apparently first appeared in the forms $\sum_{n=0}^{\infty} m^{n^2}$, $\sum_{n=0}^{\infty} m^{1/2n(n+1)}$, $\sum_{n=0}^{\infty} m^{1/2n(n+3)}$ in the work of Jakob Bernoulli. In his work on partition theory, Euler introduced a second variable ζ and studied functions of the form $\prod_{n=1}^{\infty}(1 - q^n \zeta)^{-1}$. For Euler, the primary objects were partition functions such as $\prod(1 - q^n)$, but the function $\prod(1 - q^n \zeta)^{-1}$ was considered as a function of ζ with q occurring as a parameter; after deriving identities for the function of ζ he then set $\zeta = 1$.

Jacobi made two important notational changes that turned out to be crucial for the modern development. He replaced q by $e^{\pi i \tau}$ and ζ by e^{2iz}; thus was born the theta function in its present form

$$\theta(\tau, z) = \sum e^{\pi i n^2 \tau + 2inz}.$$

The change from q to τ allowed him to formulate the "imaginary transformation" $\tau \to -1/\tau$, which together with the obvious transformation $\tau \to \tau + 2$ leads to the modular group and eventually to the modern theory of modular forms and their ramifications. (The formulation of the modular group in the variable q is complicated; see the paper by Ehrenpreis in this volume.)

In addition, Jacobi studied $\theta(\tau, z)$ as a function of z in its own right. The quasi double periodicity under $z \to z + \pi$ and $z \to z + \pi \tau$ enabled him to relate theta functions as functions of z to elliptic function theory. For Jacobi as for Euler the primary working variable was z. Of course, this theory has had far reaching generalizations to higher genera Riemann surfaces, abelian varieties, etc.

Surprisingly, theta functions made their appearance in another case of nineteenth century mathematics, namely mechanics. It was discovered by Carl Neumann and Jacobi that certain mechanical (Hamiltonian) systems could be explicitly integrated by means of theta functions. These ideas could have formed the foundation of some of the modern ideas on KdV, KP, and integrable systems in general, but the modern viewpoint seems to have been discovered without knowledge of the eighteenth century results.

When the organizing committee met to discuss the possibility of a conference on theta functions, we saw how perfectly the notation $\theta(\tau, z)$ fit into a three week conference: one week for τ, one week for z, and one week for the comma. (This conforms to the above described three aspects of theta functions that appeared in the nineteenth century.) The conference was thereby organized accordingly. The first week was devoted to the comma, that is, to the interplay of τ and z. The sections on infinite analysis, integrable systems, Kac-Moody algebras, lattice models, and physics are, roughly speaking, devoted to this interplay; the sections on Jacobi varieties, Prym varieties, and algebraic geometry emphasize the z variable. These sections form Part 1 of Volume 49. The sections on modular forms, number theory, and combinatorics emphasize the τ variable. They comprise Part 2 of Volume 49.

It was our hope in organizing the conference that the presentation of a cross section of modern work on theta functions would enable mathematicians to see where we stand now and in what directions we should go in the future.

Leon Ehrenpreis

Robert C. Gunning

Modular Forms

Proceedings of Symposia in Pure Mathematics
Volume **49** (1989), Part 2

Siegel Modular Forms and Theta Series

SIEGFRIED BÖCHERER

Herrn Prof. Dr. H. Klingen zum sechzigsten Geburtstag gewidmet

In the spirit of this conference the most important and most natural modular forms are of course those that arise from theta functions. So it seems natural to ask the following question:

"Do all modular forms arise from Θ?"

Of course this question is far too vague to have a reasonable answer (what do we mean by "all," "arise," and "Θ"?). In fact, there are several different possibilities to make this question more precise. For one of these possibilities ("polynomials in theta constants") I refer to the work of Igusa [**Ig 1**] and Salvati-Manni [**S-M 1, S-M 2**]. A second possibility will be the main subject of the present paper; it is concerned with theta series attached to positive definite quadratic forms. This version of the problem is more of a number-theoretic flavour than Igusa's (which is more geometric). The results I want to describe here were obtained by Weissauer [**We 1, We 2**] and myself by different methods. It turns out that the problem is deeply related to the multiplicative arithmetic of modular forms:

> *Those Siegel modular forms, which are linear combinations of theta series, can be characterized by properties of automorphic L-functions associated to them.*

The present paper can be considered as a modified version of [**Bö 1, Bö 2**], but it is at the same time more complete and conceptually simpler than my previous papers. In particular, the concept of a Fourier-Jacobi expansion (which was an important tool in [**Bö 1, Bö 2**]) will be avoided and we do not need to restrict ourselves to "large" weights.

I. Siegel modular forms (for details see [**Fr 2**]). Siegel's upper half-space $\mathbf{H}_n$ of degree n is the set of all n-rowed complex symmetric matrices $Z = X + iY$ with positive definite imaginary part $\mathrm{Im}(Z) = Y$. The symplectic group

---

1980 *Mathematics Subject Classification* (1985 *Revision*). Primary 11F46.

$\mathrm{Sp}(n, \mathbf{R})$ acts on $\mathbf{H}_n$ by means of

$$(M, Z) \mapsto M\langle Z \rangle := (AZ + B)(CZ + D)^{-1},$$

where, as usual, $M = \begin{pmatrix} A & B \\ C & D \end{pmatrix}$ is decomposed into n-rowed matrices A, B, C, D. For a function $f: \mathbf{H}_n \to \mathbf{C}$, an integer k, and $M \in \mathrm{Sp}(n, \mathbf{R})$ we define $f|_k M$ by

$$(f|_k M)(Z) = f(M\langle Z \rangle) j(M, Z)^{-k} \quad \text{where } j(M, Z) = \det(CZ + D)$$

is the standard automorphy factor.

The space M_n^k of Siegel modular forms of degree n and weight k consists of all holomorphic functions $f: \mathbf{H}_n \to \mathbf{C}$ satisfying

$$f|_k M = f \quad \text{for all } M \in \Gamma_n := \mathrm{Sp}(n, \mathbf{Z})$$

and an additional condition "at infinity" for $n = 1$. The subspace of cuspforms (= kernel of Siegel's ϕ-operator) will be denoted by S_n^k. By convention, we have $M_0^k = S_0^k = \mathbf{C}$.

II. Theta series. For $Z \in \mathbf{H}_n$ and an m-rowed real symmetric matrix S, which we assume to be positive definite, we define a theta series by

$$\vartheta_S^n(Z) = \sum_{G \in \mathbf{Z}^{(m,n)}} e^{\pi i \, \mathrm{trace}(G' S G)}.$$

Here G' means the transpose of G. This series has nice convergence properties and depends only on the $\mathrm{Gl}(m, \mathbf{Z})$-equivalence class of S. If S has rational entries, then ϑ_S^n will be a Siegel modular form of weight $m/2$—in general not for Γ_n, but for some subgroup of finite index. If S is even and unimodular, then ϑ_S^n represents an element of $M_n^{m/2}$. It is well known that such S exist iff m is divisible by 8. We denote by $\mathfrak{S}(m)$ the genus of all such quadratic forms and by $h(m)$ the class number of $\mathfrak{S}(m)$; in general, we don't have an exact formula for that class number, but—thanks to the work of Minkowski and Siegel—we do have a formula for the mass $\mathfrak{M}(m)$ of $\mathfrak{S}(m)$; recall that the mass is defined by

$$\mathfrak{M}(m) = \sum_{i=1}^{h(m)} \frac{1}{\varepsilon(S_i)},$$

where $S_1, \ldots, S_{h(m)}$ is a complete set of representatives of the classes of quadratic forms in $\mathfrak{S}(m)$ and $\varepsilon(S)$ is the number of (integral) units of S.

III. The problem. We can now give a precise meaning to the vague question raised at the beginning of this paper. More precisely, we want to ask three questions, which are closely related to each other. We fix n, m with m divisible by 8.

(III.a) Which modular forms $f \in M_n^{m/2}$ are finite linear combinations of the theta series ϑ_S^n with $S \in \mathfrak{S}(m)$?

(III.b) Under which conditions on n and m do we have $M_n^{m/2} = \Theta(m, n)$, where $\Theta(m, n) = \mathbf{C} - \text{span of the } \vartheta_S^n, S \in \mathfrak{S}(m)$?

(III.c) In general, the $\vartheta_{S_i}^n$, $1 \le i \le h(m)$, are not linearly independent; nevertheless we can ask: Is there a "canonical" way to express any $f \in \Theta(m, n)$ in terms of the $\vartheta_{S_i}^n$?

Problem (III.b) is known as the basis problem; for $n = 1$ and arbitrary level there are famous and far-reaching results on this problem due to Eichler [Ei], Serre and Stark [Se-St], and Waldspurger [Wa]. In the context of Siegel modular forms the basis problem was solved by Freitag [Fr 1] for "small" weights: The theory of "singular" modular forms says that

$$M_n^{m/2} = \Theta(m, n) \quad \text{for } m < n.$$

For $m = n$ there is a similar result due to Andrianov [An].

IV. Siegel's theorem. The starting point for the theory of Siegel modular forms was an identity that solves our problems (III.a) and (III.c) for a special type of modular forms, namely for Eisenstein series. The Eisenstein series in question are defined, for $Z \in \mathbf{H}_n$ and even k, by

$$E_n^k(Z, s) = \sum_{M \in \Gamma_{n,\infty} \backslash \Gamma_n} j(M, Z)^{-k} \det(\text{Im}(M\langle Z \rangle))^s,$$

where s is a complex parameter and

$$\Gamma_{n,\infty} = \left\{ \begin{pmatrix} A & B \\ C & D \end{pmatrix} \in \Gamma_n \,\middle|\, C = 0 \right\}.$$

It is well known that this series converges for $k + \text{Re}(2s) > n + 1$ and—as a function of s—it has a meromorphic continuation to the whole complex plane. The identity mentioned above is known as Siegel's main theorem [Si]; in a more general context it is called the Siegel-Weil formula. In the special case we are interested in, it may be formulated as follows:

THEOREM 0. *Suppose m is divisible by 8; then we have for all n*

(IV.1) $$\frac{\kappa(m, n)}{\mathfrak{M}(m)} \sum_{i=1}^{h(m)} \frac{1}{\varepsilon(S_i)} \vartheta_{S_i}^n(Z) = E_n^{m/2}(Z, s)|_{s=0},$$

where

$$\kappa(m, n) = \begin{cases} 1 & \text{if } m > n + 1, \\ 2 & \text{if } m \le n + 1. \end{cases}$$

We briefly recall the history of this theorem. For arbitrary genera of positive definite quadratic forms (but with the restriction $m/2 > n + 1$) it was stated by Siegel in his famous paper [Si]; then it was made more explicit by Witt [Wi] for the case of even unimodular quadratic forms. The restriction $m > 2n + 2$ was recently removed by Weissauer [We 1] and more generally by Kudla and Rallis [K-R].

Note that this theorem does not only express the Eisenstein series in terms of theta series, but it also does it in a particularly nice explicit way. We shall show below that the Eisenstein series shares this property with all eigenforms of the Hecke algebra in $\Theta(m, n)$! Siegel's theorem (IV.1) will turn out to be just a special case of a more general result, which is true for all eigenforms of the Hecke algebra. However, to prove this for degree n, we have to use (IV.1) for degree $> n$ as a main tool.

V. The basic idea: **The operator** $\Lambda_{n,r}^k$. How do we get a link between Siegel's theorem and arbitrary modular forms? The basic idea is very simple. For arbitrary r, n and k—with k divisible by 4—we define an operator

$$\Lambda_{n,r}^k : \begin{cases} S_r^k \rightarrow M_n^k \cap \Theta(2k, n), \\ f \mapsto \dfrac{\kappa(2k, n+r)}{\mathfrak{M}(2k)} \sum_{i=1}^{h(2k)} \dfrac{1}{\varepsilon(S_i)} \langle f, \vartheta_{S_i}^r \rangle \vartheta_{S_i}^n, \end{cases}$$

where $\langle\, ,\, \rangle$ is the well-known Petersson scalar product. The following remark is the key for all subsequent considerations:

REMARK. Assume that $f \in S_r^k$ is an eigenfunction of $\Lambda_{r,r}^k$ with eigenvalue $\lambda(f)$:

$$\lambda(f)f = \frac{\kappa(2k, 2r)}{\mathfrak{M}(2k)} \sum_{i=1}^{h(2k)} \frac{1}{\varepsilon(S_i)} \langle f, \vartheta_{S_i}^r \rangle \vartheta_{S_i}^r.$$

Then we have

$$f \in \Theta(2k, r) \quad \text{iff} \quad \lambda(f) \neq 0.$$

(Note that $\lambda(f) = 0$ implies that $f \perp \vartheta_{S_i}^r$ for all i.)

By this remark our problems (III.a) and (III.c) are (almost) trivially solved for eigenfunctions of the operator $\Lambda_{r,r}^k$. Two questions arise here:

(V.1) Is $\Lambda_{r,r}^k$ diagonalizable?

(V.2) What is the nature of the eigenvalues $\lambda(f)$?

To answer these questions, we introduce in a first step another operator $\Lambda_{n,r}^k(s)$, $s \in \mathbf{C}$, as follows: We start from $\mathcal{Z} = \left(\begin{smallmatrix} Z & 0 \\ 0 & W \end{smallmatrix}\right) \in \mathbf{H}_{n+r}$ with $Z \in \mathbf{H}_n$, $W \in \mathbf{H}_r$. For general s, the Eisenstein series $E_{n+r}^k\left(\left(\begin{smallmatrix} Z & 0 \\ 0 & W \end{smallmatrix}\right), s\right)$ is not a modular form in Z and W, but—at least outside the set of poles—it has the transformation properties of a modular form of weight k and degree n and r respectively and it satisfies some growth condition; so it makes sense to consider the scalar product of a cusp form $f \in S_r^k$ with $E_{n+r}^k\left(\left(\begin{smallmatrix} Z & 0 \\ 0 & * \end{smallmatrix}\right), s\right)$ for $Z \in \mathbf{H}_n$.

Now we define, for $f \in S_r^k$ with even k and any complex s (outside the set of poles of the Eisenstein series) a function $\Lambda_{n,r}^k(s)(f)$ on $\mathbf{H}_n$ by

$$\Lambda_{n,r}^k(s)(f)(Z) := \left\langle f, E_{n+r}^k\left(\left(\begin{smallmatrix} -\overline{Z} & 0 \\ 0 & * \end{smallmatrix}\right), \overline{s}\right) \right\rangle$$

$$= \int_{\mathcal{F}} f(W) \overline{E_{n+r}^k\left(\left(\begin{smallmatrix} -\overline{Z} & 0 \\ 0 & W \end{smallmatrix}\right), \overline{s}\right)} \det(U)^{k-r-1} \, dU \, dV,$$

where $\mathcal{F}$ is any fundamental domain for the action of Γ_r on $\mathbf{H}_r$ and $W = U + iV$. In general, $\Lambda_{n,r}^k(s)(f)$ is not a Siegel modular form, but an automorphic form in a more general sense. Combining Siegel's theorem with the elementary facts

(V.3)
$$\vartheta_{S_i}^{n+r}\left(\begin{pmatrix} Z & 0 \\ 0 & W \end{pmatrix}\right) = \vartheta_{S_i}^n(Z) \cdot \vartheta_{S_i}^r(W),$$

(V.4)
$$\overline{\vartheta_{S_i}^n(-\overline{Z})} = \vartheta_{S_i}^n(Z),$$

we obtain a description of the operator $\Lambda_{n,r}^k$ for k divisible by 4, which does no longer involve theta series, namely

(V.5)
$$\Lambda_{n,r}^k = \Lambda_{n,r}^k(s)|_{s=0}.$$

VI. Pullbacks of Eisenstein series: first version. We shall now give a new description of $\Lambda_{r,r}^k(s)(f)$, which does not involve Eisenstein series, but only data attached directly to f. It will turn out that the operator $\Lambda_{r,r}^k(s)$ can be characterized by Hecke operators of the following type: For an r-rowed integral matrix

$$D = \begin{pmatrix} d_1 & & 0 \\ & \ddots & \\ 0 & & d_r \end{pmatrix} \quad \text{with } d_i > 0, d_i | d_{i+1}$$

—which we shall call an elementary divisor matrix—we consider the double coset $\Gamma_r\begin{pmatrix} D & 0 \\ 0 & D^{-1} \end{pmatrix}\Gamma_r$; it is well known that such a double coset can be decomposed into finitely many left cosets:

$$\Gamma_r\begin{pmatrix} D & 0 \\ 0 & D^{-1} \end{pmatrix}\Gamma_r = \bigcup_i \Gamma_r g_i, \qquad g_i \in \mathrm{Sp}(n, \mathbf{Q}).$$

We define a Hecke operator $T(D)$ on M_r^k by the formula

$$f|T(D) = \sum_i f|_k g_i.$$

The basic theorem on $\Lambda_{r,r}^k(s)$, which occurs in different versions in the work of Garrett [**Ga 1**, **Ga 2**], Piatetski-Shapiro and Rallis [**P-S-R**], and the author [**Bö 3**], is

THEOREM 1. *For k even, $f \in S_r^k$, and $s \in \mathbf{C}$, $\mathrm{Re}(s) \gg 0$ we have*

(VI.1)
$$\Lambda_{r,r}^k(s)(f) = \gamma_r^k(s) \sum_D f|T(D)\det(D)^{-k-2s},$$

where D runs over all r-rowed elementary divisor matrices and

$$\gamma_r^k(s) = 2^{(r^2+3r)/2 - 2rs - rk + 1}(-1)^{rk/2}\pi^{r(r+1)/2}\frac{\Gamma_r(k + s - (r + 1)/2)}{\Gamma_r(k + s)}.$$

Here $\Gamma_r(s)$ is a product of ordinary gamma-factors, defined by

$$\Gamma_r(s) = \Gamma(s)\Gamma\left(s - \frac{1}{2}\right)\cdots\Gamma\left(s - \frac{r-1}{2}\right).^1$$

---

[1] The distinction between $\Gamma_r = \mathrm{Sp}(r, \mathbf{Z})$ and $\Gamma_r(s)$ will always be clear from the context.

The right-hand side of (VI.1) converges for $\mathrm{Re}(s) \gg 0$, but the left-hand side tells us that meromorphic continuation to the whole complex plane is possible.

It is well known that M_r^k has a basis consisting of eigenfunctions of all Hecke operators. For the questions (V.1) and (V.2) on $\Lambda_{r,r}^k$ this implies

COROLLARY. *The operator $\Lambda_{r,r}^k(s)$ is diagonalizable; if $f \in S_r^k$ is an eigenfunction of all the Hecke operators $T(D)$ with eigenvalues $\lambda(f, D)$, then the corresponding eigenvalue for $\Lambda_{r,r}^k(s)$ is—up to the factor $\gamma_r^k(s)$—equal to $\mathscr{L}(f, 2s + k)$, where (for $\mathrm{Re}(s) \gg 0$) $\mathscr{L}(f, s)$ is defined by the Dirichlet series*

$$\mathscr{L}(f, s) = \sum_D \lambda(f, D) \det(D)^{-s}.$$

VII. Pullbacks of Eisenstein series: second version. To get a statement for $\Lambda_{n,r}^k(s)$ similar to Theorem 1, we need the concept of Eisenstein series of Klingen type [K1]. For $f \in S_r^k$ we define (with $\mathrm{Re}(s) \gg 0$) a function on $\mathbf{H}_n$, $n \geq r$, by

$$E_{n,r}^k(f, s)(Z) = \sum_{M \in C_{n,r} \backslash \Gamma_n} f(M\langle Z \rangle^*) j(M, Z)^{-k} \left(\frac{\det(\mathrm{Im}(M\langle Z \rangle))}{\det(\mathrm{Im}(M\langle Z \rangle^*))} \right)^s,$$

where Z^* means the r-rowed matrix in the upper left corner of $Z \in \mathbf{H}_n$; the subgroup $C_{n,r}$ is defined by

$$C_{n,r} = \left\{ M = \begin{pmatrix} * & * \\ 0^{(n-r,n+r)} & * \end{pmatrix} \Big| M \in \Gamma_n \right\}.$$

This series converges for $k + 2\,\mathrm{Re}(s) > n + r + 1$ and has a meromorphic continuation to the whole complex plane. For $k > n + r + 1$ and $r \leq \mu \leq n$ the series converges at $s = 0$ and defines an element of M_n^k, which satisfies

(VII.1) $\phi^{n-\mu}(E_{n,r}^k(f, 0)) = E_{\mu,r}^k(f, 0).$

Note that for $r = n$ and $r = 0$ we don't get anything new:

(VII.2) $E_{r,r}^k(f, s) = f,$

(VII.3) $E_{n,0}^k(1, s) = E_n^k(Z, s).$

Using these Eisenstein series we can remove the restriction $r = n$ in Theorem 1. For simplicity, we only consider eigenfunctions of Hecke operators.

THEOREM 2. *Let $f \in S_r^k$ (with k even) be an eigenfunction of all Hecke operators $T(D)$ with eigenvalues $\lambda(f, D)$. Then we have*

(VII.4) $\Lambda_{n,r}^k(s)(f) = \begin{cases} 0 & \text{if } n < r, \\ \gamma_r^k(s)\mathscr{L}(f, 2s + k)E_{n,r}^k(f, s) & \text{if } n \geq r. \end{cases}$

VIII. Automorphic L-functions. Let $f \in S_r^k$, k even, be a Hecke eigenform (i.e., an eigenfunction of all Hecke operators $T(D)$); the analytic properties of the Eisenstein series $E_{2r}^k(Z, s)$ imply that the Dirichlet series $\mathscr{L}(f, s)$ has a meromorphic continuation to the whole complex plane and satisfies a functional equation relating s and $1 + 2r - s$. On the other hand—following Langlands [La]—we attach to f the (standard-) L-function $L(f, s)$, defined by the Euler product

$$L(f, s) = \prod_p \left\{ \frac{1}{(1 - p^{-s})} \prod_{i=1}^r \frac{1}{(1 - \alpha_{i,p} p^{-s})(1 - \alpha_{i,p}^{-1} p^{-s})} \right\},$$

where, for a prime p, the $\alpha_{i,p}$ are the Satake parameters of the Hecke eigenform f.

It is an important point that the Dirichlet series $\mathscr{L}(f, s)$, defined in terms of eigenvalues, and the Langlands L-function $L(f, s)$, defined by an Euler product involving the Satake parameters, are essentially the same. In [Bö 3, §5] and in [Bö 4] one can find two different proofs of the following

THEOREM 3. *Let $f \in M_r^k$ be a Hecke eigenform. Then we have (for $\text{Re}(s) \gg 0$)*

$$\zeta(s) \prod_{i=1}^r \zeta(2s - 2i) \mathscr{L}(f, s) = L(f, s - r).$$

In particular one can deduce from this a functional equation for $L(f, s)$ relating s and $1 - s$ (for details see [Bö 3]); this functional equation was also obtained by Piatetski-Shapiro and Rallis in a more general context [P-S-R]. The precise statement is

(VIII.1) $$\psi_f(1 - s) = \psi_f(s),$$

where

$$\psi_f(s) = \begin{cases} \pi^{-(2n+1)s/2} \Gamma\left(\frac{s}{2}\right) \Gamma_n\left(\frac{s+k-1}{2}\right) \Gamma_n\left(\frac{s+k}{2}\right) L(f, s) & (n \text{ even}), \\ \pi^{-(2n+1)s/2} \Gamma\left(\frac{s+1}{2}\right) \Gamma_n\left(\frac{s+k-1}{2}\right) \Gamma_n\left(\frac{s+k}{2}\right) L(f, s) & (n \text{ odd}). \end{cases}$$

In the sequel, we shall work with $L(f, s)$ instead of $\mathscr{L}(f, s)$, since $L(f, s)$ is the more natural object from the viewpoint of the theory of automorphic forms. We abbreviate the product of Riemann zeta-functions, which relates both series, by

$$\omega_r(s) := \zeta(s) \prod_{i=1}^r \zeta(2s - 2i).$$

IX. The main result. From our investigations of the operator

$$\Lambda_{n,r}^k = \Lambda_{n,r}^k(s)|_{s=0}$$

we obtain

THEOREM 4. *Let* $f \in S_r^k$ *be a Hecke eigenform and assume that* 4 *divides* k. *Then we have for all* $n \geq r$

(IX.1)
$$
\left(\gamma_r^k(s) \frac{1}{\omega_r(2s+k)} L(f, 2s+k-r) E_{n,r}^k(f,s) \right) \Big|_{s=0}
$$
$$
= \frac{\kappa(2k, n+r)}{\mathfrak{M}(2r)} \sum_{i=1}^{h(2k)} \frac{1}{\varepsilon(S_i)} \langle f, \vartheta_{S_i}^r \rangle \vartheta_{S_i}^n .
$$

This theorem contains the answer to all the problems formulated in §III (at least for cusp forms).

In view of the remark in §V and $E_{r,r}^k(f,s) = f$ we get in particular (solving problem (III.a)

THEOREM 4_1. *For* f *as above we have*

$$
f \in \Theta(2k, r) \quad \text{iff} \quad \left(\gamma_r^k(s) \frac{1}{\omega_r(2s+k)} L(f, 2s+k-r) \right) \Big|_{s=0} \neq 0.
$$

By the functional equation (VIII.1) this condition is easily seen to be equivalent to

$$
L(f, k-r) \neq 0 \qquad (k \geq r+1),
$$
$$
L(f, s) \text{ has a simple pole in } s = 1+r-k \qquad (k < r+1).
$$

Using the Euler product expansion of $L(f, s)$, which converges for $\operatorname{Re}(s) > k$, and the fact (see [Kl]) that for $k > 2n$ we have

(IX.2)
$$
M_n^k = \{ E_{n,r}^k(f, 0) \mid 0 \leq r \leq n, \ f \in S_r^k \},
$$

we can also give a solution of problem (III.b):

THEOREM 4_2. *For* $k > 2n$, k *divisible by* 4 *we have* $M_n^k = \Theta(2k, n)$.

Problem (III.c) is solved by (IX.1) in the sense that whenever a Hecke eigenform can be expressed by theta series, it can also be done by means of the equation (IX.1). Equation (IX.1) is canonical in the sense that the coefficients of $\vartheta_{S_i}^n$ on the right-hand side of (IX.1) do not depend on n but only on r (up to a possible factor 2 coming from $\kappa(2k, *)$); moreover, due to the fact that for $n \geq 2k$ the $\vartheta_{S_i}^n$ are linearly independent, the expression on the right-hand side of (IX.1) is uniquely determined by that property.

REMARKS. (1) Theorem 4 also makes sense for $r = 0$, namely, it is precisely Siegel's theorem in the version of Theorem 0. Therefore equation (IX.1) can be considered to be a generalization of Siegel's theorem.

(2) Notice that equation (IX.1) in particular implies that, if the condition

$$
\left(\gamma_r^k(s) \frac{1}{\omega_r(2s+k)} L(f, 2s+k-r) \right) \Big|_{s=0} \neq 0
$$

is satisfied, we have

(IX.3)
$$
E_{n,r}^k(f, s)|_{s=0} \text{ is a Siegel modular form}
$$

with the property

(IX.4) $\qquad \phi^\nu(E_{n,r}^k(f,s)|_{s=0}) = \dfrac{\kappa(2k,n+r)}{\kappa(2k,n-\nu+r)} E_{n-\nu,r}^k(f,s)\bigg|_{s=0}$

for $0 \le \nu \le n-r$.

Both these properties are not obvious if we are outside the range of absolute convergence of the series defining $E_{n,r}^k(f,s)$, i.e., $k \le n+r+1$. These questions of "Hecke summation" are the main subject of [**We 1**]. It should be emphasized that we cannot get the full strength of Weissauer's results here, because equation (IX.1), which is our main tool, is only helpful for the question of "stable lifting," that is, the question of whether (or not) the Eisenstein series $E_{n,r}^k(f,s)|_{s=0}$ satisfies (IX.3) and (IX.4) for all $n \ge r$.

X. Noncusp forms in general. Unfortunately we do not know whether (IX.2) also holds for all weights $k \le 2n$. To extend our results to non-cusp forms in general, we need some technical preparations. We recall that the Petersson scalar product $\langle f, g \rangle$ is defined for $f, g \in M_n^k$ with at least one of f and g a cusp form; therefore $\langle\ ,\ \rangle$ induces an orthogonal decomposition

$$M_n^k = S_n^k \perp N_n^k.$$

For $f \in M_n^k$ we denote by f_0 the "cuspidal component of f" (i.e., $f_0 \in S_n^k$ and $f - f_0 \in N_n^k$). Now we can extend the scalar product $\langle\ ,\ \rangle$ to all of M_n^k by

$$\langle\langle f, g \rangle\rangle := \sum_{\mu=0}^n \langle(\phi^\mu f)_0, (\phi^\mu g)_0\rangle,$$

and, starting with $r = n$, we inductively define subspaces $M_{n,r}^k$ of M_n^k by

$$M_{n,r}^k = \left\{ F \in M_n^k \ \bigg|\ \phi^{n-r}F \in S_r^k,\ F \perp \sum_{j=r+1}^n M_{n,j}^k \right\}.$$

In this way we get an orthogonal decomposition

(X.1.) $\qquad\qquad\qquad\qquad M_n^k = \bigoplus_{r=0}^n M_{n,r}^k.$

In particular, the restriction of ϕ^{n-r} to $M_{n,r}^k$ is injective. For details, see [**Ev 1, Ev 2**]; note that other (perhaps more familiar) definitions of $M_{n,r}^k$ require some surjectivity properties of Siegel's ϕ-operator. The relevance of the decomposition (X.1) will be evident from the following lemma.

LEMMA. *Let $F = \sum_{r=0}^n F_r$ with $F_r \in M_{n,r}^k$, be a Hecke eigenform. Then*
(a) *All F_r are Hecke eigenforms ($0 \le r \le n$).*
(b) *If k is divisible by 4, $F \in \Theta(2k,n)$ iff $F_r \in \Theta(2k,n)$ for all r.*

PROOF. Assertion (a) follows from the fact (proved in [**Ev 1**]) that each of the spaces $M_{n,r}^k$ is invariant under Hecke operators. To prove (b) let r_0 be the smallest number r with $F_r \neq 0$. It is sufficient to show that $F \in \Theta(2k, n)$ implies $F_{r_0} \in \Theta(2k, n)$. From $F \in \Theta(2k, n)$ it follows that

$$0 \neq f := \phi^{n-r_0} F = \phi^{n-r_0} F_{r_0} \in \Theta(2k, r_0) \cap S_{r_0}^k.$$

Moreover from Theorem 4 and the properties (IX.3) and (IX.4) of the Eisenstein series it follows—for this f—that

$$E_{n,r_0}^k(f, s)|_{s=0} \in M_{n,r_0}^k$$

(since $E_{n',r_0}^k(f, s)|_{s=0} \in N_{n'}^k$ for all $r_0 \leq n' \leq n$) and

$$\frac{\kappa(2k, n + r_0)}{\kappa(2k, 2r_0)} F_{r_0} = E_{n,r_0}^k(f, s)|_{s=0}.$$

The lemma justifies the restriction of the following theorem to a fixed subspace $M_{n,r}^k$.

THEOREM 5. *Suppose that 4 divides k and $F \in M_{n,r}^k$ is a Hecke eigenform. Then we have (with $f = \phi^{n-r} F$)*

$$\left(\gamma_r^k(s) \frac{1}{\omega_r(2s + k)} L(f, 2s + k - r) \right) \Bigg|_{s=0} F$$

$$= \frac{\kappa(2k, n + r)}{\mathfrak{M}(2k)} \sum_{i=1}^{h(2k)} \frac{1}{\varepsilon(S_i)} \langle\langle F, \vartheta_{S_i}^n \rangle\rangle \vartheta_{S_i}^n.$$

This theorem follows from Theorem 4, the considerations above, and the formula

(X.2) $$\langle\langle F, \vartheta_S^n \rangle\rangle = \frac{\kappa(2k, n + r)}{\kappa(2k, 2r)} \langle f, \vartheta_S^r \rangle.$$

The truth of this formula is obvious, if $F \in \Theta(2k, n)$; for $F \notin \Theta(2k, n)$ both sides of (X.2) are zero, which is obvious for the right-hand side; on the left-hand side one has to observe that it is sufficient to show that $F \perp (M_{n,r'}^k \cap \Theta(2k, n))$ for all r'.

We can of course draw similar conclusions from this theorem as we did from Theorem 4; this will be left to the reader.

One should improve Theorem 5 by replacing the factor on the left-hand side of the equation by a factor involving $L(F, s)|_{s=k-n}$ instead of $L(f, s)|_{s=k-r}$, but we do not want to go into this here.

XI. Harmonic coefficients. It is evident that we lost much information using Siegel's theorem for degree $n+r$ and specializing to $\mathscr{Z} = \begin{pmatrix} Z^{(n)} & 0 \\ 0 & W^{(r)} \end{pmatrix}$. One possibility to keep track of some part of that information is to apply certain differential operators before specializing the variable $\mathscr{Z}$. In this way theta series with harmonic coefficients arise, and we can ask the same questions as before for those more general theta series. We do not want to become involved in the theory of vector-valued modular forms here, but we should mention that it is possible to generalize the method sketched below to that class of modular forms.

A polynomial function $P: \mathbf{C}^{(m,n)} \to \mathbf{C}$ is called a harmonic form of (integral) weight $\nu \geq 1$ if

(a) $P(XA) = \det(A)^\nu P(X)$ for all $A \in \mathrm{Gl}(n, \mathbf{C})$,

(b) P is harmonic: $\sum_{i,j} \partial^2/\partial x_{ij}^2 = 0$.

Such polynomials are automatically pluriharmonic in the usual sense (see [**Fr 2**, III, §3]). The finite-dimensional vector space of all such polynomials will be denoted by $H_\nu(m, n)$. For details on $H_\nu(m, n)$—and more generally on all pluriharmonic polynomials—we refer to [**K-V**].

From now on we assume that m is even and $\nu \geq 1$. For $m \geq 2n$ there are many harmonic forms; more precisely we have for $m = 2(n+t)$, $t \geq 0$, and some constant $c \neq 0$

(XI.1) $$\dim H_\nu(m, n) \sim c\nu^{n(n+1)/2+n(2t-1)} \quad \text{for } \nu \to \infty.$$

Moreover, there is a rather explicit way to construct all of them [**Ma**]. On the other hand, for $m < 2n$ there are very few of them:

$$H_\nu(m, n) = \{0\} \quad \text{for } m < n, \text{ all } \nu \geq 1,$$
$$H_\nu(m, n) = \{0\} \quad \text{for } n \leq m < 2n \text{ and } \nu \neq 1.$$

We are interested in theta series of type

$$\vartheta_{S,P}^n(Z) = \sum_{G \in \mathbf{Z}^{(m,n)}} P(S^{1/2}G)e^{\pi i \, \mathrm{trace}(G'SGZ)},$$

where $P \in H_\nu(m, n)$, $Z \in \mathbf{H}_n$, and $S \in \mathfrak{S}(m)$; here $S^{1/2}$ is the (unique) real m-rowed positive definite symmetric matrix with $(S^{1/2})(S^{1/2}) = S$. For m divisible by 8, $m \geq n$, and $\nu \geq 1$ we define the space

$$B(m, n, \nu) = \mathbf{C} - \text{span of } \{\vartheta_{S,P}^n \mid S \in \mathfrak{S}(m), \ P \in H_\nu(m, n)\};$$

this is a subspace of S_n^k with $k = m/2 + \nu$. Of course we can now ask the same questions as in §III for the spaces $B(m, n, \nu)$ and S_n^k. In [**Bö 2**] we introduced a holomorphic differential operator $\mathscr{D}_{k,n}$ acting on functions defined on $\mathbf{H}_{2n}$; that operator, which was a linear differential operator with nonconstant coefficients, had some remarkable transformation properties (for holomorphic functions f):

(XI.2) $$(\mathscr{D}_{k,n}f)|_{k+1}M = \mathscr{D}_{k,n}(f|_k M)$$

for all $M \in \mathrm{Sp}(n, \mathbf{R}) \times \mathrm{Sp}(n, \mathbf{R}) \hookrightarrow \mathrm{Sp}(2n, \mathbf{R})$; here we embed $\mathrm{Sp}(n) \times \mathrm{Sp}(n)$ in $\mathrm{Sp}(2n)$ in the obvious way by

$$\begin{pmatrix} A & B \\ C & D \end{pmatrix} \times \begin{pmatrix} a & b \\ c & d \end{pmatrix} \mapsto \begin{pmatrix} A & 0 & B & 0 \\ 0 & a & 0 & b \\ C & 0 & D & 0 \\ 0 & c & 0 & d \end{pmatrix}.$$

For any $\nu \geq 1$ and any k we consider the operator

(XI.3) $$(\mathscr{D}_{k+\nu-1,n} \circ \cdots \circ \mathscr{D}_{k,n})|_{U=0},$$

where we decompose $\mathscr{Z} \in \mathbf{H}_{2n}$ into n-rowed submatrices

$$\mathscr{Z} = \begin{pmatrix} Z & U \\ U' & W \end{pmatrix}, \qquad Z, W \in \mathbf{H}_n.$$

If $k \geq n$ or $\nu = 1$ we can normalize that operator (XI.3) in such a way that we get an operator $\mathbf{D}^{\nu}_{k,n}$ with the property

$$\mathbf{D}^{\nu}_{k,n} = \det \left(\frac{\partial}{\partial u_{ij}} \right)^{\nu} \bigg|_{U=0} + \cdots ,$$

where $\cdots$ is a polynomial in $\partial/\partial u_{ij}$, $\partial/\partial z_{ij}$, $\partial/\partial w_{ij}$, which as a polynomial in the $\partial/\partial u_{ij}$ has total degree $< n\nu$, in other words, $\det(\partial/\partial u_{ij})^{\nu}$ is the "main term"; for $\nu = 1$ those terms of lower degree do not occur at all.

Now we define a polynomial function $Q^{\nu}_{n,k}$ on $SC^{(2n,2n)}$ (= space of symmetric $2n$-rowed complex matrices) by the formula

$$(\mathbf{D}^{\nu}_{k,n} f_T) \left(\begin{pmatrix} Z & 0 \\ 0 & W \end{pmatrix} \right) = Q^{\nu}_{n,k}(T) e^{\text{trace}(T_1 Z + T_4 W)},$$

where

$$f_T(\mathscr{Z}) = e^{\text{trace}(T\mathscr{Z})} \quad \text{and} \quad T = \begin{pmatrix} T_1 & * \\ * & T_4 \end{pmatrix}, \qquad T_1, T_4 \in SC^{(n,n)}.$$

In [**Bö 2**, Satz 15] we proved that for m even we can define an element $P^{\nu}_{m,n}$ of $H_{\nu}(m,n) \otimes H_{\nu}(m,n)$ by means of

$$P^{\nu}_{m,n}(X, Y) = Q^{\nu}_{n,m/2}((X,Y)'(X,Y)), \qquad X, Y \in \mathbf{C}^{(m,n)}.$$

This polynomial is different from zero in all the cases we are interested in (i.e., $m \geq 2n$ and all ν as well as $n \leq m < 2n$ and $\nu = 1$). Moreover we have

$$P^{\nu}_{m,n} \text{ is symmetric: } P^{\nu}_{m,n}(X, Y) = P^{\nu}_{m,n}(Y, X),$$
$$P^{\nu}_{m,n}(AX, Y) = P^{\nu}_{m,n}(X, AY) \quad \text{for all } A \in O(m, \mathbf{C}).$$

Now we recall from [**K-V**] that the natural action of the orthogonal group $O(m, \mathbf{C})$ on $H_{\nu}(m,n)$ is an irreducible representation. From all these facts we get the important property

(XI.4) $$P^{\nu}_{m,n}(X, Y) = c(m, n, \nu) \sum_{l_{\nu}} \varphi_{l_{\nu}}(X) \cdot \varphi_{l_{\nu}}(Y),$$

where $(\varphi_{l_{\nu}})$ is an orthonormal basis of $H_{\nu}(m,n)$ with respect to a $O(m, \mathbf{C})$-invariant scalar product on $H_{\nu}(m,n)$ and $c(m, n, \nu)$ is some nonzero constant. We may (and we shall) assume that the polynomials $\varphi_{l_{\nu}}$ have real coefficients.

For (almost) all formulas developed in previous sections, we have analogues involving the differential operator $\mathbf{D}^{\nu}_{k,n}$ introduced above. For example, the analogue of (V.3) is (for $r = n$, $S \in \mathfrak{S}(m)$, and ν as above)

$$(\mathbf{D}^{\nu}_{m/2,n} \vartheta^{2n}_S) \left(\begin{pmatrix} Z & 0 \\ 0 & W \end{pmatrix} \right) = c(m, n, \nu) \sum_{l_{\nu}} \vartheta^{n}_{S, \varphi_{l_{\nu}}}(Z) \vartheta^{n}_{S, \varphi_{l_{\nu}}}(W)$$

with $Z, W \in \mathbf{H}_n$.

To get analogues of the pullback formulas for Eisenstein series, we do not apply $\mathbf{D}^\nu_{m/2,n}$ to $E^{m/2}_{2n}(\mathscr{Z},s)$ but a more general[2] operator $\mathbf{D}^\nu_{m/2+s,n}$ to

$$G^{m/2}_{2n}(\mathscr{Z},s) := \sum_{M\in\Gamma_{n,\infty}\backslash\Gamma_n} j(M,\mathscr{Z})^{-k-s}\, j(M,\overline{\mathscr{Z}})^{-s},$$

which differs from $E^{m/2}_{2n}(\mathscr{Z},s)$ just by the factor $\det(\operatorname{Im}\mathscr{Z})^s$. The generalized pullback, which we want to consider, is (for $Z, W \in \mathbf{H}_n$)

$$E^{m/2,\nu}_{2n}(Z,W,s) = \mathbf{D}^\nu_{m/2+s,n}(G^{m/2}_{2n}(*,s)) \left(\begin{pmatrix} Z & 0 \\ 0 & W \end{pmatrix} \right) \det(\operatorname{Im} Z)^s \det(\operatorname{Im} W)^s.$$

By calculations similar to those in [**Bö 2, Bö 3**] we now get for any eigenform $f \in S^k_n$, $k = m/2 + \nu$, a precise analogue of (VII.4) (for $r = n$):

$$\langle f, E^{m/2,\nu}_{2n}(-\overline{Z}, *, \overline{s}) \rangle$$
$$= \gamma^k_n(s)\mathscr{C}_n\left(\frac{m}{2}+s,\nu\right)\frac{1}{\omega_n(m/2+2s)}L\left(f,\frac{m}{2}-n+2s\right)f(Z),$$

where, for $s \in \mathbf{C}$,

$$\mathscr{C}_n(s,\nu) = 2^{n\nu}C_n(-s)\cdots C_n(-s-\nu+1)$$

with

$$C_n(s) = s\left(s+\frac{1}{2}\right)\cdots\left(s+\frac{n-1}{2}\right).$$

Combining all these results again with Siegel's theorem, we easily get (in analogy to Theorem 4)

THEOREM 5. *For m divisible by 8 and any $\nu \geq 1$ (if $m \geq 2n$) or $\nu = 1$ (if $n \leq m < 2n$) we have for all Hecke eigenforms $f \in S^k_n$, $k = m/2 + \nu$:*

$$\left(\gamma^k_n(s)\mathscr{C}_n\left(\frac{m}{2}+s,\nu\right)\frac{1}{\omega_n(m/2+2s)}L\left(f,\frac{m}{2}-n+2s\right)\right)\bigg|_{s=0}f$$
$$= \frac{\kappa(m,2n)}{\mathfrak{M}(m)}\sum_{i=1}^{h(m)}\frac{1}{\varepsilon(S_i)}\sum_{l_\nu}\langle f,\vartheta^n_{S_i,\varphi_{l_\nu}}\rangle\vartheta^n_{S_i,\varphi_{l_\nu}};$$

in particular we get $f \in B(m,n,\nu)$ iff

$$\left(\gamma^k_n(s)\mathscr{C}_n\left(\frac{m}{2}+s\right)\frac{1}{\omega_n(m/2+2s)}L\left(f,\frac{m}{2}-n+2s\right)\right)\bigg|_{s=0} \neq 0.$$

Moreover (using the Euler product expansion of $L(f,s)$) we have for all ν as above

$$B(m,n,\nu) = S^{m/2+\nu}_n \quad \text{if } m > 4n.$$

REMARKS. (a) Notice that Theorem 5 is considerably stronger than Theorem 4; in particular, Theorem 5 implies that for $m > 4n$ a finite number of quadratic forms is sufficient to produce all cusp forms of all weights $k > m/2$.

---

[2]In [**Bö 2**] the operator $\mathscr{D}_{k,n}$ was introduced for any complex "weight" k.

For instance, all cusp forms of degree 1 can be constructed from one quadratic form in 8 variables (this result is also contained in [Wa]). For degree 2 all cusp forms can be obtained as linear combinations of theta series arising from the two classes of quadratic forms in $\mathfrak{S}(16)$.

(b) For $n = 1$ and $n = 2$ Theorem 4_2 was proved by Hecke [He,§7] and Ozeki [Oz] respectively by looking at generators of the ring of modular forms. Results like those of Theorem 5 however cannot be proved by looking at generators of the ring of modular forms!

XII. Final remarks. (A) If we restrict ourselves to cusp forms $f \in S_n^k$, where f is a Hecke eigenform, then our results can be summarized as follows:

> Those f that can be expressed by theta series attached to quadratic forms from $\mathfrak{S}(m)$, m divisible by 8 and $k \geq m/2$, are characterized by a condition on the behaviour of the automorphic L-function $L(f,s)$ at $s = m/2 - n$; that condition is always satisfied if $m > 4n$.

(B) We have restricted ourselves completely to modular forms of level 1 (i.e., modular forms for the full Siegel modular group). Of course one can ask the same questions also for higher levels, as it was done successfully for degree 1 by several authors [Ei, Wa]. But the method of Waldspurger, which is close to ours in some sense, uses the theory of newforms; such a theory of newforms however is at present not available for Siegel modular forms.

(C) The only case where we definitely know that the equality

$$M_n^k = \Theta(m, n), \quad m \text{ divisible by 8}$$

or

$$B(m, n, \nu) = S_n^{m/2+\nu}$$

does not hold is for $m = 2n$. In this case the space $B(2n, n, \nu)$ is a very small subspace of $S_n^{n+\nu}$ for $\nu \gg 0$, since there are not enough harmonic forms in $H_\nu(2n, n)$ (see (XI.1)).

(D) It is well known that there exist finitely many Siegel modular forms $f_1, \ldots, f_t$ of degree n, whose set of common zeros characterizes the Jacobians of curves of genus n among all principally polarized abelian varieties ("Schottky problem", see [v.G], where the ideal of all such Siegel modular forms is described precisely). Of course the modular forms $f_1, \ldots, f_t$ can be chosen to have large weight. Therefore, by the results explained in the present paper, the $f_1, \ldots, f_t$ can be chosen as linear combinations from $\bigcup_{8|m} \Theta(m, n)$ or from $\bigcup_{\nu \geq 1} B(m, n, \nu)$, where m is any fixed number divisible by 8 with $m > 4n$. This solves a problem raised by Igusa [Ig 2].

REFERENCES

[An] A. N. Andrianov, *On Siegel modular forms of genus n and weight n/2*, J. Fac. Sci. Univ. Tokyo **28** (1982), 487–503.

[Bö 1] S. Böcherer, *Über die Fourier-Jacobi-Entwicklung Siegelscher Eisensteinreihen*, Math. Z. **183** (1983), 21–46.

[Bö 2] ____, *Über die Fourier-Jacobi-Entwicklung Siegelscher Eisensteinreihen.* II, Math. Z. **189** (1985), 81–110.

[Bö 3] ____, *Über die Funktionalgleichung automorpher L-Funktionen zur Siegelschen Modulgruppe,* J. Reine Angew. Math. **362** (1985), 146–168.

[Bö 4] ____, *Ein Rationalitätssatz für formale Heckereihen zur Siegelschen Modulgruppe,* Abh. Math. Sem. Univ. Hamburg **56** (1986), 35–47.

[Ei] M. Eichler, *The basis problem for modular forms and the traces of the Hecke operators,* Lecture Notes in Math., vol. 320, Springer-Verlag, 1973, pp. 75–151.

[Ev 1] S. A. Evdokimov, *A basis of eigenfunctions of Hecke operators in the theory of modular forms of genus n,* Math. USSR-Sb. **43** (1982), 299–321.

[Ev 2] ____, Letter to the editor, Math. USSR-Sb. **43** (1982), 322 (=Correction to **[Ev 1]**).

[Fr 1] E. Freitag, *Die Invarianz gewisser von Thetareihen erzeugter Vektorräume unter Heckeoperatoren,* Math. Z. **156** (1977), 141–155.

[Fr 2] ____, *Siegelsche Modulfunktionen,* Springer-Verlag, 1983.

[Ga 1] P. B. Garrett, *Pullbacks of Eisenstein series; applications,* Automorphic Forms of Several Variables (Taniguchi Sympos., 1983), Birkhäuser, 1984, pp. 114–137.

[Ga 2] ____, *Integral representations of Eisenstein series and L-functions,* Preprint, 1987; Proc. Selberg Sympos., Oslo, 1987 (to appear).

[He] E. Hecke, *Analytische Arithmetik der positiven quadratischen Formen,* Mathematische Werke, Vandenhoeck & Ruprecht, Gottingen, 1970, pp. 789–918.

[Ig 1] J.-I. Igusa, *Theta functions,* Springer-Verlag, 1972.

[Ig 2] ____, *Problems on theta functions,* this volume.

[K-V] M. Kashiwara and M. Vergne, *On the Segal-Shale-Weil representations and harmonic polynomials,* Invent. Math. **44** (1978), 1–47.

[Kl] H. Klingen, *Zum Darstellungssatz für Siegelsche Modulformen,* Math. Z. **102** (1968), 30–43.

[Ku-Ra] S. S. Kudla and S. Rallis, *On the Weil-Siegel formula,* Preprint, 1987.

[La] R. P. Langlands, *Euler products,* Yale Univ. Press, New Haven, Conn., 1971.

[Ma] H. Maass, *Harmonische Formen in einer Matrixvariablen,* Math. Ann. **252** (1980), 133–140.

[Oz] M. Ozeki, *On basis problem for Siegel modular forms of degree* 2, Acta Arith. **31** (1976), 17–30.

[P-S-R] I. I. Piatetski-Shapiro and S. Rallis, *L-functions for the classical groups,* Preprint.

[S-M 1] R. Salvati-Manni, *On the not integrally closed subrings of the ring of the thetanullwerte,* Duke Math. J. **52** (1985), 25–33.

[S-M 2] ____, *On the not integrally closed subrings of the ring of thetanullwerte.* II, J. Reine Angew. Math. **372** (1986), 64–70.

[Se-St] J.-P. Serre and H. M. Stark, *Modular forms of weight* 1/2, Lecture Notes in Math., vol. 627, Springer-Verlag, 1977, pp. 27–67.

[Si] C. L. Siegel, *Über die analytische Theorie der quadratischen Formen,* Ann. of Math. **36** (1935), 527–606.

[v.G] B. van Geemen, *Siegel modular forms vanishing on the moduli space of curves,* Invent. Math. **78** (1984), 329–349.

[Wa] J. L. Waldspurger, *Engendrement par des séries theta de certains espaces de formes modulaires,* Invent. Math. **50** (1979), 135–168.

[We 1] R. Weissauer, *Stabile Modulformen und Eisensteinreihen,* Lecture Notes in Math., vol. 1219, Springer-Verlag, 1986.

[We 2] ____, *Stable modular forms,* Contemp. Math., no. 53, Amer. Math. Soc., Providence, R.I., 1986, pp. 535–542.

[Wi] E. Witt, *Eine Identität zwischen Modulformen zweiten Grades,* Abh. Math. Sem. Univ. Hamburg **14** (1941), 323–337.

MATHEMATISCHES INSTITUT, FREIBURG, WEST GERMANY

Proceedings of Symposia in Pure Mathematics
Volume **49** (1989), Part 2

Arithmetic Compactification
of the Siegel Moduli Space

CHING-LI CHAI

0. Introduction. This paper is intended as a survey and expository account of recent progress on the arithmetic theory of compactification of the moduli spaces of polarized abelian varieties, most importantly due to Gerd Faltings [**Fa1**, **Fa2**]. What emerged was in fact a sort of "compactification machinery" for moduli spaces coming from polarized abelian varieties (possibly with endomorphism structures). Detailed expositions can be found in the monograph [**Fa-Ch**]. In preparing this paper I have consulted notes taken from three lectures given by Faltings at Borel's seminar at the Institute for Advanced Studies from March 31 to April 14, 1987. In some senses the present paper can be regarded as an expanded version of these lectures.

This paper is organized as follows: §1 gives a review of the definitions and general properties about polarized abelian schemes, semi-abelian schemes, and their toric part, algebraic stacks, the moduli stack A_g, and complex uniformization of A_g. In §2 we explain the main technical theorem, which analyzes completely degeneration of (polarized) abelian varieties over complete noetherian normal domains in terms of multiplicative periods. In §3 we construct the arithmetic toroidal compactification of A_g, which is an over-$\mathbb{Z}$ version of Mumford's toroidal compactification. In §4 we construct the arithmetic minimal compaction, an over-$\mathbb{Z}$ version of the Satake-Bailey-Borel compactification. At the same time we explain the basic arithmetic theory of Siegel modular forms, including Koecher's principle, Fourier-Jacobi expansion, and the q-expansion principle. As a corollary we give a geometric proof of a result on p-adic monodromy, which was first proved by Ribet in the Hilbert-Blumenthal case by arithmetic method.

1. Review about abelian schemes.

1.1. *Abelian schemes and polarizations.*

1980 *Mathematics Subject Classification* (1985 *Revision*). Primary 14K10, 14K15, 11F46.

1.1.1. DEFINITION. An *abelian scheme* is a group scheme $\pi\colon A \to S$ that is smooth, proper with (geometrically) connected fibres. Notice that since A has a section over S, namely the zero section, the adverb "geometrically" can be added or omitted without affecting the definition. A basic fact is that an abelian scheme is actually a commutative group scheme.

1.1.2. A fundamental result about abelian schemes is the *theorem of the cube*, which we will formulate as follows:

THEOREM 1.1.2. *Let* $\pi\colon A \to S$ *be an abelian scheme and* $\underline{L}$ *an invertible sheaf on* A. *For any subset* $I \subseteq \{1, 2, 3\}$, *let* $m_I\colon A \times_S A \times_S A \to A$ *be the morphism "adding the ith factor for all* $i \in I$"; *for example,* $m_\varnothing\colon A \to A$ *is the zero morphism. Then the invertible sheaf*

$$\Theta(\underline{L}) = \bigotimes_{I \in \{1,2,3\}} m_I^* \underline{L}^{\otimes (-1)^{\mathrm{Card}(I)}}$$

on $A \times_S A \times_S A$ *is canonically trivial.*

1.1.3. For an abelian scheme $\pi\colon A \to S$, let $A^t = \mathrm{Pic}^0(A/S) \to S$ be the *dual abelian scheme*. Thus functorially

$A^t(T) = \{$isomorphism classes of invertible sheaves $\underline{L}$ on $A \times_S T$, rigidified

along the zero section, such that $\underline{L}_{\bar{t}}$ is algebraically

equivalent to zero for any geometric point $\bar{t}$ of $T\}$

for any scheme T over S. Here if $e\colon S \to A$ is the zero section, a *rigidification* of an invertible sheaf $\underline{L}$ along e is an isomorphism $e^* \underline{L} \xrightarrow{\sim} \mathcal{O}_S$. The identity morphism $A^t \to A^t$ corresponds to the universal rigidified invertible sheaf on $A \times_S A^t$, which is called the Poincaré sheaf and denoted by $\underline{P}_{A \times_S A^t}$. An important fact is that the dual of A^t is canonically isomorphic to A and the Poincaré sheaf for A can also be regarded as the Poincaré sheaf for A^t. We remark that the $\mathbf{G}_m$-torsor associated to the Poincaré sheaf, still denoted by $\underline{P}$, is the canonical $\mathbf{G}_m$-biextension of $A \times_S A^t$. ($\underline{P}$ has two mutually compatible partial group laws, see [SGA7 I, exposé VII, VIII] for details.)

1.1.4. Let $\underline{L}$ be an invertible sheaf on A. $\underline{L}$ defines a morphism

$$\lambda(\underline{L})\colon A \to A^t \quad \text{given by}$$
$$a \mapsto T_a^*(\underline{L}) \otimes \underline{L}^{-1} \otimes a^*(\underline{L})^{-1} \otimes e^* \underline{L}$$

for any functorial point $a \in A(Z)$ for any scheme Z over S, where $T_a\colon A \to A$ is the morphism "translation by a."

1.1.5. DEFINITION. A *polarization* of $A \to S$ is a homomorphism $\lambda\colon A \to A^t$ such that $\forall$ geometric points $\bar{s}$ of S, $\lambda_{\bar{s}} = \lambda(\underline{L}_{\bar{s}})$ for some ample invertible sheaf $\underline{L}_{\bar{s}}$ on $A_{\bar{s}}$. Equivalently, locally on S_{fppf} (resp. $S_{\mathrm{\acute{e}t}}$, resp. S_{fpqc}) λ is defined by some ample invertible sheaf. On S_{fppf} (resp. $S_{\mathrm{\acute{e}t}}$, resp. S_{fpqc}), the invertible sheaves on A giving rise to λ form a principal homogeneous space under A^t (i.e., an A^t-torsor).

1.1.6. Let $\lambda: A \to A^t$ be a polarization. Then λ is finite and faithfully flat (i.e., an isogeny). The pull-back of the Poincaré sheaf by $(\mathrm{id}_A, \lambda): A \to A \times_S A^t$ is an ample invertible sheaf $\underline{L}$ such that $\lambda(\underline{L}) = 2\lambda$. We say that λ is a *principal polarization* if λ is an isomorphism.

1.1.7. If $\underline{L}$ is a relatively ample invertible sheaf on A, then $\pi_*(\underline{L})$ is a locally free sheaf on S. The rank of $\pi_*(\underline{L})$ and the degree of $\lambda(\underline{L})$ are locally constant on S w.r.t. the Zariski topology S_{Zar}. In fact $\lambda(\underline{L}): A \to A^t$ is a polarization and $\mathrm{rk}(\pi_*(\underline{L}))^2 = \deg(\lambda(\underline{L}))$, which is a consequence of the theorem of the cube. $\lambda(\underline{L})$ is an isomorphism iff $\pi_*(\underline{L})$ is an invertible sheaf, in which case $\lambda(\underline{L})$ is a principal polarization. The pull-back of the Poincaré sheaf by $(\mathrm{id}_A, \lambda(\underline{L})): A \to A \times_S A^t$ is canonically isomorphic to $\underline{L} \otimes [-1]^*\underline{L}$, where $[n]: A \to A$ denotes the homomorphism "multiplication by n" $\forall n \in \mathbb{Z}$.

1.1.8. DEFINITION. Let $A \to S$ be an abelian scheme of relative dimension g, $1 \le n \in \mathbb{N}$. A *principal level-n-structure* is an isomorphism

$$\alpha: (\mathbb{Z}/n\mathbb{Z})^{2g} \xrightarrow{\sim} A[n],$$

where $A[n] = \mathrm{Ker}[n]$ is the subgroup of n-torsion points. Such an isomorphism exists only if n is invertible in $\mathscr{O}_S$. A lemma of Serre says that if $n \ge 3$, a polarized abelian scheme with level-n-structure has no nontrivial automorphism that preserves both the polarization and the level structure.

1.2. *Semi-abelian schemes and tori.*

1.2.1. DEFINITION. A *torus* T over a scheme S is a group scheme over S such that locally on S_{fppf} (equivalently, $S_{\text{ét}}$ or S_{fpqc}) it is isomorphic to a product of finitely many copies of the multiplicative group. T is a *split torus* if it is S-isomorphic to a product of finitely many copies of the multiplicative group.

1.2.2. For a torus T over S, $\mathrm{X}(T) = \underline{\mathrm{Hom}}_{\mathbb{Z}}(T, \mathbf{G}_m)$ will denote the character group of T. It is an $S_{\text{ét}}$ (or S_{fppf} resp. S_{fpqc}) sheaf of torsion-free $\mathbb{Z}$-modules of finite type. T is uniquely determined by $\mathrm{X}(T)$: $T = \underline{\mathrm{Hom}}_{\mathbb{Z}}(\mathrm{X}(T), \mathbf{G}_m)$. T is a split torus iff $\mathrm{X}(T)$ is constant.

1.2.3. A fundamental property of tori is that they are pretty rigid: you cannot deform them. We describe this phenomenon as follows:

THEOREM 1.2.3 [SGA3, IX, §3]. *Let S be a scheme $\mathscr{I} \subset \mathscr{O}_S$ be a sheaf of ideals such that $\mathscr{I}^2 = 0$. Let $S_0 = \mathrm{Spec}(\mathscr{O}_S/\mathscr{I})$ be the closed subscheme of S defined by $\mathscr{I}$. Let H_0 be a torus defined over S_0, and $G \to S$ a commutative group scheme smooth over S, $G_0 = G \times_S S_0$. Then*

(i) *H_0 can be uniquely lifted to a torus H over S. (Any two liftings are isomorphic by a unique isomorphism.)*

(ii) *Any homomorphism $u_0: H_0 \to G_0$ can be uniquely lifted to a homomorphism $u: H \to G$. If moreover u_0 is a closed embedding, so is u.*

1.2.4. DEFINITION. A *semi-abelian scheme* is a smooth commutative group scheme $\pi: G \to S$ with (geometrically) connected fibres, such that each fibre G_s is an extension of an abelian variety A_s by a torus T_s. We remark that

A_s and T_s are uniquely determined by G_s: T_s is the largest affine subgroup scheme of G_s and A_s is the quotient of G_s by T_s. T_s is called the *toric part* of G_s.

For a semi-abelian scheme $G \to S$ and $s \in S$, let $X(s) = X(T_s)$ denote the character group of the torus part of G_s. Let $\bar{s}$ be a geometric point lying over s. Then $X(\bar{s})$ is a free $\mathbb{Z}$-module of finite type. The Galois group $\mathrm{Gal}(\kappa(s)^{\mathrm{sep}}/\kappa(s))$ naturally acts on $X(\bar{s})$, and $X(s)$ is (or can be interpreted as) just this Galois module.

1.2.5. There are two useful results about semi-abelian schemes. The first one is about extension of homomorphisms between semi-abelian schemes:

PROPOSITION 1.2.5 [**Ra1**, IX, 1.4, p. 130]. *Let S be a noetherian normal scheme and G and H two semi-abelian schemes over S. Suppose that over a dense open subscheme U of S there is a homomorphism $\phi_U \colon G_U \to H_U$. Then ϕ_U extends (uniquely) to a homomorphism $\phi \colon G \to H$ over S.*

1.2.6. A variant of the above proposition is about extending the toric part:

PROPOSITION 1.2.6 [**Ra1**, IX, 2.4, p. 137]. *Let S be a noetherian normal scheme, U a dense open subscheme of S, and G a semi-abelian scheme over S. If H_U is a closed subgroup of G_U that is a torus over U, then the closure of H_U in G is a torus $H \to S$ contained in G.*

We remark that if we drop the assumption that S is normal in either Proposition 1.2.5 or 1.2.6, both statements become *false*.

1.2.7. Let $\pi \colon G \to S$ be a commutative group scheme. Following the notations of 1.1.2, for any subset $I \subseteq \{1, 2, 3\}$ we let $m_I \colon G \times_S G \times_S G \to G$ be the morphism "adding the ith factor for all $i \in I$." Let $\underline{L}$ be a $\mathbf{G}_m$-torsor on G, *rigidified* along the zero section. (As always we will freely switch between the two equivalent notions of $\mathbf{G}_m$-torsors and invertible sheaves.) Let

$$\Theta(\underline{L}) = \bigotimes_{I \in \{1,2,3\}} m_I^* \underline{L}^{\otimes (-1)^{\mathrm{Card}(I)}}.$$

Similarly, let

$$\Lambda(\underline{L}) = m^* \underline{L} \otimes p_1^* \underline{L}^{\otimes -1} \otimes p_2^* \underline{L}^{\otimes -1},$$

where $m, p_1, p_2 \colon G \times_S G \to G$ denote the group law and the two projections respectively. A *cubical structure* on $\underline{L}$ ($\underline{L}$ rigidified) is a trivialization $\Theta(\underline{L}) \xrightarrow{\sim} \underline{1}$ (= the trivial torsor) such that it induces on $\Lambda(\underline{L})$ a structure of symmetric biextension of $G \times_S G$ by $\mathbf{G}_m$. For instance, the theorem of the cube says that any rigidified invertible sheaf on an abelian scheme has a canonical cubical structure. In fact, any rigidified invertible sheaf on a smooth commutative group scheme $G \to S$ with connected fibres over a *normal* base scheme S such that the maximal fibres are semi-abelian varieties (e.g., a semi-abelian scheme over normal base) admits a unique cubical structure compatible with the rigidification. For details, see [**Br**, §2, Proposition 2.4, p. 16].

1.2.8. A major merit of the notion of cubical structure is that it makes $\mathbf{G}_m$-torsors a bit more rigid, so that it is easier to do descent.

by) a separated algebraic space (again simply called algebraic space) in the sense of **Artin**. This is simply the definition of algebraic spaces.

1.3.2. All notions and operations in algebraic geometry that are compatible with étale descent can be carried over to the context of algebraic stacks. For example, a quasicoherent sheaf $\underline{F}$ on **S** is given by a quasicoherent sheaf $\underline{F}_U$ on U and an isomorphism $p_1^* \underline{F}_U \xrightarrow{\sim} p_2^* \underline{F}_U$ satisfying the transitivity condition. As an easy example, if G is a finite group, $U = * = $ a (geometric) point, $R = G$, then U/R is what topologists call BG. A sheaf on U/G is (essentially by definition) just a G-set; a quasicoherent sheaf on U/G is nothing but a G-module. Similarly we can define relative schemes over algebraic stacks and abelian schemes over algebraic stacks. We can define fiber products of algebraic stacks. We can talk about flat, étale, smooth morphisms, etc. The list is very long. In fact one can work out most parts of EGA and SGA in the context of algebraic stacks.

1.3.3. Let $\mathbf{S} = (U/R)$ be a noetherian algebraic stack. There is an algebraic space S and a morphism $\mathbf{S} \to S$ such that

(i) for any algebraically closed field k, $S(k) = \mathbf{S}(k) = U(k)/R(k)$, and

(ii) any morphism from **S** to an algebraic space factors uniquely through S.

S is called the *coarse moduli space* of **S** and is obviously characterized by the second property alone. We will write U/R for this coarse moduli space. The strict local rings of S can be described as follows: for a geometric point $\bar{x}: \operatorname{Spec} k \to \mathbf{S}$, $\operatorname{Aut}(\bar{x}) = \bar{x} \times_{\mathbf{S}} \bar{x}$ is a finite group acting on $\mathscr{O}_{\bar{x},\mathbf{S}}$. Then $\mathscr{O}_{\bar{x},S} = (\mathscr{O}_{\bar{x},\mathbf{S}})^{\operatorname{Aut}(\bar{x})}$ is the $\operatorname{Aut}(\bar{x})$-invariants of $\mathscr{O}_{\bar{x},\mathbf{S}}$. Note that if $\bar{x}$ factorizes through U, then $\mathscr{O}_{\bar{x},\mathbf{S}} = \mathscr{O}_{\bar{x},U}$ and $\mathscr{O}_{\bar{x},S}$ is the ring of invariants in $\mathscr{O}_{\bar{x},U}$ under the equivalence relation induced by R.

1.3.4. Our basic example of algebraic stack is the moduli stack $\mathbf{A}_g$ classifying principally polarized abelian varieties and its variants. First we recall that, roughly speaking, a *stack* is a fibered category such that the base category is endowed with a Grothendieck topology with respect to which we can glue both the objects and morphisms of the source category.

DEFINITION 1.3.4. Let **Sch** be the big étale site of all schemes. Define $\mathbf{A}_g \to$ **Sch** to be the fibered category such that for any scheme $S \in \operatorname{Ob}(\mathbf{Sch})$, $\mathbf{A}_g(S)$ is the category of all principally polarized abelian schemes $(A \to S, \lambda: A \xrightarrow{\sim} A^t)$ with relative dimension g. Clearly $\mathbf{A}_g$ is a groupoid, i.e., all morphisms in $\mathbf{A}_g(S)$ are isomorphisms. It is also immediate to check that it is a stack.

1.3.5. To show that $\mathbf{A}_g$ is an algebraic stack, we first have to show that there is an étale covering $U \to \mathbf{A}_g$, where U is an (affine) scheme. In fact we will be able to choose U to be of finite type over $\operatorname{Spec} \mathbb{Z}$, so that not only is $\mathbf{A}_g$ quasicompact but indeed of finite type over $\operatorname{Spec} \mathbb{Z}$. There are at least two ways of doing it. The first way is to use the theory of Hilbert schemes. This theory together with the rigidity of abelian schemes tells us that there is a universal family $(A \to H, A \hookrightarrow \mathbb{P}^N/H, \lambda: A \xrightarrow{\sim} A^t)$ such that $A \to H$ is an abelian scheme of relative dimension g, λ is a principal

PROPOSITION 1.2.8. *Let S be a strictly henselien local ring, and $G \to S$ a (global) extension of an abelian scheme A by a torus T: we have an exact sequence $0 \to T \to G \to A \to 0$. Let $p: G \to A$ be the name of this arrow. Let $\underline{L}$ be a cubical invertible sheaf (i.e., invertible sheaf with cubical structure) on G. Then $\underline{L}$ can be descended to a cubical invertible sheaf on A. The set of all descent data of $\underline{L}$ is a principal homogeneous space under $\mathrm{X}(T)$ in a natural way.*

We remark here that extensions of A by a torus T are classified by $\mathrm{Hom}_{\mathbf{Z}}(\mathrm{X}(T), A^t)$.

1.2.9. We state the important semistable reduction theorem:

THEOREM 1.2.9. *Let V be a discrete valuation ring, K the fraction field of V, and G_K a semi-abelian variety over K. Then there exists a finite extension V' of V with fraction field K' such that $G_{K'} = G_K \otimes_K K'$ extends to a semi-abelian scheme over $\mathrm{Spec}(V')$. (When G_K is an abelian variety, this means that the Néron model of $G_{K'}$ contains a semi-abelian open subgroup scheme over V.)*

1.3. The moduli stack $\mathbf{A}_g$.

1.3.1. We first give an awfully inadequate review about algebraic stacks. An algebraic stack is a scheme-theoretic analogue of what topologists call "orbifold," although the notion of algebraic stacks seems to have appeared earlier. It generalizes the notion of algebraic spaces, therefore also the notion of schemes. Because its definition is somewhat technical, many people (including myself several years ago) are afraid of and do not like it. But speaking from experience, the difficulty is purely psychological and can be overcome with time. Basically all one needs in order to handle algebraic stacks is to do descends systematically. I will only give a "working definition" of it. For more details one should consult [De-Mum], see also [Ar4].

A separated algebraic stack (which we will simply call algebraic stack by abuse of language) $\mathbf{S}$ is given by a pair of schemes U, R together with two morphisms $p_1, p_2: R \to U$ such that

(i) p_1, p_2 are both étale and the morphism $(p_1, p_2): R \to U \times U$ is finite.

(ii) $(p_1, p_2): R \to U \times U$ is a categorical equivalence relation. In particular, R is a groupoid with identity.

$\mathbf{S}$ should be thought of as the quotient of U by the étale equivalence relation R. We will write $\mathbf{S} = [U/R]$.

Two quadruples $Q = (U, R, p_1, p_2)$, $\bar{Q} = (\bar{U}, \bar{R}, \bar{p}_1, \bar{p}_2)$ may define the same algebraic stack: If $U' \to U$ is an étale covering, then we get another quadruple (U', R', p_1', p_2') by pull-back, which defines the same algebraic stack. Thus Q and $\bar{Q}$ define the same algebraic stack if we can find a third quadruple $\tilde{Q}$ dominating both.

In case for a (hence all) representative quadruple $Q = (U, R, p_1, p_2)$, the morphism $(p_1, p_2): R \to U \times U$ is a closed embedding, then $\mathbf{S}$ is (represented polarization, $N = 4^g - 1$, $A \hookrightarrow \mathbb{P}^N/H$ is a (closed) projective embedding such that the pull-back of $\mathcal{O}(1)$ is a rigidified symmetric H-ample invertible sheaf on A giving rise to 4λ (so that the Hilbert polynomial of $A \hookrightarrow \mathbb{P}^N/H$ is $f(n) = 4^g \cdot n^g$). Moreover, H is quasiprojective, therefore of finite type over $\mathrm{Spec}(\mathbb{Z})$. The universal property tells us that PGL_{N+1} naturally acts on $A \to H$. Thus morally $\mathbf{A}_g$ "is" "the quotient of H by the natural PGL_{N+1}-action." To actually get an étale covering of $\mathbf{A}_g$, use an étale slice argument.

Another way is to use the method developed by M. Artin [Ar1, Ar2, Ar3, Ar5], which in some sense is more elementary. Given any geometric point $\bar{x} \colon \mathrm{Spec}\, k \to \mathbf{A}_g$, i.e., a principally polarized abelian variety A_k over an algebraically closed field k of dimension g, one first uses standard deformation theory to get the universal deformation $A^\wedge$ of A_k. $A^\wedge$ is a formal principally polarized abelian variety over $\mathrm{Spf}(R^\wedge)$, where $R^\wedge$ is a formally smooth complete noetherian local domain of dimension $n = 1 + g \cdot (g+1)/2$ of mixed characteristics with residue field k (hence isomorphic to $W(k)[\![t_1, \ldots, t_n]\!]$). By Grothendieck's formal existence theorem $A^\wedge \to \mathrm{Spec}(R^\wedge)$ is uniquely algebraizable to a principally polarized abelian scheme $A \to \mathrm{Spec}(R^\wedge)$. Then we can approximate $A \to \mathrm{Spec}(R^\wedge)$ by another principally polarized abelian scheme $A' \to \mathrm{Spec}(R)$, where R is a (strictly) henselian local ring whose formal completion is isomorphic to $\hat{R}^\wedge$. A standard limit argument gives us a principally polarized abelian scheme $A^\dagger \to \mathrm{Spec}(R^\dagger)$, where $R^\dagger$ is a subring of R of finite type over $\mathbb{Z}$ such that $\mathrm{Spec}(R) \to \mathrm{Spec}(R^\dagger)$ is étale and $A^\dagger$ is still versal at $\bar{x} \colon \mathrm{Spec}\, k \to \mathrm{Spec}(R^\dagger)$. By the openness of versality we may assume that $A^\dagger \to \mathrm{Spec}(R^\dagger)$ is versal at every point of $\mathrm{Spec}(R^\dagger)$. Therefore we have found an étale morphism $\mathrm{Spec}(R^\dagger) \to \mathbf{A}_g$ covering a neighborhood of $\bar{x}$. This shows that we can find an étale covering of $\mathbf{A}_g$. Since $\mathbf{A}_g$ is quasicompact (which follows from the theory of Hilbert schemes), we can cover $\mathbf{A}_g$ by a finite number of affine schemes as above.

1.3.6. Having found an étale covering $U \to \mathbf{A}_g$, the equivalence relation R is just $U \times_{\mathbf{A}_g} U$ (what else can it be?). Let $(A_U \to U, \lambda_U)$ be the principally polarized abelian scheme corresponding to $U \to \mathbf{A}_g$ and $p_1^*(A_U, \lambda_U) p_2^*(A_U, \lambda_U)$ be the pull-backs of (A_U, λ_U) via the two projections $p_1, p_2 \colon U \times U \to U$. Clearly $U \times_{\mathbf{A}_g} U = \underline{\mathrm{Isom}}_{U \times U}(p_1^*(A_U, \lambda_U), p_2^*(A_U, \lambda_U))$, therefore, is a scheme finite over $U \times U$, which is clearly étale over U via either projection. The groupoid structure of R is just the composition of isomorphisms. Therefore we have shown that $\mathbf{A}_g$ is indeed an algebraic stack.

1.3.7. We introduce two useful coherent sheaves on $\mathbf{A}_g$. The first one is a locally free sheaf $\underline{\Omega}$ on $\mathbf{A}_g$ of rank g. The second is just $\Lambda^g(\underline{\Omega})$, denoted by $\underline{\omega}$, therefore is an invertible sheaf. To define $\underline{\Omega}$, we will define for every $S \to \mathbf{A}_g$ a locally free rank-g sheaf $\underline{\Omega}_S$ on S that is compatible with pull-backs. This clearly suffices. A morphism $S \to \mathbf{A}_g$ corresponds to a principally polarized abelian scheme $(A \to S, \lambda)$, and we define $\underline{\Omega}_S$ to be $e^*\underline{\Omega}^1_{A/S}$, where $e = S \to A$ is the zero section. Clearly this construction is functorial in S.

1.3.8. We introduce a variant of $\mathbf{A}_g$ by considering level structures. Let $1 \leq n \in \mathbb{N}$ be an integer. Define $\mathbf{A}_{g,n}$ to be the moduli stack that classifies principally polarized abelian schemes with level-n-structure. By exactly the same method as before, $\mathbf{A}_{g,n}$ is an algebraic stack. If $n \geq 3$, Serre's lemma tells us that $\mathbf{A}_{g,n}$ is an algebraic space faithfully flat and smooth over $\mathrm{Spec}(\mathbb{Z}[1/n])$. In fact, Mumford proved in [GIT] that $\mathbf{A}_{g,n}$ is actually a quasiprojective scheme over $\mathrm{Spec}(\mathbb{Z}[1/n])$.

1.4. *Complex uniformization of* $\mathbf{A}_g$.

1.4.1. We can define the moduli stack of principally polarized abelian varieties over the category of complex analytic spaces and get a stack $\mathbf{A}_g^{\mathrm{an}} =$ the analytification of $\mathbf{A}_g$. (I propose to call it a complex analytic stack, but you may like to call it an analytic orbifold.) Everyone knows what it is:

$$\mathbf{A}_g^{\mathrm{an}} = [\mathsf{H}_g / \mathrm{Sp}_{2g}(\mathbb{Z})], \quad \text{where } \mathsf{H}_g = \{Z \in M_{g \times g}(\mathbb{C}) | Z = {}^t Z, \mathrm{Im}(Z) \gg 0\}$$

is the Siegel upper half-space of genus g, and

$$\mathrm{Sp}_{2g}(\mathbb{Z}) = \{\gamma \in \mathrm{GL}_{2g}(\mathbb{Z}) | {}^t\gamma J \gamma = J\}$$

with $J = \left(\begin{smallmatrix} 0 & I_g \\ -I_g & 0 \end{smallmatrix}\right)$, and a typical element of $\mathrm{Sp}_{2g}(\mathbb{Z})$ is $\gamma = \left(\begin{smallmatrix} A & B \\ C & D \end{smallmatrix}\right)$ such that A, B, C, D are integral $g \times g$ matrices with $A \cdot {}^t B = B \cdot {}^t A$, $C \cdot {}^t D = D \cdot {}^t C$, $A \cdot {}^t D - B \cdot {}^t C = I_g$. Thus H_g is an étale covering of $\mathbf{A}_g^{\mathrm{an}}$, and $\mathrm{Sp}_{2g}(\mathbb{Z})$ acts on H_g via the classical formula $\gamma \colon Z \mapsto (AZ + B) \cdot (CZ + D)^{-1}$ for γ as above. The étale equivalence relation $p_1, p_2 \colon R \rightarrow \mathsf{H}_g \times \mathsf{H}_g$ is given by $R = \mathsf{H}_g \times \mathrm{Sp}_{2g}(\mathbb{Z})$, where p_1 is the first projection and p_2 is the group action. The universal family of principally polarized abelian varieties A over H_g has fibre $A_Z =$ fibre over $Z \in \mathsf{H}_g = \mathbb{C}^g / (\mathbb{Z}^g + Z \cdot \mathbb{Z}^g)$. The principal polarization on A_Z is induced by the hermitian form $(u, v) \mapsto u \cdot \mathrm{Im}(Z)^{-1} \cdot v$ for $u, v \in \mathbb{C}^g$.

1.4.2. Under the uniformization $\mathbf{A}_g^{\mathrm{an}} = [\mathsf{H}_g / \mathrm{Sp}_{2g}(\mathbb{Z})]$, the locally free sheaf $\underline{\Omega}$ on $\mathbf{A}_g^{\mathrm{an}}$ is the trivial sheaf $(\mathscr{O}_{\mathsf{H}^g})^{\oplus g}$ with action of $\mathrm{Sp}_{2g}(\mathbb{Z})$ given by the 1-cocycle $(CZ + D)$. The invertible sheaf $\underline{\omega}$ on $\mathbf{A}_g^{\mathrm{an}}$ is the trivial sheaf $\mathscr{O}_{\mathsf{H}^g}$ with action of $\mathrm{Sp}_{2g}(\mathbb{Z})$ given by the 1-cocycle $\det(CZ + D)$.

1.4.3. Recall that the classical Riemann theta function is the complex holomorphic function on $\mathbb{C}^g \times \mathsf{H}_g$ given by the theta series

$$\theta(\zeta, Z) = \sum_{n \in \mathbb{Z}^g} \mathbf{e}(1/2 \cdot {}^t n \cdot Z \cdot n + {}^t n \cdot \zeta),$$

where $\mathbf{e}(\cdot) = \exp(2\pi\sqrt{-1}\cdot) \colon \mathbb{C} \rightarrow \mathbb{C}^*$ is the exponential function. $\theta(\zeta, Z)$ gives a nonzero section of an ample invertible cubical sheaf $\underline{L}_Z$ on A_Z that defines the principal polarization. The pull-back of $\underline{L}_Z$ to $\mathbb{C}^g$ is cubically trivial. Easy calculation shows that

$$\theta(\zeta + m, Z) = \theta(\zeta, Z) \quad \forall m \in \mathbb{Z}^g,$$

$$\theta(\zeta + Z \cdot m, Z) = \mathbf{e}(-1/2 \cdot {}^t m \cdot \zeta - 1/2 \cdot {}^t m \cdot Z \cdot m) \cdot \theta(\zeta, Z) \quad \forall m \in \mathbb{Z}^g.$$

Thus we see that if we denote by $\mathbf{E} = \mathbf{e}^g \colon \mathbb{C}^g \rightarrow (\mathbb{C}^*)^g$ the homomorphism "exponentiating down half of the periods $\mathbb{Z}^g \subset \mathbb{Z}^g + Z \cdot \mathbb{Z}^g$," then

(i) $\theta(\zeta, Z)$ factorizes through **E**.

(ii) $A_Z = (\mathbb{C}^*)^g / \mathbf{E}(Z \cdot \mathbb{Z}^g)$ and $\underline{L}_Z$ is defined by the 1-cocycle

$$\mathbf{e}(-1/2 \cdot {}^t m \cdot \zeta - 1/2 \cdot {}^t m \cdot Z \cdot m)$$

on $(\mathbb{C}^*)^g$. Note that this 1-cocycle has the form

$$\mathbf{e}(\text{linear term in } \zeta + \text{term constant in } \zeta).$$

(iii) As a function on $(\mathbb{C}^*)^g$, the theta series $\theta(\zeta, Z)$ assumes the form

$$\sum_\chi (\text{quadratic function } a(\chi) \text{ in } \chi) \cdot \chi,$$

where χ runs through all characters of the algebraic torus $(\mathbb{C}^*)^g$. Since $a(\chi) = \mathbf{e}(1/2 \cdot {}^t n \cdot Z \cdot n)$ if χ is $\mathbf{E}(\zeta) \to \mathbf{E}({}^t n \cdot \zeta)$, we see that *the quadratic function $a(\cdot)$ determines the (multiplicative) periods* $\mathbf{E}(Z \cdot \mathbb{Z}^g)$! Pursuing the analogue of the above in the algebraic context proves to be the key to understanding degeneration of abelian varieties.

2. Degeneration of abelian varieties.

2.1. We first fix the notation of this section. Throughout this section, R will be a *complete* noetherian *normal* local ring, m = the maximal ideal of R, K = fraction field of R, $k = R/\mathrm{m}$, $S = \mathrm{Spec}(R)$, $\eta = \mathrm{Spec}(K)$ = generic point of S, $s = \mathrm{Spec}(k)$ = the closed point of S. For each integer $n \geq 0$, let $R_n = R/\mathrm{m}^{n+1}$, $S_n = \mathrm{Spec}(R_n)$. Let $G \to S$ be a semi-abelian scheme over S, $G_n = G \times_S S_n$, and let $G^\wedge = \varinjlim_n G_n$ be the formal completion of G. We assume the generic fibre G_η *is an abelian variety* and that the *toric part of G_s is split*. (The last assumption on G_s is added only for simplicity. It can be eliminated by descent.) Let $\underline{L}$ be an invertible sheaf on G rigidified along the zero-section. We know that the rigidification already determines a unique cubical structure on $\underline{L}$. We assume that $\underline{L}_\eta$ is *ample* on G_η, which defines a *principal* polarization on G_η.

2.2. *Case when G completely degenerates.*

2.2.1. We first treat the case when the closed fibre G_s is a split torus. In fact the general case is more complicated only in notation, not in substance. The advantage of this special case is that it allows us to see the phenomenon very clearly.

By the rigidity of tori (1.2.3) G_s can be uniquely lifted to a split torus T_n over S_n for each n. The identification of the closed fibre of T_n with G_s uniquely extends to an embedding $T_n \hookrightarrow G_n$; hence $G_n = T_n$ is a split torus over S_n. Therefore $G^\wedge$ is a formal split torus, which we will also denote by $T^\wedge$. Note that $X(T^\wedge) = X(G_s) = X(G_n) \ \forall n$.

The formal cubical invertible sheaf $\underline{L}^\wedge = \underline{L}|_{G^\wedge}$ is trivial and can be descended to $\mathrm{Spf}(R)$ by Proposition 1.2.8. Choose a cubical isomorphism $\underline{L}^\wedge \xrightarrow{\sim} \mathcal{O}_{T^\wedge}$. (The descent data is unique up to a character of the formal torus $T^\wedge$, so we have to make a choice.) Since $\underline{L}_\eta$ defines a principal polarization of G_η, $\Gamma(G_\eta, \underline{L}_\eta)$ is one-dimensional over K and is generated by a section θ. Multiplying θ by an element of R if necessary, we may and do assume that θ is a nonzero global section of $\underline{L}$. On the other hand, it is clear that global sections of $\mathscr{O}_{T^\wedge}$ are m-adically convergent formal power series in the characters of $T^\wedge$ with coefficients in R. Via the (formal) trivialization $\underline{L}^\wedge \xrightarrow{\sim} \mathscr{O}_{T^\wedge}$, we can expand the global (algebraic) section θ as

$$\theta = \sum_{\chi \in X(T^\wedge)} a(\chi) \cdot \chi,$$

where $a(\chi) \in R$ and $a(\chi) \to 0$ m-adically.

2.2.2. We can now state the main result about degeneration of principally polarized abelian varieties in the completely degenerate case.

THEOREM 2.2.2. *Under the above assumptions, we have*
(i) $a(\chi) \neq 0 \ \forall \chi \in X(T^\wedge)$.
(ii) $a(\chi)$ *is multiplicatively quadratic in* χ, *i.e., the function*

$$b(\mu,\nu) = a(\mu+\nu) \cdot a(0)/a(\mu) \cdot a(\nu) \colon X \times X \to K^*$$

is bi-multiplicative.

(iii) *The function* $b(\cdot,\cdot)$ *depends only on the (principal) polarization of* G_η, *and is independent of the invertible sheaf* $\underline{L}$ *giving rise to the polarization or the choice of cubical trivialization* $\underline{L}^\wedge \xrightarrow{\sim} \mathscr{O}_{\hat{T}}$.

(iv) *If we have two semi-abelian schemes* G_1, G_2 *as above and a formal isomorphism* $\hat{\phi} \colon G_1^\wedge \xrightarrow{\sim} G_2^\wedge$ *that identifies the two functions* b_1 *and* b_2, *then* $\hat{\phi}$ *is the formal completion of a (necessarily unique) isomorphism* $\phi \colon G_1 \xrightarrow{\sim} G_2$ *preserving the principal polarization of the generic fibres.*

This theorem is one among those results that are not very hard to prove once they are correctly formulated. Indeed, all one needs to do in this case is to use the algebraic version of the addition theorem for theta functions already formulated by Mumford in [**Mum1**]. The result drops out after executing the necessary details of technicalities.

2.2.3. The function $b(\cdot,\cdot)$ defines a homomorphism $i \colon X \to (X^* \otimes_{\mathbf{Z}} \mathbf{G}_m)(K)$. So in some sense $b(\cdot,\cdot)$ provides a period subgroup of the torus $X^* \otimes_{\mathbf{Z}} \mathbf{G}_m$ and a polarization of periods. The reader should look at 2.3 and 2.4 in order to fully appreciate the above theorem. Let us make some historical remarks. It has long been understood that compactifying moduli spaces depends crucially on being able to analyze and construct degenerations over big base schemes (i.e., dim(base) is high). In many situations the degenerate fibre is *proper*, in which case one would study deformations of a degenerate fibre, construct (uni-)versal deformations and try to algebraize them. In the case of degenerate abelian varieties the degenerate fibre is not proper; this might have deterred people from thinking about it. The case when $\dim(R) = 1$ (i.e., R is a complete d.v.r.) was proved independently by Raynaud (see [**Ra**]) using rigid geometry and Mumford using his theory of algebraic theta functions (for residue characteristic $\neq 2$). In fact the general case follows immediately from the 1-dimensional case, but no one saw the right formulation and the usefulness of it until Faltings came along.

2.2.4. We remark that there is a generalization of 2.2.2 to the case where the polarization of G_η is not necessarily principal. For the notion of periods and polarization of periods in this more general case, see 2.3.3. The proof uses a family of addition formulae.

2.3. *Mumford's construction.*

2.3.1. This is about construction of semi-abelian degeneration of abelian varieties over *complete* base schemes (i.e., Spec(R) with R complete w.r.t. an ideal). Historically, this was done first for elliptic curves, namely, the famous Tate curve. It may be worthwhile to describe it briefly. In the early 1960s (maybe even earlier) Tate had the insight that if we think of a uniformized elliptic curve $\mathbb{C}/\mathbb{Z}+\mathbb{Z}\tau$ as $\mathbb{C}^*/q^{\mathbb{Z}}$ with $q = \mathbf{e}(\tau)$, then we can do the same thing p-adically. More accurately, if K is a complete discrete valuation ring, $\mathscr{O}$ is the ring of integers in K, $q \in K$ with ord(q) > 0, then one can construct an elliptic curve $E = \text{``}\mathbf{G}_m/q^{\mathbb{Z}}\text{''}$ such that $E(\bar{K}) = \bar{K}/q^{\mathbb{Z}}$ with the obvious Galois action. E has split multiplicative reduction over k = residue field of K, and any such elliptic curve arises in this way. The above quotient $\mathbf{G}_m/q^{\mathbb{Z}}$ was made in the sense of rigid geometry, which Tate developed at around the same time. This was a very exciting discovery! Pursuing the same theme, several people, including Morikawa and Tate himself, saw how to construct higher-dimensional abelian varieties over complete d.v.r. as quotients of (split) tori by multiplicative periods. Unfortunately just being able to do it over a one-dimensional base is not good enough, for although the modular curve is one-dimensional, the modular spaces of higher-dimensional (polarized) abelian varieties have dimension > 1.

2.3.2. In 1972, Mumford published a paper full of ingenuity [**Mum2**], in which he constructed a completely degenerate semi-abelian schemes over excellent normal complete base schemes. (This is actually the second paper of his in the same journal that year; the first one deals with degenerate curves.) This paper is excellently written and should be thoroughly studied by anyone serious about degeneration of abelian varieties. I will limit myself just to explain the ideas about this construction and an over-simplified formulation of his result.

Before formulating Mumford's construction, it might be instructive to describe the algebraic way of constructing Tate's curve. We start with $\mathbb{P}^1 \to$ Spec($\mathscr{O}$) with $\mathscr{O}$ a complete d.v.r. as before. It has two obvious sections: 0 and ∞. We first blow up the closed fibres (same as closed points in this case) of 0 and ∞. Then we blow up the closed fibres of the strict transform of 0 and ∞. Keep taking strict transforms of 0 and ∞ and blow up their closed fibres. These birational transforms of $\mathbb{P}^1$ stabilize in an obvious sense and we get in the limit a scheme $\tilde{P}$ locally of finite type over Spec($\mathscr{O}$). The period $q^{\mathbb{Z}}$ naturally acts on $\tilde{P}$ extending the translation action over $K^{\times}$. The point is that $q^{\mathbb{Z}}$ acts *properly discontinuously* (w.r.t. Zariski topology) on the π-adic completion $\tilde{P}^\wedge$ of $\tilde{P}$, where π is a uniformizer of $\mathcal{O}$. So we can take the quotient formal scheme $\tilde{P}^\wedge/q^\mathbb{Z}$ in the naive sense. $\tilde{P}^\wedge/q^\mathbb{Z}$ is *proper* over $\mathrm{Spf}(\mathcal{O})$. Grothendieck's formal existence theorem tells us that $\tilde{P}^\wedge/q^\mathbb{Z}$ is algebraizable. Let P be the algebraization. $P \to \mathrm{Spec}(\mathcal{O})$ is a semistable curve over $\mathrm{Spec}(\mathcal{O})$ of genus one. Let $G^\dagger = P - \{\text{singularities of the closed fibre}\}$. Let G be the dense open subscheme of $G^\dagger$ such that the generic fibre of G is the same and the special fibre of G is the connected component of $G^\dagger \otimes k$ containing $1 \in \mathbf{G}_m(k)$. Then $G^\wedge$ is canonically isomorphic to $\mathbf{G}_m^\wedge$ and $G \to \mathrm{Spec}(\mathcal{O})$ is a semi-abelian scheme whose generic fibre is an elliptic curve.

2.3.3. Let us first set the scene for Mumford's construction. Let R be a noetherian normal domain that is complete w.r.t. an ideal I, $K = $ fraction field of R, $S = \mathrm{Spec}(R)$, $\eta = \mathrm{Spec}(K)$, $S_0 = \mathrm{Spec}(R/I)$. We assume $\sqrt{I} = I$ for convenience. Let $\tilde{G}$ be a split torus of dimension g over S so that $X = X(\tilde{G})$ is a constant free abelian group of rank g. Let Y be a free abelian group of rank g, and $i: Y \to G(K)$ an injective homomorphism. We will identify Y as a subgroup of $G(K)$ and call it the *periods subgroup*.

DEFINITION 2.3.3. (a) A *polarization* of the periods Y is a homomorphism $\phi: Y \to X$ such that
   (i) $\phi(y)(z) = \phi(z)(y) \ \forall y, z \in Y$.
   (ii) $\phi(y)(y) \in I \ \forall y \in Y, \ y \neq 0$.
Notice that ϕ is injective with finite cokernel. We say that ϕ is a *principal polarization* if ϕ is an isomorphism.

(b) Let ϕ be a polarization of Y. An *ample sheaf data* on ϕ is a function $a(\cdot): Y \to K^*$ such that $a(0) = 1$,

$$a(y + z) \cdot a(0)/a(y) \cdot a(z) = \phi(z)(y) \quad \forall y, z \in Y,$$

and that there exists some nonzero element $c \in R$ such that $c \cdot a(y) \in R$ $\forall y \in Y$ and $c \cdot a(y)$ tends to zero in the I-adic topology of R. We also remark that what $a(\cdot)$ does is define an action of Y on the trivial cubical invertible sheaf on $\tilde{G}$.

2.3.4. Here is a formulation of Mumford's result:

THEOREM 2.3.4. *Notation as in 2.3.3.*

(a) *Let ϕ be a polarization of the periods group Y. Then there is a semiabelian scheme $G \to S$ and a homomorphism $\lambda_\eta: G_\eta \to G_\eta^t$ canonically associated to $(\tilde{G}, Y, \phi)$ such that*
   (i) *the formal completion $G^\wedge$ of G is canonically isomorphic to $\tilde{G}^\wedge$. In particular, $G_0 = G \times_S S_0$ is a split torus.*
   (ii) *The generic fibre G_η of G is an abelian variety. In fact for $s \in S$, G_s is an abelian variety iff $\phi(y)(y) \in \mathcal{O}_{s,s}^* \ \forall y \in Y$.*
   (iii) *$\lambda_\eta: G_\eta \to G_\eta^t$ is a polarization of $G_\eta \cdot \deg(\lambda_\eta) = [X: \phi(Y)]^2$.*

(b) *If moreover $a(\cdot): Y \to K^*$ is an ample sheaf data, then there is a rigidified cubical invertible sheaf $\underline{L} = \underline{L}_{\phi,a}$ on G such that $\underline{L}_\eta$ is ample and the polarization $\lambda(\underline{L}_\eta)$ associated to $\underline{L}_\eta$ is equal to λ_η above.*

(c) *Let $X' = Y$, $Y' = X$, $\tilde{G}'$ be the split torus with character group X'. Define $\phi': Y' \to X'$ by $\phi'(y')(x') = \phi(y)(x)$ $\forall y' = x \in X = Y$, $\forall x' = y \in Y = X'$. Then the abelian variety G'_η associated to $(\tilde{G}', Y', \phi')$ is the dual of G_η.*

(d) *If $f: Y \to X$ is a homomorphism and $a(\cdot): Y \to K^*$ is a function such that $a(0) = 1$, $a(y + z) \cdot a(0)/a(y) \cdot a(z) = f(z)(y)$ $\forall y, z \in Y$, and $\exists N > 0$ s.t. $a(y) \cdot \phi(y,y)^N \in R$ $\forall y \in Y$, then there is a rigidified cubical invertible sheaf $\underline{L}_{f,a}$ canonically associated to f and $a(\cdot)$. The association $(f, a) \to$ is additive in the obvious sense. When f is a polarization, $\underline{L}_{f,a}$ coincides with the one given by* (b).

(e) *Let $(\tilde{G}_1, Y_1, \phi_1)$ and $(\tilde{G}_2, Y_2, \phi_2)$ be two triples such that $\tilde{G}_i$ is a split torus over S, Y_i is a periods subgroup of $\tilde{G}_i(K)$, and ϕ_i is a polarization of Y_i for $i = 1, 2$. Let $\tilde{h}: \tilde{G}_1 \to \tilde{G}_2$ be a homomorphism such that $\tilde{h}(Y_1) \subset Y_2$. Then*

(i) *There exists a (necessarily unique) homomorphism $h: G_1 \to G_2$ such that the formal completion $h^\wedge$ of h is equal to $\tilde{h}^\wedge$ via $G_i^\wedge \xrightarrow{\sim} \tilde{G}_i^\wedge$.*

(ii) *If $\tilde{h}^*(\phi_2) = \phi_1$ in the sense that $\phi_1 = \tilde{h}^* \circ \phi_2 \circ \tilde{h}$ where $\tilde{h}^*: X_2 \to X_1$ is dual to $\tilde{h}$, then $\lambda_{1,\eta}$ is the pull-back of $\lambda_{2,\eta}$ in the sense that $\lambda_{1,\eta} = h_\eta^t \circ \lambda_{2,\eta} \circ h_\eta$.*

(iii) *If $a_i(\cdot)$ is a polarization of ϕ_i for $i = 1, 2$ and $a_1(\cdot)$ is the pull-back of $a_2(\cdot)$ in the sense that $a_1(y_1) = a_2(\tilde{h}(y_1))$ $\forall y_1 \in Y_1$, then $\underline{L}_{1,\eta} = h_\eta^*(\underline{L}_{2,\eta})$.*

2.3.5. The idea of Mumford's construction is basically the same as in Tate's curve case. One first constructs a "partial compactification" $\tilde{P}$ of $\tilde{G}$ that is locally of finite type over S and contains $\tilde{G}$ as a dense open subscheme. Moreover $\tilde{P}_\eta = \tilde{G}_\eta$ and the translation action of Y on $\tilde{G}_\eta$ extends to $\tilde{P}$. Y acts properly discontinuously on the formal completion $\tilde{P}^\wedge$ of $\tilde{P}$. $\tilde{P}^\wedge/Y$ is a proper formal scheme over $\text{Spf}(R)$, which can be algebraized to a proper scheme $P \to \text{Spec}(R)$. P contains a dense open subscheme G that is defined by $(P - G)^\wedge = (\tilde{P}^\wedge - (Y \cdot (\tilde{P} - \tilde{G}))^\wedge)/Y$. Then one shows that G inherits from $\tilde{G}$ a natural group law making it a semi-abelian scheme. When there is an ample sheaf data, one can carry it along first to an invertible sheaf on $\tilde{P}$ then to $\tilde{P}^\wedge/Y$ and P to get an ample invertible sheaf on P. All these are guaranteed by a set of axioms on $\tilde{P}$ that Mumford called a "relatively complete model." They always exist but G does not depend on which relatively complete model you use. I will say nothing more but urge you to read his brilliant paper.

2.4.1. Mumford stated in his paper that his construction is the keystone in the compactification of the moduli space of (2-dimensional) abelian varieties. He was certainly right. Let us elaborate it in our context: The two processes in 2.2 and 2.3 are inverse to each other (in a very precise sense). In 2.2 we stated results only for the case G_η is principally polarized; therefore we have implicitly identified the character group X with the periods group Y via $\phi: Y \xrightarrow{\sim} X$ with $\phi(y)(z) = b(y, z)$. (Of course the split torus $\tilde{G}$ involved is the split torus over S with character group X.) Therefore we have analyzed split completely degenerate principally polarized abelian varieties over complete normal base ring R in terms of periods with polarizations. Actually the same is true for arbitrary polarizations, as was remarked in 2.2.4. On the other hand, Mumford's construction tells us that starting with periods and polarizations, we can construct such degenerate abelian varieties.

Now assume that R is a normal complete noetherian local ring. Then the above two processes give a pair of functors that are quasi-inverse to each other, if we suitably formulate the relevant categories. The slogan is: Mumford's construction gives an equivalence of categories. Of course it should be kept in mind that there are several variants of this statement: one may talk about polarizations, or ample invertible sheaves, or arbitrary invertible sheaves, or just (polarizable) abelian varieties.

2.4.2. It is also important to note that Mumford's construction can be done not just over (complete noetherian normal) local rings but actually complete noetherian normal rings with bigger "souls" as well. Roughly speaking the strategy for compactifying $\mathbf{A}_g$ is first to use Mumford's construction (suitably generalized) to construct enough semi-abelian schemes whose generic fibres are principally polarized abelian varieties. One should be careful in constructing these semi-abelian schemes so that they are compatible in the sense that it is possible to glue them in the étale topology. This is where a choice of certain combinatorial data has to be made (the so-called "admissible polyhedral cone decomposition"). The actual gluing is made possible via the above equivalence of categories (suitably generalized also). We will come to more details later.

2.5. *The general (mixed) case.*

2.5.1. We keep the notation and assumptions in 2.1. Again from the rigidity of tori we see that $\forall n \geq 1$, G_n is an extension of an abelian scheme $A_n \to S_n$ by a split torus $T_n \to S_n$. Taking the inductive limit, we see that the formal (m-adic) completion $G^\wedge$ of G is an extension of a formal abelian scheme $A^\wedge \to \mathrm{Spf}(R)$ by a formal split torus $T^\wedge \to \mathrm{Spf}(R)$. Thus we have an exact sequence

$$0 \to T^\wedge \to G^\wedge \to A^\wedge \to 0$$

of commutative formal group schemes; the map $G^\wedge \to A$ will be named $f^\wedge$. As in 2.2, the formal cubical invertible sheaf $\underline{L}^\wedge$ can be descended to a formal invertible sheaf $\underline{M}^\wedge$ on $A^\wedge$ that is relatively ample on $A \to \mathrm{Spf}(R)$ after choosing a descent datum. Moreover all the descent data form a principal homogeneous space under $\mathrm{X} = \mathrm{X}(T^\wedge) = \mathrm{X}(T_s)$, so that if we twist a chosen descent datum by $\chi \in \mathrm{X}$ the effect is to change $\underline{M}^\wedge$ to $\underline{M}_\chi^\wedge = \underline{M}^\wedge \otimes \mathscr{O}_\chi^\wedge$, where $\mathscr{O}_\chi^\wedge$ corresponds to the push-forward of the $T^\wedge$-torsor $T^\wedge \to G^\wedge \to A^\wedge$ by $\chi \colon T^\wedge \to \mathbf{G}_m^\wedge$. We fix such a choice, which will not affect our result at the end. By Grothendieck's formal existence theorem, $A^\wedge \to \mathrm{Spf}(R)$ and $\underline{M}^\wedge$ are algebraizable to an abelian scheme $A \to \mathrm{Spec}(R)$ and an ample cubical invertible sheaf $\underline{M}$ on A.

2.5.2. The extension $0 \to T^\wedge \to G^\wedge \to A^\wedge \to 0$ is coded by a homomorphism $c \colon \mathrm{X} \to A'^\wedge(\mathrm{Spf}(R)) = A'(R)$. c can be used to define an extension $0 \to T \to \tilde{G} \to A \to 0$, where T is the split torus over S with character group X. The map from $\tilde{G}$ to A will be named f. Let $\underline{\tilde{L}} = f^* \underline{M}$ be the pull-back of the ample cubical sheaf on A. Denote by $\underline{M}_\chi$ the twisting of $\underline{M}$ by $\chi \in X$, which is also a descended form of $\underline{\tilde{L}}$. Clearly the formal completions of f, $\underline{\tilde{L}}$, $\underline{M}_\chi$ are just $f^\wedge$, $\underline{L}^\wedge$, and $\underline{M}_\chi^\wedge$ defined earlier.

As before choose a nonzero global section $\theta \in \Gamma(G, \underline{L}) \subset \Gamma(G^\wedge, \underline{L}^\wedge)$. Expanding this algebraic section θ against the formal structure of $G^\wedge$, we can write

$$\theta = \sum_{\chi \in X} f^* \sigma_\chi$$

with $\sigma_\chi \in \Gamma(A, \underline{M}_\chi)$, and the sum converges in the $\mathfrak{m}$-adic topology. Again we would like to produce from the σ_χ's a periods subgroup Y of $\tilde{G}(K)$, a polarization of Y, and an action of Y on $\underline{\tilde{L}}_\eta$.

$\underline{M}^\wedge$ defines a principal polarization on $A^\wedge$ since $\underline{L}_\eta$ does, therefore so does $\underline{M}$. Thus c defines a homomorphism $X \to A(S)$, which we also call c. Let Y be defined to be X as abstract group (the polarization is principal). $\forall \chi \in X$, $T^*_{c(\chi)} \underline{M}$ is canonically isomorphic to $\underline{M}_{-\chi} \otimes_R \underline{M}(c(\chi))$ (note the sign!). Using these canonical isomorphisms, the cubical structure on $\underline{M}$, the biextension structure on $\Lambda(\underline{M}) = m^* \underline{M} \otimes p_1^*(\underline{M})^{\otimes -1} \otimes p_2^*(\underline{M})^{\otimes -1}$, and the addition theorem, we can carry out everything we did in the completely split case and get the polarization. In this case, the suitable notion of principal polarization is a bi-multiplicative map $X \times X \to \Lambda(\underline{M})(K)$ lying over $c \times c$ that satisfies a suitable positivity condition. Although there is nothing really new, it requires much more complicated notation to state results correctly, and one has to exercise a lot of care to keep track of all the canonical isomorphisms involved. So I will be a coward and just leave things as they stand now. Details of the above together with generalizations to arbitrary polarizations can be found in [Fa, Ch, Fa-Ch].

The situation for Mumford's construction is the same: notation becomes pretty complicated, there are lots of canonical isomorphisms floating around, and the situation tends to be confusing. Other than these, everything is in Mumford's paper. Several people have considered this; Brylinski seems to be the first to put this case in writing. Details can be found in [Ch, Fa, Fa-Ch]. So I chicken out again. The discussion about these two processes is exactly the same as before: Mumford's construction gives an equivalence of categories. Again we refer to [Fa, Fa-Ch] for details.

3. Toroidal compactification of A_g.

3.1. *Admissible polyhedral cone decompositions.*

3.1.1. We fix a free abelian group X of rank g. $X_\mathbf{R} = X \otimes_\mathbf{Z} \mathbf{R}$ is a g-dimensional vector space with integral structure given by X. Denote by C the convex cone of all positive semidefinite symmetric bilinear forms on $X_\mathbf{R}$ whose radicals are defined over $\mathbf{Q}$. $C \subset \mathrm{Hom}_\mathbf{R}(S^2(X), \mathbf{R})$. We give $\mathrm{Hom}_\mathbf{R}(S^2(X), \mathbf{R})$ the integral structure given by $B(X) = \mathrm{Hom}_\mathbf{Z}(S^2(X), \mathbf{Z}) =$ all $\mathbf{Z}$-valued symmetric bilinear forms on X, canonically isomorphic to the second divided power $\Gamma_2(X^*)$ of X^*, where $X^* = \mathrm{Hom}_\mathbf{Z}(X, \mathbf{Z})$. Denote by

C° the interior of C, i.e., the convex open cone of all positive definite symmetric bilinear forms on $X_{\mathbf{R}}$. By definition, a GL(X)-*admissible polyhedral cone decomposition* of C is a collection $\{\sigma_\alpha\}_{\alpha\in J}$ with an indexing set J such that

(i) Each σ_α is a nondegenerate rational polyhedral cone that is open in the smallest $\mathbf{R}$-subspace of $X\otimes\mathbf{R}$ containing itself. In other words, there exist $v_1,\ldots,v_k\in B_2(x)\otimes_{\mathbf{Z}}\mathbf{Q}$ such that $\sigma_\alpha = \mathbf{R}_{>0}\cdot v_1 + \cdots + \mathbf{R}_{>0}\cdot v_k$ and σ_α does not contain any line. ($\sigma_\alpha = \{0\}$ is allowed.)

(ii) C is the disjoint union of the σ_α's.

(iii) $\{\sigma_\alpha\}$ is GL(X)-invariant under the natural action of GL(X) on $B_2(X)$ $\otimes_{\mathbf{Z}}\mathbf{R}$, and $\{\sigma_\alpha\}/\mathrm{GL}(X)$ is finite.

(iv) The closure of each σ_α is the (disjoint) union of finitely many σ_β's.

3.1.2. By classical reduction theory of positive definite quadratic forms, GL(X)-admissible polyhedral cone decomposition always exists. A modern treatment of reduction theory in the form we need can be found in chapter 2 of [SC]. In fact it is possible to choose $\{\sigma_\alpha\}$ such that each σ_α is generated by part of a $\mathbf{Z}$-basis of $B_2(X)$: $\sigma_\alpha = \mathbf{R}_{>0}\cdot v_1 + \cdots + \mathbf{R}_{>0}\cdot v_k$, where $v_1,\ldots,v_k\in B_2(X)$ and can be extended to a basis of $S^2(X^*)$. In the following we assume that we have made such a choice for psychological comfort although it is not logically necessary.

3.1.3. We take this chance to recall the notion of torus embedding. The standard reference is chapter 1 of [TE]. We remark that usually (in the documented literature) one works over an algebraically closed field, in which case one can classify torus embeddings in terms of rational partial polyhedral cone decompositions. In our case we will work over $\mathbf{Z}$ so that the classification part is not available but the construction part is intact, and that's what we need.

Associated to the (chosen and fixed) admissible polyhedral cone decomposition $\{\sigma_\alpha\}$ is a torus embedding $E\hookrightarrow \bar{E}$. Here is a description and some general properties of torus embeddings (this can be done for any rational partial polyhedral cone decomposition of any $\mathbf{R}$-vector space with $\mathbf{Z}$-structure; condition (iii) above is not relevant so far):

(i) E is the split torus over Spec($\mathbf{Z}$) with character group $Q = S^2(X)$, the symmetric quotient of $X\otimes_{\mathbf{Z}} X$. Elements of Q will be identified with linear forms on $B_2(X)\supset C$.

(ii) $\bar{E}$ is a scheme separated over Spec($\mathbf{Z}$) that contains E as a dense open subscheme. The translation action of E on itself extends to $\bar{E}$.

(iii) For each σ_α there is an affine E-invariant dense open subscheme $E(\sigma_\alpha)$ of $\bar{E}$ containing E. $\bar{E}$ is the union of various $E(\sigma_\alpha)$'s.

(iv) For each cone $\sigma = $ some σ_α,

$$E(\sigma) = \mathrm{Spec}(B_\alpha)$$

with $B_\alpha = \mathbf{Z}[q]_{q\in Q, q\geq 0 \text{ on } \sigma}$. Notice that $\{q\in Q, q\geq 0 \text{ on } \sigma\}$ is a subsemigroup of Q containing the unit of Q that generates Q. The coordinate ring B_α

of $E(\sigma)$ is just the group ring of this semigroup over $\mathbb{Z}$. When $\sigma = \{0\}$, $E(\sigma) = E$.

(v) If $\sigma_\alpha \subset \bar\sigma_\beta$ where $\bar\sigma_\beta$ is the closure of σ_β, then $E(\sigma_\alpha) \subset E(\sigma_\beta)$.

(vi) If $\bar\sigma_\alpha = \bar\sigma_\beta \cap \bar\sigma_\gamma$, then $E(\sigma_\alpha) = E(\sigma_\beta) \cap E(\sigma_\gamma)$.

(vii) $\bar E$ has a natural stratification by locally closed subschemes Z_α, $\alpha \in J$. Each stratum Z_α is an orbit of E on $\bar E$ that is canonically isomorphic to a quotient torus of E. $\sigma_\alpha \subset \bar\sigma_\beta$ iff $\overline{Z}_\alpha \supset Z_\beta$.

(viii) $\forall \alpha \in J$, $Z_\alpha \cong \mathrm{Spec}(\mathbb{Z}[q]_{q \in Q, q=0 \text{ on } \sigma})$. Z_α is a closed subscheme of $E(\sigma_\alpha)$ via $\mathbb{Z}[q]_{q \in Q, q=0 \text{ on } \sigma} \xrightarrow{\sim} B_\alpha / I_\alpha$, where $I_\alpha =$ the ideal of B_α generated by $\{q|_{q \in Q, q>0 \text{ on } \sigma}\}$.

(ix) $\bar E$ is smooth over $\mathrm{Spec}(\mathbb{Z})$ by the assumption that each σ_α is generated by part of a $\mathbb{Z}$-basis.

Notice that $\mathrm{GL}(X)$ acts on $\bar E$ by functoriality of the construction of the torus embedding.

3.2. *Local charts at the boundary.*

3.2.1. We will first consider the case "near the 0-dimensional cusp," namely the completely degenerate case. We will construct a semi-abelian scheme over the formal completion $\bar E^\wedge$ of $\bar E$ along the locally closed subscheme whose support is $\bigcup Z_\alpha$, where α runs through those with $\sigma_\alpha \subset C^\circ$. You may be troubled by what this means. There is a book by Hakim [Ha] about relative schemes (over any topos). For each cone $\sigma_\alpha \subset C^\circ$, let $R_\alpha = I_\alpha$-adic completion of B_α, $S_\alpha = \mathrm{Spec}(R_\alpha)$, $S_{\alpha,0} = \mathrm{Spec}(R_\alpha / I_\alpha \cdot R_\alpha) \cong Z_\alpha$, $K_\alpha =$ fraction field of R_α. In down-to-earth terms, a semi-abelian scheme $G' \to \bar E^\wedge$ just means that for each $\alpha \in J$, we have a semi-abelian scheme $G_\alpha \to S_\alpha$, and this assignment is compatible with all pull-backs $S_\beta \to S_\alpha$ induced by $E(\sigma_\beta) \subset E(\sigma_\alpha)$, $\sigma_\beta \subset \bar\sigma_\alpha$.

Naturally we would employ Mumford's construction to get the desired semi-abelian schemes $G'_\alpha \to S_\alpha$. (The reason we use G'_α instead of G_α will become clear later.) Clearly R_α is normal because B_α is normal excellent. Take $Y = X$, $\tilde G \to S_\alpha$ the split torus over S_α with character group X. All we need in order to define a principal polarization is to give a bi-multiplicative symmetric map $b_\alpha : X \vee X \to K_\alpha$ such that $b(\chi, \chi) \in I_\alpha \cdot R_\alpha \; \forall \chi \neq 0$, $\chi \in X$. This is very easy: just take the tautological map. Indeed b_α is defined as the composition of canonical maps $X \otimes X \to Q \hookrightarrow K_\alpha$. We recall that $Q = S^2(X)$ is canonically a quotient of $X \otimes X$. Since σ_α is assumed to lie in C°, (the image of) $\chi \otimes \chi > 0$ on σ_α if $\chi \neq 0$, hence lies in I_α. This shows that the positivity condition is satisfied. Therefore Mumford's construction gives us a completely degenerate semi-abelian scheme $G'_\alpha \to S_\alpha$ whose generic fibre is a principally polarized abelian variety.

3.2.2. We have addressed those cones $\sigma_\alpha \subset C^\circ$. What would we do for those σ_α lying in the boundary $C - C^\circ$ of C? Clearly we want to use Mumford's construction in the general (mixed) case. We have to construct a suitable base scheme first. For a general σ_α, let $X \to X_\alpha$ be the smallest quotient of X such that all elements of σ_α factorize through X_α. $B_2(X_\alpha) \otimes \mathbb{R} \subset B_2(X) \otimes \mathbb{R}$ is the rational subspace spanned by σ_α. Let $C(X_\alpha) = C \cap B_2(X_\alpha) \otimes \mathbb{R}$ be the cone of all positive semidefinite symmetric bilinear forms on X_α whose radicals are rational, and let $C(X_\alpha)^\circ$ be the interior of $C(X_\alpha)$ consisting of all positive definite symmetric bilinear forms on X_α. Then $\{\sigma_\beta | \sigma_\beta \subset B_2(X_\alpha) \otimes \mathbb{R}\}$ is a $GL(X_\alpha)$-admissible polyhedral decomposition of $C(X_\alpha)$. Let $E_\alpha \hookrightarrow \bar{E}_\alpha$ be the torus embedding associated to $\{\sigma_\beta | \sigma_\beta \subset B_2(X_\alpha) \otimes \mathbb{R}\}$. We have $E_\alpha = Q_\alpha^* \otimes \mathbf{G}_m$, $\bar{E}_\alpha$ = union of affine E_α-invariant open subschemes $E_\alpha(\sigma_\beta)$'s for $\sigma_\beta \subset C(X_\alpha)$, and $\bar{E}_\alpha$ is stratified by E_α-orbits $Z_\alpha(\sigma_\beta)$'s for $\sigma_\beta \subset C(X_\alpha)$. Notice that the indexing scheme here is not perfect, because we could have used any cone that spans the same subspace as σ_α.

The (formal completion of) torus embedding $\bar{E}_\alpha$ itself will not provide us base schemes for Mumford's construction for several reasons. For one thing the (relative) dimension is wrong: it should be $1/2 \cdot g \cdot (g+1)$, but the dimension of $\bar{E}_\alpha$ is $1/2 \cdot r \cdot (r+1)$ if $r = \mathrm{rk}(X_\alpha)$. Also we have to have an extension of an abelian scheme by a torus over the base. Therefore we should use the universal family. Let $A \to \mathbf{A}_{g-r}$ be the universal abelian scheme with principal polarization $\lambda \colon A \xrightarrow{\sim} A^t$ over $\mathbf{A}_{g-r}$. We will construct an E_α-torsor D_α over $\underline{\mathrm{Hom}}_{\mathbb{Z}}(X_\alpha, A^t)$. We always identify A^t with A, and therefore we can regard the Poincaré $\mathbf{G}_m$-torsor $P \to A \times_{\mathbf{A}_{g-r}} A^t$ as a $\mathbf{G}_m$-torsor $P \to A \times_{\mathbf{A}_{g-r}} A$. D_α is defined by requiring the push-forward of it by $[\mu \otimes \nu] \colon E_\alpha \to \mathbf{G}_m$ to be the $\mathbf{G}_m$-torsor $(c(\mu) \times c(\nu))^* P^{-1}$ for any $\mu, \nu \in X_\alpha$. Here $[\mu \otimes \nu]$ denotes the image of $\mu \otimes \nu$ in Q and $c \colon X_\alpha \to A$ is the tautological map over $\underline{\mathrm{Hom}}_{\mathbb{Z}}(X_\alpha, A^t)$. Define an $\bar{E}_\alpha$-bundle $\bar{D}_\alpha$ over $\underline{\mathrm{Hom}}_{\mathbb{Z}}(X_\alpha, A^t)$ by $\bar{D}_\alpha = D_\alpha \times^{E_\alpha} \bar{E}_\alpha$. $\bar{E}_\alpha$ has a locally closed subscheme W_α that is the union of all E_α-orbits Z_β with $\sigma_\beta \subset C(X_\alpha)^\circ$. Finally let S_α be the formal completion of $\bar{D}_\alpha$ along the locally closed subscheme $D_\alpha \times^{E_\alpha} W_\alpha$. This is the base we wanted. $\bar{D}_\alpha$ has a natural stratification by $D_\alpha \times^{E_\alpha} Z_\alpha(\sigma_\beta)$'s for $\sigma_\beta \subset C(X_\alpha)$.

By its very definition, we have an extension $\tilde{G} \to S_\alpha$ of $A \times_\mathbf{A} S_\alpha$ by $X_\alpha^* \otimes \mathbf{G}_m \to S_\alpha$ together with periods group $i \colon Y_\alpha = X_\alpha \to \tilde{G}(K_\alpha)$ and polarization of Y_α. Applying Mumford's construction, we obtain a semi-abelian scheme $G'_\alpha \to S_\alpha$ whose generic fibre is a principally polarized abelian variety.

3.3. Exercising Artin's method.

3.3.1. In 3.2 we saw how to construct semi-abelian schemes over formal schemes S_α. In other words we constructed semi-abelian schemes $G'_\mu \to S_\mu$, μ running through another indexing set J', where each S_μ has the form $S_\mu = \mathrm{Spec}(R_\mu)$, R_μ being a complete formally smooth w.r.t. ideal I_μ, $\dim(R_\mu) = 1 + 1/2 \cdot g \cdot (g+1)$. Smoothness of R_μ comes from our assumption that each σ_α is generated by part of a $\mathbb{Z}$-basis, which we made for psychological comfort. You may think that the natural thing to do is to algebraize these $G'_\mu \to S_\mu$'s. It would be very nice if we could do this directly. Unfortunately there does not seem to be any functor that can be defined in some elegant and simple way that is represented by some compactification of $\mathbf{A}_g$; therefore the usual algebraization technique cannot be applied directly. But we *can still use Artin's approximation theorem*, which is the foundation of the algebraization method developed by Artin and is more elementary in some sense.

3.3.2. For each $\mu \in J'$, write $R = R_\mu = \varinjlim_\alpha \bar{V}_\alpha$, where each $\bar{V}_\alpha \subset R$ is a finitely generated algebra over $\mathbb{Z}$. By the standard limit argument $G' = G'_\mu$ is defined over $\bar{V}_\alpha$ for some α. By Artin's approximation theorem, the morphism $\text{Spec}(R) \to \text{Spec}(\bar{V}_\alpha)$ can be approximated by $\text{Spec}(V) \to \text{Spec}(\bar{V}_\alpha)$ if we shrink $\text{Spf}(R)$ when necessary, where V is smooth and finitely generated over $\mathbb{Z}$ whose formal completion is R. Thus we obtain a semi-abelian scheme $G \to \text{Spec}(V)$ by pulling back $G' \to \text{Spec}(\bar{V}_\alpha)$. Clearly the generic fibre of G is a principally polarized abelian variety, because we just did a pull-back. The problem is that if we recomplete G, we may *not* get back G' itself.

3.3.3. We keep the notation as above. What we have to show is that if V approximates R close enough (i.e., $\text{Spec}(V) \to \text{Spec}(\bar{V}_\alpha)$ is close to $\text{Spec}(R) \to \text{Spec}(\bar{V}_\alpha)$), then we can control G so that G is very close to G' in a suitable sense. We will explain this only in the case G' is completely degenerate. The mixed case requires more notation than we have developed so far and tends to be confusing, but involves nothing really new.

In 3.3.2, we explained how to obtain $G \to \text{Spec}(V)$ by approximating R, where V is algebraic (i.e., finitely generated over $\mathbb{Z}$) and the completion of V (w.r.t. a suitable ideal defining smooth subscheme of $\text{Spec}(R)$ over $\mathbb{Z}$) is R. Let $G \to \text{Spec}(R)$ be the completion of $G \to \text{Spec}(V)$. Since we assumed that G' is completely degenerate, so is G. Let X be the character group of $G'_0 = G_0$. The main result of §2 says that G (resp. G') determines and is determined by a symmetric bi-multiplicative map $b' : \text{X} \times \text{X} \to K^*$ (resp. $b : \text{X} \times \text{X} \to K^*$) satisfying the positivity condition. The truth is

(∗)   For any given $N > 0$, if V is sufficiently close to R, then b/b' takes values in $(1 + I^N) \subset R^* =$ units of R.

The key point is that b' (modulo the nth power) (resp. b (modulo the nth power)) is already "encoded in" $G'[n] = n$-torsion points of G' (resp. $G[n]$). Since $G'[n]$ is finite over $\text{Spec}(R)$ and étale outside of $\text{Spec}(R/nR)$, we can make $G[n]$ arbitrarily close to $G'[n]$. This proves (∗). We also remark that $G \to \text{Spec}(R)$ is obtained from $G' \to \text{Spec}(R)$ by base change via an isomorphism $f : R \xrightarrow{\sim} R$ very close to identity I-adically.

3.3.4. Let $\text{Spec}(V)^\circ \subset \text{Spec}(V)$ be the largest open subscheme over which G is abelian. By studying the Kodaira-Spencer map, we can achieve that the classifying morphism $\text{Spec}(V)^\circ \to \mathbf{A}_g$ is *étale* at the generic point of $\text{Spec}(V)$ and induces an isomorphism $S^2(\Omega_{G/V}) \simeq \Omega^1_V(d \log \infty)$, provided $\text{Spec}(V) \to \text{Spec}(\bar{V})$ is sufficiently close to $\text{Spec}(R) \to \text{Spec}(\bar{V})$.

3.4. *Gluing.*

3.4.1. By the discussion in 3.3, we can construct semi-abelian schemes $G_\mu \to U_\mu = \text{Spec}(V_\mu)$ that approximate very well the semi-abelian schemes $G'_\mu \to S_\mu$'s constructed in 3.2. Adjoining to the U_μ's a collection of local charts of $\mathbf{A}_g$, we obtain an affine scheme $U = \coprod_\nu U_\nu$ and a semi-abelian scheme $G \to U$. Let $U^\circ \subset U$ be the largest open subscheme of U such that

$G^\circ = G \times_U U^\circ$ is an abelian scheme. By construction, U° is an étale covering of $\mathbf{A}_g$.

3.4.2. Define a groupoid $p_1, p_2 \colon \mathscr{R} \to U \times_{\mathrm{Spec}(\mathbb{Z})} U$ as follows: first, let

$$\mathscr{R}^\circ = \underline{\mathrm{Isom}}_{\mathrm{pol}}(\mathrm{pr}_1^* G^\circ, \mathrm{pr}_2^* G^\circ) = U^\circ \times_{\mathbf{A}_g} U^\circ.$$

We know that $\mathscr{R}^\circ$ is a groupoid under composition, which is finite over $U \times_{\mathrm{Spec}(\mathbb{Z})} U$. Then, define

$$\mathscr{R} = \text{normalization of } \mathscr{R}^\circ \to U \times_{\mathrm{Spec}(\mathbb{Z})} U.$$

Then we have the important

PROPOSITION 3.4.2. *The two projections* $\mathrm{pr}_1, \mathrm{pr}_2 \colon \mathscr{R} \to U$ *are both étale.*

From this proposition it follows that the composition morphism $\mathscr{R}^\circ \times_U \mathscr{R}^\circ \to \mathscr{R}^\circ$ extends (uniquely) to a morphism $\mathscr{R} \times_U \mathscr{R} \to \mathscr{R}$ and defines a groupoid. Therefore $(U, \mathscr{R} \to U \times_{\mathrm{Spec}(\mathbb{Z})} U)$ defines a (separated) algebraic stack.

To prove Proposition 3.4.2, one shows that the homomorphism $\phi \colon \mathrm{pr}_1^* G \to \mathrm{pr}_2^* G$ over $\mathscr{R}$ extending the tautological isomorphism over $\mathscr{R}^\circ$ is an isomorphism. A key point is that the cone σ_α's do not overlap, and (*) in 3.3.3 basically means that the cones do not change when we do this approximation. Another key point is the openness of versality: one must show that the formal completion of G_μ at a point of U_μ occurs in a $G'_\nu \to S_\nu$, corresponding to passing to a face of a given cone. The major tool is of course the main result of §2. We shall not enter into the details here and refer the readers to [Fa-Ch].

3.5. *Statement of the main result.*

THEOREM 3.5.1. *Let* $\{\sigma_\alpha\}_{\alpha \in J}$ *be a* GL(X)-*admissible polyhedral cone decomposition of* C *such that each* σ_α *is generated by part of a* $\mathbb{Z}$-*basis. Let* $\bar{\mathbf{A}}_g = \bar{\mathbf{A}}_g(\{\sigma_\alpha\}_{\alpha \in J}) = [U/\mathscr{R}]$ *be the algebraic stack constructed in 3.4. Then*

(i) $\bar{\mathbf{A}}_g$ *is smooth and proper over* $\mathrm{Spec}(\mathbb{Z})$. (*Properness is a translation of the semistable reduction theorem in view of the valuative criterion for properness and the result on degeneration in* §2.)

(ii) *There is a semi-abelian scheme* $G \to \bar{\mathbf{A}}_g$ *that extends the universal principally polarized abelian scheme over* $\mathbf{A}_g$.

(iii) $\bar{\mathbf{A}}_g \backslash \mathbf{A}_g$ *is a relative Cartier divisor over* $\mathbb{Z}$ *with normal crossings. Thus* $\bar{\mathbf{A}}_g$ *has a natural stratification "by intersection of divisors at the boundary," whose dual complex is* $\{\sigma_\alpha\}/\mathrm{GL}(X) \colon \bar{\mathbf{A}}_g = \coprod W_{[\sigma]}, [\sigma] \in \{\sigma_\alpha\}/\mathrm{GL}(X)$. *Locally in étale topology* $W_{[\sigma]}$ *looks like* $D_\sigma(\sigma)$ (*notation as in 3.2*).

(iv) *For a cone* σ *in* $\{\sigma_\alpha\}$, *the formal completion* $\bar{\mathbf{A}}_g^{/\sigma}$ *of* $\bar{\mathbf{A}}_g$ *along* $W_{[\sigma]}$ *is isomorphic to the formal algebraic stack* S_σ/Γ_σ *and* $G \to \bar{\mathbf{A}}_g^{/\sigma}$ *is isomorphic to* $(G' \to S_\sigma)/\Gamma_\sigma$ *in the notation of 3.2, where* Γ_σ *is the (finite) subgroup of elements of* $GL(X_\sigma)$ *which stabilize* $\sigma \subseteq B_2(X_\sigma) \otimes \mathbb{R}$.

(v) *Let* $G_1 \to \mathrm{Spec}(R)$ *be a semi-abelian scheme with* R *a noetherian complete local ring such that the generic fibre is a principally polarized abelian* *variety and $G_1 \times_{\mathrm{Spec}(R)} \mathrm{Spec}(R/\mathfrak{m})$ is an extension of an abelian scheme by a split torus. Then whether $G_1 \to \mathrm{Spec}(R)$ is a pull-back of $G \to \bar{A}_g$ can be read off from periods and the cone decomposition $\{\sigma_\alpha\}$. We will only formulate the case when G_1 is completely degenerate: G_1 is a pull-back iff the bilinear symmetric function b satisfies the following condition:*

(*)
$$\{v \circ b \in B_2(X) \otimes \mathbb{R}) | v \colon K^* \to \mathbb{Z} \text{ is a discrete valuation of } K\} \subset \bar{\sigma}$$
for some cone σ in $\{\sigma_\alpha\}$.

Notice that (*) is invariant under $\mathrm{GL}(X)$, therefore makes sense. Here we used the notation in 2.2. The general (mixed) case involves more notation.

3.5.2. COROLLARY. *All geometric fibres of $\bar{A}_g$ (hence also A_g) are geometrically connected.*

The generic fibre of $\bar{A}_g$ is geometrically connected by complex uniformization of A_g. 3.5.1 implies that all fibres of $\bar{A}_g$ are geometrically normal, hence geometrically unibranch. The corollary follows from these and Zariski's connectedness theorem.

3.5.3. REMARK. Theorem 3.5.1 has a variant for $A_{g,n}$. We leave it to the reader for its precise formulation, which is entirely similar. When $n \geq 3$, $\bar{A}_{g,n}$ is an algebraic space. If in addition $\{\sigma_\alpha\}_{\alpha \in J}$ satisfies a certain convexity condition as formulated in chapter 4 of [SC], then $\bar{A}_{g,n}$ is projective over $\mathrm{Spec}\,\mathbb{Z}(1/n, e^{2\pi\sqrt{-1}/n}]$. We make the obvious remark that $\bar{A}_{g,n}^{\mathrm{an}}$ is the smooth toroidal compactification of $A_{g,n}^{\mathrm{an}}$ constructed in [SC]; therefore $\bar{A}_{g,n}$ is just a $\mathbb{Z}$-version of the toroidal compactification.

4. Modular forms and minimal compactification.

4.1. *Koecher principle, q-expansion principle, and Fourier-Jacobi expansion.*

4.1.1. DEFINITION. Let $k \in \mathbb{Z}$. A *Siegel modular form f of weight k* is a global section of $\underline{\omega}^{\otimes k} = \underline{\omega}_{A_g}^{\otimes k}$. In other words, f assigns to every principally polarized abelian scheme $(A \to S, \lambda)$ of relative dimension g an element of $\Gamma(S, \underline{\omega}_{A/S}^{\otimes k})$ that is compatible with all pull-backs. Similarly a *Siegel modular form f of level n and weight k* is a global section of $\underline{\omega}_{A_{g,n}}^{\otimes k}$.

4.1.2. We *assume from now on that $g \geq 2$*, because the case $g = 1$ corresponds to the theory of modular curves and has been very intensively studied.

In 3.2.1 we constructed semi-abelian schemes $G_\alpha' \to S_\alpha$ for $\sigma_\alpha \subset C^\circ$. Let $S_\alpha^\circ \subset S_\alpha$ be the open subscheme over which G_α' is a principally polarized abelian scheme. On S_α, the sheaf $\underline{\omega}_{G'/S_\alpha}$ is canonically isomorphic to $\mathscr{O}_{S_\alpha} \otimes_{\mathbb{Z}} \Lambda^g(X)$. Given a Siegel modular form f of weight k, f evaluated at $G_\alpha' \times_{S_\alpha} S_\alpha^\circ$ is (identified with) an element of $K_\alpha \otimes \Lambda^g(X)^{\otimes k}$, where K_α is the quotient field of R_α. This is just the classical Fourier expansion of f up to a factor $(2\pi\sqrt{-1})^{gk}$, if we remember that R_α is a formal power series ring. According to the usual terminology in the theory of elliptic modular functions, we will call this element of $K_\alpha \otimes \Lambda^g(X)^{\otimes k}$ the *q-expansion* of f. It has the form

$(\sum_{q \in Q} a_q \cdot q) \cdot$(a generator of $\Lambda^g(X)^{\otimes k}$). It is easy to see that the q-expansion of f is independent of the choice of σ_α and is invariant under the natural GL(X)-action. It is also independent of the choice of admissible rational polyhedral cone decomposition. In complex analytic context, the q-expansion principle is just the usual Fourier expansion.

4.1.3. From the fact that f is regular on S_α°, we see that there exists some $q_0 \in Q$ such that $q_0 \cdot (\sum_{q \in Q} a_q \cdot q) \in R_\alpha$. The GL(X)-invariance of $\sum_{q \in Q} a_q \cdot q$ implies that in fact $a_q = 0$ unless $\langle q, C \rangle \geq 0$ (remember that $g \geq 2$). This shows

KOECHER'S PRINCIPLE ($g \geq 2$). Any global section of $\underline{\omega}^{\otimes k}$ over $\mathbf{A}_g$ extends uniquely to a global section $\underline{\omega}^{\otimes k}$ over $\bar{\mathbf{A}}_g$.

4.1.4. Using the fact that $\bar{\mathbf{A}}_g$ is smooth with geometrically connected fibres, we easily deduce the useful

q-EXPANSION PRINCIPLE. (i) The q-expansion defines a homomorphism

$$\Phi_B : \Gamma(\mathbf{A}_g \otimes_\mathbb{Z} B, \underline{\omega}^{\otimes k}) = \Gamma(\bar{\mathbf{A}}_g \otimes_\mathbb{Z} B, \underline{\omega}^{\otimes k}) \to B[\![Q \cap \check{C}]\!]^{\mathrm{GL}(X)} \otimes_\mathbb{Z} \Lambda^g(X)^{\otimes k},$$

for every $\mathbb{Z}$-algebra B. Here $\check{C} = \{q \in Q \otimes_\mathbb{Z} \mathbb{R} | \langle q, C \rangle \geq 0\}$ is the convex dual of C. Φ_B is functorial in B.

(ii) If B is a subalgebra of B', $f \in \Gamma(\bar{\mathbf{A}}_g \otimes_\mathbb{Z} B', \underline{\omega}^{\otimes k})$, and $\Phi_{B'}(f) \in B[\![Q \cap \check{C}]\!] \otimes_\mathbb{Z} \Lambda^g(X)^{\otimes k}$, then in fact f is (the image of) an element of $\Gamma(\bar{\mathbf{A}}_g \otimes_\mathbb{Z} B, \underline{\omega}^{\otimes k})$.

4.1.5. REMARK. The traditional way to define and study rational structure of modular forms is via their q-expansions. The arithmetic structure of modular varieties is then determined by that of modular forms. See Shimura's works [Sh] for (overwhelming) examples of deep arithmetic studies of modular forms along the classical line. Of course our definition of rationality coincides with the traditional one (over fields of characteristic 0), which is (part of) the statement of the q-expansion principle.

4.1.6. We can also expand a Siegel modular form f w.r.t. the semi-abelian scheme $G'_\alpha \to S_\alpha$ for general σ's. What comes out is the analogue of the so-called Fourier-Jacobi expansion. It has the property similar to the statements in the q-expansion principle.

We also remark that all the above results have obvious analogues if we put in level structures.

4.2. *The minimal compactification.*

4.2.1. THEOREM. (i) *For some* $k_0 \in \mathbb{N}$ *with* $k_0 > 1$ *(therefore for all* k *sufficiently divisible), the invertible sheaf* $\underline{\omega}^{\otimes k}$ *on* $\bar{\mathbf{A}}_g = \bar{\mathbf{A}}_g(\{\sigma_\alpha\})$ *is generated by its global sections.*

(ii) *The canonical morphism (which exists by (i))*

$$\bar{\pi} : \bar{\mathbf{A}}_g \to A_g^* = \mathrm{Proj}\left(\bigoplus_{k \geq 0} \Gamma(\bar{\mathbf{A}}_g, \underline{\omega}^{\otimes k})\right)$$

is surjective. A_g^* *is independent of the choice of* $\{\sigma_\alpha\}$.

(iii) *The graded ring $\bigoplus_{k \geq 0} \Gamma(\bar{\mathbf{A}}_g, \underline{\omega}^{\otimes k}$ is finitely generated over $\mathbb{Z}$, so A_g^* is a projective normal scheme of finite type over $\mathrm{Spec}(\mathbb{Z})$.*

(iv) *The image A_g of $\mathbf{A}_g$ under $\bar{\pi}$ is a dense open subscheme of A_g^*. The morphism $\pi = \bar{\pi}|_{\mathbf{A}_g} : \mathbf{A}_g \to A_g$ identifies A_g as the coarse moduli space of $\mathbf{A}_g$.*

(v) *A_g^* has a natural stratification: $A_g^* = A_g \amalg A_{g-1} \amalg \cdots \amalg A_1 \amalg A_0 (\cong \mathrm{Spec}\ \mathbb{Z})$. Each A_i is the coarse moduli space of $\mathbf{A}_i$ and is the image of the union of strata $W_{[\sigma]}$ with $\mathrm{rank}(X(\sigma)) = g - i$. In other words, this stratification is given by the dimension of the abelian part of $G \to \bar{\mathbf{A}}_g$. The strata $A_{g-1}, \ldots, A_1, A_0$ are called the cusps of A_g^*.*

(vi) *Let $\bar{x}_1$, $\bar{x}_2$ be two geometric points of $\bar{\mathbf{A}}_g$, and let $G_{\bar{x}_1}$, $G_{\bar{x}_2}$ be the fibres of G over $\bar{x}_1$ and $\bar{x}_2$. Then $\bar{\pi}(\bar{x}_1) = \bar{\pi}(\bar{x}_2)$ iff the abelian parts of $G_{\bar{x}_1}$ and $G_{\bar{x}_2}$ are isomorphic as principally polarized abelian varieties.*

(vii) *The completed strict local rings of A_g^* can be explicitly described. Again we will only say this for the 0-dimensional cusp A_0. In fact the completion of A_g^* along A_0 is isomorphic to $\mathrm{Spf}(\mathbb{Z}[\![Q \cap \check{C}]\!]^{\mathrm{GL}(X)})$. The general (mixed) case involves more notation.*

4.2.2. The proof of 4.2.1 is not difficult, but we will only give some ideas of the proof. The major work is (i). The quickest way is to quote Theorem 2.1 on p. 208 of [M-B], but it is also possible to prove it directly. The basic idea is that theta constants are modular forms of weight $1/2$, as manifested in the classical context by the theta transformation formula. In the algebraic context, the fact that $\underline{\omega}$ on $\mathbf{A}_g$ is up to torsion the square of the invertible sheaf whose sections are theta constants pops out from the Grothendieck-Riemann-Roch theorem. (For better understanding of this magic, see [Fa-Ch] Chap. I. §5.) Algebraically, theta constants are easier to control, and (i) follows. The proof of (vi) uses the following fact: if C is a complete nonsingular curve over an algebraically closed field k and $G \to C$ is a semi-abelian scheme such that $\deg(\underline{\omega}_{G/C}) = 0$, then G is an extension of an isotrivial abelian scheme $A \to C$ by a torus $T \to C$. (Isotrivial means that there is a finite étale covering $C' \to C$ such that $G \times_C C'$ is constant.) The other statements follow from high-tech, but not difficult, machinery.

4.2.3. REMARKS. (i) An analogue of 4.2.1 holds for $\bar{\mathbf{A}}_{g,n}$. The statement has to be mildly modified. For instance, $\bar{\mathbf{A}}_{g,n}$ is no longer geometrically connected (look at the Weil pairing) and the cusps have more components due to the level structure.

(ii) When $n \geq 3$, $\underline{\omega}$ on $\bar{\mathbf{A}}_{g,n}$ itself can be descended to $\mathbf{A}_{g,n}^*$, unlike the case for A_g^* where we have to raise $\underline{\omega}$ to some power to kill (finite) automorphisms.

(iii) $(A_g^*)^{\mathrm{an}}$ is simply the Satake-Baily-Borel compactification of $(A_g)^{\mathrm{an}}$ (cf. [Sa, B-B, Shb]).

4.3. *p-adic monodromy.*

4.3.1. We conclude this paper by an application of Theorem 4.2.1 about *p*-adic monodromy. Let us first explain the problem. We fix an algebraically closed field k of characteristic $p > 0$, and an integer $g \geq 1$. Consider the moduli stack $\mathbf{A}_g$ and the universal abelian scheme $A \to \mathbf{A}_g = \mathbf{A}$ *over* k. Let $\mathbf{A}^\circ \subset \mathbf{A}$ be the ordinary locus so that $\mathbf{A}^\circ$ is open dense in $\mathbf{A}$ and all geometric points of $\mathbf{A}^\circ$ correspond to ordinary abelian varieties and $\mathbf{A}^\circ$ is maximal w.r.t. this property. (Recall that an abelian variety A over k is ordinary iff $A[p](k) \cong (\mathbb{Z}/p\mathbb{Z})^{\dim(A)}$.) Let $A[p^\infty] = \varprojlim_n A[p^n]$ be the Barsotti-Tate group (or p-divisible group) associated to A. Over $\mathbf{A}^\circ$, we have an exact sequence

$$0 \to A[p^\infty]_{\text{tor}} \to A[p^\infty] \to A[p^\infty]_{\text{ét}} \to 0,$$

where $A[p^\infty]_{\text{tor}}$ (resp. $A[p^\infty]_{\text{ét}}$) are the toric part (resp. étale quotient) of $A[p^\infty]$. Over $\mathbf{A}^\circ$, $A[p^\infty]_{\text{ét}}$ can be identified with an étale sheaf of torsion free $\mathbb{Z}_p$-modules of rank g; therefore it corresponds to a representation

$$\rho \colon \pi_1(\mathbf{A}^\circ) \to \mathrm{GL}_g(\mathbb{Z}_p)$$

well-defined up to conjugation. A natural conjecture is that ρ is actually surjective (i.e., the p-adic monodromy is as big as possible), a question that was raised explicitly in [**Ka-La**]. In the elliptic modular case, this was first proved by Igusa [**Ig**] by analyzing local monodromy near supersingular points. In the Hilbert modular case the analogous statement was first proved by Ribet [**Ri, De-Ri**] by the arithmetic method, using Deligne's result on canonical lifting of abelian varieties over finite fields. Unfortunately it is not easy to use Ribet's method in our present case, because the target group of ρ is not commutative. We would like to indicate a proof of the surjectivity of ρ as a corollary to the theory of minimal compactification.

4.3.2. What we will actually prove is that the local monodromy of ρ near the 0-dimensional cusp w.r.t. $A^\circ \subset A^*$ already contains $\mathrm{SL}_g(\mathbb{Z}_p)$; this clearly suffices. Here A° is the image of $\mathbf{A}^\circ$ in A^*.

First of all, over $\mathbf{A}^\circ$, the Serre dual of $A[p^\infty]_{\text{tor}}$ is $A[p^\infty]_{\text{ét}}$. Therefore it suffices to show that the local monodromy of (the Cartier dual of) $A[p^n]$ over A° near the 0-dimensional cusp w.r.t. $A^\circ \subset A^*$ already contains $\mathrm{SL}_g(\mathbb{Z}/p^n\mathbb{Z})$. Let ∞ denote the 0-dimensional cusp of A^*. $A[p^\infty]_{\text{tor}} \to \mathbf{A}^\circ$ extends to a neighborhood of $\bar\pi^{-1}(\infty)$, because G is a torus over $\bar\pi^{-1}(\infty)$. By 3.5.1, the formal completion of $G \to \bar{A}$ along $\bar\pi^{-1}(\infty)$ is isomorphic to $(G' \to \bar{E}^\wedge)/\mathrm{GL}(X)$, using the notation in 3.2.1. Now we can simply read off the local monodromy of $A[p^n]$ near ∞: it is the image of $\mathrm{GL}(X)$ in $\mathrm{GL}(X/nX)$. In particular, it contains $\mathrm{SL}(X/nX) \cong \mathrm{SL}_g(\mathbb{Z}/p^n\mathbb{Z})$. Note that we have actually shown that the local monodromy of $A[p^\infty]_{\text{ét}}$ near ∞ is the closure of $\mathrm{GL}_g(\mathbb{Z})$ inside $\mathrm{GL}_g(\mathbb{Z}_p)$.

BIBLIOGRAPHY

[**Ar1**] M. Artin, *The implicit function theorem in algebraic geometry*, Papers presented at the Bombay Colloquium in Algebraic Geometry, Tata Inst., Bombay, Oxford Univ. Press, 1969, 13–34.

[**Ar2**] ——, *Algebraic approximation of structures over complete local rings*, Inst. Hautes Études Sci. Publ. Math. **36** (1969), 23–58.

[Ar3] ____, *Algebraization of formal moduli*. I, Global Analysis, Univ. of Tokyo Press-Princeton Univ. Press, 1969, pp. 21–71.

[Ar4] ____, *Versal deformations and algebraic stacks*, Invent. Math. **27** (1974), 165–189.

[Ar5] ____, *Théorème de représentabilité pour les espaces algébriques*, Presses Univ. Montréal, Montréal, 1973.

[SC] A. Ash, D. Mumford, M. Rapoport, and Y. S. Tai, *Smooth compaction of locally symmetric varieties*, Math. Sci. Press, Brookline, Mass., 1975.

[B-B] W. L. Baily and A. Borel, *On the compactification of arithmetically defined quotients of bounded symmetric domains*, Ann. of Math. (2) **84** (1966), 442–528.

[Br] L. Breen, *Fonctions thêta et théorème du cube*, Lecture Notes in Math., vol. 980, Springer-Verlag, 1983.

[Ch] C.-L. Chai, *Compactification of Siegel moduli schemes*, London Math. Soc. Lecture Notes Ser., no. 107, Cambridge Univ. Press, 1985.

[De-Mum] P. Deligne and D. Mumford, *The irreducibility of the space of curves of given genus*, Inst. Hautes Études Sci. Publ. Math. **36** (1969), 75–109.

[De-Ri] P. Deligne and K. Ribet, *Values of abelian L-functions at negative integers over totally real fields*, Invent. Math. **59** (1980), 227–286.

[Fa1] G. Faltings, *Arithmetische Kompaktifizierung des Modulraums der abelschen Varietäten*, Lecture Notes in Math., vol. 111, Springer-Verlag, 1985, pp. 321–383.

[Fa2] ____, *Course given at Princeton University*, Spring 1985.

[Fa-Ch] G. Faltings and C.-L. Chai, *Monograph on arithmetic compactification of Siegel moduli spaces and applications* (to appear).

[Hak] M. Hakim, *Schémas relatifs sur les topos annelés*, Ergeb. Math. Grenzgeb. (3), Bd. 64, Springer-Verlag, 1972.

[Ig] J.-I. Igusa, *On the algebraic theory of elliptic modular functions*, J. Math. Soc. Japan **20** (1968), 96–106.

[Ka-La] N. Katz and S. Lang, *Finiteness theorems in geometric class field theory*, L'Ens. Math. **27** (1981), 285–319.

[TE] G. Kempf, F. Knudson, D. Mumford, and B. Saint-Donat, *Toroidal embeddings*. I, Lecture Notes in Math., vol. 339, Springer-Verlag, 1973.

[M-B] L. Moret-Bailly, *Pinceaux de variété abéliennes*, Astérisque, no. 129, Soc. Math. France, Paris, 1985.

[GIT] D. Mumford and J. Fogarty, *Geometric invariant theory*, Ergeb. Math. Grenzgeb. (3), Bd. 34, second edition, Springer-Verlag, 1982.

[Mum1] D. Mumford, *On the equations defining abelian varieties*. I, II, III, Invent. Math. **1** (1966), 287–354; **3** (1967), 71–135, 215–244.

[AV] ____, *Abelian varieties*, Tata Inst. Fund. Res. Studies in Math., vol. 5, Oxford Univ. Press, 1970.

[Mum2] ____, *An analytic construction of degenerate abelian varieties over a complete ring*, Compositio Math. **23** (1972), 239–272.

[Ra1] M. Raynaud, *Faisceaux amples sur les schémas en groupes et les espaces homogènes*, Lecture Notes in Math., vol. 119, Springer-Verlag, 1970.

[Ra2] ____, *Variétés abéliennes et géométrie rigide*, Proc. Internat. Congr. Math. (Nice, 1970), Vol. 1, Gauthier-Villars, Paris, 1971, pp. 473–477.

[Ri] K. Ribet, *P-adic interpolation via Hilbert modular forms*, Proc. Sympos. Pure Math., vol. 29, Amer. Math. Soc., Providence, R. I., 1975, pp. 581–592.

[Sa] I. Satake, *On the compactification of the Siegel space*, J. Indian Math. Soc. **20** (1956), 259–281.

[Sh] G. Shimura, Shimura's many papers on families of abelian varieties and/or automorphic functions. Satake's book on symmetric domains contains a rather extensive list. The following is just a sampling of them:

    a. *Moduli of abelian varieties and number theory*, Proc. Sympos. Pure Math., vol. 9, Amer. Math. Soc., Providence, R. I., 1966, pp. 312–332.

    b. *Moduli and fibre systems of abelian varieties*, Ann. of Math. (2) **83** (1966), 294–338.

    c. *On canonical models of arithmetic quotients of bounded symmetric domains*. I, II, Ann. of Math. (2) **91** (1970), 144–222; **92** (1970), 528–549.

    d. *On the Fourier coefficients of modular forms of several variables*, Göttingen Nachr. Akad. Wiss., 1975, pp. 261–268.

    e. *On certain reciprocity-laws for theta functions and modular forms*, Acta. Math. **141** (1978), 35–71.

    [SGA3] M. Demazure and A. Grothendieck, *Schémas en groupes*. I, II, III, Lecture Notes in Math., vols. 151, 152, 153, Springer-Verlag, 1971.

    [SGA7] A. Grothendieck, *Groupes de monodromie en géométrie algébrique*, Lecture Notes in Math., vol. 288, Springer-Verlag, 1971.

PRINCETON UNIVERSITY

Proceedings of Symposia in Pure Mathematics
Volume **49** (1989), Part 2

Fourier Analysis, Partial Differential Equations, and Automorphic Functions

LEON EHRENPREIS

Chapter I. Introduction

Our study of theta functions, and automorphic functions in general, may be said to be rooted in the construction of Euler whereby he introduced a variable ζ into series like $\vartheta_3(q) = \sum q^{n^2}$ to obtain

$$(1.1) \qquad \vartheta_3(q, \zeta) = \sum q^{n^2} \zeta^n.$$

By this means properties of $\vartheta_3(q) = \vartheta_3(q, 1)$ could be studied by examining the behavior of $\vartheta_3(q, \zeta)$ as a function of ζ. This interrelation between the behaviors of ϑ_3 on $q = 1$ and on $\zeta = 1$ is an instance of a process we call *intertwining*.

This idea of Euler forms the cornerstone of our present work. In its ultimate form it involves the embedding of the functions or equations one is studying into a space of more dimensions in which other important structures and equations appear. Then the interplay, or intertwining, of these structures and equations allows us to gain much information about the individual structures and equations.

One can consider this passage of increasing the number of variables or equations as the construction of a *hierarchy*. This is studied in detail in [10]. Actually, hierarchy is one of the two methods we use here; the other can be termed *basis*, meaning that we search for a basis in which our equation can be managed.

The present work can be considered as introductory in that we shall not develop any of our constructs to completion, but we shall content ourselves with several illustrations. I am preparing several articles to detail the concepts of this paper.

1980 *Mathematics Subject Classification* (1985 Revision). Primary 11FXX, 22EXX, 35CXX; 35G15, 42BXX.
Work supported by NSF 33-1807-101.

Several ideas used in this article have their source in my book [6], which will be referred to as FA.

To put (1.1) in the framework of Fourier analysis we require the insight of Jacobi, who used variables $\zeta = e^{2iz}$ and $q = e^{i\pi\tau}$ in place of ζ and q. Thus we write

$$(1.2) \qquad \vartheta_3(\tau, z) = \sum e^{\pi i n^2 \tau + 2inz}.$$

Euler's construction is reflected in two crucial properties of $\vartheta_3(\tau, z)$:

(a) $\vartheta_3(\tau, z)$ satisfies the heat equation

$$(1.3) \qquad \frac{\partial \vartheta_3}{\partial \tau} = -\frac{i\pi}{4} \frac{\partial^2 \vartheta_3}{\partial z^2};$$

(b) ϑ_3 is quasi doubly periodic

$$(1.4) \qquad \begin{aligned} \vartheta_3(\tau, z + \pi) &= \vartheta_3(\tau, z), \\ \vartheta_3(\tau, z + \pi\tau) &= e^{-i\pi\tau - 2iz} \vartheta_3(\tau, z). \end{aligned}$$

Property (a) uses the fact that the series for $\vartheta_3(q)$ has only exponents n^2, the coefficients being irrelevant. But Property (b) uses the defining relation for squares $(n + 1)^2 = n^2 + 2n + 1$ and the fact that all the coefficients of q^{n^2} in $\vartheta_3(q)$ are equal.

To get back to our philosophy, which entails examining $\vartheta_3(0, z)$, we observe that $\vartheta_3(0, z)$ has a central significance in the Poisson summation formula

$$(1.5) \qquad \vartheta_3(0, z) = \sum e^{2inz} = \sum \delta_{2\pi n},$$

where δ_a is the Dirac delta function at the point a. This evaluation of $\vartheta_3(0, z)$ has as a standard consequence the modular relation for $\vartheta_3(\tau, 0)$

$$(1.6) \qquad \vartheta_3(\tau, 0) = (-i\tau)^{-1/2} \vartheta_3(-1/\tau, 0).$$

REMARK. Theta functions crop up in many areas of mathematics as this conference is designed to demonstrate. In most cases the theta function appears as a sum over all points in a lattice, perhaps satisfying some congruence condition. But sometimes, as in combinatorics, we obtain sums over lattice points belonging to a proper cone. For example, the natural generating function for the Rogers–Ramanujan identities (see [10]) is a sum over nonnegative integers. This seems to account for the difficulty in proving that this function is automorphic.

We can understand the above idea from the viewpoint of *nonlinear Fourier analysis*. Linear Fourier analysis examines functions in terms of the linear exponentials $\exp(ix \cdot \hat{x})$, where x is a variable in R^k and $\hat{x}$ is a dual variable. For nonlinear Fourier analysis we change our notation. We let $_1x$ be a variable in R^k. Then for each $j \geq 1$ we let $_jx$ be an enumeration of the monomials in R^k. The linear space spanned by $_jx$ is the space of homogeneous polynomials on R^k of degree j. This space is denoted by $_jR^k$. The dimension of $_jR^k = \binom{k+j-1}{j}$. Let τ^i be a dual variable so that $\tau^j \cdot {}_jx$ represents the general homogeneous polynomial of degree j on R^k. We write x for $({}_1x, \ldots, {}_Jx)$ and $\tau = (\tau^1, \ldots, \tau^J)$.

For any function $f({}_1x)$ we define (formally) its nonlinear Fourier transform of degree J by

$$(1.7) \qquad F(\tau^1, \ldots, \tau^J) = \int f({}_1x) \exp(i\tau^1 \cdot {}_1x + \cdots + i\tau^J \cdot {}_Jx) \, dx.$$

The nonlinear Fourier transform F satisfies a great number of heat-like and other partial differential equations. These are discussed in Chapter II.

In particular, $\vartheta_3(\tau, z)$ is the nonlinear Fourier transform of degree 2 of $f = \sum \delta_n$. (Actually there is a factor 2 which we ignore.) In this sense we may say that we have reversed Euler's idea since $\vartheta_3(\tau, z)$ is built from $\vartheta_3(0, z)$.

The most simple-minded generalization of $\vartheta_3(\tau, z)$ is to start with $f = \sum \delta_{{}_1n}$ where $\{{}_1n\}$ represents the lattice points in R^k. The resultant function is denoted by $\Theta(\tau^1, \ldots, \tau^J)$. Thus

$$(1.8) \qquad \Theta(\tau^1, \ldots, \tau^J) = \sum_{{}_1n} e^{i\tau^1 \cdot {}_1n + \cdots + i\tau^J \cdot ({}_1n)^J}.$$

Here $({}_1n)^j$ is the point whose $_jx$ coordinates are the appropriate products of the coordinates of $_1n$. Thus if the lth coordinate of $_jx$ is

$$(_jx)_l = {}_1x_1^{l_1} \cdots {}_1x_k^{l_k},$$

where $l_1 + \cdots + l_k = j$, then

$$(1.9) \qquad (_1n)_l^j = {}_1n_1^{l_1} \cdots {}_1n_k^{l_k}.$$

Thus we regard j as a power in $(_1n)^j$ and as a superscript in τ^j.

The relation between Fourier transform on the τ^1 axis and analogs of (1.6) on the τ^j axis, which is another instance of intertwining, can be thought of as a general form of the Weil representation (see [34]). For $J = 2$ it bears a close connection with a transformation introduced by Appell [1] for the heat equation. This is clarified in Chapter IV.

There is an important distinction between the cases $J = 2$ and $J > 2$. For $J = 2$ we find that there is a transformation like (1.6), which is essentially equivalent to the (linear) Fourier invariance of $\Theta(\tau^1, 0)$. Equation (1.6) is of a particular form that we term *geometric*, meaning that is built from transformations of τ (independent variable) and multiplication of Θ (dependent variable). Put in other terms, a geometric transformation has a very simple form on the basis $\{\delta_\tau\}$, namely $\delta_\tau \to a(\tau)\delta_{b(\tau)}$). Thus it is like a multiplier representation. For $J > 2$ no such geometric transformations exist.

Nevertheless, there are many linear transformations that leave, say, $\Theta(0, 0, \tau^3)$ invariant. For, the isotropy group in the space of linear transformations of any function is large. We can even distinguish one such transformation that characterizes this Θ function in analogy with Hamburger's theorem for the zeta function or the result of Ehrenpreis–Kawai concerning the uniqueness of $\sum \delta_n$ (see [16]).

However, it is the intrinsic "beauty" of geometric transformations that has set apart the case $J = 2$ from $J > 2$.

What can we obtain from the theta function? We have already mentioned how Fourier invariance on the τ^1 axis can be translated to an invariance on the τ^j axis. This depends on having an explicit expression for the fundamental solution of the heat equation. This is, in fact, one of the standard ways of deriving the modularity (1.6) of ϑ_3 from the Fourier invariance (1.5).

Because Θ satisfies many partial differential equations, it is determined by its values, or the values of it and some derivatives on certain subsets of $\sum_j R^k$. If o_1 and o_2 are two such subsets, then we might be able to find an *explicit* expression for some kind of "fundamental solution" e that enables one to express the solution on o_1 in terms of the solution on o_2. Since the restrictions of Θ to o_1 and o_2 can be expressed by the definition (1.8), this gives certain integral identities for Θ. In Chapter VI below we make this explicit; in the simplest situation of $k = 2$, $J = 2$ these identities, for suitable o_1, o_2, reduce to the Hecke-Siegel theory of periods of Eisenstein series (see [33]).

Sets like o_1 and o_2 are called *parametrization sets* for the system of differential equations; they are discussed in detail in Chapter II. Let me make some heuristic remarks here.

There are several different kinds of parametrization sets o and corresponding parametrization problems:

(a) o that are of the natural dimension associated to the system of partial differential equations.

(b) o whose dimension is less than this natural dimension;

(c) o whose dimension is greater than this natural dimension.

(d) o that are preserved by some nontrivial subgroup of the conformal group of the system of differential equations. [The conformal group is the group of spacial transformations that preserves the kernel of the equation up to a multiplicative factor (see [8]).]

(e) If the system of equations has constant coefficients, then the variety V of common zeros of the Fourier transforms of the operators is a parametrization set of a different kind since all solutions are Fourier transforms over V by the Fundamental Principle of FA. The dimension of V is as in (a).

For the heat equation for $k = 1$ the t and x axes are examples of (a). Appell's idea (mentioned above) shows that the t axis is an example of (d); this example provides us with a new insight into the Weil representation.

As we shall see later, the time axis for the wave equation is an example of (b). Such parametrization sets play a crucial role in our work. [This time axis is also an illustration of (d).]

For $1 < j < k$ the τ^j axis is an example of (c). Another important situation occurs when there are two different types of parametrization sets of type (d). Then we can interplay the two subgroups. An example of this occurs for the wave operator. The parametrization sets in question are the time axis and the hyperboloids [and also (e)]. This situation contains as special cases many instances of separation of variables and also Roger Howe's theory of dual pairs (see [21]).

The above "intertwining" between o_1 and o_2 can be regarded from the general viewpoint of *bases*. Thus the fact that there are Cauchy-like problems with data given on o_1 and o_2 means that the δ functions of the points on o_1 (and some "normal derivatives") form a basis for the dual of the space of solutions of the system of partial differential equations; a similar property holds for o_2.

The fundamental solution gives us a way of comparing these bases, which we call intertwining. Actually one needs a slight variation of the fundamental solution, which we call a *null solution*, which is a solution having special conditions placed on its support so that suitable integrals will converge. The formalism goes as follows: Suppose we can find null solutions $K(\lambda, \tau)$ depending on the parameter λ so that for λ varying in some suitable set the restrictions of $K(\lambda, \tau)$ to o_j form a dense set of Cauchy data for $j = 1$ and $j = 2$. Then clearly the relation

$$(1.10) \qquad \int K(\lambda, \tau_1)\mu_1(\tau_1)\,d\tau_1 = \int K(\lambda, \tau_2)\mu_2(\tau_2)\,d\tau_2$$

for linear functions $\mu_j(\tau_j)\,d\tau_j$ on the Cauchy data on o_j implies

$$(1.11) \qquad \int F(\tau_1)\mu_1(\tau_1)\,d\tau_1 = \int F(\tau_2)\mu_2(\tau_2)\,d\tau_2$$

for all solutions F.

This idea is often used in reverse. Namely, we start with μ_1, then solve (1.10) for μ_2 if we can. Then apply this solution to (1.11). It seems that a related idea dates back to the profound work of L. J. Rogers [29] on theta functions.

In particular we apply (1.11) to $F = \Theta$ to obtain generalizations of the Hecke-Siegel theory.

Note that this construction is useful only if we can find specific pairs μ_1, μ_2 satisfying (1.11). The construction of such pairs is referred to as "explicit calculation" (see Chapters V and VI).

The nonlinear Fourier transform was introduced by the author and Paul Malliavin in [17]. The specific form used in the present work was developed in [15] to study the Radon transform on algebraic varieties. We shall see that it explains why the above results of Hecke and Siegel are expressed in terms of certain types of zeta functions; these zeta functions measure the number of lattice points on various algebraic varieties.

The above constructions are obviously closely related to the group $G = \mathrm{SL}(k, R)$ and to the discrete subgroup $\Gamma = \mathrm{SL}(k, Z)$. In particular we started with the fundamental representation ρ of G on R^k, and then we took ρ^j, which is the jth symmetric power of ρ. This is the representation on $_jR^k$. Using the above inner product between τ and x, we transfer the action of G to τ.

For the theta function Θ we started with the Γ invariant measure $\sum {}_1n$ on R^k. We then formed its image $\sum({}_1n)^j$ in $_jR^k$. Finally Θ could be considered as the linear Fourier transform in $R = \sum {}_jR^k$ of $\sum_{nj} \delta_{({}_1n)^j}$.

Obvious generalizations that occur to us are

(a) use other Lie groups,

(b) use other fundamental representations and direct sums of representations,

(c) use more general discrete subgroups,

(d) use "initial distributions" other than $\sum \delta_{{}_1n}$.

In later sections we shall examine various aspects of these generalizations, except that we shall not deal with (c) here. Actually the generalization we work with most is (d). In particular, instead of starting with $\sum \delta_{{}_1n}$ in the $_1x$ axis, we can start with a discrete orbit of Γ in the $_jx$ axis. Here G acts on $_jR$ by the symmetric power ρ^j as noted above. In Chapter II we discuss these discrete orbits or, more generally, a discrete Γ invariant set Δ, and how to intertwine δ_Δ so as to obtain interesting Γ invariant functions.

The interesting objects we obtain are constructed by

(1) starting with a discrete Γ invariant set Δ,

(2) applying various Fourier transforms,

(3) solving Cauchy or other boundary value problems.

We shall apply the term *Eisenstein* to this approach. The point is that (1) usually begins by picking a point p_0 whose isotropy group G_{p_0} in G has a large intersection $\Gamma_{p_0} = \Gamma \cap G_{p_0}$ with Γ. Then $\Delta = \Gamma p_0$, so we form the Eisenstein series

$$(1.12) \qquad \sum_{p \in \Delta} \delta_p = \sum_{\gamma \in \Gamma/\Gamma_{p_0}} \delta_{\gamma p_0}.$$

Thus the starting object, namely δ_{p_0}, is invariant under a continuous group G_{p_0}.

In contrast to this, *Poincaré* studied series that started with functions f_0 that might be invariant by a subgroup Γ_0 of Γ. Then form the Poincaré series

$$(1.13) \qquad \sum_{\gamma \in \Gamma/\Gamma_0} \gamma f_0.$$

The difference between Eisenstein series and Poincaré series is most clearly illustrated in the upper half-plane $z = x + iy$. The Eisenstein series is

$$(1.14) \qquad E(z, s) = \sum_{\gamma \in \Gamma/\Gamma_0} (\mathrm{Im}\, \gamma z)^s = \sum \frac{y^s}{|cz + d|^s}.$$

The Poincaré series is

(1.15) $$P(z) = \sum_{\gamma \in \Gamma/\Gamma_0} \gamma e^{2\pi i z}.$$

(In this case we could use a multiplier representation for Γ.) The subgroup $\Gamma_0 = \{(\begin{smallmatrix} 1 & n \\ 0 & 1 \end{smallmatrix})\}$ in both cases. But, whereas for the Eisenstein series the function Im z is invariant under a continuous group N, for the Poincaré series the function $\exp(2\pi i z)$ is invariant only under the subgroup Γ_0 of N.

Of course, the difference between Poincaré and Eisenstein series is subtle and both are formed by similar principles.

We have emphasized the use of various bases, which are chosen for their convenience. Now bases are concerned with solutions of equations by *sums*. There is another way of solving equations, namely by a form of *iteration* or *products*. The contrast between these two ideas occurs most simply for the equation $f' - f = 0$. Euler's (iteration) method leads to the solution

(1.16) $$f(x) = \lim(1 + x/N)^N,$$

while power series lead to $\sum x^n/n!$.

When applied to partial differential equations, Euler's method leads to the Wiener–Feynmann path integral (see [9]). The simplest application of the basis method yields the Cauchy–Kowalewski theorem. In Chapter VII we give a more subtle example; the equality of the two methods has as a consequence a new setting for Selberg's trace formula.

There is another way to look at the sum versus product methods. When we have an invariance property such as $Tf = f$, then there is a natural basis associated, namely the "creation basis" (see [10] for other applications). This means that we start somewhere, say δ_x and "create" other elements $\{T^n \delta_x\}$. (We have to take many x.)

More generally, if $Tf = \lambda f$, then the creation basis for T is convenient as it is easy to derive equations for the expansion coefficients for f in this basis. We refer to this as the "dynamic" approach as it moves from one basis element to another. This idea is close to the product idea expressed above.

There is also a "stagnant approach", which is somewhat dual to the above. This works best when T is part of a continuous group. Then, to analyse $Tf = f$ only those group characters or representation functions are relevant that are invariant under the subgroup $\{T^n\}$.

We shall see in Chapter IV how equations (1.4) are used in a dynamic fashion to derive (1.6). Also the contrast of dynamic and stagnant will appear in our discussion of the Poisson Summation Formula.

It is important that the equation we are studying involves only the first power of T. When higher powers occur, then special methods can often be applied. The difficulty in the celebrated Rogers-Ramanujan identities occurs because the natural equation related to these identities is of second order. However it is possible to use creation bases to solve the equation (see [10] for details). Another possibility, which is also discussed in that paper, is to reduce the higher-order equation to a first-order "hierarchy" using a variant of Infeld–Hull factorization.

It is interesting to note that while the Wiener-Feynmann path integral is quantum-mechanical in nature since the integral is over all paths, Selberg's trace formula involves only geodesics, which may be regarded as classical paths. A variation of this idea, whereby we integrate over some *incomplete* paths that are defined group theoretically, is of interest for the construction of elliptic functions. For example, if $k = 2$ and $J = 2$, then the indefinite integral over a ray in the τ^2 axis of a suitable null solution is the inverse of the elliptic function $cn(z, k)$ where k depends on the ray. This is the origin of the relation between $\mathrm{SL}(2, R)$ and elliptic functions.

In higher rank groups we meet a new situation; for we deal with the "indefinite integral" of a closed l form. This indefinite integral is defined on some sort of loop space (see Chapter VII).

For a somewhat different form of the product method let us go back to, e.g., $\vartheta_3(\tau, z)$ for τ fixed but $\neq 0$. Then $\vartheta_3(\tau, z)$ is a quasiperiodic function of z. We have already remarked how bases lead from this to modularity. The multiplicative idea is to consider the zeros of ϑ_3 or, what is the same thing, the singularities of $\log \vartheta_3$ thought of as a function of z.

It is classical that ϑ_3 has simple zeros at the half period $\pi/2 + \pi\tau/2$ and all points congruent under the additive period group. Thus

$$(1.17) \qquad \frac{\vartheta_3'(z)}{\vartheta_3(z)} = \sum \frac{1}{z - \pi(n + 1/2) - \pi(m + 1/2)\tau}.$$

The right side of (1.17) is clearly invariant under translation by π or by $\pi\tau$.

The double periodicity of the singularities of $\log \vartheta_3(z)$ does not entail the double periodicity of $\vartheta_3(z)$. The variation of $\vartheta_3(z)$ from double periodicity is often referred to as *scattering*. An analysis of the scattering associated to (1.17) is made in Chapter V where it leads, again, to the modularity (1.6). A deeper study of this scattering leads to a nonlinear ordinary differential equation for the "nullwerte" $\vartheta_3(\tau, 0)$. This equation was discovered by Jacobi [22] by a somewhat different method.

The general idea of studying a function defined by its singularities stands at the beginning of this article, for we defined Θ by using the Cauchy data $\sum \delta_{\backslash n}$ on the τ^1 axis. Another aspect of this concept appears in the singular series of Hardy, Littlewood, and Ramanujan. Thus $\vartheta_3(\tau, z)$ is described by its singularities at the rational points of the real τ axis. For automorphic forms of higher weight this is not always true due to the presence of cusp forms. The situation is clarified in Chapter V.

There is a third aspect of the dichotomy between additive and multiplicative theory. Up to now the main function we have used is the exponential function, which is related to the additive group. Thus, for example, $\vartheta(\tau, 0)$ is an exponential series. Since the definitive work of Hecke [18], we understand that it is significant to replace the "additive basis" $\exp(ix\hat{x})$ by the "multiplicative basis" x^s (for $x > 0$). This change of basis is accomplished by the Mellin transform, which when applied to $\vartheta_3(\tau, u)$ yields (see [4, vol. I, p. 30])

$$(1.18) \qquad \Gamma\left(\frac{s}{2}\right) F(u,s) = \Gamma\left(\frac{s}{2}\right) \sum \frac{u^{n-1}}{n^s}.$$

For s a positive integer this is the polylogarithm.

The multiplicative structure of n^s would lead us to expect that there is an iteration related to these functions. Perhaps it would not be amiss to say that this iteration is the "multiplicative iteration" related to decomposition into prime factors and so leading naturally to adelization.

CHAPTER II. PARAMETRIZATION PROBLEMS; INTERTWININGS A AND B

II.1. Parametrization problems. The standard treatment of parametrization problems goes roughly as follows: We are given a manifold M and a system of differential operators $\vec{D} = \{D_j\}$ of arbitrary order. (For the present article we can assume that M is some euclidean space.) We are given submanifolds $M_1 \supset M_2 \supset \cdots \supset M_r$, and for each l we are given a finite set of differential operators $\{\delta_l^i\}$, which can be thought of as "normal derivatives" to M_l. (We always set $\delta_l^1 = $ identity.) Then the parametrization problem is expressed in terms of the map

$$(2.1) \qquad P: f \to [\vec{D}f, \{\delta_l^i f|_{M_l}\}].$$

Thus we can enquire as to whether P is injective or whether P is surjective on suitable function spaces. In the simplest case when there is only one D_j, we can deal with one M_l, which is a hypersurface. The reader can then recognize the standard Cauchy and Dirichlet problems. But when there is more than one D_j, we may have to have more than one M_l (compare Chapter IX of [6]).

Instead of considering scalar functions f, one could deal equally well with vector functions. In this case $\vec{D}$ becomes a matrix $\boxed{D}$ and it is possible that $M_1 = M$. There are no essential modifications.

The true subtlety comes from trying to diminish the dimension of the M_l. This can sometimes be accomplished at the expense of using infinitely many $\{\delta_l^k\}$. Let me give two illustrations of this idea, which is developed in detail in [7]. (Originally I called this parametrization problem the "Watergate problem" because I developed it at the time of the Watergate hearings. For the wave equation one example of this problem is using M_1 as the time axis. This means that one person makes all the observations, such as the listening to telephones at Watergate, and then knows everything.)

THEOREM 2.1. *Let*

$$(2.2) \qquad \square = \frac{\partial^2}{\partial t^2} - \frac{\partial x^2}{\partial x_1^2} - \cdots - \frac{\partial^2}{\partial x_n^2}$$

be the wave operator. For each q let $\{h_m^q\}$ be a suitably normalized basis for the harmonic polynomials in x of degree q. Then the parametrization map

$$(2.3) \qquad f \rightarrow \left[\Box f, \left\{ h_m^q \left(\frac{\partial}{\partial x} \right) f \Big|_{x=0} \right\} \right]$$

is injective. Moreover it is surjective on suitable spaces.

This theorem is taken from [7]. We shall not discuss the "suitable spaces" as this would take us too far afield and is not important for us. Let us content ourselves with the remark that we must impose growth conditions in q. In the simplest case of $n = 2$, we can choose

$$(2.4) \qquad h_m^q = h_{\pm}^q = (\partial/\partial x_1 + i\partial/\partial x_2)^q.$$

These constitute the "normal derivatives" to the t axis.

The t axis can be thought of as the fixed set of the compact group K, which is the rotation group about the t axis. K is a maximal compact subgroup of $G = $ Lorentz group $\mathrm{SO}(1, n)$. Taking any other maximal compact subgroup would not affect matters significantly. However, taking another real form $\tilde{K}$ of K does lead to interesting results. We shall meet many examples of this in the sequel, so let me start with the simplest case when we replace the t axis by the x_1 axis. Suppose first that $n = 2$, and write x, y for x_1, x_2.

THEOREM 2.2. *For $n = 2$ the parametrization map*

$$(2.5) \qquad f \rightarrow \left[\Box f, \left\{ \mathscr{F}(\hat{r}^{is} e^{is\hat{\theta}}) f \Big|_{t=y=0} \right\} \right]$$

is injective and it is surjective on suitable spaces.

Here $\hat{r}$ represents the hyperbolic distance in the $\hat{t}, \hat{y}$ plane, and $\hat{\theta}$ is the hyperbolic angle in this plane. $\mathscr{F}$ represents the Fourier transform in the t, y variables. Thus $\hat{r}^{is} \exp(is\hat{\theta})$ represents the hyperbolic analog of $\hat{r}^q \exp(\pm iq\hat{\theta})$, which appears in (2.4). [Since $q \geq 0$ but s is an arbitrary real number we need $\pm$ in (2.4) but not in (2.5).]

In (2.3), we used the functions h_m^q, which, for fixed q, represent a basis for the representation of K on the unit sphere in x space or, more importantly, are the Fourier transforms in x of a basis for the representation of K on the unit sphere in $\hat{x}$ space. In fact, the same property holds for $\hat{r}^{is} \exp(is\hat{\theta})$ when K is replaced by $\tilde{K}$.

The general theory is very complicated and can only be phrased in the language of harmonicity as in [7]. The main idea is that we need some *fixed* set of functions such as $\{h_m^q\}$ or $\hat{r}^{is} \exp(is\hat{\theta})$ whose restrictions to the generic orbits of K or $\tilde{K}$ (acting on the Fourier transform spaces) on the variety of zeros of $\hat{\Box}$ parametrize all functions on those orbits, and which degenerate "nicely" when the orbits degenerate. In the above cases these orbits were $\hat{r} = $ constant so the fact that $\{h_m^q\}$ form a basis is classical.

The above illustrates the main aspect of harmonicity: one basis suffices for all orbits.

Let us examine the general setting of this approach: Suppose the D_j (variables x) have constant coefficients, and let $\vec{\hat{D}}$ represent the Fourier transform of $\vec{D}$. Let V be the algebraic variety in $\hat{x}$ space of common zeros of the $\hat{D}_j$. (According to FA, V might be a multiplicity variety, but we shall assume that it is an ordinary algebraic variety.) Let T be a linear subspace, and write $x = (t, y)$. We want to set up a parametrization problem on T.

What we need is some fixed set of functions $\{h_m^q\}$, which we call *harmonic*, whose restrictions to each set

$$(2.6) \qquad\qquad V_{\hat{t}_0} = \{\hat{t} = \hat{t}_0\} \cap V$$

parametrize a suitable set of functions on $V_{\hat{t}_0}$.

How are these harmonic functions constructed? In the cases we study the T axis is the fixed set for some group G_T that is a subgroup of a linear group G leaving V invariant. Thus G_T acts on the sets $\hat{t} =$ constant. The most favorable situation occurs as follows: The homogeneous polynomial invariants, say $\vec{i} = (i_1, \ldots, i_u)$, of G_T acting on y are *strongly independent* meaning that the set $\vec{i} = $ const is generically of codimension u, or, equivalently, that the Koszul complex associated to $\vec{i}$ has trivial cohomology.

A function $h(y)$ is called *harmonic* if $\vec{i}(\partial/\partial y)h = 0$. Under the above conditions the restrictions of the harmonics to the generic orbits o_c of G_T, which are defined by $\vec{i} = c$, parametrize all polynomials on o_c. (A polynomial on o_c is the restriction to o_c of a polynomial in y.) The functions h_m^q referred to above are a basis for the homogeneous harmonic polynomials of degree q.

The relation of this to the variety V is that the $\hat{y}$ plane is "lifted" to the variety V by raising the orbit $\vec{i} = c$ to the level $\hat{t} = c$. Thus the dimension of the $\hat{t}$ axis is u. Of course we need to know that the sets $V_{\hat{t}_0}$ of (2.6) are the orbits of G_T.

All this works nicely when G_T is compact. But when G_T is not compact, then polynomial harmonics do not suffice. One needs to deal with general homogeneous harmonics with suitable growth conditions. This explains the contrast between Theorems 2.1 and 2.2.

In the noncompact case we can go one step deeper. The point is that the Cauchy data on the T axis provides us with a meaning of the *restriction* of a solution F to the T axis, and of the Cauchy data of F to the T axis. (The restriction corresponds to the Cauchy datum corresponding to δ^1.) But sometimes F may not have an obvious restriction to the T axis so we have to use another device.

The simplest case of this lack of restrictions occurs if we take for F a function whose Fourier transform $\hat{F}$ (which lives on V) is constant on the orbits of G_T, which is not compact. We cannot integrate $\hat{F}$ over such an orbit so as to obtain $F|_T$. Restricting F to the T axis is like restricting δ_0 to the origin.

We shall follow a procedure that is, at its barest level, due to Hecke (or perhaps Poisson). Suppose we are given a discrete subgroup Γ_T of G_T such that G_T/Γ_T is compact. Then we shall not be interested in all Cauchy data but only those Cauchy data that correspond to Γ_T invariant harmonics.

For example, if we are in the situation of Theorem 2.2, then we could consider only those s for which $\exp(is\hat\theta)$ is periodic corresponding to Γ_T. This is the case that Hecke studied in connection with real quadratic fields (see [20]). The point is that for periodic functions it is better to use Fourier series than Fourier integrals.

In the general situation when G_T is a real reductive Lie group, we can define those Cauchy data relative to Γ_T by using harmonics whose restrictions to a generic G_T orbit define the harmonic analysis on G_T/Γ_T. We call these Γ_T *invariant harmonics*.

We obtain meaningful results when we define the Γ_T *restriction* to T by integrating $\hat F$ against Γ_T invariant harmonics over a fundamental domain for Γ_T acting on a generic orbit of G_T.

REMARK. An equivalent definition can be given in terms of residues of a suitable meromorphic function; other formulations are possible. In particular, the idea of Hecke and Siegel corresponds to regarding $\hat F$, which lives on V, as a function on the $\hat y$ axis obtained by projecting V. Then one could take the ordinary Fourier transform $\tilde F$ of $\hat F$ in $\hat y$. But this is Γ_T invariant, so we can form its Fourier coefficients on the orbits of Γ_T. This differs but formally from our construction. We note in passing that $\tilde F$ is the Fourier transform of the Cauchy datum of F on the y axis corresponding to δ^1 =identity (see Chapter IX of FA).

The harmonic polynomials that define the Γ_T restriction will be called *grossencharactere*.

THEOREM 2.3. *Any suitable Γ_T invariant solution F is uniquely determined by the Γ_T restriction of its Γ_T invariant harmonic Cauchy data. The Γ_T invariant Cauchy problem is well posed.*

One can go further and consider the case when the orbits of Γ_T are not compact. When the orbits have finite measure, there is a good harmonic expansion theory, which has its origins in the work of Selberg and Langlands. But when the measure of the orbits is infinite, we do not know how to proceed. (If there is a discrete series, then we can make sense of Γ_T restrictions corresponding to harmonics in the discrete series.)

The theory just presented works very nicely when T is linear. It is possible to extend the scope of this method somewhat beyond the linear case. What we really used was some general form of *separation of variables*. Fourier transform was significant because it gave a good "product splitting" of a basis into functions of t and functions of y. What is really needed is a basis for all functions of x of the form $\varphi_\lambda(t)\psi_\mu(y)$ so that $\vec D\varphi_\lambda\psi_\mu = 0$ if and only if (λ, μ) belongs to some algebraic variety V.

Given such a product basis, we can, under suitable conditions, set up an analog of the above-described harmonicity. This aspect of separation of variables will not appear in the present work.

Rather, our treatment of nonlinear T will be based upon some variation of fundamental solutions. In the case of the wave operator, or, more generally when $\vec{D}$ is a single operator D, and T is of codimension 1, we search for a fundamental solution e meaning that

$$(2.7) \qquad\qquad De = \delta$$

such that when we translate e by x (denote this translate by e_x) then {support of $e_x\} \cap T$ is compact for a "large" set of x. This is a rare situation and corresponds to some sort of hyperbolicity. Also, as the wave equation illustrates, we may have to use various fundamental solutions for various regions of x.

Even if {support $e_x\} \cap T$ is not compact we can proceed. What is crucial is that the restriction of e_x to T or, what is the same thing, the product

$$(2.8) \qquad\qquad e_x \delta_T$$

should be defined. The existence of this product is best formulated in terms of wave front sets; I shall not discuss this here.

When the product (2.8) is defined, then we can apply D to $e_x \delta_{>T}$, where $> T$ refers to the half-space that contains x. Then

$$(2.9) \qquad\qquad De_x \delta_{>T} = \delta_x + P_T,$$

where P_T is a distribution supported by T. Applying (2.9) to suitable solutions of $Df = 0$ gives

$$(2.10) \qquad\qquad f(x) = P_T \cdot f,$$

which is a parametrization problem.

Unfortunately, fundamental solutions can exist only when there is only one D_j because of Hartogs's phenomenon (see [12]). Thus we have to deal with parametrization problems by systematically reducing to the case of a single D_j when we are able to.

There is another level of subtlety in parametrization problems; this comes from *increasing* rather than decreasing the dimension of M_l. To compensate for this, the data on M_l are no longer arbitrary functions, but they satisfy differential equations. We shall meet such examples in Chapter VI when we study nonlinear Fourier transforms in the space defined by the quadratic variable.

Actually, a classical case of this occurs in the study of the restrictions of holomorphic functions of several variables to subvarieties. These restrictions must satisfy $\bar{\partial}_b$ equations.

These parametrization problems are of the type (c) of Chapter I.

II.2 Intertwinings A and B. Parametrization problems lead naturally to the idea of *intertwining*. (We shall introduce other types of intertwining later, so we refer to this as *Parametric Intertwining* or *Intertwining* A.) Suppose that we have two parametrization problems. For simplicity we assume that they are each related to single manifolds M^1 and M^2 respectively. (Recall that in the general case a parametrization is related to a nested sequence of manifolds $M_1 \supset M_2 \supset \cdots \supset M_r$.) Suppose that there is an action of a group G on M^1, that is, either a geometric action or a linear action on the functions on M^1. Then the differential equation allows us to transform this action to M^2.

To see how this is done, let $g \in G$ and let f be a suitable function on M^1. Call F the solution of the parametrization problem whose data on M^1 is $(f, 0, 0, \ldots, 0)$; that is, the restriction of F to M^1 is f and the restriction of the "normal derivatives" is zero.

Since G acts on M^1, it acts on the space of such solutions F. Hence G acts on their parametrization data on M^2. Thus, from an action of G on M^1 we have derived an action on suitable (vector-valued) functions on M^2. Of course, it is important to determine which parametrization data on M^2 correspond to data of the form $(f, 0, 0, \ldots, 0)$.

If G acts on vector-valued functions on M^1, then the same procedure intertwines this action to M^2.

Let me give some illustrations of intertwining that will play a role in what follows.

The simplest example occurs for the heat equation in one space variable, which we write in the form (differing from our usual notation)

$$(2.11) \qquad\qquad \frac{\partial F}{\partial t} = \frac{\partial^2 F}{\partial x^2}.$$

For parametrization sets we choose $M^1 = x$ axis and $M^2 = t$ axis. The parametrization problems we treat are the Cauchy problems with data on M^1 or on M^2. Thus for M^1 the only operator δ is identity while for M^2 we have (identity, $\partial/\partial x$).

G is the cyclic group of order 4 whose generator is $\mathcal{F}$ = Fourier transform on functions of x [so $(\mathcal{F}^2 f)(x) = f(-x)$]. Our task is to determine the intertwining of $\mathcal{F}$ from M^1 to M^2.

The most direct way to attack this problem is to start with two functions f_0, f_1, which represent Cauchy data on the t axis, and follow the diagram:

$$(2.12) \qquad
\begin{array}{ccccc}
(f_0, f_1) & \overset{\mathrm{CP}}{\to} & F(t, x) & \to & F(0, x) = g(x) \\
\downarrow W & & & & \mathrm{FT} \downarrow \\
[\tilde{F}(t,0), \tilde{F}_x(t,0)] & \leftarrow & \tilde{F}(t,x) & \overset{\mathrm{CP}}{\Leftarrow} & \hat{g}(x)
\end{array}$$

Here CP represents solving the Cauchy problem, FT is Fourier transform, and W is the desired map.

A little care must be taken since the CP is not well posed on $t = 0$ unless some growth restrictions in x are imposed on $F(t, x)$ or, what is the same thing, regularity conditions in t. (This is clarified in Chapter IX of [6].) Let us start with f_0, f_1 being exponential, say

$$(f_0, f_1) = [\exp(i\hat{t}_0 t), \exp(i\hat{t}_1 t)]$$

with $\hat{t}_0 \hat{t}_1$ nonvanishing. (The case of $\hat{t}_0 \hat{t}_1 = 0$ presents only minor difficulties.) Then we find easily, using the ideas of Chapter IX of [6]

$$(2.13) \qquad F(t, x) = e^{it\hat{t}_0} \cosh(i\hat{t}_0)^{1/2} x + e^{it\hat{t}_1} \frac{\sinh(i\hat{t}_1)^{1/2} x}{(i\hat{t}_1)^{1/2}}.$$

Continuing in (2.12),

$$(2.14) \qquad g(x) = F(0, x) = \cosh(i\hat{t}_0)^{1/2} x + [\sinh(i\hat{t}_1)^{1/2} x]/(i\hat{t}_1)^{1/2}.$$

In order to apply the Fourier transform, we assume that $(i\hat{t}_0)^{1/2}$ and $(i\hat{t}_1)^{1/2}$ are both pure imaginary, say are $i\alpha_0, i\alpha_1$. This assumption does not affect the computation because such Cauchy data give rise to a dense set. Thus

$$(2.15) \qquad \hat{g}(x) = \frac{1}{2}(\delta_{\alpha_0} + \delta_{-\alpha_0}) + \frac{1}{2i\alpha_1}(\delta_{\alpha_1} - \delta_{-\alpha_1}).$$

$$(2.16) \qquad \begin{aligned} \tilde{F}(t, x) &= \frac{1}{2\sqrt{t}}[e^{-(x-\alpha_0)^2/t} + e^{-(x+\alpha_0)^2/t}] \\ &+ \frac{1}{2i\alpha_1\sqrt{t}}[e^{-(x-\alpha_1)^2/t} - e^{-(x+\alpha_1)^2/t}]. \end{aligned}$$

Hence

$$(2.17) \qquad W(e^{i\hat{t}_0 t}, e^{i\hat{t}_1 t}) = \left[\frac{1}{t^{1/2}} e^{-i\hat{t}_0/t}, \frac{2}{it^{3/2}} e^{-i\hat{t}_1/t} \right].$$

Finally we obtain by linearity

$$(2.18) \qquad W(f_0, f_1) = \left[\frac{1}{t^{1/2}} f_0 \left(\frac{-1}{t} \right), \frac{2}{it^{3/2}} f_1 \left(\frac{-1}{t} \right) \right].$$

(We have "renormalized" some unimportant constants to be 1.)

We conclude

THEOREM 2.4. *The Fourier transform in x corresponds to inversion in t times suitable multipliers as in (2.18).*

This bears a close relation to the Weil (or Segal-Shale-Weil) representation. We shall return to this point in Chapter IV below. The factor i in the right side of (2.18) accounts for the fact that $W^4 = $ identity but $W^2 \neq$ identity.

REMARK. The transformation $t \to 1/t$ times multipliers is geometric. Spatial transformations α like $t \to 1/t$ satisfy identities of the form $\alpha(fg) = \alpha(f)\alpha(g)$. One might wonder as to what the analog of this "distributivity" is for the Fourier transform in x. The answer can be found by careful analysis of the above argument. To no surprise it is the fact that Fourier transform interchanges multiplication and convolution.

There is no great difficulty in extending this result to higher-dimensional x space (still one dimension for t). As in the case of the wave equation discussed above, we must give infinitely many derivatives on the t axis; the normal derivatives are of the form $h(\partial/\partial x)$, where h is a (suitably normalized) spherical harmonic in x. The function F is now expressed using Bessel functions. We find again that the Fourier transform induces the geometric transformation $t \rightarrow -1/t$ times various powers of t on the Cauchy data on the t axis.

One can make the same computation for the wave equation. The intertwining to the t axis of the Fourier transform in x can be expressed in terms of the fundamental solution for the wave equation. Thus in the case of odd number of space dimensions, the component of the Cauchy data corresponding to the spherical harmonic h gets mapped into the Fourier transform times a suitable spherical harmonic.

The above computation for the heat equation works because we are using the quadratic Fourier transform. This is apparent in the passage from (2.15) to (2.16). If we had used a higher order of nonlinearity in the Fourier transform, meaning a higher order equation like $D \equiv \partial/\partial t = \partial^{2l}/\partial x^{2l}$, then we could compute W explicitly in terms of fundamental solutions for D. The result would be an integral kernel; but it is a geometric transformation only for $l = 1$.

We can regard parametric intertwining from a somewhat different point of view. If $m_0^1 \in M^1$, then under suitable conditions we should be able to express the value of a function f on M^1 at m_0^1 in terms of its values on M^2. This can usually be accomplished by means of a fundamental solution as we discussed above. Thus we should have a formula of the type (formally)

$$(2.19) \qquad F(m_0^1) = \int K(m_0^1, m^2) F(m^2)\, dm^2$$

for some suitable kernel K. (F is the solution of $\vec{D}F = 0$ having Cauchy Data $(f, 0, \ldots, 0)$ on M^1. Also the right side of (2.19) may involve all data of F on M^2.)

There is a second intertwining that bears certain similarities with (2.19). We shall refer to it as *Geometric Intertwining* or *Intertwining* B. To define it we suppose that M^1 and M^2 are orbits under the action of a continuous linear group G. Pick base points m_0^1 and m_0^2, and let H^1 and H^2 be their respective isotropy groups in G. Then we can think of the points $m^1 \in M^1$ as the cosets gH^1, which we write as $g(m^1)H^1$.

We want to go from gH^1 to an object on $M^2 = G/H^2$. Thus we must integrate on the right with respect to H^2. This sends

$$(2.20) \qquad gH^1 \rightarrow gH^1H^2.$$

On G/H^2 the set gH^1H^2 can be thought of as $gH^1m_0^2$. Thus we form the H^1 orbit of the base point m_0^2 and then translate by $g = g(m^1)$ on G/H^2.

Put in different terms, if we start with a function $f(m^2)$ we obtain the intertwining

$$(2.21) \qquad If(m^1) = \int f[g(m^1)h^1 m_0^2] \, dh^1.$$

The dual formulation is that the δ function of m^1 corresponds to the δ function of the orbit $g(m^1)H^1 m_0^2$ on M^2.

We think of $g(m^1)H^1 m_0^2$ as a cycle on M^2. Thus the point m^1 corresponds to this geometric cycle.

The relation of (2.21) to (2.19) is rather simple. Since m_0^1 is fixed by H^1, if the differential equation $\vec{D}F = 0$ is G invariant, then we should expect that $K(m_0^1, m^2)$, thought of as a function of m^2, is H^1 invariant, because the value $F(m_0^1)$ is H^1 invariant. More generally we expect a formula of the sort

$$(2.22) \qquad K(m^1, m^2) = \int \alpha(m^1, u)\delta_u \, du.$$

Here $\{u\}$ are the orbits on M^2 of the isotropy group $H(m^1)$ of m^1, and δ_u is the (suitably normalized) $H^2(m^1)$ invariant measure on u. Since

$$H(m^1) = g(m^1)H^1 g^{-1}(m^1),$$

we see that the kernel in (2.21) corresponds to the case when only

$$u = H(m^1)g(m^1)m_0^2$$

appears, meaning that $\alpha(m^1, u) \, d\mu$ has support at this particular u.

Actually, a similar idea appears at several places in our work. We often represent an invariant function or operator by a not necessarily unique integral formula; we want the representation to be invariant. In spirit this is like the condition that a cocycle be a coboundary, or it is like a Tauberian condition. All these conditions are designed to say that properites of invariance, solution of equations, convergence, etc. hold only if they hold for obvious reasons.

Although parametric intertwining as formulated in (2.19) appears to be more general than geometric intertwining as given by (2.21), it is often essentially the same thing. For if we are dealing with a δ_{m^1} that is H^1 invariant, then we expect the kernel K of (2.19) to be H^1 invariant, meaning that it should be of the form (2.22).

Now comes the main point: The differential form $\alpha(m^1, u)\delta_u \, du$ is generally a closed form. This means that we can reduce the integral in the right side of (2.22) to a single orbit, thus obtaining a formula like (2.21). We shall meet several instances of this idea below.

Both intertwinings A and B will usually be applied in the following situation. The group G has an interesting discrete subgroup Γ. Let p_0 be a point in M^1 whose Γ orbit is discrete. Then we form

$$(2.23) \qquad \mu = \sum_{\gamma \in \Gamma/\Gamma_0} \delta_{\gamma p_0},$$

where Γ_0 is the isotropy group of p_0 in Γ. (More generally we can start with $\partial \delta_{p_0}$, where ∂ is a constant coefficient partial differential operator.)

We then intertwine μ from M^1 to M^2. Often we also decompose under the commuting ring of G. We thus obtain an interesting Γ invariant object on M^2.

CHAPTER III.   NONLINEAR FOURIER TRANSFORM;
THE HEAT AND PLÜCKER EQUATIONS; INTERTWINING C

III.1. Nonlinear Fourier transform and Radon transform. We introduced the nonlinear Fourier transform of order J in (1.7). Let me explain briefly its relation to the Radon transformation as that represents one of its origins. Let j be fixed and fix a ray $A = \{a\tau_0^j\}$ in τ^j space. The restriction of F to A is

$$(3.1) \qquad F(0, \ldots, a\tau_0^j, 0, \ldots, 0) = \int f(_1 x) e^{i a \tau_0^j \cdot_j x} \, dx.$$

Note that $\tau_0^j \cdot_j x$ is some fixed homogeneous polynomial of degree j. Let us make a change of variables by using $\tau_0^j \cdot_j x$ as one coordinate and $y = k - 1$ coordinates on the surface $\tau_0^j \cdot_j x = c = $ constant. Then we can rewrite (3.1) in the form

$$(3.2) \qquad F(0, \ldots, a\tau_0^j, \ldots, 0) = \int dc \int f(c, y) e^{iac} \, đy$$

for a suitable measure $đy$. Thus the restriction of F to A corresponds, via Fourier transform, to the integrals of f on the sets $\tau_0^j \cdot_j x = $ constant, that is, to the Radon transform on these nonlinear varieties for a suitable measure $đy$.

Instead of taking one homogeneous polynomial $\tau_0^j \cdot_j x$ we can take several independent $\tau_1^j \cdot_j x, \ldots, \tau_l^j \cdot_j x$. The same argument shows that the restriction of F to the linear space generated by the τ_m^j corresponds, via Fourier transform, to the integrals over the sets $\tau_1^j \cdot_j x = c_1, \ldots, \tau_l^j \cdot_j x = c_l$.

In particular we have

PROPOSITION 3.1. *The restriction of* Θ *to a linear space spanned by independent* τ_i^j *in* τ^j *space represents the counting function for the number of lattice points on sets of the form*

$$(3.3) \qquad \tau_1^j \cdot_j x = c_1, \ldots, \tau_l^j \cdot_j x = c_l.$$

Proposition 3.1 makes sense when the number of lattice points is finite. When the number is infinite, we can sometimes extract finite answers by various techniques. In particular, when the sets (3.3) are orbits of a Lie group H, then there might be a discrete subgroup Γ_0 of H that acts on these lattice points. In such a case it makes sense to count the number of orbits of Γ_0 in the set of lattice points in (3.3) if the number of such orbits is finite. We think of Γ_0 as a "unit group." We shall meet instances of this idea below.

There is an interesting case of Proposition 3.1 that corresponds to matrix multiplication. Let $k = mk'$ and regard R^k as the direct sum of m copies of $R^{k'}$. We write the points in R^k in the form $x = (y^1, \ldots, y^m)$, where y^i has k' components. Let B be a symmetric k' by k' matrix. Think of x as an m by k' matrix. We set

$$(3.4) \qquad\qquad y^i \cdot_B y^{i'} = \sum_{jj'} y^i_j B_{jj'} y^{i'}_{j'},$$

so $y^i \cdot_B y^{i'}$ are the components of xBx^t. In particular if $B = I$ then $y^i \cdot_I y^{i'}$ is the usual inner product $y^i \cdot y^{i'}$.

The $y^i \cdot_B y^{i'}$ are quadratic functions on R^k, which we may write in the form $\tilde{\tau}^2_{ii'} \cdot_2 x$. Thus, according to Proposition 3.1 the restriction of F to the space spanned by the $\tilde{\tau}^2_{ii'}$ corresponds to the integrals of f on the set of matrices x for which xBx^t takes a fixed value.

In particular, if B is integral then the Θ function tells the number of representations of a given symmetric m by m matrix by B. The determination of these numbers is the main problem in the arithmetical theory of quadratic forms.

The variable τ^2_{ij} lives naturally in the Siegel upper half-space H_m. As in the case of $m = 1$, for $B = I$ the function $\Theta(\tau^2_{ij})$ is a Siegel modular form. In fact, $\Theta(\tau^2_{ij})$ is the usual theta function (nullwerte).

III.2. The heat and Plücker equations. The theta function Θ like any other nonlinear Fourier transform satisfies many identities. In the first place there are differential equations. The simplest of these are the *heat equations* (which might be more properly called "Schrödinger" equations)

$$(3.5) \qquad\qquad i^{j-1} \frac{\partial F}{\partial \tau^j_l} = \frac{\partial^j F}{\partial (\tau^1)^j_l}.$$

Here $\partial(\tau^1)^j_l$ is a jth-order operator in the τ^1 variables defined analogously to (1.9).

In addition to the heat equations there are Plücker equations. These are consequences of the heat equations. Now, the heat equations arise from expressing $_j x_l$ in terms of the $_1 x$ coordinates, namely, as functions of $_1 x$ we have $_j x_l = (_1 x)^j_l$. The Plücker equations arise, more generally, when there is an identity amongst various $_j x_l$ thought of as functions of $_1 x$. For example, if $j = 2$ then

$$(3.6) \qquad\qquad [_2 x_{(1,2)}]^2 = [_2 x_{(1,1)}][_2 x_{(2,2)}]$$

since both sides of (3.6) are $(_1 x_1)^2 (_1 x_2)^2$. This leads to the differential equation

$$(3.7) \qquad\qquad \frac{\partial^2 F}{\partial (\tau^2_{(1,2)})^2} = \frac{\partial^2 F}{(\partial \tau^2_{1,1})(\partial \tau^2_{(2,2)})},$$

which is the three-dimensional wave equation.

In addition to differential equations that hold for all nonlinear Fourier transforms there are equations that generalize the second equation of (1.4) (the first is trivial), which we refer to as *periodicity relations*. These apply to Θ because the sum is over all lattice points. As an example of such relations we have, for $k = 1$,

$$(3.8) \qquad \Theta(\tau^1 + 3\tau^3 + 2\tau^2, \tau^2 + 3\tau^3, \tau^3) = e^{-i\tau^3 - i\tau^2 - i\tau^1} \Theta(\tau^1, \tau^2, \tau^3).$$

The reader will have no difficulty in formulating the general relations; they are just a restatement of the binomial theorem.

In case $J = 2$ we are in the situation of (1.4). Then (1.4) asserts that for fixed $\tau = \tau^2$ the function $\vartheta_3(\tau, z) = \Theta(\tau^1, \tau^2)$ is a doubly periodic function of $z = \tau^1$. For $J = 3$ and τ^3 fixed the function $\Theta(\tau^1, \tau^2, \tau^3)$ satisfies 4 periodicity relations, namely periodicity in $\tau^1 \to \tau^1 + 2\pi$, $\tau^2 \to \tau^2 + 2\pi$, and the quasi-periodicity (1.4) (with $\tau = \tau^2$ and $z = \tau^1$ and π replaced by 2π), and also (3.8).

THEOREM 3.1. *The heat equations and the periodicity conditions determine Θ up to a constant multiple*

We postpone a proof of Theorem 3.1 to Chapter IV.

There are some important differences amongst the variables $\tau^1, \tau^2, \tau^3, \dots$. Thus for $\operatorname{Im} \tau^2 > 0$ and $\tau^3, \dots$ real, the Θ function is an entire function of τ^1. In fact, if $\operatorname{Im} \tau^{2j} > 0$, then Θ is an entire function of $\tau^1, \dots, \tau^{2j-1}$. But no similar statement holds for τ^{2j+1}.

Let us go back to the nonlinear Fourier transform; we want to look at it from a different viewpoint. The theory of FA shows that the solutions of the heat equations (3.5) correspond to Fourier transforms on the algebraic variety V

$$(3.9) \qquad\qquad\qquad (_jx)_l = (_1x)_l^j.$$

When we parametrize points of V by the $_1x$ coordinate, such a Fourier transform is exactly a nonlinear Fourier transform.

Of course, the Plücker relations, say on the $_jx$ components, are necessary conditions for numbers $(_jc)_l$ to be of the form $(_1c)_l^j$ for numbers $_1c$. Moreover the projection of V on the $_jx$ coordinate is a j-fold covering. This is seen as follows: Start with x_1^j which is clearly invariant under multiplication by a jth root of unity, say $x_1 \to \zeta x_1$. Once ζ and the $(_jx)_l$ are given then the x_i are determined from $x_1^{j-1} x_i$. Moreover, this determination is compatible with the Plücker relations because we can write, for $\sum l_p = j$

$$(x_1)^{j(j-l_1-1)}(x_1^{l_1} x_2^{l_2} \cdots x_k^{l_k}) = (x_1^{j-1} x_2)^{l_2} \cdots (x_1^{j-1} x_k)^{l_k}$$

which means that any monomial of degree j is a rational expression in $(x_1)^j$ and the $x_1^{j-1} x_i$.

Finally we think of $x_1 \to \zeta x_1$ as being a group of order j acting on V. The invariants are generated by x_1^j so $1, x_1, \dots, x_1^{j-1}$ form a basis for the harmonics. Using our above ideas this leads to

THEOREM 3.2. *The Cauchy problem of the type (c) as for the heat equations can be well posed on the τ^j axis. One must prescribe j data, for which the differential operators can be chosen, for example, as*

$$(3.10) \qquad [\text{identity}, \partial/\partial\tau^j_{1,\ldots,1}, \partial^2/\partial(\tau^j_{1,\ldots,1})^2, \ldots, \partial^{j-1}/\partial(\tau^j_{1,\ldots,1})^{j-1}].$$

The data all satisfy the Plücker relations.

Theorem 3.2 is a parametrization problem of the type (C) as formulated in Chapter I above in that the data themselves satisfy equations.

Let us return to the situation we were concerned with in the discussion following Proposition 3.1. We write $R^k = (R^{k'})^m$. Instead of considering the whole quadratic Fourier transform, we consider its restriction to the space spanned by the $\tilde{\tau}^2_{ii'}$ and by τ^1. We write $\tau^2 = \hat{y}^1, \ldots, \hat{y}^m$ in conformity with $x = (y^1, \ldots, y^m)$.

Write $\widetilde{F}$ for this restriction of the quadratic Fourier transform. Then $\widetilde{F}$ satisfies the heat equations

$$(3.11) \qquad i\frac{\partial\widetilde{F}}{\partial\tilde{\tau}^2_{ii'}} = \sum_{j>j'}\frac{\partial^2\widetilde{F}}{\partial\hat{y}^i_j\partial\hat{y}^{i'}_{j'}}B_{jj'}.$$

In the present situation there are usually infinitely many τ^1 for given $\tilde{t}^2$. The case of interest is when $m \le k'$. Prescribing $\tilde{\tau}^2$ means prescribing the lengths and angles between m vectors in $R^{k'}$. The set of such figures is an orbit of the orthogonal group of the matrix B. When B is definite, we are in the favorable case of the Cauchy problem, while if B is indefinite (but nondegenerate), then we have to deal with the more subtle Cauchy problem as described in Chapter II.

If we choose an arithmetic subgroup Γ_T of the orthogonal group G_T so that G_T/Γ_T has finite measure, then we can set up the Γ_T restriction.

THEOREM 3.3. *The Γ_T restriction defines a well-posed Cauchy problem on the Γ_T invariant solutions of (3.11).*

III. 3. Intertwining C. Let $G = \text{SL}(k, R)$. Then G acts on the $_1x$ axis. Hence by symmetric product G acts on all of the variety V of (3.9). As G acts on each symmetric product, it acts on $_jV$, which is the projection of V on the $_jx$ axis. Via Fourier transform these actions of G give actions of G on various spaces of solutions of heat or Plücker equations.

More generally, let V be any algebraic variety and G a linear group that preserves V. Then by Fourier transform G acts on the space of solutions of the system of equations

$$(3.12) \qquad \vec{D}F = 0,$$

where V is defined by $\vec{\hat{D}} = 0$. (The inner product used in the Fourier transform determines the nature of the action of G on τ. We shall usually deal with a G invariant inner product.)

Let M be a parametrization set for (3.12). Then, via the action of G on the solutions F we obtain an action of G on functions (or vector-valued functions) on M.

We refer to this transferance of the action of G from V to M as *Fourier Intertwining* or *Intertwining* C.

The most important usage for us of Fourier intertwining occurs in the following context. Let Γ be a discrete subgroup of G, and let $p_0 \in V$ be a (real) point whose orbit under Γ is discrete. Call Γ_0 the isotropy group of p_0 in Γ. Then the measure

$$(3.13) \qquad \mu = \sum_{\gamma \in \Gamma/\Gamma_0} \delta_{\gamma p_0}$$

is Γ invariant. Hence so is its Fourier transform. Thus we have a way of going from discrete orbits on V to Γ-invariant objects on M.

Instead of starting with a point or rather with δ_{p_0}, we could start with a differential operator, using the action of G induced on differential operators. Thus we start with $\partial \delta_{p_0}$, where ∂ is a differential operator with constant coefficients. Our primary object is

$$(3.14) \qquad \mu = \sum_{\gamma \in \Gamma/\Gamma_0} (\gamma \partial) \delta_{\gamma p_0},$$

where Γ_0 is now the subgroup of Γ that leaves both p_0 and ∂ invariant.

The Fourier transform of μ is

$$(3.15) \qquad \hat{\mu}(x) = \sum (\gamma \hat{\partial}) e^{i \gamma p_0 \cdot x}.$$

Usually we follow this intertwining by a decomposition under the commuting ring of G. Actually we should use the commuting ring of Γ. This involves some new ideas concerning Hecke operators that we do not yet understand.

PROBLEM. What operates commute with Γ but do not commute with all of G.

Of course, we can give an abstract answer in terms of double cosets, but this is unsatisfactory.

CHAPTER IV.   THE POISSON SUMMATION FORMULA;
THE FUNCTIONAL EQUATION FOR THETA, AND THE WEIL REPRESENTATION

IV.1. The Poisson summation formula. We wish to study the nullwerte equation

$$(4.1) \qquad \vartheta_3(\tau) = (-i\tau)^{1/2} \vartheta_3(-1/\tau) = S\vartheta_3(\tau).$$

Our first task is to analyse the stroke of genius that led Jacobi (and probably Gauss) to replace the variable q by τ where $q = \exp(i\pi\tau)$. How can we formulate (4.1) in terms of q rather than its logarithm?

Perhaps the main thrust of the passage from q to τ is that it allows us to break the forced periodicity $\tau \to \tau + 2$ that occurs when we think of only integral powers of q. This symmetry could also be broken if we consider

q^s instead of q^n. Putting things in other terms, we think of the integers as embedded in the reals so we can apply real analysis.

Why is it so important to drop the integrality of s or the periodicity? The answer is that S does not preserve periodicity. Thus S needs to be formulated in a space without this symmetry. However we must postulate that when applied to the sum of terms that make up ϑ_3, symmetry is restored.

We want to compute S in terms of q. Since ϑ_3 is expressed as a series involving $\{q^n\}$ we shall express S in terms of the Mellin transform. It is actually more conducive to our notation to express S in terms of Fourier transforms, meaning in terms of the basis $\{\exp(i\tau\hat{\tau})\}$ rather than in terms of the basis $\{q^{is}\}$; there is no difficulty in transferring from the Fourier basis to the Melin basis.

The important point is that we want to compute S in a basis other than $\{\delta_\tau\}$ which is relevant for geometric transformations.

Thus we assume, formally, that S is given by an integral kernel L; that is, if

$$(4.2) \qquad f(\tau) = \int \hat{f}(\hat{\tau})e^{i\tau\hat{\tau}}\,d\hat{\tau},$$

then

$$(4.3) \qquad \begin{aligned} (-i\tau)^{-1/2}f(-1/\tau) &= (-i\tau)^{-1/2}\int \hat{f}(\hat{\tau})e^{-i\hat{\tau}/\tau}\,d\hat{\tau} \\ &= \int e^{i\tau\hat{y}}\,d\hat{y}\int L(\hat{y},\hat{\tau})\,\hat{f}(\hat{\tau})\,d\hat{\tau} \\ &= \int \hat{f}(\hat{\tau})\,d\hat{\tau}\int e^{i\tau\hat{y}}L(\hat{y},\hat{\tau})\,d\hat{\tau}. \end{aligned}$$

The Fourier inversion of (4.3) gives

$$(4.4) \qquad \begin{aligned} L(\hat{y},\hat{\tau}) &= \int e^{-i\tau\hat{y}-i\hat{\tau}/\tau}(-i\tau)^{-1/2}\,d\tau \\ &= \int (-i\tau)^{-1/2}e^{-i\hat{y}(\tau+\hat{\tau}/\hat{y}\tau)}\,d\tau \\ &= \lambda(\hat{\tau}/\hat{y})^{1/4}K_{1/2}[2i(\hat{y}\hat{\tau})^{1/2}] \\ &= \tilde{\lambda}\hat{y}^{-1/2}e^{-2i(\hat{y}\hat{\tau})^{1/2}} \end{aligned}$$

for suitable constants $\lambda, \tilde{\lambda}$ as follows from Bateman [4, vol. II, p. 82 (24) and p. 10 (42)]. Here $K_{1/2}$ is a Bessel function. (Strictly speaking, the formulae in Bateman do not apply as the parameters we use are on the boundary of the region of validity; in any case, the formula we use is readily verified directly.) Thus the geometric aspect of S that appears in the τ basis gives way to an integral kernel in the q^s basis.

In case f is periodic, $\hat{f} = \sum a_n\delta_{\pi n}$ so the condition that S fix f is, by (4.4),

$$(4.5) \qquad \sum a_n L(\hat{y},\pi n) = \sum a_n\delta_{\hat{y}=\pi n}.$$

The set of L satisfying (4.5) is very large. The Poisson summation formula is easily seen to be equivalent to saying that the L of (4.4) satisfies (4.5) when $\{a_n\}$ is the characteristic function of the squares.

We can put things in another way. The isotropy group, in the group of operators, of any function is very large. (4.5) is the condition that the operator with integral kernel L preserves the periodic function f. Condition (4.5) is a sort of Poisson summation formula for L.

Thus Jacobi's geometric condition, which is the expression of S in the basis $\{\delta_\tau\}$, corresponds to the Poisson summation condition (4.5). As mentioned above, it is important that we changed from q to τ to have a reasonable geometric formulation. The beauty of the complex variable approach to modular forms depends on the fact that S can be expressed as a geometric transformation.

Hamburger's theorem, as extended by Kawai and myself, shows how to characterize those L that have special forms (see [16]).

PROBLEM 4.1. Find other interesting kernels that satisfy (4.5).

A *Poisson Summation Formula* (PSF) means an intertwining of two distributions of "small support." In its simplest form the PSF asserts

$$(4.6) \qquad\qquad \widehat{\sum \delta_n} = \sum \delta_n.$$

(We have normalized 2π to be 1.) The sums are taken over all lattice points.

Our first approach to PSF is an *operator approach*. It is a simple matter to prove that (4.6) holds up to a multiplicative factor. For $T = \sum \delta_n$ is characterized by

(a) T is invariant under translation by 1 in any coordinate, that is, $T(x_i \to x_i + 1) = T$.

(b) T is a measure that has its support at the integers so that $(\sin x_i)T = 0$ for any i.

In the language of Chapter I, (a) is dynamic and (b) is static.

Note that (a) and (b) are interchanged by the Fourier transform. [Actually by setting $2\pi = 1$ we have a slight contradiction since the Fourier transform of $x \to x + 1$ is multiplication by $\exp(ix)$ so invariance under $x \to x + 1$ corresponds to being annihilated by $\sin(x/2)$; this matter is readily rectified.]

Of course, it remains to calculate one constant, which is the ratio of the right and left sides of (4.6). This can be accomplished in various manners, for example by applying (4.6) to a simple function like the characteristic function of a cube.

When we apply the PSF to the function $\exp(i\tau x^2)$ (for $k = 1$) we obtain the functional equation for the theta function

$$(4.7) \qquad\qquad \vartheta_3(\tau, 0) = (-i\tau)^{1/2} \vartheta_3 \left(-\frac{1}{\tau}, 0 \right).$$

The above proof can be said to emphasize the transformation $\tau \to \tau + 2$ rather than $\tau \to -1/\tau$. When emphasizing the latter transformation it seems that the basis τ^{is} seems more reasonable since $\tau \to -1/\tau$ acts as a permutation, i.e. is geometric, in this basis. In this basis $\vartheta_3(\tau)$ becomes the zeta function. This viewpoint culminates with the deep ideas of Hecke [**18**] on the relation between automorphic forms and Dirichlet series satisfying suitable functional equations.

It is important to note that the basis $\{\tau^{is}\}$ differs considerably from $\{q^{is}\}$. In the former basis $\tau \to -1/\tau$ is geometric while in the latter it is given by an integral kernel.

Up to now we have dealt mostly with the nullwerte equation (4.1). Is there a "natural" passage from $\vartheta_3(\tau)$ to $\vartheta_3(\tau, z)$? To make this question meaningful we remark that in [**8**] we show how to pass from the Bessel function J_0 to the hierarchy $\{J_n\}$. A similar idea applies to the Legendre functions and some other hypergeometric functions. In [**10**] we discuss how a combinatorial analog leads to a "Rogers hierarchy."

Thus we might try to start with the equations $\vartheta_3(\tau + 2) = \vartheta_3(\tau)$ and $(-i\tau)^{-1/2}\vartheta_3(-1/\tau) = \vartheta_3(\tau)$ and "derive" some hierarchy, that is $\vartheta_3(\tau, z)$, from them. However, this does not seem likely because these equations are very far from determining $\vartheta_3(\tau)$. We must impose some growth condition at infinity to obtain uniqueness or even finite dimensionality.

One might search for some equation that "almost" determines $\vartheta_3(\tau)$ and from which the hierarchy appears naturally. Jacobi (see [**22**]) has discovered an ordinary differential equation satisfied by the nullwerte $\vartheta_3(\tau)$. Jacobi's equation is

$$(4.8) \qquad \begin{aligned} (y^2 y''' - 15yy'y'' + 30y'^3)^2 + 32(yy'' - 3y'^2)^3 \\ = y^{10}(yy'' - 3y'^2)^2. \end{aligned}$$

Equation (4.8) was rediscovered by me using scattering theory (see [**13**]). Unfortunately, both proofs use the variable z to derive (4.8). (See Section V.1 below for more details.)

Even if we could derive (4.8) without using z, we do not see how to construct z from (4.8).

PROBLEM 4.2. Derive (4.8) without using z. Use (4.8) to give a natural construction of z.

A solution to this problem might prove useful in defining (directly) a z variable for automorphic forms other than theta functions. (One could express the automorphic form in terms of theta functions and thereby introduce z.)

We want to determine to what extent the equations (1.4) determine $\vartheta_3(\tau, z)$ (compare Theorem 3.1). The first equation in (1.4) expresses the periodicity of ϑ_3 in z. Thus we can write

$$(4.9) \qquad \vartheta_3(\tau, z) = \sum a_n(\tau)e^{2inz}.$$

Impose the second equation:

$$\vartheta_3(\tau, z + \pi\tau) = \sum a_n(\tau)e^{2\pi in\tau}e^{2inz}$$
$$= \sum a_n(\tau)e^{-i\pi\tau}e^{2i(n-1)z}.$$

(4.10)

This means that

(4.11) $$a_n(\tau)e^{2\pi in\tau} = a_{n+1}(\tau)e^{-i\pi\tau},$$

that is,

(4.12) $$a_{n+1}(\tau) = a_n(\tau)e^{2\pi i(n+1/2)\tau}.$$

By iteration this gives

(4.13) $$a_n(\tau) = e^{\pi in^2\tau}a_0(\tau).$$

This is the entire content of equations (1.4) except that a_0 is periodic.

To determine ϑ_3 uniquely we need a condition to guarantee that a_0 is a constant. The simplest such condition is the heat equation (actually the Schrödinger equation)

(4.14) $$\frac{\partial \vartheta_3}{\partial \tau} = -\frac{i\pi}{4}\frac{\partial^2 \vartheta_3}{\partial z^2}.$$

Applying (4.14) to (4.9) and using (4.13) yields $a_0' = 0$, that is, $a_0 = $ constant.

The same method gives a complete proof of Theorem 3.1.

The functional equation (4.7) (up to a constant) is an easy consequence of this, as it is readily verified that the right side of (4.7) satisfies the heat equations and (1.4).

What do the heat equations mean for the nullwerte $\vartheta_3(\tau)$? We want to eliminate z from these equations. We write (4.14) in the form

(4.15) $$\left[\frac{\partial}{\partial z} - \lambda_0 \left(\frac{\partial}{\partial \tau}\right)^{1/2}\right]\vartheta_3 = 0.$$

The square root $(\partial/\partial\tau)^{1/2}$ is defined most simply by

(4.16) $$\left(\frac{\partial}{\partial \tau}\right)^{1/2} e^{\hat{\tau}\tau} = (\hat{\tau})^{1/2}e^{\hat{\tau}\tau}$$

and $\lambda_0^2 = 4i/\pi$. (Compare the paper by Kawai in these proceedings.)

We can now rewrite equations (1.4) in the form [recall that $\exp(a\,d/dx)$ is translation by a]

(4.17) $$\left[\exp \pi\lambda_0 \left(\frac{d}{d\tau}\right)^{1/2} - I\right]\vartheta_3(\tau) = 0,$$

$$\left[\exp \pi\lambda_0\tau : \left(\frac{d}{d\tau}\right)^{1/2} - \exp(-i\pi\tau)\right]\vartheta_3(\tau) = 0.$$

Here we have written $\tau\colon (d/d\tau)^{1/2}$ to indicate that we order powers of this operator by commuting all powers of τ to the left and all differentiations to the right.

REMARK. In some sense we can regard equation (4.14) as a "rationalization" of (4.17).

Actually one can see directly that (4.17) characterizes $\vartheta_3(\tau)$ up to a constant multiple. For the first equation implies that the support of $\hat{\vartheta}_3$ is contained in the squares of the integers. The second equation then, as in the above proof of Theorem 3.1, gives a recursion relation for the value at each integer.

One can search for a direct link between (4.17) and Jacobi's imaginary transformation $\tau \to -1/\tau$. Denote by A and B the operators that occur in (4.17) so ϑ_3 is characterized (up to a constant multiple) by $A\vartheta_3 = B\vartheta_3 = 0$. Denote by S the operator $f(\tau) \to (-i\tau)^{-1/2}f(-1/\tau)$. The passage from (4.17) to (4.1) would be accomplished if we could solve

PROBLEM 4.3. Find operators α and β so that

(4.18) $$\alpha A + \beta B = S - I.$$

If we used the Fourier base for functions, then the computations (4.3)ff. show that $S - I$ is an integral operator whose kernel is essentially

$$\exp[-2\pi i(xy)^{1/2}].$$

The possibility of solving (4.18) depends on the fact that the null space for $S - I$ contains the simultaneous null space of A and B, which, by the above, is $\{a\vartheta_3(\tau)\}$. The vanishing of $S - I$ at $\vartheta_3(\tau)$ has been shown to be exactly the Poisson summation formula.

Of course, if one could solve (4.18) directly, then this would give a new proof of the Poisson summation formula.

Although we do not know how to solve Problem 4.3 we can find β on the kernel of A and α on the kernel of B.

The above may be thought of as operator aspects of PSF. We now give a setting in ergodic theory.

The cornerstone of the PSF is the one-dimensional theorem asserting that $\sum \delta_n$ is its own Fourier transform (when things are suitably normalized). The lattice points in R^k can be obtained from those on the τ^1 axis by applying the modular group Γ. We shall investigate how to use this to derive the PSF for general k from $k = 1$.

For simplicity of notation we shall explain the case $k = 2$. In particular we write $\tau^1 = x, y$. By the one-dimensional PSF, it follows that

(4.19) $$\sum \widehat{\delta_{n,0}} = \sum \delta_{x=n}.$$

Here we have taken the two-dimensional Fourier transform, and $\delta_{x=n}$ is the δ function of the line $x = n$. We want to apply Γ to (4.19). Since Γ commutes with Fourier transform, we should expect

(4.20) $$\sum_{n,\gamma} \widehat{\delta_{\gamma(n,0)}} = \sum \delta_{\gamma(x=n)}.$$

We sum over Γ/Γ_0 where $\Gamma_0 = \{(\begin{smallmatrix} 1 & j \\ 0 & 1 \end{smallmatrix})\}$ is the isotropy group of $(n,0)$ in Γ.

While the left side of (4.20) looks like $2\sum \widehat{\delta_{n,m}}$, the right side seems far from $2\sum \delta_{n,m}$. In fact, it would seem that the right side should be the uniform measure on the plane.

The point is that the origin is a fixed point of Γ. Thus the left side is not $2\sum \widehat{\delta_{n,m}}$ but we have to add something like $\infty \hat{\delta}_{00}$. Thus the PSF for the plane asserts that the right side is $\infty\, dx\, dy + 2\sum \delta_{n,m}$.

To make sense of this, we have to appeal to ergodic theory. The standard Birkhoff ergodic theorem, which asserts that time averages equal space averages for a suitable transformation T on a probability space, is usually written

(4.21) $$\lim \frac{1}{2N+1} \sum_{-N}^{N} T^n f = \int f(x)\, dx.$$

We rewrite (4.21) in the form

(4.22) $$\sum_{-N}^{N} T^n = (2N+1)\, dx + o(N).$$

We want to improve the term $o(N)$ in (4.22). The best thing to do would be to find an asymptotic formula in N for the left side. The least we can do is improve $o(N)$ to a smaller power.

The simplest illustration of this latter improvement occurs in the question of equidistribution on the circle (normalized to be reals mod 1). Let θ_0 be a fixed number and let $Tf(\theta) = f(\theta + \theta_0)$. Writing the Fourier series for f we deduce easily

PROPOSITION 4.1. *Let θ_0 be an irrational number that is not a Liouville number, meaning that for l sufficiently large there are no large integers p, q satisfying*

(4.23) $$|p\theta_0 + q| < q^{-l}.$$

Then

(4.24) $$\sum_{-N}^{N} T^n - (2N+1)\, d\theta = O(1)$$

when the left side of (4.24) is applied to C^∞ functions on the circle.

Thus although we cannot improve (4.22) for all functions, we can improve it for C^∞ functions.

Now the ergodic theorem depends on the natural order of the integers n in (4.22). But the group Γ is not amenable so it does not have a natural ordering.

For $\gamma \in \Gamma$, denote by $|\gamma|$ the maximum of the modulus of the matrix entries of any $\gamma' \in \gamma \Gamma_0$. Call $|N|$ the number of $\gamma \in \Gamma/\Gamma_0$ with $|\gamma| \leq N$. We have

THEOREM 4.2. *When applied to functions in* $L_1 \cap \hat{L}_1$

$$(4.25) \qquad \overbrace{\sum_{|\gamma| \leq N} \sum_n \delta_{\gamma(n,0)}} = (|N| - 2)\, dx\, dy + 2 \sum \delta_{n,m} + O(1).$$

PROOF. If K is any square, then for N large enough it is clear that $\{\gamma(n,0)\}$ for all n and $|\gamma| \leq N$ covers the lattice points in K twice, plus gives the origin $|N| - 2$ times, plus gives some lattice points outside K twice. If K is sufficiently large and $f \in L_1 \cap \hat{L}_1$ then the contribution to the left side of (4.25) of the points outside K is arbitrarily small. Thus the left side of (4.25) is $2 \sum \hat{f}(n, m) + (|N| - 2)\hat{f}(0, 0) + \varepsilon$. By the PSF this is exactly the right side.

REMARK. Theorem 4.2 shows how to derive the ergodic result (4.25) from the PSF. It is possible to give a direct proof that the right side of (4.20) gives the right side of (4.25) [when we sum for $|\gamma| \leq N$ in (4.20)]. But we can prove this only for f which are Schwartz functions. This gives a proof that the PSF for $k = 1$ implies it for $k > 1$. This proof will be given elsewhere as well as applications to certain exotic PSF.

For another aspect of the functional equation for the theta function, let us reexamine Theorem 2.3. We showed that the Fourier transform in x corresponds to the map W given by (2.18).

IV.2. The Appell transformation and the Weil representation. In [1] Appell studied the most general geometric transformation that preserves solutions of the heat equation $\partial/\partial t - \partial^2/\partial x^2$. He found that the only nontrivial transformations are given by

$$(4.26) \quad F(t, x) \to \frac{C}{\sqrt{t-\alpha}} e^{-(x-\beta)^2/4(t-\alpha)} F\left[-\frac{k^2}{t-\alpha} + \alpha', \frac{k(x-\beta)}{t-\alpha} + \beta'\right].$$

Here $C, k, \alpha, \beta, \alpha', \beta'$ are arbitrary constants. In particular the analog of inversion of the wave equation is

$$(4.27) \qquad F(t, x) \xrightarrow{A} \frac{1}{\sqrt{\pm t}} e^{-x^2/4t} F\left(-\frac{1}{t}, \frac{x}{t}\right).$$

If we denote by $\tilde{F}(t, x)$ the right side of (4.27), the Cauchy data of $\tilde{F}$ on the t axis is

$$(4.28) \qquad \left[\frac{1}{\sqrt{\pm t}} F\left(-\frac{1}{t}, 0\right), \frac{1}{t\sqrt{\pm t}} F_x\left(-\frac{1}{t}, 0\right)\right].$$

With the proper definition of square root, it is seen that (4.28) agrees with W. In any case it is easily seen that $\tilde{F}$ of (2.16) is AF with F as in (2.13) [except for some constants that we have normalized]. Thus we have

PROPOSITION 4.3. *The Appell inversion A corresponds to the Fourier transform on the x axis.*

Proposition 4.3 leads to a new interpretation of the Weil representation of SL(2, R) or, more precisely, of the metaplectic group (see [34]). For, the crucial point of the Weil representation is the correspondence between inversion in t and the Fourier transform in x. The intertwining to the x axis of the other generators of the metaplectic group acting on the Cauchy data on the t axis is readily computed.

In Chapter III we discussed a general type of heat equation. In particular, Theorem 3.3 set up a Cauchy problem for the heat equation (3.11). In case $m = 1$ this is just the usual heat equation in several space variables. For $m > 1$ the t axis should be considered as contained in the Siegel upper half-plane. As such there are natural multiplier representations of the symplectic modular group.

There is no difficulty in extending the above Appell idea to this general situation; we must change from scalar to matrix notation. (However I do not know if this defines the most general geometric transformation preserving solutions of the heat equation.)

The above ideas lead to

THEOREM 4.4. *The Weil representation on the x axis corresponds, using intertwining via the heat equation (3.11), to a multiplier representation of the metaplectic group on Cauchy data on the t axis.*

For each fixed harmonic in x (i.e., for each component of the Cauchy data on the t axis) we get a different multiplier. This is already apparent from (2.18).

We can use Proposition 4.3 (and its extension to the symplectic group) in the following manner. (We again, for simplicity, start with the case $k = 1$.) (f_0, f_1) is the Cauchy data on the t axis of an Appell-invariant solution F of the heat equation if and only if $F(0, x)$ is Fourier invariant. In particular, if $f_1 = 0$ and

$$(4.29) \qquad f_0(t) = \sum a_n e^{int}$$

so that

$$(4.30) \qquad F(t, x) = \sum a_n e^{int} \cos \sqrt{n} x$$

and

$$(4.31) \qquad \widehat{F(0, \hat{x})} = \sum a_n \delta_{\pm\sqrt{n}},$$

the invariance of $F(0, x)$ under Fourier inversion is

$$(4.32) \qquad \sum a_n \cos \sqrt{n} x = \sum a_n \delta_{\pm\sqrt{n}}.$$

That this is equivalent to

$$(4.33) \qquad a_n = \begin{cases} c & \text{if } n \text{ is a square,} \\ 0 & \text{otherwise} \end{cases}$$

can be given a direct proof.

Let us go back to the second equation of (1.4) (but we drop the factor π for convenience). If F is a solution of the heat equation, then by the Fundamental Principle of FA we can express F in the form (formally) as a quadratic Fourier transform

$$(4.34) \qquad F(t, x) = \int \hat{f}(\hat{x}) e^{-it\hat{x}^2 - 2ix\hat{x}} \, d\hat{x}.$$

If F satisfies the second equation of (1.4), then we expect that

$$
\begin{aligned}
(4.35) \qquad F(t, x) &= \int \hat{f}(\hat{x}) e^{-it(\hat{x}^2 + 2\hat{x} + 1) - 2ix(\hat{x} + 1)} \, d\hat{x} \\
&= \int \hat{f}(\hat{x} - 1) e^{-it\hat{x}^2 - 2ix\hat{x}} \, d\hat{x}.
\end{aligned}
$$

Thus the second equation (1.4) formally states that $\hat{f}$ is invariant under $\hat{x} \to \hat{x} - 1$. The first equation says that the support of $\hat{f}$ is **Z**.

This means that (1.3) and (1.4) are the formal analogs of (a) and (b) discussed at the beginning of this section.

We have thus shown

PROPOSITION 4.5. *For solutions F of the heat equation the following are equivalent*:

A. *Appell invariance*;

B. *properties* (a) *and* (b) *for* $F(0, x)$;

C. *Fourier invariance of $F(0, x)$ together with the first equation of* (1.4);

D. *both equations of* (1.4);

E. *the first equation of* (1.4) *and the invariance of the Cauchy data on the t axis under the inversion described above.*

Naturally there are analogs for higher-dimensional symplectic groups.

There is another aspect of the Appell transform and Weil representation that is of interest. Unfortunately, I understand this only in the special case $k = 2$.

We consider the complete quadratic Fourier transform. We use the notation x, y for a point in R^2, and $t = x^2$, $u = xy$, $v = y^2$. Then the Plücker equation is

$$(4.36) \qquad \Box F \equiv \frac{\partial^2 F}{\partial u^2} - \frac{\partial^2 F}{\partial t \, \partial v} = 0,$$

which is the three-dimensional wave equation.

For the three-dimensional wave equation there is one nontrivial geometric transformation preserving solutions; this is inversion in the unit hyperboloid (see [14] for details). In our notation this inversion is given as follows: Denote

$$(4.37) \qquad r^2 = tv - u^2$$

so r is the Minkowski distance. The unit hyperbola is $r = 1$, and inversion in the unit hyperbola is

$$(4.38) \qquad\qquad c: F(t, u, v) \rightarrow r^{-1} F(t/r^2, u/r^2, v/r^2).$$

The inversion c preserves solutions of the wave equation.

We think of c as the analog of the inversion $t \rightarrow -1/t$ for the usual heat equation. However we have not succeeded in finding the action of c on the Cauchy data on $t = u = v = 0$, (that is, the x, y plane) which we call the Weil inversion. More precisely

PROBLEM 4.4. Find analogs of c for the other Cauchy data for the heat equation so that we can compute the effect of their inversion on the Cauchy data on $t = u = v = 0$.

The inversion c has added interest. For it is shown in the work of Baily-Ehrenpreis-Wells [3] that solutions of the wave equation extend to the conformal compactification $\tilde{M}$ of Minkowski space M. The inversion c sends the origin into a point $c(0, 0, 0) \in \tilde{M}$. Thus the answer to Problem 4.4 would be the transformation sending restrictions to $(0, 0, 0)$ of solutions of the heat equation to restrictions to $c(0, 0, 0)$.

On the other hand, it is not difficult to compute the effect of the Fourier transform in x, y on solutions of the heat equation. But this does not seem to lead to anything of interest.

CHAPTER V.   THE PSF AND THE FUNCTIONAL EQUATION FOR THETA—
SCATTERING THEORY; SINGULAR SERIES; ZETA FUNCTIONS, AND c FUNCTIONS

V.1. Scattering theory. In the previous chapter we discussed the operator and ergodic approaches to the PSF and the theta function. In this chapter we discuss the Hecke relation of the theta function to Dirichlet series (the zeta function), and also the zeros and singularities of theta functions.

We begin with the relation of the theta function to scattering theory. To understand how things work, let us begin with the function $\sin \pi z$. It is important that we consider $z = x + iy$ as a complex variable. Then, in terms of distributions in the x, y plane

$$(5.1) \qquad\qquad \frac{\partial^2}{\partial \bar{z} \partial z} \log \sin \pi z = \sum_{n=-\infty}^{\infty} \delta_n.$$

The distribution $\sum \delta_n$ is invariant under translation $z \rightarrow z + 1$. However, $\sin \pi z$ satisfies

$$(5.2) \qquad\qquad \sin \pi(z + 1) = -\sin \pi z.$$

We think of the singularity $\sum \delta_n$ as the "scattering data." Then the usual consequence of an invariance property of scattering is that the solution of an equation such as

$$(5.3) \qquad\qquad \bar{\partial} \partial \log f = \text{scattering data}$$

with f satisfying suitable growth conditions "almost" satisfies the invariance of the scattering data. The deviation from true invariance is measured by a "potential," which is a cocycle in a suitable sense. Of course it is important that the operator $\bar{\partial}\partial$ be invariant.

For the case of theta functions, we shall deal with ϑ_4 rather than ϑ_3. Here

$$(5.4) \qquad \vartheta_4(\tau, z) = \sum (-1)^n e^{i\pi n^2 \tau + 2inz}.$$

It is readily verified that ϑ_4 vanishes when $z = \pi\tau/2$ and at all points congruent to $\pi\tau/2$ modulo the period parallelogram. This means that

$$(5.5) \qquad \bar{\partial}\partial \log \vartheta_4(\tau, z) = \sum \delta_{\pi\tau/2 + m\pi + n\pi\tau}.$$

The right side of (5.5) is doubly periodic so we expect that some solution of

$$(5.6) \qquad \bar{\partial}\partial \log f = \sum \delta_{\pi\tau/2 + m\pi + n\pi\tau}$$

should be almost doubly periodic. Since the zeros of f are prescribed,

$$(5.7) \qquad f(z + m\pi + n\pi\tau) = \varphi(m, n; z) f(z),$$

where φ is an entire function with no zeros.

We search for an f with some sort of minimal growth. Since the number of zeros of f in a circle of radius r is about πr^2, the theory of entire functions predicts that the smallest f should be an entire function of order 2. Then the cocycle φ can be adjusted so as to be the exponential of a linear function that is trivial when $n = 0$. We then arrive at equation (1.4). Note that the quasi-periodicity factor is a cocycle on Z.

The above idea leads to the famous Jacobi triple product with $\zeta = e^{2iz}$

$$(5.8) \qquad \vartheta_4(z) = \prod (1 - q^{2n})(1 - q^{2n-1}\zeta)(1 - q^{2n-1}/\zeta).$$

This formula plays a central role in partition theory. The reason for this is that the theta series is subject to linear analysis and appears in connection with many linear problems. But it is the product that has combinatorial significance.

It should be pointed out that, from a slightly different viewpoint, the scattering idea appears, even in higher dimension, in the work of Weierstrass, Appell, Poincaré, and Siegel on abelian functions (see, e.g., [31]).

Our novelty of the scattering theory approach is that it gives a natural setting to find equations for potentials; these equations say that the potentials are cocycles. In fact, by a rather complicated modification of the above we can derive equation (4.8) for the nullwerte as a cocycle condition. This work will appear elsewhere.

V.2. Singular series and almost mean periodicity. In its simplest form the PSF is an example of a distribution with discrete support whose Fourier transform has discrete support. There is another instance of a transformation going from discrete support to discrete support; this is the singular series of

Hardy-Littlewood-Ramanujan. I wish to give a slightly different perspective on the singular series.

Let us begin with functions in the unit disc of the complex plane. We have, of course

$$(5.9) \qquad \frac{1}{1-z} = \sum_{n\geq 0} z^n = \mathscr{P} \sum_{n=-\infty}^{\infty} z^n = \mathscr{P}(\delta_{\theta=0}).$$

Here $\mathscr{P}$ refers to projection on holomorphic functions and $\delta_{\theta=0}$ is considered as a distribution on the circle. Equation (5.9) means that, when applied to holomorphic functions, the distributions $1/(1-z)$ and $\delta_{\theta=0}$ have the same values.

From (5.9) we derive the "singular series" expression

$$(5.10) \qquad \sum \frac{c_{p/q}}{1 - e^{-2\pi ip/q} z} = \mathscr{P} \sum c_{p/q} \delta_{\theta=p/q}.$$

We can give a new interpretation of the right side of (5.10). For the "discreteness" of the rational numbers p/q means that $\sum c_{p/q} \delta_{p/q}$ is the Fourier series transform of an almost periodic (ap) function. That is,

PROPOSITION 5.1. *A function $f(z) = \sum_{n\geq 0} a_n z^n$ that is holomorphic in the unit disc has a singular series representation in the form (5.10) if and only if the sequence $\{a_n\}_{n\geq 0}$ has an extension $\{\tilde{a}_n\}_{n=-\infty}^{\infty}$ that is ap with spectrum $\{p/q\}$.*

I shall not discuss here the exact type of almost periodicity. Let me mention that a relation between singular series and almost periodicity was explored by Kac [23], Kac, van Kampen, and Wintner [24] among others. However the introduction of almost mean periodic functions below, which is necessary for the deeper ideas, in novel to this work.

We shall presently discuss how to characterize this almost periodicity directly in terms of $\{a_n\}$.

Now (5.10) represents only the simplest singular series. To go beyond we have to start with $(1 - z)^{-\alpha}$ for various α, which may be integers or half integers. (Even more general α must be considered.) We shall assume α is an integer; the half-integer case differs only formally in that differential operators become pseudodifferential operators. For example, (5.9) is replaced by

$$(5.11) \qquad \frac{1}{(1-z)^2} = \sum_{n\geq 0} (n+1) z^n = \mathscr{P}(\delta_0' + \delta_0).$$

This leads to

$$(5.12) \qquad \sum \frac{c_{p/q}}{(1 - e^{-2\pi ip/q} z)^2} = \mathscr{P} \left[\sum c_{p/q} (\delta_{p/q}' + \delta_{p/q}) \right].$$

The right side of (5.12) is no longer related to ap functions but rather to *almost mean periodic* (amp) functions. Such functions are the analogs of ap functions with multiple spectrum. They arise in several connections; the theory is exposed in detail in [11]. The main idea is that ap functions h almost satisfy translation invariance for many translations, meaning

$$(5.13) \qquad\qquad (T_\lambda - I)h(x) = h(x + \lambda) - h(x)$$

is small for "many" λ. On the other hand, amp functions almost satisfy many convolution equations

$$(5.14) \qquad\qquad S_\mu * h \quad \text{small}$$

for "many" S_μ. In the case we treat here we can assume that S_μ are higher-order difference operators.

THEOREM 5.2. *A necessary and sufficient condition that f admit a singular series*

$$(5.15) \qquad\qquad f(z) = \sum_{p/q,l} \frac{c^l_{p/q}}{(1 - e^{-2\pi i p/q} z)^l}$$

is that $\{a_n\}$ admit an amp extension to $n \in \mathbf{Z}$, with rational spectrum.

Actually when $\{a_n\}$ admits an amp extension, then this extension can be expressed directly in terms of $\{a_n\}$. For (5.9) and (5.11) (and similarly for larger l) show that the extension is unique. This uniqueness can be thought of as a sharpening of Bohr's uniqueness for ap functions since we are using the projection $\mathscr{P}$ and also we are using amp functions. In [11] we discuss the case of infinitely many l. In this case we need a condition like

$$(5.16) \qquad\qquad a_n = O(\exp(n^\gamma))$$

with $\gamma < 1/2$. In fact uniqueness does not hold for $\gamma = 1/2$. All this is related to work of Radamacher [27] on functions like the partition function.

One of the most important examples of singular series occurs in the analytic theory of quadratic forms (see [32] for details). There it is shown that the average of theta functions of quadratic forms belonging to a fixed genus (the analytic genus invariant) is a singular series; this is the content of the Main Theorem of Minkowski and Siegel. From our point of view this means that the average of restrictions to various lines of the quadratic Fourier transform of the sum of δ functions at the lattice points is the projection of a "discrete-like" object.

In order to place this within the framework of PSF, we apply the ideas that began with Ono [26] and culminated with the work of Tamagawa. These show that the natural setting for this study is by means of adelization. For, upon adelization the distinction between class and genus disappears. Moreover the rationals become discrete. Having done this the Main Theorem can be thought of as a PSF.

The adelization process will be detailed elsewhere. The reader might object on grounds that our approach is analytic. However, the "analysis" is very algebraic since we don't need differential operators but only their solutions, which are Fourier transforms on algebraic varieties.

Let me point out that the condition in Theorem 5.2 on $\{a_n\}$ can be expressed directly in terms of congruence properties. Thus it is not surprising that one of the main techniques used by Siegel in [32] for the proof of the Main Theorem is to show congruence properties of the coefficients of the analytic genus invariant.

V.3. The Hecke theory. Our next project involves the change of basis that was so successfully promulgated by Hecke. This involves the change from exponentials to powers, which changes theta functions to zeta functions. One must take a little caution because we usually consider τ as real and now we need imaginary τ; in fact $\tau = iy$ with $y > 0$. But this does not give any major difficulty.

In its simplest form we change from $\vartheta_3(\tau, z)$ via the Mellin transform in τ to (an unimportant multiple of)

$$(5.17)\qquad \Gamma(s/2)F(s, x) = \Gamma(s/2)\sum_1^\infty \frac{x^{n-1}}{n^s}.$$

From our point of view the term n^s should really be written $(n^2)^{s/2}$. Moreover, the sum should be for $n = (-\infty, \infty)$. We shall meet this point below. The function F has many interesting specializations. For example, $F(2, s)$ is known as Euler's dilogarithm; it has found interesting applications in topology.

We must make an important observation that applies to all further considerations regarding Dirichlet series. The theta functions, e.g., $\vartheta_3(\tau, z)$, have a term corresponding to $n = 0$. This does not appear in the zeta function. The Mellin transform of the term corresponding to $n = 0$ is $\delta_{s=0}$ on the imaginary s axis. This term does not appear explicitly in the Dirichlet series. Yet it appears in a somewhat different form. For when we deal with Dirichlet series in the region $\operatorname{Re} s$ large, the effect of the term corresponding to $n = 0$ is to introduce a pole at $s = 1$.

All this is, of course, clarified in Hecke's work [18].

At the end of Chapter IV we discussed various uniqueness properties of ϑ_3. The main point is that the integrality of the exponents plus the heat equation plus an invariance property determine ϑ_3 up to a constant multiple. Hecke's idea is that the invariance property is geometric for the "nullwerte" $F(s, 1)$ relating the values $F(s, 1)$ and $F(1 - s, 1)$ (in a much more general setting). How about the other properties?

It is easily seen that the heat equation becomes

$$(5.18)\qquad F(s - 1, x) = \left(x\frac{d}{dx} + 1\right)F(s, x).$$

[Actually, since n^s is really $(n^2)^{s/2}$ the exact analog of the heat equation is $F(s - 2, x) = (x\frac{d}{dx} + 1)^2 F(s, x)$. Thus (5.18) is a "square root" of the heat equation, which makes some things easier.]

Thus what we need to describe are

(a) Integrality condition,
(b) Appell transformation,
(c) Weil representation,
(d) Quasi double periodicity.

Usually the integrality condition is postulated. Then Hamburger's theorem states that F is uniquely determined (up to constant multiple) by the inversion $s \to 1 - s$ on the nullwerte. Hamburger's proof, in fact, is similar to the arguments of Chapter IV in that it eventually proves that the Dirichlet series coefficients of $F(s, 1)$ are invariant under translation.

However, we can formulate the first equation in (1.4) as

$$(5.19) \qquad F(s, xe^{2\pi i}) = F(s, x).$$

Actually, in terms of nullwerte we can write it as [compare (a) following (4.6)]

$$(5.20) \qquad [\exp(2\pi i e^{\partial/\partial s}) - I]F(s, x) = 0.$$

The operator $\exp(2\pi i \exp(\partial/\partial s))$ can be represented as additive convolution with $(2\pi i)^s \Gamma(s)$.

Let us pass to the Appell transformation (b). We do not know how to take the Mellin transform of the Appell transform in general, but for solutions of the heat equation we get an interesting answer. For the exponential

$$(5.21) \qquad e^{itu^2 + 2izu} \xrightarrow{\text{Appell}} t^{-1/2} e^{-iu^2/t - 2izu/t - iz^2/t}$$

$$\xrightarrow{\text{Mellin}} (u + z)^{1-s} \Gamma,$$

where Γ represents a Γ function times a simple exponential. This means that for a solution of the heat equation

$$(5.22) \qquad G(t, z) = \int e^{itu^2 + 2izu} \hat{g}(u) \, du,$$

the Mellin transform of the Appell transform of G is

$$(5.23) \qquad \Gamma \int (u + z)^{1-s} \hat{g}(u) \, du.$$

This result, when applied to the theta function, is the functional equation for the Hurwitz zeta function (see Bateman [4, vol. I, p. 26(6)])

$$(5.24) \qquad \zeta(s, v) = 2(2\pi)^{s-1} \Gamma(1 - s) \sum_{n=1}^{\infty} n^{s-1} \sin\left(2\pi nv + \frac{1}{2}\pi s\right).$$

When dealing with the Weil representation (c), we must consider $n \in (-\infty, \infty)$. A simple calculation shows that, up to simple factors, the Fourier transform in x takes the solution $u^{-s} x^{u-1}$ of (5.18) into $(\log x + u)^{-s}$. Thus the PSF again gives (5.24).

(d) does not seem to lead to anything of interest.

A point of great interest is the Mellin transform of the singular series. The difficulty is that we are now taking the Mellin transform of $\vartheta_3(iy, 0)$ with

$y > 0$. The singular series involves sums of terms of the form $(\tau - p/q)^{-\alpha}$. Thus we must evaluate sums of the form: Gaussian times

$$(5.25) \qquad \int_0^\infty (iy - p/q)^{-\alpha} y^{is} \frac{dy}{y}.$$

PROBLEM 5.1. Can one use the singular series to give a viable representation of the zeta function?

V.4. c functions. Our final discussion of this chapter concerns the relation of PSF to what Harish-Chandra called c functions. Let us begin with PSF for the x, y plane. We take the Mellin transform of the radial part of the formal Fourier transform of $\sum \delta_{mn}$. Thus we form

$$(5.26) \qquad \begin{aligned} &\int r^{is} e^{ir(m\cos\theta + n\sin\theta)} \frac{dr}{r} \, d\theta \\ &= \Gamma(is) e^{\pi s/2} \int (m\cos\theta + n\sin\theta)^{-is} \, d\theta. \end{aligned}$$

Now,

$$(5.27) \qquad m\cos\theta + n\sin\theta = |(m, n)| \cos(\theta + \theta_0),$$

where $\theta_0 = \arg(m + in)$. Thus the right side of (5.24) equals

$$(5.28) \qquad \Gamma(is) e^{\pi s/2} (m^2 + n^2)^{-is/2} \int \cos^{-is}\theta \, d\theta.$$

The integral in (5.28) is known as a c function; in this case it is $\beta(\frac{1}{2}, \frac{1}{2} - is)$. Roughly speaking this means that it is the integral of some power of a group character (or a simple combination of group characters).

To understand the direction we are going, let me give a second example. We consider the quadratic Fourier transform on the x, y plane. Start with the lattice points (m, n) and take their quadratic images (m^2, mn, n^2), which belong to the light cone Γ. We ask the following question:

PSF for Γ. Is the linear Fourier transform of $S = \sum \delta_{m^2, mn, n^2}$ when restricted to Γ equal to S (perhaps plus some simple terms)?

To put this question in perspective, let us consider a two-dimensional analog. Thus we consider the light cone Γ^2 in the x, y plane. The lattice points on Γ^2 are (n, n) and $(n, -n)$. If we count the origin twice, then the ordinary one-dimensional PSF says that

$$(5.29) \qquad \begin{aligned} \sum \widehat{\delta_{m,m}} &= \sum \delta_{x+y=n}, \\ \sum \widehat{\delta_{m,-m}} &= \sum \delta_{x-y=n}. \end{aligned}$$

The restriction of $\delta_{x\pm y=2n}$ to Γ^2 is $\delta_{n,\pm n}$ except for $n = 0$. Thus the PSF holds for the cone Γ^2 (except for a factor of 2).

But for the cone $\Gamma = \Gamma^3$ the situation is somewhat more complicated. Using the parametrization of Γ by the quadratics on R^2, we take our cue from the first example to write points on Γ in the form $r^2(\cos^2\theta, \cos\theta\sin\theta, \sin^2\theta)$. The "natural", i.e. $SL(2, R)$ invariant, inner product in τ^2 is (see Section VI.1)

(5.30) $$(x, y, z) \cdot (x', y', z') = xz' + x'z - 2yy'.$$

Thus

(5.31) $$\widehat{\sum \delta_{m^2, mn, n^2}}\Big|_\Gamma = \sum \exp[ir^2(n^2\cos^2\theta - 2mn\cos\theta\sin\theta + m^2\sin^2\theta)].$$

To check whether the PSF holds we decompose under the Mellin transform (and Fourier series in θ). This idea uses the fact that the light cone is of the form KA^+ in the usual notation of semi-simple Lie groups. This means that the K invariant part of the Mellin (i.e. A^+ Fourier) transform of (5.31) is

(5.32)
$$\int \exp[ir^2(n^2\cos^2\theta - 2mn\cos\theta\sin\theta + m^2\sin^2\theta]r^{is}\frac{dr}{r}\,d\theta$$

$$= \Gamma(is/2)e^{\pi s/4}\int (n^2\cos^2\theta - 2mn\cos\theta\sin\theta + m^2\sin^2\theta)^{-is/2}\frac{dr}{r}\,d\theta$$

$$= \Gamma(is/2)e^{\pi s/4}(m^2 + n^2)^{-is/2}\int \cos^{-is}\theta\,d\theta,$$

which we see by writing $(m, n) = |m, n|(\cos\theta', \sin\theta')$.

On the other hand, the Mellin transform of the radial part of $\sum \delta_{m^2, mn, n^2}$ is $\sum (m^2 + n^2)^{is-1}$ by (5.30).

Thus when we change $s \to 2s$ in (5.28) so as to obtain the Gaussian zeta function $\zeta_i(s)$ when summing, we see that the combination of elementary factors does not agree with those of (5.32). Since PSF holds for the plane, it follows that PSF does not hold for the light cone Γ.

The same calculation in general shows

PROPOSITION 5.3. *PSF does not hold for the cone Q, which is the quadratic image of R^k.*

The point is that the elementary factors for Q differ from those for R^k. However, if we adelize our constructions then the elementary factors become trivial. This will be shown elsewhere. As a consequence

THEOREM 5.4. *PSF holds for the adelized Q.*

By using the fundamental solution for the heat equation to express the values of Θ on the τ^2 axis in terms of those on Q, we obtain the singular series for Θ. This is clarified in Section VI.1 below.

CHAPTER VI. THE THEORY OF HECKE-SIEGEL, EISENSTEIN, AND POINCARÉ

VI.1. Eisenstein series. The Hecke-Siegel theory (see [32] for a clear discussion) shows how to express the periods of Eisenstein series over geodesic cycles defined by units in a real quadratic field in terms of the zeta function of this field.

To fit this into our framework, we have to add a new type of parametrization problem. Up to now we have emphasized parametrization problems on linear spaces. Such parametrization problems have the advantage of simplicity: they can be analysed by harmonic function theory, which is essentially algebra. On the other hand, the nonlinear parametrization problems seem to involve analysis.

To see how things work, let us start with the quadratic Fourier transform on R^2. Writing the kernel of the Fourier transform as usual,

$$\exp(ix\hat{x} + iy\hat{y} + itx^2 + iuxy + ivy^2),$$

every quadratic Fourier transform satisfies the wave equation in t, u, v space (i.e., $\hat{x} = \hat{y} = 0$)

$$(6.1) \qquad\qquad \Box F \equiv \left(\frac{\partial^2}{\partial t \partial v} - \frac{\partial^2}{\partial u^2}\right) F = 0.$$

The group $G = \mathrm{SL}(2, R)$ acts on t, u, v as the symmetric product of the 2-dimensional representation. [We can also regard this as a representation of $\mathrm{SO}(1, 2)$.] This means that we identify (t, u, v) with the symmetric matrix $X = \begin{pmatrix} t & u \\ u & v \end{pmatrix}$ and $g \in G$ acts on such matrices as

$$(6.2) \qquad\qquad\qquad X \to gXg'.$$

The orbits of G are

(a) (*Half*) *of hyperboloid of two sheets.* This is the orbit of a positive definite X (or $-X$). The orbit consists of all symmetric Y with $\det Y = \det X$ and $\operatorname{sgn} \operatorname{tr} Y = \operatorname{sgn} \operatorname{tr} X$. The isotropy group of such an X is the unit group of the associated quadratic form. Thus such an orbit can be identified with G/K, where K is a maximal compact subgroup of G.

In particular, consider the orbit of I. The isotropy group is

$$(6.3) \qquad\qquad K = \left\{ \begin{pmatrix} \cos\theta & \sin\theta \\ -\sin\theta & \cos\theta \end{pmatrix} \right\}.$$

We can identify the orbit of I with the upper half-plane. The precise identification is given as follows: Write $z = \xi + i\eta$ for a point in the upper half-plane. In terms of group coordinates

$$(6.4) \qquad\qquad z = \xi + i\eta = \begin{pmatrix} \eta^{1/2} & \xi\eta^{-1/2} \\ 0 & \eta^{-1/2} \end{pmatrix} i,$$

where G acts by fractional linear transformation.

The corresponding point in Minkowski space is

$$(6.5) \qquad \begin{pmatrix} \eta^{1/2} & \xi\eta^{-1/2} \\ 0 & \eta^{-1/2} \end{pmatrix} \begin{pmatrix} \eta^{1/2} & 0 \\ \xi\eta^{-1/2} & \eta^{-1/2} \end{pmatrix}$$
$$= \begin{pmatrix} \eta + \xi^2\eta^{-1} & \xi\eta^{-1} \\ \xi\eta^{-1} & \eta^{-1} \end{pmatrix}.$$

Thus a point $(t, u, v) \in$ orbit of I is given by

(6.6) $$t = \eta + \xi^2 \eta^{-1}, \quad u = \xi \eta^{-1}, \quad v = \eta^{-1}.$$

(b) (*Half*) *of open light cone.* The light cone is defined by $\det X = 0$. The halves are $\operatorname{sgn} \operatorname{tr} X > 0$ or $\operatorname{sgn} \operatorname{tr} X < 0$. The isotropy group of $\left(\begin{smallmatrix} 1 & 0 \\ 0 & 0 \end{smallmatrix}\right)$ is given by

(6.7)
$$\begin{pmatrix} g_{11} & g_{12} \\ g_{21} & g_{22} \end{pmatrix} \begin{pmatrix} 1 & 0 \\ 0 & 0 \end{pmatrix} \begin{pmatrix} g_{11} & g_{21} \\ g_{12} & g_{22} \end{pmatrix}$$
$$= \begin{pmatrix} g_{11}^2 & g_{11} g_{21} \\ g_{11} g_{21} & g_{21}^2 \end{pmatrix} = \begin{pmatrix} 1 & 0 \\ 0 & 0 \end{pmatrix}.$$

Thus $g_{11} = \pm 1$ and $g_{21} = 0$. This means that g is an upper triangular matrix with diagonal $\pm I$. In the usual notation for semisimple Lie groups, the isotropy group is MN.

The points of the positive light cone are exactly the quadratic transform of R^2, that is, the points

(6.8) $$X = \begin{pmatrix} x^2 & xy \\ xy & y^2 \end{pmatrix}.$$

(c) *Hyperboloid of one sheet.* This is the orbit of X with $\det X < 0$. Let us compute the isotropy group of $\left(\begin{smallmatrix} 1 & 0 \\ 0 & -1 \end{smallmatrix}\right)$. The equation is

(6.9)
$$\begin{pmatrix} g_{11} & g_{12} \\ g_{21} & g_{22} \end{pmatrix} \begin{pmatrix} 1 & 0 \\ 0 & -1 \end{pmatrix} \begin{pmatrix} g_{11} & g_{21} \\ g_{12} & g_{22} \end{pmatrix}$$
$$= \begin{pmatrix} g_{11}^2 - g_{12}^2 & g_{11} g_{21} - g_{12} g_{22} \\ g_{11} g_{21} - g_{12} g_{22} & g_{21}^2 - g_{22}^2 \end{pmatrix} = \begin{pmatrix} 1 & 0 \\ 0 & -1 \end{pmatrix}.$$

From this we conclude that the isotropy group consists of

(6.10) $$\tilde{K} = \left\{ \pm \begin{pmatrix} \cosh \zeta & \sinh \zeta \\ \sinh \zeta & \cosh \zeta \end{pmatrix} \right\}.$$

$\tilde{K}$ is (except for the $\pm$) another real form of K. Actually in this case $\tilde{K} = A$, where A is the diagonal subgroup; but this is an accident of $\operatorname{SL}(2, R)$.

It is clear that $\det X$ is the basic G invariant polynomial. Actually, from the general viewpoint of symplectic groups it is better to use the equivalent quadratic form

(6.11) $$-\operatorname{tr}(XJXJ) = r^2(X),$$

where J is, as usual, the symplectic form

(6.12) $$J = \begin{pmatrix} 0 & 1 \\ -1 & 0 \end{pmatrix}.$$

Let Γ denote the subgroup $\operatorname{SL}(2, \mathbf{Z})$. We want to find some interesting points p_0 whose orbit under Γ is discrete. It is standard that every point in G/K has a discrete Γ orbit, so these points are uninteresting.

For the light cone we can easily pass to the linear action on the $\tau^1 = (\mathbf{x}, \mathbf{y})$ plane.

PROPOSITION 6.1. *The Γ orbit of τ^1 is discrete if and only if $\xi = \mathbf{x}/\mathbf{y}$ is rational. If ξ is irrational, then the orbit of τ^1 is dense in S^1.*

PROOF. It is clear that if ξ is rational (including infinity) then $\Gamma\tau^1$ is discrete.

Suppose that ξ is irrational. By Kronecker's theorem we can find integers c, d so that $c\mathbf{x} + d\mathbf{y} = \varepsilon$ is arbitrarily small and $\varepsilon \neq 0$. By dividing by (c, d) we make ε even smaller, so we may assume that $(c, d) = 1$. Thus (c, d) forms the second row of a matrix in Γ. This means that we can assume that $\mathbf{y}$ is arbitrarily small.

Next let $\tilde{\tau}^1$ be another point in S^2. Then we can find $\tilde{\gamma} \in \Gamma$ so that the second coordinate of $\tilde{\gamma}\tilde{\tau}^1$ is arbitrarily small. Finally we observe that

$$(6.13) \qquad \begin{pmatrix} 1 & m \\ 0 & 1 \end{pmatrix}\begin{pmatrix} \mathbf{x} \\ \varepsilon \end{pmatrix} = \begin{pmatrix} \mathbf{x} + m\varepsilon \\ \varepsilon \end{pmatrix}.$$

This means that we can find a supertriangular matrix $\tilde{\tilde{\gamma}}$ that makes τ^1 arbitrarily close to $\tilde{\gamma}\tilde{\tau}^1$. This proves the result.

REMARK 1. The second coordinate of $\gamma\tau^1$ is $c\mathbf{x} + d\mathbf{y}$, where $(c, d) = 1$. By Proposition 6.1 the set of second coordinates is dense in the line. This represents a sharpening of Kronecker's theorem since Kronecker does not use $(c, d) = 1$. It seems rather difficult to give a direct proof of this sharpened Kronecker theorem.

REMARK 2. Proposition 3.1 was proven by F. I. Mautner and the author about 1960. Our original method involved using Vinogradov's result on the representation of a large odd number as a sum of three primes to prove the sharpened Kronecker theorem mentioned in Remark 1 and then analysing the proof in great detail. The above proof was found a little later. Since that time this idea of proof has been observed (independently) by Dani [5], who have used it for other examples.

We can formulate the general

PROBLEM. Let G_0 be a real semisimple Lie group, P a parabolic with nilradical N, and Γ_0 an arithmetic subgroup of G_0. Let $p \in G_0/N$ be chosen so that the isotropy Γ_0^p of p in Γ_0 is trivial. Is the orbit $\Gamma_0 p$ dense in G_0/N?

It should be observed that it is most difficult to prove $\Gamma_0 p$ is dense when P is a maximal parabolic since then N is minimal so G_0/N is largest and the condition that Γ_0^p be trivial is weakest.

REMARK. The analog of Proposition 3.1 is false if we replace G/N by other homogeneous spaces such as G/A. In this case some orbits are discrete, some are dense, and some are neither discrete nor dense. This can be readily understood by examining geodesics on the Riemann surface whose fundamental group is Γ, since we are really studying the double coset space $\Gamma \backslash G/A$. Although the geodesic flow (i.e., right action of A) on $\Gamma \backslash G$ is ergodic, some nonclosed geodesics may behave in a nondense fashion.

Satake has pointed out to me that the opposite situation, namely the discreteness of Γ_0^p, occurs only if N/Γ_0^p is compact.

Finally we want to find the discrete orbits on G/A. We can think of this from a somewhat different point of view. For we are really interested in the double coset space $\Gamma\backslash G/A$. This can be interpreted as the action of A on $\Gamma\backslash G$. This action is the geodesic flow.

It is readily verified that discrete orbits of Γ on G/A correspond to closed geodesics on $\Gamma\backslash G$. These are of two types

(1) *Compact geodesics.* These occur for points p_0 whose isotropy group G_{p_0} has an infinite intersection Γ_{p_0} with Γ. In the upper half-plane they correspond to circles orthogonal to the real axis whose endpoints are conjugates in a real quadratic number field.

(2) *Noncompact closed geodesics.* These must flow to the cusp in the fundamental domain for Γ. In the upper half-plane they correspond to geodesics whose endpoints are rational (including ∞).

We are now ready for explicit expressions for intertwinings. The first intertwining that we study is Intertwining C (see Chapter III) given by the Fourier transform. We start with the discrete "lattice" on the light cone consisting of the quadratic transform of the lattice in R^2. Thus we form

$$(6.14) \qquad \hat{\mathscr{E}} = \sum \delta_{m^2, mn, n^2}.$$

Note that $\hat{\mathscr{E}}$ is the sum of the δ functions of infinitely many Γ orbits, namely the quadratic transforms of $\Gamma(m, 0)$ with $m \geq 0$.

Next we form the Fourier transform $\mathscr{E}$ of $\hat{\mathscr{E}}$ using the G invariant inner product defined by $\det X$, namely

$$(6.15) \qquad X \cdot \tilde{X} = tv' + t'v - 2uu'.$$

(Actually we should take $\frac{1}{2}$ of the right side but that is unimportant.) Thus

$$(6.16) \qquad \mathscr{E} = \sum e^{itn^2 + ivm^2 - 2iumn}.$$

We regard $\mathscr{E}$ as a *generating Eisenstein series.* We think of $\mathscr{E}$ as an Eisenstein series rather than a Poincaré series because the points $(m, 0)$ are fixed by all of N rather than by $N \cap \Gamma$ alone.

According to our ideas of Chapter III we must decompose $\mathscr{E}$ under scalar multiplication; this requires forming the Mellin transform in scalar multiplication. We handle the constant term $n = m = 0$ à la Hecke (see Chapter V). We are led to

$$(6.17) \qquad \alpha \sum (tn^2 + vm^2 - 2umn)^{-is},$$

where α is an elementary factor. We now compute (6.17) on G/K (this makes sense since it is homogeneous). In terms of upper half-plane coordinates

(6.6), we obtain

$$(6.18) \qquad E(\xi, \eta; s) = \sum [(\eta + \xi^2 \eta^{-1}) n^2 + \eta^{-1} m^2 - 2\xi \eta^{-1} mn]^{-is}$$

$$= \eta^{is} \sum [(\eta^2 + \xi^2) n^2 + m^2 - 2\xi mn]^{-is}$$

$$= \eta^{is} \sum [(\xi n - m)^2 + (\eta n)^2]^{-is}$$

$$= \sum \frac{\eta^{is}}{|(\xi + i\eta)n - m|^{2is}}.$$

(We can change $-m$ to m in (6.18) because we sum over all m.)

Thus we have shown

THEOREM 6.2. *The Fourier transform of the sum of the δ functions at the quadratic transform of the lattice points in R^2 has as its Mellin transform the usual Eisenstein series.*

How about Intertwining A applied to this same distribution? This means that we solve the parametrization problem with data $\hat{\mathscr{E}}$ on the forward light cone. The explicit formula can be given using the fundamental solution for the wave equation. It was made explicit for the first time by d'Adhémar. d'Adhémar's construction was taken up in detail by Marcel Riesz in [28]. The fundamental solution for the wave equation in an odd number m of dimensions is

$$(6.19) \qquad e = \begin{cases} \gamma_m r^{2-m} & \text{in retrograde cone,} \\ 0 & \text{outside,} \end{cases}$$

where

$$(6.20) \qquad \gamma_m = (-1)^{(m+1)/2} \frac{1}{\pi \omega_{m-2}}$$

with ω_m the surface area of the m sphere.

D'Adhémar's explicit evaluation $u(P)$ for a solution u of the wave equation at the point P inside the forward light cone is

$$(6.21) \qquad u(P) = u(0) - \frac{1}{\pi} \int_{S^P} ds \int_0^{R_0} \frac{du}{dR} R^{-1/2} dR.$$

Here S^P is the intersection of the positive light cone with the retrograde light cone translated by P and $R = r^2$, where r is the Minkowski distance from P to the point on the positive light cone lying below S^P. In particular $R = 0$ on S^P and $R = R_0 = \|P\|^2$ at the origin.

Thus the solution of the wave equation at a point P with data $\hat{\mathscr{E}}$ on the light cone is given by a sum over the points (m^2, mn, n^2) that lie in the interior of the intersection of the forward light cone with the retrograde cone based at P. In particular this representation holds on the time axis (K invariant axis). It is reminiscent of the singular series discussed in Chapter V. However there is an important difference, for there is a jump in the representation whenever the point P on the time axis is such that the retrograde cone based at P meets a point (m^2, mn, n^2). The variation from the singular series is measured by the c function for the light cone.

Nevertheless when the theory is adelized then we get an actual singular series. This is, in fact, a formulation of the main theorem of Minkowski-Siegel on quadratic forms (see [32]). These remarks should clarify the corresponding discussion in Chapter V.

VI.2. The Hecke-Siegel theory and Poincaré duality. The Hecke-Siegel theory deals with the relation between two different types of parametrization manifolds. In Chapter II we discussed the parametrization problem on various time axes. In particular, let T_K be the K fixed axis and let $T_{\check{K}}$ be an axis that is fixed by a conjugate of A whose intersection $\check{K} \cap \Gamma = \Gamma_{\check{K}}$ is the unit group in a real quadratic number field Q.

We have shown in Chapter II that the $\Gamma_{\check{K}}$ restriction to $T_{\widetilde{K}}$ of the quadratic Fourier transform of $\sum \delta_{m,n}$, which is the linear Fourier transform of $\mathscr{E}$, consists of the theta functions with grossencharactere for Q, say $\vartheta_Q(\chi)$.

We can find another expression for these theta functions using the fundamental solution e for the wave operator $\square$. Thus we can express the values of $\vartheta_Q(\chi)$ in terms of its values on any other parametrization surface. We choose for this surface (half) the unit hyperboloid of two sheets, which we have identified with G/K.

Note that the points P on $T_{\check{K}}$ are $\check{K}$ fixed. Thus the translate of e to P is $\check{K}$ fixed; hence so is its intersection with G/K. The orbit of the point $i \in G/K$ under $\check{K}$ is a semicircle orthogonal to the real axis with endpoints $\omega, \overline{\omega}$, which are conjugates in Q. The orbits of the other points $\xi + i\eta \in G/K$ are semicircles with the same endpoints.

From the structure (6.19) of e we conclude that $\vartheta_Q(I)$ can be expressed as an integral of the form

(6.22)
$$\vartheta_Q(I) = \int \lambda(\tilde{r}) \, d\tilde{r} \int \mathscr{E}_{\check{K}} \, d\theta.$$

Here $d\theta$ is the $\check{K}$ invariant measure on the semicircle and $\tilde{r}$ parametrizes the semicircles in question. $\hat{\mathscr{E}}_{\check{K}}$ differs from $\hat{\mathscr{E}}$ in that we pick one lattice point from each equivalence class $\mod \Gamma_{\check{K}}$ in accordance with the ideas presented in Chapter II on $\Gamma_{\check{K}}$ restriction.

We can now use an idea that, in essence, goes back to Eli Cartan but that in this connection was observed by Kudla and Millson [25]: We can use a form of Green's theorem to transfer the integrals of $\mathscr{E}$ to a fixed semicircle, say the semicircle through the point i. This is an instance of the general principle relating "invariant" and "closed".

To formulate things precisely would require a long diversion and will appear elsewhere. We content ourselves with a heuristic formulation of *Poincaré duality*.

Before going into the details let me make a remark that is important for what follows: I shall phrase it as *Fourier duality implies Poincaré duality*.

Poincaré duality deals with the question of associating an analytic object to a geometric object. The simplest geometric object is a point; Fourier duality assigns an analytic object to a point. Thus, for example, if the point is the origin, then

$$(6.23) \qquad \delta_0 \cdot f = f(0) = \int \hat{f}(\hat{x}) \, d\hat{x}$$

shows that the measure (or n form) $d\hat{x}$ in the dual space corresponds to the origin in the sense that integrating $\hat{f}$ times this n form corresponds to δ_0.

In terms of x itself we write

$$(6.24) \qquad \delta_0 \cdot f = \int f(x) \, dx \int e^{ix \cdot \hat{x}} \, d\hat{x},$$

so the n form $(\int \exp(ix \cdot \hat{x}) \, d\hat{x}) \, dx$ corresponds to δ_0.

Instead of taking δ_0 we can take δ_L, where L is a linear variety say of dimension m in R^N. Then the Fourier representation assigns $\delta_{\hat{L}^\perp}$ in the Fourier transform space. We can think of $\delta_{\hat{L}^\perp}$ as an $N - m$ form.

The above implies, in particular, that convolution with $\delta_{\hat{L}^\perp}$ corresponds to multiplication by δ_L, that is, to restriction to L.

So far all is general. But suppose that we convolve $\delta_{\hat{L}^\perp}$ with a function $\hat{f}$ supported on an algebraic variety V. This convolution

$$(6.25) \qquad (\delta_{\hat{L}^\perp} * \hat{f})(\hat{y}) = \text{measure of } \hat{f} \text{ on } \hat{L}^\perp + \hat{y}.$$

This is the Fourier transform of restriction of f to L. But, since support $\hat{f} \subset V$ the function f satisfies a system $\vec{D}$ of partial differential equations. Let M be a parametrization set for $\vec{D}$. Then we can evaluate f in terms of $f|_M$ (and some derivatives). In particular we can evaluate $f|_L$ in terms of $f|_M$ (and some derivatives). But $f|_L$ expresses the measure of $\hat{f}$ on $\{\hat{L}^\perp + y\}$. Hence this measure can be evaluated in terms of $f|_M$. In order for this to work we need to

Find the explicit evaluation of $f|_L$ in terms of $f|_M$.

To put this in the framework of Poincaré duality, suppose that $\hat{L}^\perp$ is fixed (pointwise) by a linear group $\hat{H}$ that preserves V and acts on M. Then we can expect that the kernel $B(l, m)$ representing the passage from $f|_L$ to $f|_M$ is H invariant in the m variable. As we shall see below, this means that $B(l,)$ defines a closed form. This closed form is, up to a constant, dual to the geometric object O_H, which is a suitable orbit of H on M.

Thus we started with the H invariant set L and ended up with the geometric object O_H. Note that under suitable circumstances O_H is essentially the Intertwining B (Geometric Intertwining) defined in Chapter II.

In case the function f also has an invariance property, then $f|_M$ defines a closed form. Poincaré duality now says:

General Hecke-Siegel Principle: We can evaluate $f|_L$, that is, the measure of $\hat{f}$ on $\hat{L}^\perp + y$, in terms of the integrals of f on O_H.

We shall now give several illustrations of the Hecke-Siegel Principle.

Let us give some precision to the above:

Let M be a Riemannian manifold of dimension $m + n$. (In practice M is a symmetric space.) Let B be a Lie group that acts smoothly on M such that the generic orbits X of B are of dimension m. We denote by dx a suitably normalized B invariant measure on X.

Our main assumption is that we can find a "complement" Y to X, meaning that $M = X \times Y$, and the sets $y = \text{const}$ are the orbits of B. We denote by dy a suitable measure on Y. We assume that Y is connected.

We regard dx as an m form and dy as an n form on M.

We also assume that there is a discrete subgroup Γ_B of B, and we set $X/\Gamma_B = c$. In most of our application c is compact. For some purposes we also want a discrete group Γ of isometries of M such that M/Γ has finite volume and $\Gamma_B = \Gamma \cap B$. In general, c is a cycle on M/Γ.

For most of our applications the sets $x = \text{const}$ are the orbits of another group $\tilde{B}$. Moreover, if we choose dy as a $\tilde{B}$ invariant measure, then $dx \wedge dy$ is the volume element of M defined by the metric.

We want to compute the Poincaré dual of c on M/Γ_B. We think of c as lying on some fixed set $y = y_0$. Recall that a Poincaré dual of c is a closed n form α such that

$$(6.26) \qquad \int_c \beta = \int_{M/\Gamma_B} \alpha \wedge \beta$$

for any closed m form β. In case we are dealing with noncompact c, then β has to be small at infinity. If α is harmonic, then it is called *the* Poincaré dual of c.

THEOREM 6.3. *Let f be a B invariant function on M. Then, for $f \in L_1(Y, dy)$, $\alpha = f(y) \, dy$ is a closed n form that, up to a constant, is a Poincaré dual form to c on $M/\Gamma_B = (X/\Gamma_B) \times Y$.*

PROOF. It is clear that α is closed. Let β be any closed m form on $X \times Y$ that is Γ_B invariant. We want to compute the integral of $\alpha \wedge \beta$ over M/Γ_B. We use coordinates $x_1, \ldots, x_m$ on X and $y_1, \ldots, y_n$ on Y. We assume our coordinate system is chosen so that $dx = dx_1 \wedge \cdots \wedge dx_m$ and $dy = dy_1 \wedge \cdots \wedge dy_n$. Since α involves only dy, the only part of β that survives in $\alpha \wedge \beta$ is the term $h(x, y) \, dx$. Thus

$$(6.27) \qquad \int \alpha \wedge \beta = \int f(y) \, dy \int h(x, y) \, dx.$$

The inner integral can be thought of as an integral over c. We want to show that it is independent of y. Now, c is $y = y_0$ and the inner integral can also be regarded as the integral of β over $y = \text{const}$. Since the sets $y = \text{const}$ are deformations of each other and β is closed, we obtain the invariance of the integral.

Hence

$$(6.28) \qquad \int \alpha \wedge \beta = \left(\int_c \beta \right) \int f(y) \, dy = \text{const} \int_c \beta,$$

which is the desired result.

Theorem 6.3 is used to compute the periods of certain closed forms on c. The point is that Theorem 6.3 allows us to transform the integration from c to all of $(X/\Gamma_B) \times Y$. Using the unfolding lemmas described below such an integral becomes an integral over all of M. Now, M is thought of as an orbit of a group G acting in R^N or as a parametrization set for a G invariant differential equation. Thus suitable integrals over M represent intertwinings.

We now present two unfolding lemmas.

LEMMA 6.4. *Let $\mathscr{D}$ be a fundamental domain for Γ on M. If h is Γ invariant and f is Γ_B invariant, then*

$$(6.29) \qquad \int_{\mathscr{D}} h(m) \sum_{\gamma \in \Gamma_B \backslash \Gamma} (\gamma f)(m) \, dm = \int_{(X/\Gamma_B) \times Y} h(m) f(m) \, dm.$$

Here X/Γ_B is a fundamental domain for the action of Γ_B on X.

We note that $(\gamma f)(m)$ is defined to be $f(\gamma m)$ so that $(\gamma \gamma')f = \gamma'(\gamma f)$. This accounts for the appearance of $\Gamma_B \backslash \Gamma$ rather than Γ / Γ_B in the left side of (6.29).

PROOF. It is clear that $(X/\Gamma_B) \times Y$ is a fundamental domain for the action of Γ_B on M. Now, consider $\bigcup \gamma \mathscr{D} = \tilde{M}$, where the union is taken over any set of representatives for $\Gamma_B \backslash \Gamma$. $\tilde{M}$ is also a fundamental domain for Γ_B since

$$M = \bigcup_{\gamma \in \Gamma} \gamma \mathscr{D} = \bigcup_{\gamma_0 \in \Gamma_B} \gamma_0 \tilde{M}.$$

For $\gamma_0 \tilde{m} = \gamma_0' \tilde{m}'$ means $\gamma_0 \gamma p = \gamma_0' \gamma' \tilde{p}$ for some $p, \tilde{p} \in \mathscr{D}$ and some γ, γ' that are representatives for $\Gamma_B \backslash \Gamma$. But this contradicts the definition of fundamental domain.

Thus the left side of (6.29) is the integral of hf over $\tilde{M}$. But hf is Γ_B invariant, so the integral over any two fundamental domains is the same. This establishes Lemma 6.4.

For our second unfolding lemma, let Γ' be a subgroup of Γ.

LEMMA 6.5. *Suppose h is of the form $h = \sum_{\gamma \in \Gamma' \backslash \Gamma} \gamma h_0$, where h_0 is Γ' invariant and, as before, f is Γ_B invariant. Suppose moreover that Γ_B acts freely on $\Gamma' \backslash \Gamma$ (on the right) meaning that no $\gamma_0 \in \Gamma_B$, $\gamma_0 \neq 1$ has a fixed point. Then the integrals in (6.29) are equal to*

$$(6.30) \qquad \int_M f(m) \sum_{\gamma \in \Gamma' \backslash \Gamma / \Gamma_B} (\gamma h_0)(m) \, dm.$$

PROOF. For each $m \in M$ there is a unique $\gamma_0 \in \Gamma_B$ so that $\gamma_0 m \in X\backslash\Gamma_B \times Y = \mathscr{D}_{\Gamma_B}$. Hence

$$(6.31) \qquad \int_M f(m) \sum_{\gamma \in \Gamma'\backslash\Gamma/\Gamma_B} (\gamma h_0)(m)\, dm = \sum_{\gamma_0 \in \Gamma_B} \int_{\gamma_0 \mathscr{D}_{\Gamma_B}}$$

$$= \int_{\mathscr{D}_{\Gamma_B}} f(m) \sum_{\gamma \in \Gamma'\backslash\Gamma/\Gamma_B} \sum_{\gamma_0 \in \Gamma_B} (\gamma h_0)(\gamma_0 m)\, dm$$

$$= \int_{\mathscr{D}_{\Gamma_B}} f(m) \sum_{\gamma \in \Gamma'\backslash\Gamma} (\gamma h_0)(m)\, dm$$

$$= \int_{\mathscr{D}_{\Gamma_B}} h(m) f(m)\, dm$$

because Γ_B acts freely on $\Gamma'\backslash\Gamma$; so when γ goes through a set of representatives, for $\Gamma'\backslash\Gamma/\Gamma_B$, then $\gamma\gamma_0$ goes uniquely through a set of representatives for $\Gamma'\backslash\Gamma$ as γ_0 goes through Γ_B.

Lemma 6.5 is established.

For our first application we set $B = \tilde{K}$, $h = \mathscr{E}$, and f the kernel for the parametrization problem (Cauchy Problem) with data on G/K, that is, f is the λ of (6.22). We also set $\Gamma' = \Gamma_N = \Gamma\cap N$. Thus λ is $\tilde{K}$ invariant so $\lambda(\tilde{r})\, d\tilde{r}$ defines a Poincaré dual to the cycle $c = (\tilde{K} \cdot i)/\Gamma_{\tilde{K}}$. On the other hand, $\hat{\mathscr{E}}$ is obtained by summing $\sum \delta_{j,0}$ over $\Gamma_N\backslash\Gamma$. Since $\delta_{j,0}$ is N invariant, it follows that, in the usual $z = \xi + i\eta$ coordinates lifted by quadratic transform, the one-form

$$(6.32) \qquad \mathscr{E}^1 = \sum_{\gamma \in \Gamma_N\backslash\Gamma} \gamma \left(\sum \widehat{\delta_{j^2,0,0}}\, d\xi \right)$$

is closed since Γ preserves closedness of forms.

By our above results

The period of $\mathscr{E}^1$ over the cycle c is, up to a constant, the theta function given by (6.22), except that $\mathscr{E}_{\tilde{K}}$ is replaced by $\mathscr{E}^1_{\tilde{K}}$, which differs little from it.

Thus far we have dealt with the theta function. There is no difficulty in applying the Mellin transform to the above. The Mellin transform changes $\mathscr{E}^1_{\tilde{K}}$ to the Eisenstein series and changes the theta functions to zeta functions.

If instead of taking periods we take Fourier series coefficients on $\tilde{K}/\Gamma_{\tilde{K}}$, then we obtain zeta functions with grossencharactere.

This gives a new approach to the *Hecke-Siegel theory* (see [32]). In particular the fact that the *Fourier coefficients of the Eisenstein series are* (essentially) *zeta functions with grossencharactere* is purely conceptual.

We can go further. Instead of taking time axes of the form $T_{\tilde{K}}$, where $\tilde{K} \cap \Gamma = \Gamma_{\tilde{K}}$ is infinite, we can take the opposite case in which $\Gamma_{\tilde{K}} = \{I\}$. For example we can take the u axis, that is, the ray through $\begin{pmatrix} 0 & 1 \\ -1 & 0 \end{pmatrix}$ whose isotropy group is the diagonal group A. Since $A \cap \Gamma = \pm I$ (with $-I$ not differing essentially from $+I$), the cycle in question is the η (imaginary) axis in the upper half-plane.

Since $A/A \cap \Gamma \sim A$ the analog of taking Fourier coefficients on $\tilde{K}/\Gamma_{\tilde{K}}$ becomes the Mellin transform along the imaginary axis. By the above this can be expressed in terms of what could be called the ζ function of the quadratic form tv, as the intersections of the light cone with the planes orthogonal to the u axis are of the form $tv = \text{const.}$

Working out the details gives the Mellin transform of the *Kronecker limit formula* (see [33]), which expresses the *constant term in the Taylor series expansion at $s = 1$ of the Eisenstein series in terms of log of the eta function*. The point is that the Mellin transform of $\log \eta$ on the imaginary axis is expressed as a quotient of ζ functions times a simple factor.

Actually the above is the first Kronecker limit formula. To go to the second Kronecker limit formula, one has to replace the upper half-plane G/K by G itself and consider the analog of the Eisenstein series corresponding to the characters of K. For this we must use the 4-dimensional representation of G that is the direct sum of 2 copies of the 2-dimensional representation. (A study of the function theory of this representation is given in [8].)

VI.3. Hyperbolic Eisenstein series. Thus far the central role in our discussion has been played by $\hat{\mathscr{E}}$, which is, essentially, the sum of δ functions at the lattice points of the light cone $= G/MN$. The work of Kudla and Millson starts with the sum of the lattice points on $G/\tilde{K}$. This leads to *hyperbolic Eisenstein series* rather than the usual nilpotent Eisenstein series.

There is an important difference between $G/\tilde{K}$ and G/MN. In the latter case we decompose the representation of G on functions on G/MN using the Mellin transform. This is meaningful because A normalizes MN. But no such possibility exists for $G/\tilde{K}$.

However for $G/\tilde{K}$ there exists the *Bessel decomposition*. The point is that we are now dealing with an equation of the form $(\Box - 1)f = 0$. For the operator $\Box$, which is homogeneous, we decomposed f into homogeneous parts. Now, there are two ways to deal with the operator $\Box - 1$

(a) Observe that $\Box - 1$ has *two* degrees of homogeneity. Thus if we decompose f into its homogeneous parts, say $\check{f}(s)$, then the equation $(\Box - 1)f = 0$ implies a differential relation between $\check{f}(s)$ and $\check{f}(s + 2)$. This is the method pursued by Kudla and Millson. It is predicated upon using a simple decomposition (Mellin transform) in r and then a more complicated equation in the parameter θ on the hyperboloids.

(b) The more classical approach is to use a complicated equation in r and then a simple equation in θ. This equation in r takes into account the two degrees of homogeneity; in our case it is the Bessel equation. Thus we express f as an integral of terms of the form $K_s(r)f_s(\theta)$. Here K_s is a Bessel function and f_s is an eigenfunction of the Laplacian on the hyperboloids.

In particular f_0 and f_1 are harmonic. Thus, if we start with $\hat{f}$ being the δ function of a $\tilde{K}$ invariant point, then Theorem 6.3 and the unfolding lemmas the corresponding f_0 and f_1 give, upon summation over Γ, *the* Poincaré dual of the cycle associated to $\tilde{K}/\Gamma \cap \tilde{K}$ on $\mathscr{D}$ (up to a constant). The corresponding view of *the* Poincaré dual using (a) was discovered by Kudla and Millson [25], who also computed the constant.

We can also use the Hecke-Siegel theory to compute the periods of hyperbolic Eisenstein series and express them in terms of Dirichlet-like series except that the exponential, which is crucial in Dirichlet series, is replaced by the Bessel functions.

This theory and its extension to higher rank groups will be the object of another study. The crucial point in the computation is the explicit fundamental solution.

CHAPTER VII.   PRODUCT METHODS:
THE SELBERG TRACE FORMULA AND ELLIPTIC FUNCTIONS

VII.1. The Selberg trace formula. We use the notation of the previous chapter. We start with G/K, which we can identify with H_1, which is the positive half of the unit hyperboloid of two sheets. We consider the "flow" of this hyperboloid defined by the wave equation. This means that for each $\tau \in T_K$ we consider the positive half H_τ of the hyperboloid passing through τ. The Cauchy problem for $\square$ defines a one-parameter group of transformations on pairs of functions on H_τ.

We start with data of the form $(f, 0)$ on H_1. Call F the solution of the Cauchy problem with data $(f, 0)$ on H_1. Denote by $U(\tau)$ the transformation

$$(7.1) \qquad\qquad U(\tau)f = F|_{H_\tau}.$$

Let $h(\tau)$ be any nice function on T_K with $h(\tau) = h(1/\tau)$. We form

$$(7.2) \qquad\qquad \mathscr{T}(h) = \int h(\tau)\operatorname{Trace} U(\tau)\, d\tau.$$

Variations of $\mathscr{T}(h)$ play the central role in Selberg's trace formula.

To understand this point, let us observe that we can relate a K invariant function $\tilde{h}$ on $G/K = H_1$ to h. This can be done in two equivalent manners:

(a) Solve the (exotic) Cauchy problem with data $(h, 0, \dots, 0, \dots)$ on T_K. This solution is K invariant; its restriction to H_1 is $\tilde{h}$.

(b) For each $\tau \in T_K$ call μ_τ the kernel defining the explicit solution at τ of the Cauchy problem for $\square$ with data $(f, 0)$ on H_1. This means that

$$(7.3) \qquad\qquad F(\tau, 0) = \int_{H_1} \mu_\tau(w) f(w)\, dw.$$

Here dw is the G invariant measure on H_1. It is clear that $\mu_\tau(w)$, considered as a function of w, is K invariant. We now define $\tilde{h}$ as that K invariant function for which $F(\tau, 0) = h(\tau)$.

It is clear that both (a) and (b) lead to the same result.

The relation between h and $\tilde{h}$ is that of Abel transform.

The next point to observe is

$$(7.4) \qquad\qquad U(\tau)f = \mu_\tau * f,$$

the convolution being taken for K invariant functions on G with H_1 and H_τ being identified with G/K as usual. This means that we can write (formally)

$$(7.5) \qquad \mathcal{T}(h) = \int h(\tau)\mathrm{Tr}(\mu_\tau * \quad)\,d\tau$$

$$= \mathrm{Tr}\int [(h(\tau)\mu_\tau)* \quad]\,d\tau = \mathrm{Tr}\,\tilde{\tilde{h}} * \quad.$$

$\tilde{\tilde{h}}$ can be determined as follows: μ_τ is constructed in the usual fashion from a fundamental solution. But, actually, it is better to regard μ_τ as the Cauchy data of a null solution η_τ whose support on H_τ is the point $T_k \cap H_\tau$. (See Section VII.2 for more details on null solutions.) This means that $\tilde{\tilde{h}}$ is the Cauchy data of $\int h(\tau)\eta_\tau\,d\tau$. It is easy to express the values of this solution on the T_K axis; this restriction to T_K is an "Abel like" transform of h. Then solving the exotic Cauchy problem with data on T_K evaluates $\tilde{\tilde{h}}$ in terms of h.

(7.5) shows that $\mathcal{T}$ defined by (7.2) is the same as the trace considered by Selberg.

We shall say only a few words about the evaluation of the trace. In the first place we start with the space of f that are Γ invariant on G/K. Here Γ may be any discrete subgroup; we shall first assume $\Gamma\backslash G/K$ is compact and Γ has no elliptic elements. In order to use Γ, we sum $\tilde{\tilde{h}}$ over Γ. Using the unfolding lemmas of Chapter VI, this has the effect of replacing the integral defining the convolution in (7.4) by an integral over $\mathcal{D}$ = fundamental domain for Γ on G/K.

As usual we evaluate the trace by using two different bases.

(A) Using the basis φ_{λ_j} of eigenfunctions of the Laplacian on $\Gamma\backslash G/K$ or, what is the same thing, functions on $\Gamma\backslash G/K$ belonging to a fixed representation space of G, we obtain

$$(7.6) \qquad\qquad \mathcal{T}(h) = \sum \check{h}(\lambda_j).$$

Here $\check{h}$ is the Mellin transform of h.

(B) Using the basis of δ functions of points in $\Gamma\backslash G/K$, we evaluate the trace in terms of diagonal elements (fixed points). The calculations depend on the explicit formula for μ_τ that is, the explicit fundamental solution; we obtain Selberg's result [30]

$$(7.7) \qquad \frac{A(\mathcal{D})}{2\pi} \int_{-\infty}^{\infty} \tanh \pi r h(r) r\,dr$$

$$+ 2\sum_{\{P\}_\Gamma} \sum_{k=1}^{\infty} \frac{\log N(P)}{(N\{P\}^{k/2} - (N\{P\})^{-k/2}} h(k \log N\{P\}).$$

P are the primitive hyperbolic classes in Γ, and $N(P)$, which is the norm of P, is the largest eigenvalue or, equivalently, the length of the geodesic on $\Gamma\backslash G/K$ defined by P.

Selberg has extended his formula to the case that $\Gamma\backslash G$ is of finite volume and allows Γ to have fixed points. Many generalizations have been made to the higher-dimensional case; (see [2]).

Our main point is to show how Selberg's formula fits into our general ideas on intertwining and parametrization. From this point of view we can also consider the Cauchy problem on the $T_{\tilde{K}}$ axis. Of course $\Gamma\backslash G/\tilde{K}$ is not a manifold; nevertheless we can study the motion under the analog of $U(\tau)$ of some classes of functions on G that are left Γ and right $\tilde{K}$ invariant. We can also deal with such questions in higher rank groups.

Although such a discussion is beyond the scope of this work, let us content ourselves with the

Computational Principle. The computation of the explicit analog of (7.7) depends on the explicit formula for one fairly general class of solutions of the system of partial differential equations. (For rank 1 the class in question is the fundamental solutions.)

VII.2. Elliptic functions and their generalization. The final subject we deal with here is that of Jacobian elliptic functions. Our starting point is the expression for the Jacobian function $y = sn(u, k)$

$$(7.8) \qquad u = \int_0^y (1 - t^2)^{-1/2}(1 - k^2 t^2)^{-1/2}\, dt.$$

In particular the complete elliptic integral is given by

$$(7.9) \qquad K = \int_0^{\pi/2} (1 - t^2)^{-1/2}(1 - k^2 t^2)^{-1/2}\, dt$$
$$= \frac{\pi}{2} F\left(\frac{1}{2}, \frac{1}{2}; 1; k^2\right) = \frac{\pi}{2} P_{-1/2}(1 - 2k^2).$$

Thus $K(k)$ is a Legendre function.

The Legendre functions appear in group representations from another point of view; they are the spherical functions. From our point of view they are the restrictions to H_1 of homogeneous solutions of $\square$ of order s that are K invariant. We can obtain such functions by taking the Fourier transform of r^{is} on the light cone. Up to elementary factors this leads to the expression

$$(7.10) \qquad P_s(\cosh 2\zeta) = \int_{-\pi}^{\pi} [\cosh 2\zeta - \sinh 2\zeta \cos 2\theta]^s\, d\theta.$$

P_s is the Legendre function.

To verify (7.10) we recall the analogous computation in (6.18). We see that the essential term in the computation is

$$\int [(g_\zeta I) \cdot (kn_0)]^s \, dk$$

(7.11)
$$= \int \left[\begin{pmatrix} e^{2\zeta} & 0 \\ 0 & e^{-2\zeta} \end{pmatrix} \cdot \begin{pmatrix} \cos^2 \theta & -\cos \theta \sin \theta \\ -\cos \theta \sin \theta & \sin^2 \theta \end{pmatrix} \right]^s \, d\theta$$

$$= \int [e^{2\zeta} \sin^2 \theta + e^{-2\zeta} \cos^2 \theta]^s \, d\theta$$

$$= \int [\cosh 2\zeta - \sinh 2\zeta \cos 2\theta]^s \, d\theta.$$

Here

$$n_0 = \begin{pmatrix} 1 & 0 \\ 0 & 0 \end{pmatrix} \quad \text{and} \quad g_\zeta = \begin{pmatrix} e^\zeta & 0 \\ 0 & e^{-\zeta} \end{pmatrix}$$

and the inner product is as in Chapter VI.

Thus the Dirichlet representation (7.10) appears naturally in our framework. How about the Euler representation (7.9)?

To clarify this point, note that if we start with any solution f of the wave equation and decompose f under homogeneity in r and under the characters of K, then we obtain functions $f(s, \chi)$ that are solutions of associated Legendre equations in $x = \cosh \zeta$. (This separation of variables is dealt with extensively in [8].) What is the starting f that leads to the Euler representation?

The answer is (almost) $e =$ fundamental solution with singularity at I. Actually, e is not a solution, so we must really take

(7.12)
$$n = e^+ - e^-,$$

where $e^\pm$ are the forward (retrograde) fundamental solutions. n is called a *null solution* of $\Box$ because its Cauchy data on G/K has support at a point.

From the explicit formula for $e^\pm$, it follows that the homogeneous part of n is given (formally) by

(7.13)
$$\int \frac{r^\alpha}{(1 + r^2 - 2r \cosh 2\zeta)^{1/2}} \frac{dr}{r} = \frac{1}{\sqrt{2}} \int \frac{r^{\alpha - 1/2}}{((r + r^{-1})/2 - \cosh 2\zeta)^{1/2}} \frac{dr}{r}$$

$$= \sqrt{2} \int \frac{r^{\alpha - 1/2}}{(t - x)^{1/2}(t^2 - 1)^{1/2}} \, dt$$

with $x = \cosh 2\zeta$; here we have put $t = (r + r^{-1})/z$. If $\alpha = 1/2$, the term involving r drops out. Finally if we set $2u^2 = 1 - t$, we obtain

$$= c \int \frac{du}{(1 - u^2)^{1/2}(1 - x - 2u^2 x)^{1/2}}$$

$$= \frac{c}{(1 - x)^{1/2}} \int \frac{du}{(1 - u^2)^{1/2}(1 - u^2[2x/(1 - x)])^{1/2}},$$

which agrees with (7.8).

We conclude

THEOREM 7.1. *The inverse of the elliptic function $sn(u, k)$ is given, up to a simple factor, as the indefinite integral of $r^{1/2}$ times the null solution of $\square$ restricted to the ray defined by ζ where $k^2 = 2 \cosh 2\zeta / (1 - \cosh 2\zeta)$.*

One might enquire as to what is the extension of Theorem 7.1 to higher rank groups G. Our first task is to search for algebraic-like solutions f of a suitable system of partial differential equations. In [7] we discuss the construction of null solutions, which are often algebraic-like.

Given such an f, a suitable decomposition under a commuting group A of G leads to spherical-like functions. However because of the algebraic nature of f it makes sense to construct *indefinite integrals* over the orbits of A.

The indefinite integral of a closed one-form ω defined on a manifold M depends on

(1) a base point p_0,

(2) a path γ joining p_0 to a variable point p.

Then

$$(7.14) \qquad W(p) = \int_{\gamma}^{p}{}_{p_0} \omega$$

is defined on the universal covering of M. Actually it is not the universal covering of M that is relevant but rather the universal homology covering $\tilde{M}$, or, what is the same thing, the maximal abelian covering.

Suppose now that ω is a closed two-form. The point p_0 should be replaced by a fixed curve and p by a curve with the same endpoints as p_0 such that p is homologous to p. Now γ is a two cycle that bounds $p - p_0$ and $W(p)$ is defined as before.

It was pointed out to me by S. Cappel that we may replace p_0 by a point, considered as a degenerate curve. Thus p is a loop based at p_0 that is homologous to zero.

$W(p)$ is thus naturally defined on the universal abelian covering $\tilde{M}$ of this loop space.

Using the idea of variation of p, we can give a natural manifold structure on $\tilde{M}$.

PROBLEM 7.1. Study the "abelian functions" $W(p)$ related to semisimple Lie groups of rank > 1.

BIBLIOGRAPHY

1. P. Appell, *Sur l'équation $\partial^2 z/\partial x^2 - \partial z/\partial y = 0$ et la théorie de la chaleur*, J. Math. (4) **VIII** (1892), 187–215.

2. J. Arthur, *The trace formula for noncompact quotient*, Proc. Internat. Congr. Math., Warsaw, 1983, Vol. 2, Polish Scientific Publishers, Warsaw and Elsevier, Amsterdam, 1984, p. 849.

3. T. Bailey, L. Ehrenpreis, and R. O. Wells, *Weak solutions of the massless field equations*, Proc. Roy. Soc. London Ser. A **384** (1982), 403–425.

4. Bateman Manuscript Project, *Higher Transcendental Functions*, Vols. 1, 2, 3, McGraw-Hill, New York, 1953.

5. J. S. Dani and S. G. Dani, *Density properties of orbits under discrete groups*, J. Indian Math. Soc. **39** (1975), 189–217.

6. L. Ehrenpreis, *Fourier Analysis in Several Complex Variables*, Wiley-Interscience, New York, 1970.

7. _____, *Harmonic functions* (to appear).

8. _____, *Hypergeometric functions*, Algebraic Analysis, Vol. I, Academic Press, New York, 1988, pp. 85–128.

9. _____, *Conditionally convergent functional integrals and partial differential equations*, Proc. Internat. Congr. Math., Stockholm, 1962, Institut Mittag-Leffler, Sweden, 1963, pp. 337–338.

10. _____, *The Rogers-Ramanujan identities* (to appear).

11. _____, *Almost mean periodic functions* (to appear).

12. _____, *A new proof and an extension of Hartogs' theorem*, Bull. Amer. Math. Soc. **67** (1961), 507–509.

13. _____, *Scattering theory, elliptic functions, and the Cousin problem* (to appear).

14. _____, *Conformal geometry*, Differential Geometry, Birkhäuser, Basel, 1983, pp. 73–88.

15. _____, *The Radon transform and its ramifications* (to appear).

16. L. Ehrenpreis and T. Kawai, *Poisson's summation formula and Hamburger's theorem*, Res. Inst. Math. Sci. Kyoto **18** (1982), 413–426.

17. L. Ehrenpreis and P. Malliavin, *Fourier analysis on nonconvex sets*, Sympos. Mat. Inst. Nazionale de Alta Mat., Vol. XII, 1971.

18. E. Hecke, *Über die Bestimmung Dirichletscher Reihen durch ihre Functionalgleichung*, Math. Ann. **112** (1936), 664–699 (Collected Works, pp. 591–626).

19. _____, *Über die Zetafunktion beliebiger algebraischer Zahlkörpen*, Nachrichten Göttingen, 1917, pp. 77–89 (Collected Works, pp. 159–171).

20. _____, *Eine neue Art von Zetafunktionen, Part* 1, Math. Z. **1** (1918), 357–376 (Collected Works, pp. 215–234, *Part* 2, Math. Z. **6** (1920), 11–51 (Collected Works, pp. 249–290).

21 R. Howe, *θ-series and invariant theory*, Proc. Sympos. Pure Math., Vol. 33, Part I, Amer. Math. Soc., Providence, R.I., 1979, pp. 275–285.

22. K. Jacobi, *Über die differentialgleichung, welcher de Reihen* $1 \pm 2q + 2q^4 \pm 2q^9 + etc.$, *genüge leisten*, 1847 (Collected Works, pp. 173–190).

23. M. Kac, *Almost periodicity and the representation of integers as sums of squares*, Amer. J. Math. **62** (1940), 122–126.

24. M. Kac, E. R. van Kampen, and A. Wintner, *Ramanujan sums and almost periodic functions*, Amer. J. Math. **62** (1940), 107–114.

25. S. Kudla and J. Millson, *Harmonic differentials and closed geodesics on a Riemann surface*, Invent. Math. **54** (1979), 193–211.

26. T. Ono, *Sur une propriété arithmétique des groupes algébriques commutatifs*, Bull. Soc. Math. France, **85** (1957), 307–323.

27. H. Rademacher, *On the expansion of the partition function in a series*, Ann. of Math. (2) **44** (1943), 416–422.

28. M. Riesz, *L'intégrale de Riemann-Liouville et le problème de Cauchy*, Acta Math. **81** (1949), 1–223.

29. L. J. Rogers, *Second memoir on the expansion of certain infinite products*, Proc. London Math. Soc. (1) **25** (1894), 318–343.

30. A. Selberg, *Harmonic analysis and discontinuous groups in weakly symmetric Riemannian spaces with applications to Dirichlet series*, J. Indian Math. Soc. **20** (1956), 47–88.

31. C. L. Siegel, *Lectures on analytic functions of several complex variables*, Inst. Adv. Stud., Princeton, NJ., 1948–1949.

32. _____, *Lectures on the analytic theory of quadratic forms*, Inst. Adv. Stud., Princeton, NJ, 1934–1935.

33. _____, *Lectures on advanced analytic number theory*, Tata Inst., Bombay, 1961.

34. A. Weil, *Sur certains groups d'opérateurs unitaires*, Acta Math. **111** (1964), 143–211.

TEMPLE UNIVERSITY

Proceedings of Symposia in Pure Mathematics
Volume **49** (1989), Part 2

Problems on Theta Functions

JUN-ICHI IGUSA

At the Summer Institute we explained three problems on classical theta functions. This paper is based on the five-page memo we prepared for that lecture. We have expanded it so that the paper becomes as self-contained as possible.

1. Notation and some results. We use the standardized notation $\mathbf{Z}$, $\mathbf{Z}_p$ for the rings of rational, p-adic integers and $\mathbf{R}$, $\mathbf{C}$ for the fields of real, complex numbers. If m, n are positive integers, we denote by $M_{m,n}$ the affine space of $(m \times n)$-matrices; in particular we put $M_{n,n} = M_n$. We also denote by GL_n the general linear group of degree n, i.e., the group of invertible matrices in M_n. If the entries are restricted, e.g., to $\mathbf{Z}$, we write $M_{m,n}(\mathbf{Z})$, etc. We take a positive integer g and denote by Sp_{2g} the symplectic group of degree $2g$. If σ is in M_{2g}, the condition on σ to be in Sp_{2g} is

$$
{}^{\mathrm{t}}\sigma \begin{pmatrix} 0 & 1_g \\ -1_g & 0 \end{pmatrix} \sigma = \begin{pmatrix} 0 & 1_g \\ -1_g & 0 \end{pmatrix},
$$

in which ${}^{\mathrm{t}}\sigma$ is the transpose of σ and 1_g is the identity matrix of degree g. We express σ as a square matrix of degree 2 in four entry matrices a, b, c, d from M_g. We denote by $\mathscr{S}_g$ the *Siegel upper half-space* of degree g. If $\tau = {}^{\mathrm{t}}\tau$ is in $M_g(\mathbf{C})$, the condition on τ to be in $\mathscr{S}_g$ is $\mathrm{Im}(\tau) > 0$, which means that the imaginary part of τ is positive definite. We know that $\mathrm{Sp}_{2g}(\mathbf{R})$ acts transitively on $\mathscr{S}_g$ as

$$
\sigma \cdot \tau = (a\tau + b)(c\tau + d)^{-1}, \qquad \sigma = \begin{pmatrix} a & b \\ c & d \end{pmatrix}.
$$

In general if Γ is a subgroup of $\mathrm{GL}_n(\mathbf{Z})$, then for any positive integer l we denote by $\Gamma(l)$ the kernel of the homomorphism $\mathrm{GL}_n(\mathbf{Z}) \to \mathrm{GL}_n(\mathbf{Z}/l\mathbf{Z})$ restricted to Γ. In particular if $n = 2g$ and $\Gamma = \mathrm{Sp}_{2g}(\mathbf{Z})$, we write $\Gamma_g(l)$ instead of $\mathrm{Sp}_{2g}(\mathbf{Z})(l)$; the condition on σ in $\mathrm{Sp}_{2g}(\mathbf{Z})$ to be in $\Gamma_g(l)$ is $\sigma \equiv 1_{2g} \bmod l$. If s is any square matrix, we denote by $\mathrm{diag}(s)$ the column vector with the ith

---

1980 *Mathematics Subject Classification* (1985 *Revision*). Primary 10D20, 14K25.
This work was partially supported by the National Science Foundation.

diagonal entry of s as its ith entry for every i. We denote by $\Gamma_g(l, 2l)$ the sub-set of $\Gamma_g(l)$ defined by the additional condition that $\mathrm{diag}(a^t b) \equiv \mathrm{diag}(c^t d) \equiv 0$ mod $2l$. We know that $\Gamma_g(l, 2l)$ is a subgroup of $\Gamma_g(l)$ and that it is normal in $\Gamma_g(1)$ if l is even. By a *modular group* we understand any subgroup Γ of $\Gamma_g(1)$ containing $\Gamma_g(l)$ for some l; in particular $\Gamma_g(l)$ and $\Gamma_g(l, 2l)$ are modular groups.

We take m from $\mathbf{Z}^{2g} = M_{2g,1}(\mathbf{Z})$ with m' and m'' in $\mathbf{Z}^g$ as its top and bottom entry vectors; further we take τ from $\mathscr{S}_g$, z from $\mathbf{C}^g$ and use the notation $\mathbf{e}(t) = \exp(2\pi i t)$ for every t in $\mathbf{C}$. Then the *theta function* of characteristic m is defined as

$$\theta_m(\tau, z) = \sum_{\xi \in \mathbf{Z}^g} \mathbf{e} \left\{ \frac{1}{2} \cdot {}^t\left(\xi + \frac{m'}{2} \right) \tau \left(\xi + \frac{m'}{2} \right) + {}^t\left(\xi + \frac{m'}{2} \right) \left(z + \frac{m''}{2} \right) \right\}.$$

It is a holomorphic function on $\mathscr{S}_g \times \mathbf{C}^g$ and

$$\theta_{m+2n}(\tau, z) = (-1)^{{}^t m' n''} \theta_m(\tau, z)$$

for every n in $\mathbf{Z}^{2g}$. Therefore we shall assume that the entries of m are 0, 1. Since

$$\theta_m(\tau, -z) = (-1)^{{}^t m' m''} \theta_m(\tau, z),$$

by counting the number of solutions of ${}^t m' m'' \equiv 0$ mod 2 we see that there are

$$\begin{cases} 2^{g-1}(2^g + 1) = 3, 10, 36, \ldots \text{even}, \\ 2^{g-1}(2^g - 1) = 1, 6, 28, \ldots \text{odd} \end{cases}$$

theta functions for $g = 1, 2, 3, \ldots$. Accordingly we call m itself even or odd.

If m is even, then the holomorphic function θ_m on $\mathscr{S}_g$ defined by

$$\theta_m(\tau) = \theta_m(\tau, 0)$$

is different from the constant 0, and it is called a *Thetanullwert* or a theta constant. Furthermore the ring $\mathbf{C}[\theta]$ generated over $\mathbf{C}$ by $2^{g-1}(2^g + 1)$ of them is called the *ring of Thetanullwerte*.

If Γ is a modular group contained in $\Gamma_g(4)$, then for any nonnegative k in $1/2 \cdot \mathbf{Z}$ we define a *modular form* of weight k relative to Γ as a holomorphic function f on $\mathscr{S}_g$ satisfying

$$f(\sigma \cdot \tau) = \det(c\tau + d)^k f(\tau)$$

for every σ in Γ and τ in $\mathscr{S}_g$. We eliminate the sign-ambiguity in $\det(c\tau+d)^{1/2}$ by the condition that $f(\tau)/\theta_0(\tau)^{2k}$ is invariant under $\tau \to \sigma \cdot \tau$. This is based on the fact that

$$\theta_0(\sigma \cdot \tau)^2 = \mathbf{e}(\tfrac{1}{4} \cdot \mathrm{tr}(d - 1_g)) \det(c\tau + d)\theta_0(\tau)^2$$

for every σ in $\Gamma_g(2)$, cf. [5, II, pp. 229–230]; hence $(\theta_0(\sigma \cdot \tau)/\theta_0(\tau))^2 = \det(c\tau + d)$ for every σ in $\Gamma_g(4)$. If k is an integer, we can take any modular group as Γ. In the case where $g = 1$ we have to add the following condition: We take any σ from $\mathrm{Sp}_2(\mathbf{Z})$ and put

$$(\sigma^{-1} \cdot f)(\tau) = (c\tau + d)^{-k} f(\sigma \cdot \tau);$$

the sign-ambiguity in $(c\tau + d)^{1/2}$ can be eliminated, but that is not necessary. The fact is that we can expand $(\sigma^{-1} \cdot f)(\tau)$ into a Fourier series as

$$(\sigma^{-1} \cdot f)(\tau) = \sum_{n \in \mathbf{Z}} c_n \mathbf{e}(n\tau/2l),$$

and the condition is that $c_n = 0$ for all $n < 0$ and for all σ. The set of all modular forms of weight k relative to Γ forms a vector space $A(\Gamma)_k$ over $\mathbf{C}$. A part of the fundamental theorem then states that the set of finite sums of elements of $A(\Gamma)_k$ for all k forms an integrally closed subring $A(\Gamma)$ of the ring of all holomorphic functions on $\mathscr{S}_g$, and it is finitely generated over $\mathbf{C}$. Furthermore $A(\Gamma)$ is the direct sum of $A(\Gamma)_k$ for all k, $A(\Gamma)_0 = \mathbf{C}$, and $A(\Gamma)_k A(\Gamma)_{k'} \subset A(\Gamma)_{k+k'}$ for all k, k', i.e., $A(\Gamma)$ is a graded ring over $\mathbf{C}$; cf. Cartan [2, 17–11]. In particular $A(\Gamma)_k$ is finite dimensional for all k. We call $A(\Gamma)$ the *ring of modular forms* relative to Γ. We observe that $\mathbf{C}[\theta]$ is also a graded ring over $\mathbf{C}$ in the ring of all holomorphic functions on $\mathscr{S}_g$ with the $\mathbf{C}$-span of the Thetanullwerte as its homogeneous part of degree $1/2$.

Further references. Basic theorems on theta functions were obtained in the last century, cf., e.g., Krazer [14]. A foundation of the theory of modular forms was given by Siegel [22]; cf. also Koecher [13].

2. Fundamental lemma and problem. If we write characteristics as row vectors, then we have the following classical identities:

$$g = 1, \quad \sum (c\tau + d)^{-4} = \frac{1}{2} \cdot \{\theta_{00}(\tau)^8 + \theta_{01}(\tau)^8 + \theta_{10}(\tau)^8\},$$

$$g = 2, \quad \sum \det(c\tau + d)^{-4} = \theta_{0000}(\tau)^8.$$

In the first series the summation is extended over the set of pairs of relatively prime integers c, d choosing one from $\pm(c, d)$. If for any modular group Γ we denote by Γ_∞ its subgroup defined by $c = 0$, then we can say that the summation is taken over $\Gamma_1(1)_\infty \backslash \Gamma_1(1)$. Similarly in the second series the summation is taken over $\Gamma_2(1, 2)_\infty \backslash \Gamma_2(1, 2)$. On the LHS we have elements of $A(\Gamma_1(1))_4$, $A(\Gamma_2(1, 2))_4$, and on the RHS we have elements of $\mathbf{C}[\theta]$ for $g = 1, 2$. There are many identities of this kind. We have found that they can be derived from the following theorem; cf. [5, I, p. 241; 6, p. 396]:

THEOREM 1. *In general $A(\Gamma_g(4, 8))$ is the integral closure of $\mathbf{C}[\theta]$ within its quotient field; if $g = 1, 2$, then $A(\Gamma_g(4, 8)) = \mathbf{C}[\theta]$.*

We also have the following geometric supplement; cf. [9, p. 385]: If $A = \bigoplus_{k \geq 0} A_k$ is any finitely generated graded entire ring over $A_0 = \mathbf{C}$, the set of homogeneous maximal ideals of A other than $\bigoplus_{k > 0} A_k$ has the structure of a complex projective variety, which we denote by $\mathrm{proj}(A)$. In particular $\mathrm{proj}(A(\Gamma_g(4, 8)))$ and $\mathrm{proj}(\mathbf{C}[\theta])$ are defined. Furthermore, since $A(\Gamma_g(4, 8))$ is integral over $\mathbf{C}[\theta]$, we have a surjective morphism $\mathrm{proj}(A(\Gamma_g(4, 8))) \to \mathrm{proj}(\mathbf{C}[\theta])$. The fact is that this morphism is *bijective*. We might mention that $\mathrm{proj}(A(\Gamma_g(4, 8)))$ is the standard compactification of the quotient variety

$\Gamma_g(4,8)\backslash \mathscr{S}_g$, cf. [2], while proj(C[$\theta$]) can be identified with the closure of the image of $\Gamma_g(4,8)\backslash \mathscr{S}_g$ in the complex projective space $\mathbf{P}_d(\mathbf{C})$ for $d = 2^{g-1}(2^g + 1) - 1$ under the map $\tau \to (\theta_m(\tau))_m$.

We might point out that in the above theorem the *abstractly defined ring* $A(\Gamma_g(4,8))$ is identified with the integral closure of the *explicitly constructed ring* C[θ]. We might also mention that once $A(\Gamma_g(4,8))$ is known, then $A(\Gamma)$ for any modular group Γ containing $\Gamma_g(4,8)$ can be determined as we have explained with examples in the case where $g = 2$; cf. [6, pp. 397–405].

As for the difference of $A(\Gamma_g(4,8))$ and C[θ], since the morphism

$$\text{proj}(A(\Gamma_g(4,8))) \to \text{proj}(\mathbf{C}[\theta])$$

is bijective, it has to be reflected by some difference of their infinitesimal structure. We shall therefore consider Nullwerte of derivatives of theta functions. One type of such objects has appeared in history and it is as follows: We take g odd characteristics $m_1, \ldots, m_g$ and put

$$D(M) = (-\pi)^{-g}\partial(\theta_{m_1}, \ldots, \theta_{m_g})/\partial(z_1, \ldots, z_g)|_{z=0}, \qquad M = (m_1 \cdots m_g).$$

If $m_i = m_j$ for some $i \neq j$, clearly $D(M) = 0$. Salvati Manni showed that if $m_1, \ldots, m_g$ are distinct, then $D(M) \neq 0$; cf. [18]. At any rate $D(M)$ is an element of $A(\Gamma_g(4,8))_{g/2+1}$. Therefore if we denote by C[θ, D] the ring generated over C[θ] by all such $D(M)$, then we have the following situation:

$$\mathbf{C}[\theta] \subset \mathbf{C}[\theta, D] \subset A(\Gamma_g(4,8)).$$

As we have stated, three rings coincide if $g = 1, 2$. The fact is that C[θ] $\neq$ C[θ, D] if $g \geq 3$; we shall talk about it in the next section. As for C[θ, D] and $A(\Gamma_g(4,8))$, they are different if $g \geq 6$; the reason is as follows:

We have

$$\partial^k \theta_m(\tau, z)/\partial z_{j_1} \cdots \partial z_{j_k}|_{z=0} = (\pi i)^k \cdot \sum_{\xi \in m' + 2\mathbf{Z}^g} e(\tfrac{1}{4} \cdot {}^t\xi m'')$$

$$\cdot \xi_{j_1} \cdots \xi_{j_k} e\{\tfrac{1}{8} \cdot \text{tr}(\xi {}^t\xi \tau)\}$$

for all $1 \leq j_1, \ldots, j_k \leq g$ and for all $k \geq 0$. Therefore elements of C[θ, D] have expansions of the following form:

$$\sum_{\xi, \eta, \ldots} c_{\xi, \eta, \ldots} e\{\tfrac{1}{8} \cdot \text{tr}((\xi {}^t\xi + \eta {}^t\eta + \cdots)\tau)\}$$

with $c_{\xi, \eta, \ldots}$ in C depending on $\xi, \eta, \ldots$, in which $\xi, \eta, \ldots$ run over some cosets in $\mathbf{Z}^g/2\mathbf{Z}^g$. On the other hand if we introduce an equivalence relation in the set $S = \{s\}$ of all positive semidefinite symmetric integer matrices of degree g as $s \sim {}^tusu$ with u in $\text{GL}_g(\mathbf{Z})(4)$ and choose a complete set of representatives $R = \{r\}$, then elements of $A(\Gamma_g(4,8))$ have expansions of the following form:

$$\sum_{r \in R} c_r \cdot \sum_{s \sim r} e\{\tfrac{1}{8} \cdot \text{tr}(s\tau)\}$$

with c_r in C. Furthermore every r occurs in some such expansion.

We now use the fact that every s in S can be written as

$$s = \xi^t \xi + \eta^t \eta + \cdots$$

with $\xi, \eta, \ldots$ in $\mathbf{Z}^g$ if and only if $g \leq 5$. In fact for $g \geq 6$ if we define s_0 as

$$^t x s_0 x = 2 \left(\sum_{1 \leq i \leq 6} x_i^2 - \sum_{1 \leq i \leq 4} x_i x_{i+1} - x_3 x_6 \right),$$

then s_0 cannot be written as $\xi^t \xi + \eta^t \eta + \cdots$; cf. [9, p. 394]. Therefore $C[\theta, D]$ and $A(\Gamma_g(4, 8))$ are different for $g \geq 6$. Furthermore the above argument shows that the situation remains the same even if we replace $C[\theta, D]$ by the subring of $A(\Gamma_g(4, 8))$ generated over C by polynomials in the Nullwerte of all partial derivatives of theta functions. We might mention that s_0 above for $g = 6$ is the Cartan matrix of type E_6; that the if-part can be proved by a reduction theory (unpublished); and that we have actually compared the local rings of $\text{proj}(C[\theta, D])$ and $\text{proj}(A(\Gamma_g(4, 8))$ at the points defined by the homomorphism

$$\Phi \colon f \to \lim_{\lambda \to \infty} f(i\lambda 1_g).$$

At any rate the *first problem* is to find a finite set of generators of $A(\Gamma_g(4, 8))$ over $C[\theta]$. If the set is nicely defined, the result should be useful, e.g., to solve the second problem, which we shall explain in the next section.

Further references. The first part of Theorem 1, called a "fundamental lemma," was obtained by using the theory of compactifications. This theory was developed by Siegel [23], Satake [19], Baily [1], and Cartan [2] already quoted. Later in a book on theta functions from Springer we gave another proof based on Mumford's theory [16], and derived the fundamental theorem from that. The expansions of modular forms into Fourier series appeared in many works of Siegel. We could have used more general "Fourier-Jacobi series" by Piatetski-Shapiro [17, Chapter IV, §15].

3. Multiplicity-one theorem and the generalized Jacobi formula. In the case where $g = 1$ Jacobi used the following identity:

$$D\left(\begin{smallmatrix}1\\1\end{smallmatrix}\right)(\tau) = -\pi^{-1}\partial\theta_{11}(\tau, z)/\partial z|_{z=0} = \theta_{00}(\tau)\theta_{01}(\tau)\theta_{10}(\tau)$$

to determine "constants," i.e., those which are independent of z, in his theory of theta functions. There is an interesting history about how mathematicians, including Riemann and Frobenius, later tried to generalize this Jacobi formula. In the following we shall explain the up-to-date status of this topic.

If we denote by X the C-span of all $D(M)$ and by Y the C-span of all

$$P(N) = \theta_{n_1} \cdots \theta_{n_{g+2}}, \qquad N = (n_1 \cdots n_{g+2}),$$

then X, Y are subspaces of $Z = A(\Gamma_g(4, 8))_{g/2+1}$. If by a generalized Jacobi formula we understand any linear relation between the $D(M)$'s and $P(N)$'s, then a search for such a formula naturally leads us to the investigation of X

and Y. We have found that a rather remarkable relation exists between them. If we denote by O_n the orthogonal group of the sum of n squares and by $\mathbf{F}_p$ the finite field with p elements, then it can be stated as follows; cf. [11]:

THEOREM 2. *Suppose that g is even and let $\Gamma_g(1)$ act on Z as*

$$(\sigma^{-1} \cdot f)(\tau) = \det(c\tau + d)^{-g/2-1} f(\sigma \cdot \tau);$$

then X and Y are $\Gamma_g(1)$-stable and the corresponding subrepresentations share one and only one irreducible representation ρ. Furthermore all $D(M)$'s and $P(N)$'s are weight vectors of $\Gamma_g(2)$, i.e.,

$$\sigma \cdot D(M) = \psi_M(\sigma)D(M), \qquad \sigma \cdot P(N) = \psi_N(\sigma)P(N)$$

for every σ in $\Gamma_g(2)$ and there are some common weights; if $\psi = \psi_M = \psi_N$ is any one of them, every such common weight of $\Gamma_g(2)$ in X and Y can be written as ψ^σ for some σ in $\Gamma_g(1)$, in which $\psi^\sigma(\sigma_0) = \psi(\sigma\sigma_0\sigma^{-1})$ for every σ_0 in $\Gamma_g(2)$. If we put

$$\Gamma_\psi = \{\sigma \in \Gamma_g(1); \psi^\sigma = \psi\},$$

the representations of Γ_ψ in X and Y share one and only one irreducible representation χ, which is of degree 1, and ρ becomes the representation of $\Gamma_g(1)$ induced by χ:

$$\rho = \mathrm{Ind}_{\Gamma_\psi \uparrow \Gamma_g(1)} \chi.$$

Finally if we denote by G_n, S_n the images of $O_n(\mathbf{Z}_2)$, $O_n(\mathbf{Z})$ in $O_n(\mathbf{F}_2)$ and by dx the Haar measure on $O_n(\mathbf{Z}_2)$ normalized as $\mathrm{vol}(O_n(\mathbf{Z}_2)) = [G_n : S_n]$, then there exists a degree 1 character ε of the inductive limit of $O_n(\mathbf{Z}_2)$ as $n \to \infty$ such that

$$\begin{cases} f_{X,\psi} = \int_{O_g(\mathbf{Z}_2)} \varepsilon(x) \det(x)D(Mx)\,dx = \sum_{G_g/S_g} \varepsilon(x)\det(x)D(Mx), \\ f_{Y,\psi} = \int_{O_{g+2}(\mathbf{Z}_2)} \varepsilon(x)P(Nx)\,dx = \sum_{G_{g+2}/S_{g+2}} \varepsilon(x)P(Nx) \end{cases}$$

span the representation spaces of χ in X and Y.

We add the following explanation to clarify the theorem: First, there exists a degree 1 character ε_n of $O_n(\mathbf{Z}_2)$ for each n with the property that

$$\varepsilon_n(x) = \varepsilon_{n+1}\begin{pmatrix} x & 0 \\ 0 & 1 \end{pmatrix}$$

for every x in $O_n(\mathbf{Z}_2)$, which can be uniquely characterized by its properties, e.g., $\varepsilon_n = 1$ on $O_n(\mathbf{Z})$; the sequence $\{\varepsilon_1, \varepsilon_2, \dots\}$ defines the character ε in the theorem. Second, $\psi_M = \psi_N$ is equivalent to

$$M^t M - N^t N \equiv \begin{pmatrix} 0 & 1_g \\ 1_g & 0 \end{pmatrix} \bmod 2,$$

(*)

$$\mathrm{diag}(M^t M - N^t N) \equiv 0 \bmod 4.$$

If we denote by δ_g the element of $M_g(\mathbf{Z})$ defined by $\delta_{ij} = 1$ for $i \leq j$, $\delta_{ij} = 0$ for $i > j$ and by 0 the zero in $M_{g,1}(\mathbf{Z})$, then

$$(M_0 \quad N_0) = \begin{pmatrix} 1_g & 1_g & 0 & 0 \\ \delta_g & 0 & \delta_g & 0 \end{pmatrix}$$

satisfies (∗) with ≡ replaced by =. Third, for any g and for any $(M \quad N)$ satisfying (∗) with the understanding that the columns of M and N are odd and even, the RHS of $f_{X,\psi}$ and $f_{Y,\psi}$ are both defined; we shall denote them by f_M and f_N. More precisely $D(Mx)$ is well defined as $D(Mx')$ for any x' in $M_g(\mathbf{Z})$ satisfying $x \equiv x'$ mod 4; if $\bar{x}$ is the image of x in G_g, then $\varepsilon(x)\det(x)D(Mx)$ depends only on $\bar{x}S_g$ and the summation for f_M is with respect to $\bar{x}\mathscr{S}_g$. The situation is similar for f_N. Finally, if $\delta'_{k,1}$ represents 1 or 0 according as $k \equiv 1$ mod 4 or otherwise, then

$$[G_n : S_n] = (1/n!) \prod_{1 < k \leq n} \{2^{k-2} + 2^{k/2-1} \cdot \mathrm{Im}(e(k/8)) - \delta'_{k,1}\}$$

$$= 1, 1, 1, 1, 1, 2, 8, 64, 960, \ldots$$

for $n = 1, 2, \ldots, 9, \ldots$; in particular $f_M = D(M)$ if and only if $g \leq 5$.

The theorem shows that $\dim(X \cap Y)$ is either 0 or equal to $\deg(\rho) = [\Gamma_g(1) : \Gamma_\psi]$; in the second case $X \cap Y$ gives the common representation space of ρ in X and Y. The fact is that if $X \cap Y \neq \{0\}$ for some g, then we have $f_{M_0} = f_{N_0}$ and $f_M = \pm f_N$ for all $(M \quad N)$ satisfying (∗). Furthermore if $g \geq 2$ and if $(M_0^* \quad N_0^*)$ denotes the $(M_0 \quad N_0)$ for $g - 1$, then $f_{M_0} = f_{N_0}$ implies $f_{M_0^*} = f_{N_0^*}$. We call $f_M = \pm f_N$, i.e.,

$$\sum_{G_g/S_g} \varepsilon(x)\det(x)D(Mx) = \pm \sum_{G_{g+2}/S_{g+2}} \varepsilon(x)P(Nx)$$

a *generalized Jacobi formula*; any linear relation between the $D(M)$'s and $P(N)$'s is a consequence of this. The *second problem* is to show, if true, that $X \cap Y \neq \{0\}$ for all g. This is known to be true up to $g = 5$, i.e., in the case where $f_M = D(M)$; cf. Fay [3]. Furthermore $D(M)$ is in $\mathbf{C}[\theta]$ if and only if it is of the form f_M and in particular $D(^t(1_g \quad 1_q))$ is not in $\mathbf{C}[\theta]$ for any $g \geq 3$.

We shall recall the following fact to clarify the meaning of the second problem: We say that a symmetric integer matrix s is even if $\mathrm{diag}(s) \equiv 0$ mod 2. Suppose that s is a positive-definite even matrix of even degree p; let l denote the smallest positive integer such that ls^{-1} is even and $h(x)$ a polynomial in the entries x_{ij} of a variable $(p \times g)$-matrix x satisfying

$$\mathrm{tr}(^t(\partial/\partial x)s^{-1}(\partial/\partial x))h(x) = 0, \qquad h(x\gamma) = \det(\gamma)^q h(x),$$

in which $\partial/\partial x$ is the $(p \times g)$-matrix with $\partial/\partial x_{ij}$ as its entries, for some q and for every γ in $GL_g(\mathbf{R})$. Then for any ξ_0 in $ls^{-1}M_{p,g}(\mathbf{Z})$ we introduce, after Maass [15], the following "theta series with harmonic coefficients":

$$\theta_{\xi_0}(\tau; s, h) = \sum_{\xi \equiv \xi_0 \bmod l} h(\xi)e\left\{\frac{1}{2l^2} \cdot \mathrm{tr}(^t\xi s\xi\tau)\right\},$$

in which τ is in $\mathscr{S}_g$. The fact is that for fixed s, h they span a $\Gamma_g(1)$-stable subspace of $A(\Gamma_g(l))_{p/2+q}$. If we denote this space by $\Theta(s, h)$, we can write $X = \Theta(4 1_g, \det)$ and $Y = \Theta(4 1_{g+2}, 1)$. Therefore the pair (X, Y) is not only historical, it is just about the simplest pair of spaces of the form $\Theta(s, h)$ that can be compared. Yet we do not seem to have any effective method to answer the simple question whether or not $X \cap Y \neq \{0\}$.

Further references. In Bull. Amer. Math. Soc. **6** (1982), pp. 170–173 we explained a short history about generalizations of Jacobi's formula with references including the meaning of Fay's extraordinary paper quoted above.

4. Schottky's equation and problem. We recall that the moduli space $\mathcal{M}_g$ of curves of genus g is embedded in the quotient variety $\Gamma_g(1)\backslash\mathcal{S}_g$, which is quasiprojective; that

$$\operatorname{codim}(\mathcal{M}_g) = \tfrac{1}{2} \cdot (g-2)(g-3) = 0, 0, 1, \dots$$

for $g = 2, 3, 4, \dots$; and that Schottky gave an element J of $\mathbf{C}[\theta]_8$ satisfying $J = 0$ on $\mathcal{M}_4$; cf. [20]. It has been shown relatively recently that $J = 0$ defines the closure of $\mathcal{M}_4$ in $\Gamma_4(1)\backslash\mathcal{S}_4$; cf. [10]. We shall recall Schottky's J and explain another expression for J.

In general we identify the set of 2^{2g} characteristics with $\mathbf{F}_2^{2g}$; in order to avoid any confusion we denote by mn the characteristic that corresponds to $m + n$ in $\mathbf{F}_2^{2g}$. We put

$$e(m, n) = (-1)^{{}^t m' n'' - {}^t m'' n'}$$

and call a subspace N of $\mathbf{F}_2^{2g}$ totally isotropic if $e(m, n) = 1$ for every m, n in N; this implies that $g_0 = g - \dim(N) \geq 0$. A coset mN by such a subspace N is called even if all elements of mN are even; the number of even cosets is $2^{g_0-1}(2^{g_0} + 1)$. Therefore if $g = 4$ and $\dim(N) = 3$, there are 3 even cosets $m_1 N, m_2 N, m_3 N$; and if

$$r_i = \prod_{m \in m_i N} \theta_m$$

for $i = 1, 2, 3$, then

$$J = r_1^2 + r_2^2 + r_3^2 - 2(r_2 r_3 + r_3 r_1 + r_1 r_2).$$

Schottky also proved that J is independent of N; cf. [20, pp. 345–348].

On the other hand if s is any positive-definite symmetric integer matrix of degree k, then Siegel introduced its *analytic class invariant* f_s as

$$f_s(\tau) = \sum_\xi \mathbf{e}\{\tfrac{1}{2} \cdot \operatorname{tr}({}^t \xi s \xi \tau)\},$$

in which τ is in $\mathcal{S}_g$ and ξ runs over $M_{k,g}(\mathbf{Z})$; cf. [21, p. 587]. By definition f_s is holomorphic on $\mathcal{S}_g$, and it depends only on the equivalence class of s under $s \sim {}^t u s u$ for u in $\mathrm{GL}_k(\mathbf{Z})$. If s is even, i.e., $\operatorname{diag}(s) \equiv 0 \bmod 2$, and $\det(s) = 1$, then k is necessarily a multiple of 8 and f_s becomes an element of $A(\Gamma_g(1))_{k/2}$. Furthermore if h_k denotes the class number of such matrices, then $h_8 = 1$ by Mordell, $h_{16} = 2$ by Witt, and $h_{24} = 24$ by Niemeier. Actually Witt gave a sequence $\{s_k\}$ for $k = 8, 16, 24, \dots$, in which s_8 is the Cartan matrix of type E_8 and $s_8 \oplus s_8$, s_{16} form representatives of the two classes for $k = 16$; cf. [24]. In that paper he also showed that $(f_{s_8})^2 = f_{s_{16}}$ for $g = 1, 2$

and conjectured the same for $g = 3$. We have found that this follows from another expression for J; cf. [8]:

THEOREM 3. *In the case where* $g = 4$ *the homogeneous ideal of* $A(\Gamma_g(1))$ *defining* $\mathscr{M}_g$ *in* $\Gamma_g(1)\backslash\mathscr{S}_g$ *is generated by Schottky's* J; *furthermore*

$$(f_{s_8})^2 - f_{s_{16}} = 2^{-2}3^2 5 \cdot 7J.$$

A remarkable feature of this theorem is that something geometric is related to something arithmetic. At any rate it motivates the following *third problem*: Do members of a minimal basis for the homogeneous ideal of $A(\Gamma_g(1))$ defining $\mathscr{M}_g$ in $\Gamma_g(1)\backslash\mathscr{S}_g$ have similar expressions? The emphasis is on the explicitness of the expressions.

Further references. In [4] Freitag gave another proof to the irreducibility of the Schottky equation $J = 0$. The Witt conjecture was first proved independently by Kneser [12] and ourselves [7]; there we derived it from the fact that $A(\Gamma_g(1))_8$ is one-dimensional for $g = 1, 2, 3$.

REFERENCES

1. W. L. Baily, *Satake's compactification of V_n*, Amer. J. Math. **80** (1958), 348–364.
2. H. Cartan, *Fonctions automorphes*, Séminaire E.N.S. (1957–58).
3. J. Fay, *On the Riemann-Jacobi formula*, Göttingen Nachrichten (1979), 61–73.
4. E. Freitag, *Die Irreduzibilität der Schottky relation (Bemerkungen zu einem Satz von J. Igusa)*, Arch. Math. (Basel) **40** (1983), 255–259.
5. J. Igusa, *On the graded ring of theta-constants*, Amer. J. Math. **86** (1964), 219–246; II, ibid. **88** (1966), 221–236.
6. ____, *On Siegel modular forms of genus two. II*, Amer. J. Math. **86** (1964), 392–412.
7. ____, *Modular forms and projective invariants*, Amer. J. Math. **89** (1967), 817–855.
8. ____, *Schottky's invariant and quadratic forms*, E. B. Christoffel Sympos., Birkhäuser, 1981, pp. 352–362.
9. ____, *On the variety associated with the ring of Thetanullwerte*, Amer. J. Math. **103** (1981), 377–398.
10. ____, *On the irreducibility of Schottky's divisor*, J. Fac. Sci. Univ. Tokyo **28** (1982), 531–545.
11. ____, *Multiplicity one theorem and problems related to Jacobi's formula*, Amer. J. Math. **105** (1983), 157–187.
12. M. Kneser, *Lineare Relationen zwischen Darstellungsanzahlen quadratischer Formen*, Math. Ann. **168** (1967), 31–39.
13. M. Koecher, *Zur Theorie der Modulformen n-ten Grades. I*, Math. Z. **59** (1954), 399–416.
14. A. Krazer, *Lehrbuch der Thetafunktionen*, Teubner, Leipzig, 1903; Chelsea, New York, 1970.
15. H. Maass, *Spherical functions and quadratic forms*, J. Indian Math. Soc. **20** (1956), 117–162.
16. D. Mumford, *On the equations defining abelian varieties. I–III*, Invent. Math. **1** (1966), 287–354; **3** (1967), 75–135, 215–244.
17. I. I. Piatetski-Shapiro, *Geometry of classical domains and theory of automorphic functions*, Fizmatgiz, Moscow, 1961. (Russian)
18. R. Salvati Manni, *On the not identically zero Nullwerte of Jacobians of theta functions with odd characteristics*, Adv. in Math. **47** (1983), 88–104.
19. I. Satake, *On the compactification of the Siegel space*, J. Indian Math. Soc. **20** (1956), 259–281.

20. F. Schottky, *Zur Theorie der Abelschen Functionen von vier Variabeln*, Crelles J. **102** (1888), 304–352.

21. C. L. Siegel, *Über die analytische Theorie der quadratischen Formen*, Ann. Math. **36** (1935), 527–606; Werke I, 326–405.

22. ____, *Einführung in die Theorie der Modulfunktionen n-ten Grades*, Math. Ann. **116** (1939), 617–657; Werke II, 97–137.

23. ____, *Zur Theorie der Modulfunktionen n-ten Grades*, Comm. Pure Appl. Math. **8** (1955), 677–681; Werke III, 223–227.

24. E. Witt, *Eine Identität zwischen Modulformen zweiten Grades*, Abh. Math. Sem. Hamburg **14** (1941), 323–337.

THE JOHNS HOPKINS UNIVERSITY

Proceedings of Symposia in Pure Mathematics
Volume **49** (1989), Part 2

On the Fourier Coefficients of Cusp Forms
Having Small Positive Weight

MARVIN I. KNOPP

I. Introduction. The earlier work [3] carried over to the range $\frac{8}{3} < s < 4$ the standard calculation (as singular series) of the Fourier coefficients of the entire modular form $\theta^s(\tau)$, where

$$(1.1) \qquad \theta(\tau) = \theta_3(0|\tau) = \sum_{-\infty}^{\infty} e^{\pi i n^2 \tau} = 1 + 2 \sum_{1}^{\infty} e^{\pi i n^2 \tau},$$

the classical theta-function. As $\theta(\tau)$ is a modular form of weight $1/2$ on the modular subgroup Γ_θ (of index 3 in the modular group $\Gamma(1) = \mathrm{SL}(2, Z)$), the weight $k = s/2$ of $\theta^s(\tau)$ lies in the range $\frac{4}{3} < k < 2$ when $\frac{8}{3} < s < 4$. The article [3] is best understood as a partial response to the circumstance that, while the Fourier coefficients of entire automorphic forms and cusp forms of weights larger than 2 (and, in some cases, of weight 2 as well) on groups of finite hyperbolic area were derived long ago by Petersson [8], a method was lacking for handling the difficulties, arising from failure of absolute convergence, inherent in the situation when the weight is less than 2. This was the case despite evidence that the Fourier coefficients have the same structure (at least when $1 < k < 2$) as in the case $k > 2$.

The approach in [3] is an elaboration of the method of Maass [6] and Siegel [12], beginning with a modification of the Eisenstein series through introduction of the Hecke convergence factor [2, §2] and proceeding by analytic continuation with respect to the new complex variable thus introduced. (See §§III–IV for the specifics.) During my work on [3] it became evident that this technique need not be restricted to the entire modular forms (not cusp forms) $\theta^s(\tau)$ of "small" weight $s/2$ (i.e., $\frac{4}{3} < \frac{s}{2} < 2$), that—appropriately modified— it should in fact be applicable as well to all forms (including cusp forms of weight k in the small range, $\frac{4}{3} < k < 2$, on all groups of finite hyperbolic area. The present article presents a preliminary sketch of the method as it

1980 *Mathematics Subject Classification* (1985 *Revision*). Primary 11F30.
Work supported by a National Science Foundation grant at Temple University.

applies to cusp forms of these small weights, with respect to the full modular group $\Gamma(1)$. The generalization to general discrete groups of finite hyperbolic area acting on the upper half-plane $\mathscr{H}$ will be furnished in a future article, as will certain important details that are omitted here.

These details relate to Selberg's Kloosterman zeta-function [11], which plays a major role in [3]; of particular importance are its growth properties [1] and its exceptional poles (those lying in the open interval $(\frac{1}{2}, 1)$) or, equivalently, the exceptional eigenvalues of the Laplacian connected with the appropriate weight k. In the present work the exceptional eigenvalues acquire an even greater prominence. Those results (concerning the eigenvalues and the growth of the Kloosterman zeta-function) which are to be provided in a later paper will be described fully at the appropriate points in §II and §IV below.

II. Convergence of the series for the Fourier coefficients. Appropriating the familiar expression for the Fourier coefficients of the (parabolic) Poincaré series of Petersson, defined for weight $k > 2$ and the group $\Gamma(1)$ by

$$(2.1) \qquad G_\nu(\tau; k, v) = G_\nu(\tau) = \sum\sum_{(c,d)=1} \frac{\exp\{2\pi i(\nu + \kappa)M\tau\}}{v(M)(c\tau + d)^k}, \qquad \operatorname{Im}\tau > 0,$$

we introduce

$$(2.2) \qquad a_m(\nu, k, v) = 4\pi i^{-k} \sum_{c=1}^{\infty} c^{-1} A_{c,\nu}(m, v) L(c; m, \nu, k, \kappa),$$

but now without the restriction $k > 2$. In (2.1), and elsewhere in this article, branches are determined according to the convention

$$z^u = |z|^u e^{iu \arg z}, \qquad -\pi \le \arg z < \pi,$$

for complex z and real u. In (2.2) $A_{c,\nu}(m, v)$ is the "generalized Kloosterman sum" given by the expression

$$(2.3) \qquad \sum_{\substack{0 \le -d < c \\ (d,c)=1}} \bar{v}(M_{c,d}) \exp\left\{\frac{2\pi i}{c}[(m + \kappa)d + (\nu + \kappa)a]\right\},$$

and

$$(2.4) \quad L(c; m, \nu, k, \kappa) = \left(\frac{m + \kappa}{\nu + \kappa}\right)^{(k-1)/2} J_{k-1}\left[\frac{4\pi}{c}(m + \kappa)^{1/2}(\nu + \kappa)^{1/2}\right].$$

In (2.1)–(2.3), $M = M_{c,d}$ is any element of $\Gamma(1)$ of the form $\left(\begin{smallmatrix} a & b \\ c & d \end{smallmatrix}\right)$, ν is an arbitrary integer, m is an integer ≥ 0, and v is a *multiplier system* (MS) for $\Gamma(1)$ in the weight k. That is to say, v is a function defined on the *matrix group* $\Gamma(1)$, satisfying the "consistency condition"

$$(2.5) \qquad v(M_1 M_2)(c_3\tau + d_3)^k = v(M_1)(M_2)(c_1 M_2\tau + d_1)^k(c_2\tau + d_2)^k,$$

where $M_i = \begin{pmatrix} * & * \\ c_i & d_i \end{pmatrix} \in \Gamma(1)$, for $i = 1$ and 2, and $M_1 M_2 = \begin{pmatrix} * & * \\ c_3 & d_3 \end{pmatrix}$. The number κ is determined from v by

$$v(S) = e^{2\pi i \kappa}, \quad 0 \leq \kappa < 1, \quad S = \begin{pmatrix} 1 & 1 \\ 0 & 1 \end{pmatrix},$$

and J_{k-1} is the Bessel function of the first kind, which may be defined for complex x by the power series

(2.6) $$J_{k-1}(x) = \sum_{p=0}^{\infty} \frac{(-1)^p (x/2)^{k+2p-1}}{p! \Gamma(k+p)}.$$

It should be noted that when $v + \kappa < 0$ the Bessel function J_{k-1} in (2.4) may be rewritten as the modified Bessel function I_{k-1}, with a positive real argument.

We show that the proof given in [3, §II.2] for convergence of the singular series in the range $\frac{8}{3} < s \leq 4$ (see [3, §I.1] for the definition) can be applied almost intact to establish convergence of the series (2.2) defining $a_m(v, k, v)$, as long as $\frac{4}{3} < k \leq 2$. Thus, we quote the exposition of [3, 170–173] for the major burden of the proof.

Inserting the definition (2.4) of L and the power series expression (2.6) for J_{k-1} into the right-hand side of (2.2), we obtain
(2.7)

$$a_m(v, k, v) = \frac{2i^{-k}(2\pi)^k (m+\kappa)^{k-1}}{\Gamma(k)} \sum_{c=1}^{\infty} A_{c,v}(m, v) c^{-k}$$

$$+ 4\pi i^{-k} \left(\frac{m+\kappa}{v+\kappa} \right)^{(k-1)/2} \sum_{c=1}^{\infty} \frac{A_{c,v}(m, v)}{c}$$

$$\times \sum_{p=1}^{\infty} \frac{(-1)^p (2\pi/c)^{k+2p-1} (m+\kappa)^{(k-1)/2+p} (v+\kappa)^{(k-1)/2+p}}{p! \Gamma(k+p)}.$$

The first series occurring here obviously converges absolutely for $k > 2$ and, furthermore, for $k > 2$ it is the value at $w = k/2$ of Selberg's Kloosterman zeta-function, defined for $\operatorname{Re} w > 1$ by the Dirichlet series

(2.8) $$Z(w; m, v, v, \Gamma(1)) = Z_m(w) = \sum_{c=1}^{\infty} A_{c,v}(m, v) c^{-2w}.$$

Selberg [11] has continued $Z_m(w)$ to a function meromorphic in the entire w-plane, regular in particular for $\operatorname{Re} w > \frac{1}{2}$ with the possible exception of a finite set of simple poles on the real segment $\frac{1}{2} < w < 1$ (the exceptional poles). Also, recently, Goldfeld and Sarnak [1, Theorem 1] have obtained the growth estimate
(2.9)

$$Z(w; m, v, v, \Gamma(1)) = O\left(\frac{(|m|+1)|w|^{1/2}}{\operatorname{Re} w - 1/2} \right), \quad |\operatorname{Im} w| \geq 1, \ \operatorname{Re} w > \frac{1}{2},$$

uniformly in m. (In the presence of exceptional poles, of course, this estimate cannot hold near the real line, $\operatorname{Im} w = 0$.)

Using the estimate (2.9) I showed in [3, §II.2] that, when $\frac{4}{3} < k \leq 2$, the series $\sum_{c=1}^{\infty} A_{c,\nu}(m, v)c^{-k}$ converges to $Z_m(k/2)$. (However, see the corrections to the proof indicated in the following paragraph.) Thus, convergence of the series in (2.2) will be established once we have proved convergence of the second term on the right-hand side of (2.7). This, in fact, is the easy part of the proof, and we show readily that the double sum converges absolutely when $k > 0$. For, denoting this double sum $\mathscr{S}$ and applying the trivial estimate $|A_{c,\nu}(m, v)| \leq c$, we have

$$|\mathscr{S}| \leq \{2\pi(m + \kappa)\}^{k-1} \sum_{c=1}^{\infty} c^{-k-1} \sum_{p=1}^{\infty} \frac{\{4\pi^2(m + \kappa)(\nu + \kappa)\}^p}{p!\Gamma(k + p)}.$$

Absolute convergence for $k > 0$ is immediate from this inequality, so the series (2.2) converges for $\frac{4}{3} < k \leq 2$.

We take this opportunity to rectify an error and an oversight in important facets of the proof of convergence of the series $\sum_{c=1}^{\infty} A_{c,\nu}(m, v)c^{-k}$ (for $\frac{4}{3} < k \leq 2$) given in [3, pp. 170–173]. The error occurs in the first paragraph of p. 172, where it is asserted that $Z_m(w)$ has no poles in the half-plane $\operatorname{Re} w > \frac{1}{2}$, i.e., no exceptional poles. This claim, a restatement of the content of the paragraph labelled (b) on p. 170, is neither correct nor—as it turns out—necessary to the proof. It is based upon a misconception of what Roelcke [10, pp. 49-51] (and, later, others) proved concerning the exceptional eigenvalues of the hyperbolic Laplace operator. The statements on p. 170 do hold, in fact, for weight $k = 0$, but I am interested there (and here) in the range $\frac{4}{3} < k \leq 2$.

This error is easily corrected, however, by turning to the more recent result of Roelcke [9, Theorem 5.5], which asserts that in weight k the spectrum of the Laplacian is contained in the half-line $[(|k|/2)(1 - |k|/2), \infty)$. Since $k > 0$ here, this means that there are no eigenvalues $\lambda < (k/2)(1-(k/2))$. On the other hand, by [11, pp. 10–12] this implies that $Z_m(w)$ has no poles in $[k/2, 1)$, provided that $\nu+\kappa \geq 0$. (Selberg states this in [11] only when $m+\kappa \geq 0$, but his approach carries over to the case $\nu + \kappa \geq 0$, $m + \kappa < 0$ as well, as we shall show in detail in a future, more comprehensive, article). When $\nu + \kappa < 0$, by contrast, $Z_m(w)$ may in fact have a pole at $w = k/2$. If such a pole does occur, the analysis we carry out here is invalid. For this reason *we impose the restriction* $\nu + \kappa \geq 0$ *until* §VI.(ii), where we invoke a condition sufficient—and necessary, as it turns out—to prevent the occurrence of a pole at $w = k/2$ in the case $\nu + \kappa < 0$. (When $\nu + \kappa < 0$, we are dealing with holomorphic modular forms with a pole at $i\infty$; the assumption $\nu + \kappa > 0$ restricts the discussion to cusp forms through §VI.(i).)

Returning briefly to the exposition in the first paragraph of [3, p. 172], we let $w_1, \ldots, w_q$ denote the (necessarily finite number of necessarily simple) poles of $Z_m(w)$ in $(1/2, k/2)$. Then, in place of (2.8) of [3, p. 172], we obtain

$$(2.10) \qquad J(x, T) = Z_m\left(\frac{k}{2}\right) + I_1 + I_2 + I_3 + 2\sum_{j=1}^{q} \frac{\rho_j(m, \nu)}{(2w_j - k)} x^{2w_j - k},$$

where $\rho_j(m, \nu)$ is the residue of $Z_m(w)$ at the pole w_j. (Note that $s/2$ of [3] is replaced by k here.) Since $2w_j - k < 0$ for $1 \leq j \leq q$, the additional finite sum drops out as $x \to +\infty$. Thus, the final result [3, (2.11)] remains unaltered.

The oversight occurs at several later stages of the proof [3, §III, §IV], where it is assumed, without comment, that the estimate (2.2) of [3] ((2.9) here) holds near the real line. In fact, here (and in [3]) we need the estimate in the entire half-plane $\operatorname{Re} w \geq k/2 - \delta$, for some $\delta > 0$, and the proof in [1] does yield this since $Z_m(w)$ is free of poles in $[k/2, 1)$. The details will be given in a future publication.

While we are noting errors in [3] it seems worthwhile to point out that the display three lines above (3.2) in [3, p. 174] is badly mangled. It should read

$$\lim_{K \to \infty} \sum_{n=1}^{\infty} e^{\pi i n \tau} n^{s/2-1} (\Sigma(K)) = \sum_{n=1}^{\infty} e^{\pi i n \tau} n^{s/2-1} \sum_{c=1}^{\infty} \mathscr{S}(c, n) c^{-s/2}.$$

III. The nonanalytic Poincaré series. We are interested primarily in the properties of the exponential series

$$(3.1) \qquad P_\nu(\tau; k, v) = P_\nu(\tau) = 2e^{2\pi i (\nu + \kappa)\tau} + \sum_{m+\kappa > 0} a_m(\nu, k, v) e^{2\pi i (m+\kappa)\tau},$$

with $a_m(\nu, k, v)$ given by (2.2) for $\frac{4}{3} < k \leq 2$. In particular, we would like to show that P_ν is holomorphic in $\mathscr{H}$, the upper half-τ-plane, that it can be rewritten as the Poincaré series $G_\nu(\tau; k, v)$ defined by (2.1)—but now not absolutely convergent, since $\frac{4}{3} < k \leq 2$—and that P_ν is a modular form on $\Gamma(1)$, of weight k and MS v.

For $k \leq 2$ it is far from clear in what sense the Poincaré series (2.1) converges (if at all), so we begin by studying the "nonanalytic Poincaré series"

$$(3.2) \qquad G_\nu(\tau | z; k, v) = G_\nu(\tau | z) = \sum_{(c,d)=1} \sum \frac{\exp\{2\pi i (\nu + \kappa)M\tau\}}{v(M)(c\tau + d)^k |c\tau + d|^z},$$

obtained from (2.1) by introduction of the Hecke convergence factor $|c\tau + d|^z$ in the denominator. In (3.2), as in (2.1), $\operatorname{Im} \tau > 0$. Eventually z can be thought of as an arbitrary complex number, but in order to guarantee absolute convergence of the double series (3.2) we assume initially that $\operatorname{Re} z > 2 - k$. Uniform convergence of the series of absolute values implies that $G_\nu(\tau | z)$ is holomorphic (in the variable z) in the half-plane $\operatorname{Re} z > 2 - k$ and, as a function of $\tau \in \mathscr{H}$, it satisfies the transformation formulae

$$(3.3) \qquad G_\nu(V\tau | z) = v(V)(c\tau + d)^k |c\tau + d|^z G_\nu(\tau | z),$$

for all $V = \begin{pmatrix} * & * \\ c & d \end{pmatrix} \in \Gamma(1)$.

When $\nu > 0$, it is always possible to recapture the function $G_\nu(\tau)$ by continuing $G_\nu(\tau | z)$ analytically, in the variable z, into a half-plane $\operatorname{Re} z > -\delta$, for some $\delta > 0$ (§IV). In this case ($\nu > 0$) we show as well (§V) that $G_\nu(\tau | 0) = P_\nu(\tau)$, and it follows that $P_\nu(\tau)$ is a cusp form of weight k and MS

v, with respect to $\Gamma(1)$. When $\nu < 0$, the function $G_\nu(\tau|z)$ has a pole at $i\infty$, in the variable τ, and in this case it cannot always be shown that $P_\nu(\tau) = G_\nu(\tau|0)$ or, indeed, even that $G_\nu(\tau|z)$ exists at $z = 0$, since a modular form with the prescribed pole may not exist. This is the underlying reason for the technical problems that, as we already pointed out in §II, may occur when $\nu < 0$. We shall discuss this point further in the second half of §VI, the first part of which is given over to a demonstration that every cusp form on $\Gamma(1)$, of weight k, $\frac{4}{3} < k < 2$, is a linear combination of the $P_\nu(\tau)$, with $\nu > 0$.

For fixed z in $\operatorname{Re} z > 2 - k$, (3.3) with $V = S = \left(\begin{smallmatrix} 1 & 1 \\ 0 & 1 \end{smallmatrix}\right)$, implies that

$$(3.4) \qquad G_\nu(\tau + 1|z) = v(S)G_\nu(\tau|z) = e^{2\pi i \kappa} G_\nu(\tau|z).$$

Thus, as a function of $\tau \in \mathscr{H}$, $G_\nu(\tau|z)$ has an exponential (Fourier) expansion in the variable $e^{2\pi i x}$, with $\tau = x + iy$. (Later, we shall write $z = \sigma + it$.) The first step in carrying out the analytic continuation of $G_\nu(\tau|z)$ is the derivation of the coefficients in this expansion, in a more or less explicit form. With this in mind, rewrite the definition (3.2) as

$$(3.5) \quad G_\nu(\tau|z) = 2e^{2\pi i(\nu+\kappa)\tau} + 2\sum_{c=1}^{\infty} \sum_{\substack{d=-\infty \\ (d,c)=1}}^{\infty} \frac{\exp\{2\pi i(\nu + \kappa)M\tau\}}{v(M)(c\tau + d)^{k+z/2}(c\bar\tau + d)^{z/2}}.$$

For fixed $c \in Z^+$, transform the inner sum on d by putting $d = h + cm$, with $0 \le -h < c$, $(h,c) = 1$, and $m \in Z$; since

$$M\tau = \frac{a\tau + b}{c\tau + d} = c^{-1}\left(a - \frac{1}{c\tau + d}\right) = c^{-1}\left(a - \frac{1}{c\tau + h + cm}\right),$$

the inner sum becomes, after insertion of the exponential series,

$$c^{-z-k} \sum_{\substack{0 \le -h < c \\ (h,c)=1}} \bar v(M_{c,h}) \exp\{2\pi i(\nu + \kappa)a/c\}$$

$$\times \sum_{m=-\infty}^{\infty} e^{-2\pi imx}\left(\tau + \frac{h}{c} + m\right)^{-k-z/2}\left(\bar\tau + \frac{h}{c} + m\right)^{-z/2}$$

$$\times \sum_{p=0}^{\infty} \frac{1}{p!}\left\{\frac{-2\pi i(\nu + \kappa)}{c^2(\tau + h/c + m)}\right\}^p.$$

(As before, $M_{c,h} = \left(\begin{smallmatrix} * & * \\ c & h \end{smallmatrix}\right) \in \Gamma(1)$.) The double sum on m and p clearly converges absolutely, so we may interchange to obtain

$$c^{-z-k} \sum_{m=-\infty}^{\infty} \frac{1}{p!}\left\{\frac{-2\pi i(\nu + \kappa)}{c^2}\right\}^p \sum_{\substack{0 \le -h < c \\ (h,c)=1}} \bar v(M_{c,h})e^{2\pi i(\nu+\kappa)a/c}$$

$$(3.6)$$

$$\times \sum_{m=-\infty}^{\infty} e^{-2\pi imx}\left(\tau + \frac{h}{c} + m\right)^{-k-p-z/2}\left(\bar\tau + \frac{h}{c} + m\right)^{-z/2}.$$

At this point we need a generalization of the Poisson summation formula, a consequence of the Lipschitz summation formula (see [7, pp. 208–212]). It has been derived in the following form by John H. Hawkins:

$$(3.7) \quad i^{\alpha-\beta} \sum_{l=-\infty}^{\infty} (\tau+l)^{-\alpha}(\bar{\tau}+l)^{-\beta} e^{-2\pi il\kappa} = \sum_{n=-\infty}^{\infty} p_{n+\kappa}(y,\alpha,\beta) e^{2\pi i(n+\kappa)x},$$

where $\tau = x + iy$, $y > 0$, $\mathrm{Re}(\alpha + \beta) > 1$, and

(3.8)

$$p_{n+\kappa}(y,\alpha,\beta) = \begin{cases} \dfrac{2\pi\Gamma(\alpha+\beta-1)}{\Gamma(\alpha)\Gamma(\beta)}(2y)^{1-\alpha-\beta} & \text{if } n = \kappa = 0, \\[2ex] \dfrac{(2\pi)^{\alpha+\beta}(n+\kappa)^{\alpha+\beta-1}}{\Gamma(\alpha)\Gamma(\beta)} e^{-2\pi(n+\kappa)y} \sigma(4\pi(n+\kappa)y,\alpha,\beta) \\[1ex] \qquad\qquad\qquad\qquad\qquad \text{if } n+\kappa > 0, \\[2ex] \dfrac{(2\pi)^{\alpha+\beta}(n+\kappa)^{\alpha+\beta-1}}{\Gamma(\alpha)\Gamma(\beta)} e^{2\pi(n+\kappa)y} \sigma(-4\pi(n+\kappa)y,\beta,\alpha) \\[1ex] \qquad\qquad\qquad\qquad\qquad \text{if } n+\kappa < 0. \end{cases}$$

Here,

$$(3.9) \qquad \sigma(\eta,\alpha,\beta) = \int_0^\infty (u+1)^{\alpha-1} u^{\beta-1} e^{-\eta u}\, du,$$

the notation of Siegel [12].

We apply (3.7), (3.8)—with τ replaced by $\tau + h/c$, $\alpha = k + p + z/2$, and $\beta = z/2$—to transform (3.6); this yields

$$G_\nu(\tau|z) - 2e^{2\pi i(\nu+\kappa)\tau} = 2i^{-k-p} \sum_{c=1}^{\infty} c^{-z-k} \sum_{p=0}^{\infty} \frac{1}{p!} \left\{ \frac{-2\pi i(\nu+\kappa)}{c^2} \right\}^p$$

(3.10)

$$\times \sum_{\substack{0 \le -h < c \\ (h,c)=1}} \bar{v}(M_{c,h}) e^{2\pi i(\nu+\kappa)a/c} \{\Sigma_1 + \Sigma_2\},$$

with

$$\Sigma_1 = \frac{(2\pi)^{k+p+z}}{\Gamma(k+p+z/2)\Gamma(z/2)} \sum_{n=0}^{\infty} (n+\kappa)^{k+p+z-1} e^{2\pi i(n+\kappa)\tau}$$

$$\times \sigma(4\pi(n+\kappa)y, k+p+z/2, z/2) e^{2\pi i(n+\kappa)h/c},$$

(3.11)

$$\Sigma_2 = \frac{(2\pi)^{k+p+z}}{\Gamma(k+p+z/2)\Gamma(z/2)} \sum_{n=1}^{\infty} (n-\kappa)^{k+p+z-1} e^{2\pi i(-n+\kappa)\bar{\tau}}$$

$$\times \sigma(4\pi(n-\kappa)y, z/2, k+p+z/2) e^{2\pi i(-n+\kappa)h/c}.$$

In order to avoid the extra term that would arise from the special case $n = \kappa = 0$ of (3.8), we have assumed $k \ne 2$, i.e., $\frac{4}{3} < k < 2$, in the derivation of (3.11). ($\kappa = 0$ occurs only for k an even integer.)

The calculation of the Fourier expansion for $G_\nu(\tau|z)$ now requires only the interchange of summations in (3.10). This leads to

(3.12)
$$G_\nu(\tau|z) - 2e^{2\pi i(\nu+\kappa)\tau}$$
$$= 2i^{-k}\frac{(2\pi)^{k+z}}{\Gamma(z/2)}\sum_{n=0}^{\infty}(n+\kappa)^{k+z-1}e^{2\pi i(n+\kappa)\tau}\sum_{p=0}^{\infty}\frac{\{-4\pi^2(n+\kappa)(\nu+\kappa)\}^p}{p!\Gamma(k+p+z/2)}$$
$$\times \sigma(4\pi(n+\kappa)y, k+p+z/2, z/2)Z_n(z/2+k/2+p)$$
$$+ 2i^{-k}\frac{(2\pi)^{k+z}}{\Gamma(z/2)}\sum_{n=1}^{\infty}(n-\kappa)^{k+z-1}e^{-2\pi i(n-\kappa)\bar{\tau}}$$
$$\times \sum_{p=0}^{\infty}\frac{\{-4\pi^2(n-\kappa)(\nu+\kappa)\}^p}{p!\Gamma(k+p+z/2)}$$
$$\times \sigma(4\pi(n-\kappa)y, z/2, k+p+z/2)Z_{-n}(z/2+k/2+p)$$

where $Z_n(w)$ is Selberg's Kloosterman zeta-function. (See (2.8) and the text following.) The Fourier expansion (3.12), valid thus far in $\mathrm{Re}\,z > 2-k$, will provide the means to carry out the required analytic continuation of $G_\nu(\tau|z)$ into an open half-plane containing $z = 0$.

In order to justify the interchange of summations in the final step of the proof of (3.12), we prove absolute convergence, for $\sigma = \mathrm{Re}\,z > 2-k$, of the triple sum $\sum_{c=1}^{\infty}\sum_{p=0}^{\infty}\sum_{n=-\infty}^{\infty}$. Trivially, $|A_{c,\nu}(n,v)| \le c$, so it suffices to consider

$$\mathscr{S}_1 = \sum_{c=1}^{\infty}c^{-\sigma-k+1}\sum_{p=0}^{\infty}\left(\frac{2\pi|\nu+\kappa|}{c^2}\right)^p\frac{1}{p!}\frac{(2\pi)^{k+p+\sigma}}{|\Gamma(k+p+z/2)|}$$
$$\times \sum_{n=0}^{\infty}(n+\kappa)^{k+p+\sigma-1}e^{-2\pi(n+\kappa)y}|\sigma(4\pi(n+\kappa)y, k+p+z/2, z/2)|$$

and

$$\mathscr{S}_1 = \sum_{c=1}^{\infty}c^{-\sigma-k+1}\sum_{p=0}^{\infty}\left(\frac{2\pi|\nu+\kappa|}{c^2}\right)^p\frac{1}{p!}\frac{(2\pi)^{k+p+\sigma}}{|\Gamma(k+p+z/2)|}$$
$$\times \sum_{n=1}^{\infty}(n-\kappa)^{k+p+\sigma-1}e^{-2\pi(n-\kappa)y}|\sigma(4\pi(n-\kappa)y, z/2, k+p+z/2)|.$$

We derive the absolute convergence of $\mathscr{S}_1$ in some detail; the proof for $\mathscr{S}_2$ is virtually the same.

Since $p \ge 0$, $c \ge 1$, and $\sigma+k-1 > 1$, it follows that $\sum_{c=1}^{\infty}c^{-\sigma-k+1-2p} < \infty$. Next, by (3.9),

$$|\sigma(4\pi(n+\kappa)y, k+p+z/2, z/2)| \le \mathscr{I}$$
$$= \int_0^{\infty}(u+1)^{k+p+\sigma/2-1}u^{\sigma/2-1}e^{-4\pi(n+\kappa)yu}\,du,$$

with $\mathscr{I}$ clearly a convergent integral because $\sigma/2 - 1 > (1-k/2) - 1 \ge -1$. Write $\mathscr{I} = \mathscr{I}_1 + \mathscr{I}_2 = \int_0^1 + \int_1^{\infty}$ and for convenience assume $\sigma = \mathrm{Re}\,z < k$.

(This assumption is harmless, as $k > 1$ implies $k > 2 - k$.) Since $\sigma/2 - 1 < k/2 - 1 < 0$ and $k + p + \sigma/2 - 1 > 0$,

$$\mathscr{I}_2 \leq 2^{k+p+\sigma/2-1} \int_0^\infty u^{k+p+\sigma/2-1} e^{-4\pi(n+\kappa)yu}\, du$$

$$= \frac{1}{2} \frac{\Gamma(k+p+\sigma/2)}{\{2\pi(n+\kappa)y\}^{k+p+\sigma/2}}.$$

Similarly, since $\sigma > 2 - k > 0$,

$$\mathscr{I}_1 \leq 2^{k+p+\sigma/2-1} \int_0^1 u^{\sigma/2-1}\, du = 2^{k+p+\sigma/2}\sigma^{-1}.$$

It follows that

$$\mathscr{I}_1 \leq K_1 \sum_{p=0}^\infty \frac{\{8\pi^2(\nu+\kappa)\}^p}{p!|\Gamma(k+p+z/2)|} \sum_{n=0}^\infty (n+\kappa)^{k+p+\sigma-1} e^{-2\pi(n+\kappa)y}$$

$$+ K_2 \sum_{p=0}^\infty \frac{\{2\pi(\nu+\kappa)\}^p \Gamma(k+p+\sigma/2)}{p!|\Gamma(k+p+z/2)|y^p} \sum_{n=0}^\infty (n+\kappa)^{(\sigma/2)-1} e^{-2\pi(n+\kappa)y}$$

$$= \mathscr{S}_1^{(1)} + \mathscr{S}_1^{(2)},$$

where K_1 and K_2 depend upon $\sigma = \operatorname{Re} z$ and $y = \operatorname{Im} \tau$.

By the Lipschitz summation formula [5],

$$\sum_{n=0}^\infty (n+\kappa)^{k+p+\sigma-1} e^{-2\pi(n+\kappa)y} = \frac{\Gamma(k+p+\sigma)}{(2\pi)^{k+p+\sigma}} \sum_{q=-\infty}^\infty e^{2\pi iq\kappa}(y+qi)^{-k-p-\sigma};$$

on the other hand,

$$\left| \sum_{q=-\infty}^\infty (y+qi)^{-k-p-\sigma} e^{2\pi iq\kappa} \right| \leq y^{-k-p-\sigma} + 2\sum_{q=1}^\infty (y^2+q^2)^{-(k+p+\sigma)/2}$$

$$\leq y^{-k-p-\sigma} + 2\left\{ 1 + \int_1^\infty u^{-k-p-\sigma}\, du \right\}$$

$$\leq Ky^{-p} + 4,$$

with K a constant depending upon $y = \operatorname{Im}\tau$ and $\sigma = \operatorname{Re} z$. Thus,

$$\mathscr{S}_1^{(1)} \leq K_1' \sum_{p=0}^\infty \frac{\{4\pi(\nu+\kappa)\}^p \Gamma(k+p+\sigma)}{p!|\Gamma(k+p+z/2)|}(y^{-p}+4).$$

In $\mathscr{S}_1^{(2)}$, the inner sum on n is independent of p and obviously convergent, so without the Lipschitz summation formula, we conclude

$$\mathscr{S}_1^{(2)} = K_2' \sum_{p=0}^\infty \frac{\{2\pi(\nu+\kappa)y^{-1}\}^p}{p!} \frac{\Gamma(k+p+\sigma/2)}{|\Gamma(k+p+z/2)|}.$$

Hence, since $\Gamma(u)$ is monotone increasing for $u \geq \frac{3}{2}$ (say) and since $k + p + \sigma/2 > 3/2$, we find that

$$(3.13) \qquad \mathscr{S}_1 \leq K_3 \sum_{p=0}^\infty \frac{[4\pi(\nu+\kappa)y^{-1}]^p + [4\pi(\nu+\kappa)]^p}{p!|\Gamma(k+p+z/2)|} \Gamma(k+p+\sigma).$$

But Stirling's formula and an elementary calculation imply that

$$\frac{\Gamma(k+p+\sigma)}{|\Gamma(k+p+z/2)|} \sim p^{\sigma/2} \quad \text{as } p \to \infty,$$

so that, by (3.13), $\mathscr{S}_1$ is absolutely convergent.

The proof of absolute convergence of $\mathscr{S}_2$ is entirely analogous, the only difference a minor one in the estimation of the integral
(3.14)

$$\left|\sigma\left(4\pi(n-\kappa)y, \frac{z}{2}, k+p+\frac{z}{2}\right)\right| = \left|\int_0^\infty (u+1)^{z/2-1}u^{k+p+z/2-1}e^{-4\pi(n-\kappa)yu}\,du\right|$$

$$\leq \frac{\Gamma(k+p+\sigma/2)}{\{4\pi(n-\kappa)y\}^{k+p+\sigma/2}}.$$

Other changes are equally minor; we omit the details.

IV. Analytic continuation of $G_\nu(\tau|z)$. For the purpose at hand, we return to the Fourier expansion (3.12) of $G_\nu(\tau|z)$ (valid in $2-k < \operatorname{Re} z < k$), now denoting by $\mathscr{F}_1$ and $\mathscr{F}_2$ the two terms on the right-hand side.

(i) The term $\mathscr{F}_2$ is the simpler of the two. By Selberg's analytic continuation, coupled with Roelcke's Theorem 5.5 of [9], for all p the function $Z_{-n}(z/2+k/2+p)$, which occurs as a factor of $\mathscr{F}_2$, is holomorphic in an open half-plane containing $\operatorname{Re} z \geq 0$. Indeed, from [11, pp. 10–12] we infer that this half-plane, say $\operatorname{Re} z > -\delta$, $0 < \delta < \frac{1}{3}$, can be chosen independent of n. (The same δ can be chosen as well for $Z_n(z/2+k/2+p)$, a factor in $\mathscr{F}_1$.) The other significant factor of $\mathscr{F}_2$,

$$\sigma\left(4\pi(n-\kappa)y, \frac{z}{2}, k+p+\frac{z}{2}\right) = \int_0^\infty (u+1)^{z/2-1}u^{k+p+z/2-1}e^{-4\pi(n-\kappa)yu}\,du,$$

is clearly holomorphic as long as $k+p+\sigma/2 > 0$. But $p \geq 0$ and $k > 4/3$, so this integral is holomorphic in z for $\sigma = \operatorname{Re} z > -\delta$, since $\delta \leq \frac{1}{3}$. It follows that each summand in $\mathscr{F}_2$ is holomorphic in $\sigma = \operatorname{Re} z > -\delta$.

To prove $\mathscr{F}_2$ holomorphic in $\sigma > -\delta$, then, it suffices to show uniform convergence of the double sum (on n and p) in compact subsets of $\sigma > -\delta$. Thus assume $-\delta + \varepsilon \leq \sigma = \operatorname{Re} z \leq A$ and $|t| = |\operatorname{Im} z| \leq B$, with $\varepsilon > 0$, $0 < A < k$, and $B > 0$. In place of $\mathscr{F}_2$, consider the series of absolute values
(4.1)

$$\sum_{n=1}^\infty (n-\kappa)^{k+\sigma-1}e^{-2\pi(n-\kappa)y} \sum_{p=0}^\infty \frac{\{4\pi^2(n-\kappa)(\nu+\kappa)\}^p}{p!|\Gamma(k+p+z/2)|}$$

$$\times |\sigma(4\pi(n-\kappa)y, z/2, k+p+z/2)|\,|Z_{-n}(z/2+k/2+p)|.$$

Since σ and t are bounded, Stirling's formula yields

$$\frac{\Gamma(k+p+\sigma/2)}{|\Gamma(k+p+z/2)|} \leq Ke^{(t/2)\arg(k+p+z/2)}\left(\frac{k+p+\sigma/2}{|k+p+z/2|}\right)^{k+p+\sigma/2}$$

$$\leq Ke^{\pi B/2},$$

where K depends only upon A, B, δ, and ε. Thus the series (4.1) is majorized by

$$(4.2) \quad K^* \sum_{n=1}^{\infty} (n-\kappa)^{A/2-1} e^{-2\pi(n-\kappa)y} \sum_{p=0}^{\infty} \frac{\{\pi(\nu+\kappa)y^{-1}\}^p}{p!} |Z_{-n}(z/2+k/2+p)|,$$

where K^* depends only upon A, B, δ, ε, and $y = \operatorname{Im} \tau$. It remains to estimate $|Z_{-n}(z/2+k/2+p)|$; we shall return to this point at the end of the section.

(ii) $\mathscr{F}_1$ is slightly more difficult to treat because the integral

$$\sigma(4\pi(n+\kappa)y, k+p+z/2, z/2)$$

has a pole (of order one, as we shall see) at $z = 0$. This problem is overcome easily by means of integration by parts, which gives, for $\sigma = \operatorname{Re} z > 2 - k$,

$$
\begin{aligned}
\sigma(4\pi(n+\kappa)y, &k+p+z/2, z/2) \\
&= 8\pi(n+\kappa)yz^{-1} \int_0^{\infty} (u+1)^{k+p+z/2-1} u^{z/2} e^{-4\pi(n+\kappa)yu}\, du \\
&\quad - \int_0^{\infty} (u+1)^{k+p+z/2-2} u^{z/2} e^{-4\pi(n+\kappa)yu}\, du \\
&\quad - 2z^{-1}(k+p-1) \int_0^{\infty} (u+1)^{k+p+z/2-2} u^{z/2} e^{-4\pi(n+\kappa)yu}\, du.
\end{aligned}
$$

(4.3)

Thus, $\sigma(4\pi(n+\kappa)y, k+p+z/2, z/2)$ is analytic in $\sigma = \operatorname{Re} z > -2$, except for a simple pole at $z = 0$. But $\mathscr{F}_1$ also contains the factor $1/\Gamma(z/2)$, which has a simple zero at $z = 0$, thus cancelling the pole. Since the factor

$$Z_n(z/2+k/2+p)$$

is holomorphic for $\sigma = \operatorname{Re} z > -\delta$, each summand in $\mathscr{F}_1$ is holomorphic in $\sigma > -\delta$.

We turn to the question of uniform convergence of the double sum $\mathscr{F}_1$ in the compact subset of $\sigma > -\delta$ defined by $-\delta + \varepsilon \le \sigma \le A$, $|t| \le B$. It suffices to consider the series of positive terms

(4.4)

$$
\begin{aligned}
\sum_{n=0}^{\infty} (n+\kappa)^{k+\sigma-1} & e^{-2\pi(n+\kappa)y} \sum_{p=0}^{\infty} \frac{\{4\pi^2(n+\kappa)(\nu+\kappa)\}^p}{p!|\Gamma(k+p+z/2)|} \left| Z_n\left(\frac{z}{2}+\frac{k}{2}+p\right) \right| \\
&\times \left\{ (n+\kappa) \int_0^{\infty} (u+1)^{k+p+\sigma/2-1} u^{\sigma/2} e^{-4\pi(n+\kappa)yu}\, du \right. \\
&\quad \left. + (p+1) \int_0^{\infty} (u+1)^{k+p+\sigma/2-2} u^{\sigma/2} e^{-4\pi(n+\kappa)yu}\, du \right\}.
\end{aligned}
$$

Denoting the two integrals occurring here as I_1 and I_2 (left to right), we note that

$$I_2 \le I_1 \le 2^{k+p+\sigma/2-1} \left\{ \left(1+\frac{\sigma}{2}\right)^{-1} + \frac{\Gamma(k+p+\sigma)}{[4\pi(n+\kappa)y]^{k+p+\sigma}} \right\},$$

so that the series (4.4) is majorized by

$$\sum_{n=0}^{\infty}(n+\kappa)^{k+\sigma-1}e^{-2\pi(n+\kappa)y}\sum_{p=0}^{\infty}\frac{\{4\pi^2(n+\kappa)(\nu+\kappa)\}^p}{p!|\Gamma(k+p+z/2)|}\left|Z_n\left(\frac{z}{2}+\frac{k}{2}+p\right)\right|$$

$$\times 2^{k+p+\sigma/2}\left\{\left(1+\frac{\sigma}{2}\right)^{-1}+\frac{\Gamma(k+p+\sigma)}{[4\pi(n+\kappa)y]^{k+p+\sigma}}\right\}(n+1)(p+1).$$

Thus, (4.4) is bounded above by the sum of the two series

$$K_1\sum_{n=0}^{\infty}(n+1)^{k+\sigma}e^{-2\pi(n+\kappa)y}$$

$$\times\sum_{p=0}^{\infty}\frac{\{8\pi^2(n+\kappa)(\nu+\kappa)\}^p(p+1)}{p!|\Gamma(k+p+z/2)|}\left|Z_n\left(\frac{z}{2}+\frac{k}{2}+p\right)\right|$$

and

$$K_2\sum_{n=0}^{\infty}e^{-2\pi(n+\kappa)y}\sum_{p=0}^{\infty}\frac{\{2\pi(\nu+\kappa)y^{-1}\}^p(p+1)}{p!}$$

$$\times\frac{\Gamma(k+p+\sigma)}{|\Gamma(k+p+z/2)|}\left|Z_n\left(\frac{z}{2}+\frac{k}{2}+p\right)\right|,$$

where K_1 and K_2 depend only upon A, δ, ε, y, k, and κ. By Stirling's formula,

$$|\Gamma(k+p+z/2)|\geq K_3(p+1)^{p+1/2}e^{-p}\quad\text{and}\quad\frac{\Gamma(k+p+\sigma)}{|\Gamma(k+p+z/2)|}\leq K_4 p^{\sigma/2},$$

with K_3 and K_4 dependent upon these same parameters. (We have used $\sigma\leq A<k\leq 2$.) Thus, to complete the analytic continuation it is sufficient to prove uniform convergence, in the compact set, of the series (4.2), and of

$$(4.5)\quad\sum_{n=0}^{\infty}(n+1)^{k+A}e^{-2\pi(n+\kappa)y}\sum_{p=0}^{\infty}\frac{\{8\pi^2 e(n+\kappa)(\nu+\kappa)\}^p}{p!(p+1)^{p-1/2}}\left|Z_n\left(\frac{z}{2}+\frac{k}{2}+p\right)\right|$$

and

$$(4.6)\quad\sum_{n=0}^{\infty}e^{-2\pi(n+\kappa)y}\sum_{p=0}^{\infty}\frac{\{2\pi(\nu+\kappa)y^{-1}\}^p}{p!}(p+1)^{A/2+1}\left|Z_n\left(\frac{z}{2}+\frac{k}{2}+p\right)\right|.$$

(iii) We must now estimate the Kloosterman zeta-functions Z_n, Z_{-n} that occur in (4.2), (4.5), and (4.6). For $p\geq 1$ the upper bound is elementary, since then

$$\frac{\sigma}{2}+\frac{k}{2}+p\geq\frac{3}{2}+\frac{\varepsilon}{2},$$

so that

$$\left|Z_n\left(\frac{z}{2}+\frac{k}{2}+p\right)\right|\leq\zeta(2+\varepsilon)<\frac{\pi^2}{6},$$

for all integers n. For $p=0$, on the other hand, we require the growth estimate (2.9), which is valid in the half-plane $\operatorname{Re} w\geq k/2-\delta$, uniformly in m (here replaced by n). (See the penultimate paragraph of §II.) In this half-plane,

$$\left| Z_n \left(\frac{z}{2} + \frac{k}{2} \right) \right| \leq K(|n| + 1) \frac{|z/2 + k/2|^{1/2}}{(\sigma/2 + k/2 - 1/2)},$$

where k depends only upon ν and k (*not* on n). This now implies uniform convergence of the series (4.2), (4.5), and (4.6) in the compact set, since $\sigma/2 + k/2 - 1/2 \geq \varepsilon/2$ there.

V. The function $G_\nu(\tau|0)$ as exponential series. We next establish that $P_\nu(\tau; k, v)$, defined by (3.1), is holomorphic in $\mathcal{H}$ and that, furthermore,

$$P_\nu(\tau; k, v) = G_\nu(\tau|0; k, v),$$

the analytic continuation of $G_\nu(\tau|z; k, v)$ evaluated at $z = 0$. From these facts it will follow as well that $P_\nu(\tau; k, v)$ is a holomorphic modular form of weight k and MS v on $\Gamma(1)$. This, in turn (§VI), will yield exact formulae for the Fourier coefficients of modular forms of weight k, $\frac{4}{3} < k < 2$, on $\Gamma(1)$.

That $P_\nu(\tau)$ is analytic in $\mathcal{H}$ follows from a simple growth estimate on the Fourier coefficients $a_m(\nu, k, v)$ defined by (2.2). Applying (2.9) for $\mathrm{Re}\, w \geq k/2 - \delta$, we shall prove

PROPOSITION 1. *Suppose $\frac{4}{3} < k < 2$. If $\nu + \kappa > 0$, $a_m(\nu, k, v) = O(m^k)$, $m \to +\infty$.*

PROOF. Write the series in (2.7) as
(5.1)

$$\Sigma_1 + \Sigma_2 = \frac{(m + \kappa)^{k-1}}{\Gamma(k)} \sum_{c=1}^{\infty} A_{c,\nu}(m, v) c^{-k} + \left(\frac{m + \kappa}{\nu + \kappa} \right)^{(k-1)/2} \sum_{c=1}^{\infty} \frac{A_{c,\nu}(m, v)}{c}$$

$$\times \sum_{p=1}^{\infty} \frac{(-1)^p \{2\pi(m + \kappa)^{1/2}(\nu + \kappa)^{1/2} c^{-1}\}^{k+2p-1}}{p! \Gamma(k + p)}.$$

By [3, pp. 170–173], $\Sigma_1 = (m + \kappa)^{k-1}(\Gamma(k))^{-1} Z_m(k/2)$ so (2.9) implies $\Sigma_1 = O(m^k)$. On the other hand, the inner sum on p in Σ_2 can be written as

$$J_{k-1}\{4\pi(m + \kappa)^{1/2}(\nu + \kappa)^{1/2} c^{-1}\} - \{\Gamma(k)\}^{-1}\{4\pi(m + \kappa)^{1/2}(\nu + \kappa)^{1/2}\}^{k-1},$$

where J_{k-1} is the Bessel function of the first kind. If $\nu + \kappa > 0$, the alternating sum on p in (5.1) is dominated by the first term. So, by the standard estimate $J_{k-1}(x) \ll x^{k-1}$, $x \to +\infty$, we have

$$\Sigma_2 \ll m^{(k-1)/2} \sum_{c \ll m^{1/2}} (m^{1/2} c^{-1})^{k-1} + m^{(k-1)/2} \sum_{c \gg m^{1/2}} (m^{1/2} c^{-1})^{k-1} \ll m^{k/2}.$$

We now turn to the evaluation of $G_\nu(\tau|z; k, v)$ at $z = 0$. In §IV we expressed $G_\nu(\tau|z)$, defined initially by (3.2), as the Fourier series (3.12). We then observed that the second double sum in (3.12) (except for the factor $2i^{-k}(2\pi)^{k+z}\{\Gamma(z/2)\}^{-1}$, this is $\mathscr{F}_2$) is holomorphic at $z = 0$, so $\mathscr{F}_2 = 0$ for $z = 0$, since $1/\Gamma(z/2) = 0$ there.

On the other hand, we required integration by parts to effect the analytic continuation of $\mathscr{F}_1$, the first double sum in (3.12). This replaced the integral $\sigma(4\pi(n+\kappa)y, k+p+z/2, z/2)$ by the three integrals on the right-hand side of (4.3). But, again because of the factor $1/\Gamma(z/2)$, when $z=0$ only those two terms (the first and third) in (4.3) survive which contain the factor $1/z$. Since the residue of $\Gamma(z/2)$ at $z=0$ is 2, we have

$$G_\nu(\tau|0) - 2e^{2\pi i(\nu+\kappa)\tau} = 2i^{-k}(2\pi)^k \sum_{n=0}^{\infty}(n+\kappa)^{k-1}e^{2\pi i(n+\kappa)\tau}$$

$$\times \sum_{p=0}^{\infty} \frac{\{-4\pi^2(n+\kappa)(\nu+\kappa)\}^p}{p!\Gamma(k+p)} Z_n\left(\frac{k}{2}+p\right)$$

$$\times \left\{ 4\pi(n+\kappa)y \int_0^\infty (u+1)^{k+p-1}e^{-4\pi(n+\kappa)yu}\, du \right.$$

$$\left. -(k+p-1)\int_0^\infty (u+1)^{k+p-2}e^{-4\pi(n+\kappa)yu}\, du \right\}.$$

Integrating the second integral by parts (in effect, reversing the earlier integration by parts), we find that the expression in { } equals 1. Thus (5.2)

$$G_\nu(\tau|0) - 2e^{2\pi i(\nu+\kappa)\tau} = 2i^{-k}(2\pi)^k \sum_{n=0}^{\infty}(n+\kappa)^{k-1}e^{2\pi i(\nu+\kappa)\tau}$$

$$\times \sum_{p=0}^{\infty} \frac{\{-4\pi^2(n+\kappa)(\nu+\kappa)\}^p}{p!\Gamma(k+p)} Z_n\left(\frac{k}{2}+p\right).$$

However, by (2.7),

$$a_m(\nu,k,v) = 2i^{-k}(2\pi)^k(m+\kappa)^{k-1}\{\Gamma(k)\}^{-1}Z_m\left(\frac{k}{2}\right)+\mathscr{S},$$

where $\mathscr{S}$ is a double sum the absolute convergence of which has been verified in the paragraph immediately following (2.9). Replacing $a_m(\nu,k,v)$ by (2.7) in the definition (3.1) of $P_\nu(\tau;k,v)$ and then interchanging the summations on c and p in $\mathscr{S}$, we find that $P_\nu(\tau;k,v)$ is equal to the expression on the right-hand side of (5.2). Therefore, $P_\nu(\tau) = G_\nu(\tau|0)$, as claimed.

To show that $P_\nu(\tau)$ is a modular form of weight k and MS v on $\Gamma(1)$, we simply recall that, by absolute convergence, the nonanalytic Poincaré series $G_\nu(\tau|z)$ satisfies the transformation formulae (3.3) under $\Gamma(1)$. By the analytic continuation of $G_\nu(\tau|z)$ into $\sigma = \mathrm{Re}\, z > -\delta$ (§IV), the two sides of (3.3) are holomorphic there (in z) for fixed $\tau \in \mathscr{H}$. Thus, by the identity principle

$$(5.3) \qquad G_\nu(V_\tau|0) = v(V)(c\tau+d)^k G_\nu(\tau|0), \qquad V = \begin{pmatrix} * & * \\ c & d \end{pmatrix} \in \Gamma(1).$$

Since $P_\nu(\tau) = G_\nu(\tau|0)$, (5.3) shows that $P_\nu(\tau)$ is a cusp form on $\Gamma(1)$, of weight k and MS v.

REMARK. By a widely known result of Hecke, it now follows that $a_m(\nu, k, v)$ $= O(m^{k/2})$, $m \to +\infty$. Since $\Sigma_2 = O(m^{k/2})$, $m \to +\infty$, the proof of Proposition 1 implies that $Z_m(k/2) = O(m^{1-k/2})$, $m \to +\infty$. This is a measurable improvement over (2.9), but, of course, only for $w = k/2$.

Following the approach of [3, §III.1] we could show, as well, that the Poincaré series (2.1) converges for $\frac{4}{3} < k \le 2$ and in fact is equal to $P_\nu(\tau)$. We refer the reader to [3, pp. 173–175] for details.

VI. The Fourier coefficients of modular forms for small weights.

(i) *Cusp forms.* We show here that any holomorphic cusp form on $\Gamma(1)$, of weight k in the interval $(\frac{4}{3}, 2)$ ("small" weight), is a finite linear combination of the functions $P_\nu(\tau)$. This, in turn, provides an explicit calculation of the Fourier coefficients of cusp forms of these weights, as linear combinations of the coefficients $a_m(\nu, k, v)$ defined by (2.2). The method is the standard one—modified to circumvent the convergence problems arising for these small weights—based upon the behavior of the Poincaré series (here, $P_\nu(\tau) = G_\nu(\tau|0)$) with respect to the inner product on the finite-dimensional Hilbert space of cusp forms of weight k and MS v (Proposition 2).

For two such cusp forms f, g on $\Gamma(1)$, recall, the inner product is defined by

$$(6.1) \qquad (f, g) = \int\!\!\int_{\mathcal{R}} f(\tau)\overline{g(\tau)} y^k \frac{dx\, dy}{y^2}, \qquad \tau = x + iy,$$

where $\mathcal{R}$ is a fundamental region for $\Gamma(1)$. Exponential vanishing of f and g at the cusp $i\infty$ guarantees absolute convergence of the integral in (6.1).

PROPOSITION 2. *Suppose* $\nu + \kappa > 0$ *and* $f(\tau)$ *is a cusp form of weight k and MS v on* $\Gamma(1)$. *Then,*

$$(6.2) \qquad\qquad (f, P_\nu) = 2b_\nu \Gamma(k-1)\{4\pi(\nu + \kappa)\}^{1-k},$$

where $f(\tau) = \sum_{n+\kappa>0} b_n e^{2\pi i(n+\kappa)\tau}$.

This result is standard for $k > 2$ (see, for example, [4, pp. 286–287]). Since $k < 2$ here, we must modify the proof, replacing $P_\nu(\tau)$ by the function $F_\nu(\tau|z) = y^{z/2} G_\nu(\tau|z)$. It follows from (3.3) that, as a function of τ, F satisfies the same transformation formulae as does a holomorphic modular form of weight k and MS v:

$$(6.3) \qquad F_\nu(M\tau|z) = v(M)(c\tau + d)^k F_\nu(\tau|z), \qquad M = \begin{pmatrix} * & * \\ c & d \end{pmatrix} \in \Gamma(1),$$

and this allows us to employ the usual proof of (6.2) that applies for $k > 2$ to derive the

LEMMA. *With* $f(\tau)$ *as in Proposition 2,* $\nu + \kappa > 0$, *and* $\sigma = \operatorname{Re} z > 2 - k$,

$$(6.4) \qquad (F_\nu, f) = 2\overline{b}_\nu \Gamma(k - 1 + z/2)\{4\pi(\nu + \kappa)\}^{1-k-z/2}.$$

REMARK. The proof requires absolute convergence of the series (3.2) defining $G_\nu(\tau|z)$; hence the assumption $\sigma > 2 - k$.

To prove Proposition 2, we note first that the right-hand side of (6.4) is holomorphic in $\sigma > -\delta$ (in fact, in $\sigma > 2 - 2k$), so (6.2) will follow immediately by letting $z = 0$ in (6.4) and invoking the identity principle, provided that we can establish that the left-hand side of (6.4) is holomorphic in $\sigma > -\delta$. One approach is to consider difference quotients of the function $L(z) = (F_\nu(\cdot|z), f)$. By linearity

$$\frac{L(z+h) - L(z)}{h} = \left(\frac{F_\nu(\cdot|z+h) - F_\nu(\cdot|z)}{h}, f \right).$$

On the other hand, for fixed $\tau \in \mathcal{H}$, $F_\nu(\tau|z)$ is holomorphic in $\sigma > -\delta$ (§IV), and, for fixed z in $\sigma > -\delta$, it vanishes exponentially in τ as $y = \operatorname{Im} \tau \to +\infty$. (The proof of this can be extracted from the discussion of the series (4.2), (4.5), and (4.6) in §IV.) This makes available the Lebesgue dominated convergence theorem to interchange the limit and integral, so that

$$\lim_{h \to 0} \frac{L(z+h) - L(z)}{h} = \left(\frac{d}{dz} F_\nu(\cdot|z), f \right),$$

for $\operatorname{Re} z > -\delta$. This proves Proposition 2.

With Proposition 2, we can invoke the standard line of reasoning (given, for example, in [4, p. 289]), which shows that every cusp form of weight k and MS v on $\Gamma(1)$ is a finite linear combination of the $P_\nu(\tau)$. Note that this does not prove the existence of cusp forms for any weight or MS, since $P_\nu(\tau)$ can be identically 0 for all $\nu + \kappa > 0$.

(ii) *Modular forms with a pole at $i\infty$*. The analysis we have carried out above rests in a crucial way upon the fact that $Z_m(w)$ is holomorphic in $\operatorname{Re} z > k/2 - \delta$, $\delta > 0$. As mentioned in §II, this is always true for $\nu + \kappa \geq 0$, but may no longer be the case when $\nu + \kappa < 0$ (i.e., when $\nu < 0$). This follows directly from the simple observation

$$(6.5) \qquad Z(w; m, \nu, v, \Gamma(1)) = \overline{Z}(\overline{w}; -m - 1, -\nu - 1, \overline{v}).$$

In connection with (6.5) we note that since v is a MS for weight k, $\overline{v}$ is connected with weight $-k$ and—more to the point—with weight $2 - k$ as well.

Now, in [11, pp. 10–11] Selberg observes that the hyperbolic Laplacian has the eigenvalue $(k/2)(1 - k/2)$ if there exists a nontrivial cusp form of weight k and MS v. He further points out [11, p. 12] that for $\nu \geq 0$ the eigenvalue $(k/2)(1 - k/2)$ produces a pole of $Z(w; m, \nu, v, \Gamma(1))$ at the point $w = 1 - k/2$, provided a certain other condition (which need not concern us here) is satisfied. Replacing k by $2 - k$ and v by $\overline{v}$, we find that the existence of a nontrivial cusp form on $\Gamma(1)$ of weight $2 - k$ and MS $\overline{v}$ may lead to a pole of $Z(w; m, \nu, \overline{v}, \Gamma(1))$ at $w = 1 - (2 - k)/2 = k/2$, if $\nu \geq 0$. Assuming $\nu < 0$ in (6.5), so that $-\nu - 1 \geq 0$, we see that the right-hand side possibly has a pole at $w = k/2$ if there is a nontrivial cups form of weight $2 - k$, and MS $\overline{v}$, with respect to $\Gamma(1)$. The same is then true of the left-hand side, namely, of $Z(w; m, \nu, v, \Gamma(1))$ itself, with $\nu < 0$.

In order to avoid this pole, we make the assumption that

(6.6)   there are no nontrivial cusp forms on $\Gamma(1)$ of weight $2 - k$ and MS $\bar{v}$.

Under this hypothesis the entire analysis given here can be carried out to show that for any $\nu < 0$, $P_\nu(\tau)$ is a holomorphic form on $\Gamma(1)$ of weight k and MS v, with the pole term $\exp\{2\pi i(\nu + \kappa)\tau)\}$ at $i\infty$. (Proposition 1 is replaced for $\nu + k < 0$ by the estimate $a_m(\nu, k, v) = O(m^{k/2 - 3/4}e^{\gamma\sqrt{m}})$, $m \to +\infty$, where $\gamma = 4\pi|\nu + \kappa|^{1/2}$. This is still sufficient to show that $P_\nu(\tau)$ is holomorphic in $\mathscr{H}$.) But taking a linear combination of the $P_\nu(\tau)$ then implies the existence of such a modular form with arbitrarily prescribed principal part at $i\infty$. Therefore, the Petersson gap theorem for automorphic forms [8a, Theorem 1, pp. 460–461], a generalization of the Weierstrass gap theorem on compact Riemann surfaces, implies (6.6). It follows that (6.6) is both a necessary and a sufficient condition for carrying out the analysis done here, when $\nu < 0$. At a later time we shall examine this circumstance in greater detail.

REFERENCES

1. D. Goldfeld and P. Sarnak, *Sums of Kloosterman sums*, Invent. Math. **71** (1983), 243–250.

2. E. Hecke, *Theorie der Eisensteinschen Reihen höherer stufe und ihre Anwendung auf Funkctionentheorie und Arithmetik*, Mathematische Werke, Vandenhoeck und Ruprecht, Göttingen, 1959, pp. 461–486.

3. M. Knopp, *On the Fourier coefficients of small positive powers of $\theta(\tau)$*, Invent. Math. **85** (1986), 165–183.

4. J. Lehner, *Discontinuous groups and automorphic functions*, Math. Surveys, no. 8, Amer. Math. Soc., Providence, R.I., 1964.

5. R. Lipschitz, *Untersuchung der Eigenschaften einer Gattung von unendlichen Reihen*, J. Reine Angew. Math. **105** (1889), 127–156.

6. H. Maass, *Konstruktion ganzer Modulformen halbzahliger Dimension mit θ-Multiplikatoren in einer und zwei Variablen*, Abh. Math. Sem. Hansische Univ. **12** (1938), 133-162.

7. ____, *Lectures on modular functions of one complex variable*, Tata Institute of Fundamental Research, Bombay, 1964.

8. H. Petersson, *Über die Entwicklungskoeffizienten der automorphen Formen*, Acta Math. **58** (1932), 169–215.

8a. ____, *Über eine Metrisierung der automorphen Formen und die Theorie der Poincaré-schen Reihen*, Math. Ann. **117** (1940), 453–537.

9. W. Roelcke, *Das Eigenwertproblem der automorphen Formen in der hyperbolischen Ebene*. I, Math. Ann. **167** (1966), 292–337.

10. ____, *Über die Wellengleichung bei Grenzkreisgruppen erster Art*, Sitzungsber. Heidelb. Akad. Wiss. Math.-Natur. Kl. **4** (1956).

11. A. Selberg, *On the estimation of Fourier coefficients of modular forms*, Theory of Numbers, Proc. Sympos. Pure Math., vol. 8, Amer. Math. Soc., Providence, R.I., 1965, pp. 1–15.

12. C. L. Siegel, *Die Funktionalgleichungen einiger Dirichletscher Reihen*, Math. Z. **63** (1956), 363–373.

TEMPLE UNIVERSITY

THE INSTITUTE FOR ADVANCED STUDY

Proceedings of Symposia in Pure Mathematics
Volume **49** (1989), Part 2

Intersection Numbers of Cycles
on Locally Symmetric Spaces
and Fourier Coefficients
of Holomorphic Modular Forms
in Several Complex Variables

JOHN J. MILLSON

This paper is an expanded version of the lecture delivered at the conference, incorporating results obtained later at the Max Planck Institute. It is an exposition of joint work with Steve Kudla. For some time we have been studying the relationship between two types of cohomology classes for arithmetic quotients of the symmetric spaces attached to orthogonal and unitary groups. The first type of cohomology class has a geometric description as the Poincaré dual classes to natural cycles on the above arithmetic quotients. These cycles are themselves unions of arithmetic quotients of totally geodesic subsymmetric spaces associated to smaller orthogonal or unitary groups. They generalize the classical Hurwitz correspondences and the cycles in the Hilbert modular surfaces considered by Hirzebruch-Zagier [6]. The second type of cohomology class has an analytic description in terms of automorphic forms on the above arithmetic quotients constructed using the theta correspondence. This correspondence is realized by an integral transformation with kernel a theta function defined on a product of two locally symmetric spaces. The new development that led to this paper is the discovery of a new method to use the Cauchy-Riemann equations in the general theory of the theta correspondence. Our method is based on a study of the double complex of relative Lie algebra cohomology with values in the oscillator representation associated to a dual reductive pair. We use such algebraic considerations to show that the $\bar{\partial}$-operator applied to one variable of the kernel is an exact differential form in the other variable. Combining this result

---

1980 *Mathematics Subject Classification* (1985 *Revision*). Primary 32N05.
Partially supported by National Science Foundation Grant DMS-85-01742.

with Stokes's theorem, we are able to deduce that the transforms of *closed* cuspidal differential forms are *holomorphic* Siegel or hermitian modular forms. The above phenomenon concerning the $\bar{\partial}$-operator would be explained by the degeneration at E_1 of the spectral sequence described in §3. The rest of this introduction is devoted to describing how the generating function for the intersection numbers of a fixed finite cycle C with the members of the family of locally finite cycles

$$\{C_\beta \colon \beta \in M_n(\mathcal{O}), \beta^* = \beta, \ \beta \geq 0 \text{ at all infinite places}\}$$

described in §2 is a Siegel (or hermitian) modular form. Our method of proof is to use the theta correspondence to construct a holomorphic modular form, which is then shown to have the above generating function as its Fourier series. Our results are the analogues of those of Hirzebruch-Zagier [6] for the Hilbert modular subgroups of $O(2,2)$.

Let $\mathscr{k}$ be a totally real number field of degree r, let $\mathcal{O}$ be the ring of integers in $\mathscr{k}$, and let V be an oriented vector space of dimension m over $\mathscr{k}$. Let R be the set of archimedean completions of $\mathscr{k}$. Let $(\ ,\ )$ be a quadratic form on V with signature (p,q) at one archimedean completion of V and positive definite at the others. Let β be a symmetric $n \times n$ matrix with entries in $\mathcal{O}$ that is positive semi-definite at all archimedean completions of $\mathscr{k}$. Then as described in §2 we can construct a cycle C_β in the arithmetic quotient $M = \Gamma \backslash D$ of the symmetric space D of $O(p,q)$. Here Γ is a congruence subgroup of the group of units of $(\ ,\ )$. The cycle C_β is of dimension $(p-t)q$ where $t = \operatorname{rank} \beta$. If q is even there exists an invariant q-form e_q, the Euler form, on D (see Theorem 4.4 for the definition). We define $e_q = 0$ if q is odd. Suppose η is a closed rapidly decreasing $(p-n)q$ form on M (if M is compact all η are considered to be rapidly decreasing). If $\operatorname{rank} \beta = t$, then $\dim C_\beta = (p-t)q$ and $\deg \eta \wedge e_q^{n-t} = (p-t)q$ and we can take the period $\int_{C_\beta} \eta \wedge e_q^{n-t}$. We define a generating function $P(\tau)$ for these periods with $\tau \in \mathfrak{h}_n'$ (here $\mathfrak{h}_n$ denotes the Siegel upper half-space of genus n). We let β run through the lattice $\mathscr{L}$ of symmetric $n \times n$ matrices with entries in $\mathcal{O}$ in the product of r copies of the $n \times n$ symmetric matrices. We will employ the following notation throughout. If $z \in \mathbf{C}$, we will abbreviate $\exp(2\pi i z)$ to $e(z)$. If $z = (z_\lambda) \in M_n(\mathbf{C})^r$, then $e_*(\beta z)$ will denote $e(\frac{1}{2} \operatorname{tr} \sum_{\lambda \in R} \lambda(\beta) z_\lambda)$. We can now define $P(\tau)$:

$$P(\tau) = \sum_{t=0}^{n} \sum_{\substack{\beta \in \mathscr{L} \\ \operatorname{rank} \beta = t \\ \beta \geq 0}} \left(\int_{C_\beta} \eta \wedge e_q^{n-t} \right) e_*(\beta \tau).$$

REMARK. If q is odd the sum is only over β of rank n.
We then have the following theorem.

THEOREM 1. $P(\tau)$ *is a holomorphic modular form of weight* $m/2$ *for a suitable congruence subgroup of* $\operatorname{Sp}_n(\mathcal{O})$. *If q is odd, $P(\tau)$ is a cusp form.*

There is also a homology version of Theorem 1. Let C be a finite (compact) nq-cycle in M. Then the Poincaré dual cohomology class to C has a compactly supported representative. Substituting this form for η in the above theorem, we obtain the following analogue of Hirzebruch-Zagier [6]. Define $I(\tau)$ for $\tau \in \mathfrak{h}_n^r$ by

$$I(\tau) = \sum_{t=0}^{n} \sum_{\substack{\beta \in \mathcal{L} \\ \text{rank } \beta = t \\ \beta \geq 0}} C \cdot (C_\beta \cap e_q^{n-t}) e_*(\beta \tau).$$

Here $\cdot$ denotes the intersection product of cycles and $\cap$ is the cap-product between cohomology and homology.

THEOREM 1 (*bis*). *$I(\tau)$ is a holomorphic modular form of weight $m/2$ for a suitable congruence subgroup of* $\mathrm{Sp}_n(\mathcal{O})$. *If q is odd, $I(\tau)$ is a cusp form.*

We have a corresponding theorem for the symmetric spaces of the unitary groups $U(p, q)$. Let F be a totally imaginary extension of the field $\mathscr{k}$ above, let $\mathscr{O}$ be the integers in F, and let V be an oriented vector space of dimension $m = p + q$ over F. Let R be a cross-section for the action of complex conjugation on the archimedean completions of F. Let $(\, , \,)$ be a hermitian form on V with signature (p, q) at one complex conjugacy class of archimedean completions of F and positive definite at the others. Let β be a hermitian $n \times n$ matrix over $\mathscr{O}$ that is positive semi-definite at all archimedean completions of F. Again we have cycles C_β for each such matrix β in arithmetic quotients $M = \Gamma \backslash D$ of the symmetric space D of $U(p, q)$. Here Γ is a congruence subgroup of the group of units of $(\, , \,)$. The cycle C_β is of complex dimension $(p - t)q$ where $t = \mathrm{rank}\, \beta$. For all q there exists an invariant $2q$-form c_q, the qth Chern form, on D (see Theorem 4.5 for the definition of c_q).

Now suppose η is a closed rapidly decreasing $2(p - n)q$ form on M. If M is compact, all η are considered to be of rapid decrease. If rank $\beta = t$ we can take the period $\int_{C_\beta} \eta \wedge c_q^{n-t}$. We define the generating function $P(\tau)$ for these periods as β runs through the lattice $\mathscr{L}$ of hermitian $n \times n$ matrices with entries in $\mathscr{O}$ in the product of r copies of the hermitian $n \times n$ matrices. We will use the symbol $U(n, n)$ to denote the isometry group of the standard split skew-hermitian form $\langle \, , \, \rangle$ on $\mathbf{C}^{2n}$. We have for $e_1, e_2, \ldots, e_{2n}$ the standard basis

$$\left\langle \sum_{j=1}^{2n} z_j e_j, \sum_{j=1}^{2n} w_j e_j \right\rangle = \sum_{j=1}^{n} (z_j \bar{w}_{n+j} - z_{n+j} \bar{w}_j).$$

We let $\mathbf{C}\mathfrak{h}_n$ be the symmetric space of $U(n, n)$, the space of complex $n \times n$ matrices $\tau = u + iv$ with positive definite skew-hermitian part (i.e., u is hermitian, iv is skew-hermitian, and v is positive definite). Let $\tau \in (\mathbf{C}\mathfrak{h}_n)^r$.

We define

$$P(\tau) = \sum_{t=0}^{n} \sum_{\substack{\beta \in \mathscr{L} \\ \text{rank } \beta = t \\ \beta \geq 0}} \left(\int_{C_\beta} \eta \wedge c_q^{n-t} \right) e_*(\beta \tau).$$

We have the following theorem.

THEOREM 2. $P(\tau)$ *is a holomorphic modular form of weight m for a suitable congruence subgroup of the group of $\mathscr{O}$-points of $U(n, n)$.*

We have a corresponding theorem in terms of intersection numbers of cycles. Let C be a finite (compact) $2nq$-cycle. Define $I(\tau)$ for $\tau \in (\mathbf{C}\mathfrak{h}_n)^r$ by

$$I(\tau) = \sum_{t=0}^{n} \sum_{\substack{\beta \in \mathscr{L} \\ \text{rank } \beta = t \\ \beta \geq 0}} C \cdot (C_\beta \cap c_q^{n-t}) e_*(\beta \tau).$$

THEOREM 2 (bis). $I(\tau)$ *is a holomorphic modular form of weight m for a suitable congruence subgroup of the group of $\mathscr{O}$-points of $U(n, n)$.*

In order to make a true generalization of the results of Hirzebruch-Zagier [6] one is forced to give up the hypothesis that η has rapid decrease. This creates enormous complications. First the integral defining $\theta_\varphi(\eta)$ (see §3 for the meaning of this symbol) might not converge. Even if it converges, the automorphic form $\theta_\varphi(\eta)$ will no longer necessarily be holomorphic; there will be a formula for $\bar{\partial}\theta_\varphi(\eta)$ involving an integral over the "boundary" of $\Gamma \backslash D$ (recall that $\bar{\partial}$ applied to the integrand defining $\theta_\varphi(\eta)$ was not zero but was exact). Finally, cycles associated to the isotropic vectors will contribute to the Fourier coefficients. It seems likely that Eisenstein cohomology will make an appearance at this point as suggested in [8, p. 309]. Cogdell [3] has analyzed the situation in the case M is a finite-volume quotient of the 2-ball in terms of the desingularization of the Baily-Borel compactification. It will be interesting to see if there is a compactification that will allow a similar analysis in the general case.

In conclusion we should mention the work of Tong and Wang along parallel lines [14–18]. It is a pleasure to thank F. Hirzebruch and J. Schwermer for their hospitality at the Max Planck Institute where our recent work was done. Finally, we would like to thank the conference organizers for the opportunity to present our results.

1. Special cycles. In what follows $D = G/K$ will be a Riemannian symmetric space with G a semisimple Lie group having no compact factors and $K \subset G$ a maximal compact subgroup. We let $\Gamma \subset G$ be a torsion-free lattice and $M = \Gamma \backslash D$ be the associated symmetric space. Let $\pi : D \to M$ be the covering projection. Now suppose σ_1 is an isometry of D of order 2 such that

$$\sigma_1 \Gamma \sigma_1 = \Gamma.$$

Let D_1 be the fixed point set of σ_2 in D, and let Γ_1, G_1, K_1 be the fixed-point sets of σ_1 in Γ, G, and K respectively. Then D_1 is a totally geodesic subsymmetric space of D isomorphic to G_1/K_1. We let $M_1 = \Gamma_1 \backslash D_1$ be the corresponding locally symmetric space. We assume that Γ has been chosen so that M_1 is orientable (this can present a problem, but we have been able to deal with it in the cases of interest here). The restriction of π to D_1 induces a map $j_1 \colon M_1 \to M$. For a proof of the next lemma see [1].

LEMMA 1.1. *j_1 is a proper embedding onto a totally geodesic submanifold of M.*

We call the locally finite cycle represented by the image of j_1 a *special cycle*. If M is compact, then M_1 will also be compact. The image of j_1 is a component of the fixed-point set of σ_1 acting on M.

Raghunathan and I were led to consider such cycles because of a remarkable property they satisfy. *They come in complementary pairs.* In fact this is true only under the further condition (usually satisfied in practice) that there exist "rational points" on D_1. By this we mean a point $x \in D_1$ such that the associated Cartan involution θ satisfies $\theta \Gamma \theta \subset \Gamma$. In this case we put $\sigma_2 = \theta \sigma_1$ and define D_2 to be the fixed-point set of σ_2. If M is compact, the image M_2 of D_2 in M is compact. If the intersection number of M_1 and M_2 is nonzero, neither is a boundary.

EXAMPLE.

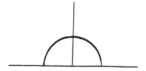

Here D is the upper half-plane and σ_1 and σ_2 are reflections in the y-axis and unit circle respectively.

Using the above idea, Raghunathan and I were able to give many examples of nonvanishing cohomology groups for locally symmetric spaces [12].

2. Special cycles in orthogonal and unitary locally symmetric spaces. In this paper we will be concerned only with the cases in which D is the symmetric space of $O(p,q)$ or $U(p,q)$. In this case we can take advantage of the projective structure of D to better understand special cycles. Let ℓ be a totally real field (respectively, totally imaginary extension of a totally real field) and $\mathcal{O}$ be the integers in ℓ. Let V_ℓ be an oriented vector space over ℓ and $L \subset V_\ell$ an $\mathcal{O}$ lattice. Let $(\ ,\ )$ be a nondegenerate quadratice (respectively, hermitian form) on V_ℓ, which is integral ($\mathcal{O}$-valued) on L and has signature (p,q) at one archimedean completion of ℓ and is positive definite at all other archimedean completions of ℓ. We let V denote the completion of V_ℓ at the place where the signature of $(\ ,\ )$ is (p,q). Let Γ denote a torsion-free congruence subgroup of the subgroup of $GL(V)$ preserving L and $(\ ,\ )$. If

$U \subset V$ is an oriented rational subspace such that $(,)|U$ is nondegenerate, we will construct special cycles $C_U \subset \Gamma \backslash D$, where D is the symmetric space of the isometry group G of $(,)$. Thus G is isomorphic to $O(p, q)$ (respectively, $U(p, q)$). Since $(,)|U$ is nondegenerate, we have a direct sum decomposition $V = U + U^\perp$. Recall that we may consider D as the open subset of the Grassmannian $\mathrm{Gr}_q(V)$ consisting of those q-planes Z such that $(,)|Z$ is negative definite. We may now define a subset $D_U \subset D$ by

$$D_U = \{Z \in D : Z = Z \cap U + Z \cap U^\perp\}$$

and let $C_U = \pi(D_U)$.

LEMMA 2.1. C_U is a special cycle.

PROOF. Let σ_1 be the involution of V that is $+1$ on U and -1 on $U^\perp$. Then D_U is the fixed point set of σ_1 on D.

We should explain how the orientations of the vector spaces U and V lead to an orientation of C_U in case $(,)|U$ is positive definite. Choose a basepoint $Z_0 \in D$ and choose an orientation of Z_0, which we fix once and for all. Propagate the orientation of Z_0 continuously to orient all other $Z \in D$. Since $T_Z(D)$ is canonically isomorphic to $Z^* \otimes Z^\perp$, we have now oriented D. Now if $Z \in D_U$, its normal fiber in D is canonically isomorphic to $Z^* \otimes U$. Thus D_U is oriented. By Millson-Raghunathan [12, Proposition 4.1], we may choose Γ so that the orientation of D_U descends for any U as above.

The signature of $(,)|U$ plays an important role in understanding the nature of the class C_U. The case in which r or s is zero is of special importance. In case $s = 0$, then U is a positive r plane (so $r \leq p$) and

$$D_U = \{Z \in D : Z \subset U^\perp\}.$$

In case $r = 0$, then X is a negative s-plane (so $s \leq q$) and

$$D_U = \{Z \in D : Z \supset X\}.$$

We say such special cycles are of definite type. We say the other cycles are of mixed type.

EXAMPLE. $D = \mathbf{H} \times \mathbf{H}$. In this case the definite cycles correspond to quotients of linearly embedded upper half-planes. The mixed cycles correspond to products of two geodesics; they are totally real geodesic tori.

Given U as above, we let D'_U be the symmetric space of $O(U)$ and D''_U be the symmetric space of $O(U^\perp)$.

LEMMA 2.2. $D_U \cong D'_U \times D''_U$.

The above product decomposition is of importance only in the mixed case. We have a corresponding decomposition of cycles $C_U \cong C'_U \times C''_U$. For later use we define some reducible cycles of positive type parametrized by the lattice $\mathscr{L}$ of symmetric (respectively, hermitian) $n \times n$ matrices over $\mathscr{O}$ that are positive semidefinite at each archimedean completion of ℓ. Let β be such a matrix and $\mathcal{Q}_\beta \subset V^n$ be the subset defined by

$$\mathcal{Q}_\beta = \{(x_1, x_2, \ldots, x_n) \in V^n : (x_i, x_j) = \beta_{ij}, \ 1 \leq i \leq j \leq n\}.$$

Assume first that β has rank n. Then $\mathcal{Q}_\beta$ is a single G-orbit, and $\mathcal{Q}_\beta \cap L^n$ consists of a finite number of Γ-orbits. We choose representatives $\{Y_1, \ldots, Y_l\}$ for these Γ orbits, so $Y_j \in L^n$ for $j = 1, 2, \ldots, l$. Let $U_j = \operatorname{span} Y_j$. We then define C_β by

$$C_\beta = \sum_{j=1}^{l} C_{U_j}.$$

We give C_{U_j} the orientation determined by Y_j. Then C_β is a locally finite cycle such that each irreducible component has dimension $(p - n)q$ (resp. $2(p - n)q$).

Now suppose rank $\beta = t$, where t is strictly less than n. In this case $\mathcal{Q}_\beta$ no longer consists of a single G-orbit but it contains a single closed G-orbit, which we now describe. We let $\mathcal{Q}_\beta^c$ denote the subset of $\mathcal{Q}_\beta$ consisting of those $Y = \{x_1, \ldots, x_n\} \in V^n$ which satisfy

(i) dim span $Y = t$,

(ii) $(\ , \ ) | \operatorname{span} Y$ is nondegenerate.

Then $\mathcal{Q}_\beta^c$ is a single G-orbit, and $\mathcal{Q}_\beta^c \cap L^n$ consists of a finite number of Γ-orbits. Again we choose Γ-orbit representatives $\{Y_1, Y_2, \ldots, Y_{l(t)}\}$ and let $U_j = \operatorname{span} Y_j$. We then define C_β as before by

$$C_\beta = \sum_{j=1}^{l(t)} C_{U_j}.$$

We give C_{U_j} the orientation obtained by refining Y_j to a basis starting at the left. Then C_β is a locally finite cycle such that each irreducible component has dimension $(p - t)q$ (resp. $2(p - t)q$). In the second case C_β is an algebraic cycle. In order that C_β be nonzero it is best to impose a congruence condition in the definition of $\mathcal{Q}_\beta$. The necessary modifications are explained in [9].

3. The Cauchy-Riemann equations and the theta correspondence. In this section we discuss a cohomological version of the theta correspondence. We will be particularly interested in proving integrals of cohomological type depending on a symplectic (or split unitary parameter) are holomorphic. It is convenient to take the viewpoint of continuous cohomology in this section. A basic reference for continuous cohomology is Borel-Wallach [2].

Let G denote the identity component of $O(p, q)$ (respectively, $U(p, q)$) and G' denote $\operatorname{Sp}_n(\mathbf{R})$ (respectively, $U(n, n)$). Let $\tilde{G}'$ be the nontrivial 2-fold cover of G'. We consider the continuous cohomology groups $H_{ct}^{\cdot}(G, \mathcal{S}(V^n))$. Here G operates on $\mathcal{S}(V^n)$, the complex-valued Schwartz functions on V^n, by the action ρ given by

$$\rho(g)\varphi(x) = \varphi(g^{-1}x).$$

Recall that by the van Est theorem the above cohomology groups may be realized as the de Rham cohomology of the complex $C^{\cdot}$, d whose ith cochain group is given by

$$C^i = (\mathscr{A}^i(D) \otimes \mathscr{S}(V^n))^G.$$

Here the superscript G denotes the G-invariants and $\mathscr{A}^i(D)$ denotes the smooth differential i-forms on D. The differential d is defined on a decomposable $\nu \otimes \varphi$ by

$$d(\nu \otimes \varphi) = (d\nu) \otimes \varphi,$$

where the d on the right-hand side is the usual exterior differential. We will identify a continuous cohomology class with a class of closed differential forms on D with values in $\mathscr{S}(V^n)$. If φ is such a differential form, we let $[\varphi]$ denote its cohomology class. We observe that we have an isomorphism given by evaluation at a point $z_0 \in D$:

$$(\mathscr{A}^i(D) \otimes \mathscr{S}(V^n))^G \cong (\Lambda^i \mathfrak{p}^* \otimes \mathscr{S}(V^n))^K.$$

Here K is the maximal compact subgroup of G that is the isotropy subgroup of z_0 and $\mathfrak{p}$ is the orthogonal complement of $\mathfrak{k}$, the complexified Lie algebra of K, in $\mathfrak{g}$ the complexified Lie algebra of G, for the Killing form on $\mathfrak{g}$.

The group $\tilde{G}'$ operates on $\mathscr{S}(V^n)$ by the oscillator (or Weil) representation ω, see [10] or [19]. This action commutes with the action ρ of G, and hence $\tilde{G}'$ operates on $H_{\mathrm{ct}}^i(G, \mathscr{S}(V^n))$. We let $\tilde{K}$ denote the maximal compact subgroup of $\tilde{G}'$ lying over $U(n) \subset \mathrm{Sp}_n(\mathbf{R})$ (respectively, $U(n) \times U(n) \subset U(n,n)$). We let $\mathfrak{p}'$ denote the orthogonal complement to $\mathfrak{k}'$, the complexification of the Lie algebra of $\tilde{K}'$, in $\mathfrak{g}'$ the complexified Lie algebra of G' for the Killing form of $\mathfrak{g}'$. Let let D' be the symmetric space of $\tilde{G}'$. Then D' is Hermitian symmetric. Since we may identify $\mathfrak{p}'$ with the complexified tangent space to D' at the identity coset, we have a splitting of $\mathfrak{p}'$ into the holomorphic and antiholomorphic tangent spaces

$$\mathfrak{p}' = \mathfrak{p}^+ + \mathfrak{p}^-.$$

We observe that $\mathfrak{p}^-$ acts on $H_{\mathrm{ct}}^{\cdot}(G, \mathscr{S}(V^n))$ by the action of ω on the coefficients.

DEFINITION. We will say cohomology class $[\varphi] \in H_{\mathrm{ct}}^{\cdot}(G, \mathscr{S}(V^n))$ is holomorphic if it is annihilated by $\mathfrak{p}^-$ under the action described above.

We will now make a short digression to give another definition of a holomorphic class. We will assume that all classes $[\varphi]$ considered are $\tilde{K}'$ finite, that is, $[\varphi]$ lies in a smallest finite-dimensional subspace of the continuous cohomology that is invariant under $\tilde{K}'$. Let us denote the representation of $\tilde{K}'$ on this subspace by σ, and let $\mathscr{E}$ denote the corresponding $\tilde{G}'$ homogeneous bundle over D'. We may then consider the double complex $C^{\cdot,\cdot}$, $d \otimes \bar{\partial}$ where

$$C^{p,q} = (\mathscr{A}^p(D) \otimes \mathscr{A}^{0,q}(D', \mathscr{E}) \otimes \mathscr{S}(V^n))^{G \times G'}$$

equipped with the differential $d \otimes \bar{\partial}$, where d is as described before and $\bar{\partial}$ is the usual $\bar{\partial}$ operator on forms in D' with values in $\mathscr{E}$. We find that $\varphi \in C^{i,0}$, and one may reformulate our previous definition as follows.

DEFINITION. $[\varphi] \in H^i_{ct}(G, \mathscr{S}(V^n))$ is holomorphic if it is a cocycle at $E_1^{i,0}$ in the spectral sequence associated to filtering the double complex by q; that is,

$$F^q C^n = \bigoplus_{\substack{p+q'=n \\ q' \geq q}} C^{p,q'}.$$

It is standard that $[\varphi]$ may be identified with an element of $E_1^{i,0}$, see [5, p. 442].

The above spectral sequence appears to be of considerable significance for the study of the theta correspondence from the point of view of differential geometry.

In case $[\varphi]$ is holomorphic we will write $\bar{\partial}[\varphi] = 0$. We emphasize that this equation does not mean $\bar{\partial}\varphi = 0$ but only that $\bar{\partial}\varphi = d\psi$ for some $\psi \in C^{i-1,1}$ in the above double complex.

We now assume that φ is an element of $(\mathscr{A}^i(D) \otimes \mathscr{S}(V^n))^G$ such that

(i) φ is $\tilde{K}'$ finite, and

(ii) $[\varphi]$ is holomorphic.

Given such a φ we now construct an integral operator Λ from closed differential forms on M to holomorphic sections of the vector bundle $\mathscr{E}$ over $M' = \Gamma' \backslash D'$, where Γ' is a lattice in G' that we now describe. We will henceforth usually abuse notation and use φ to denote the form and its class.

We recall that there is a remarkable distribution Θ, the theta distribution, on $\mathscr{S}(V^n)$, which is the sum of Dirac delta distributions centered at points of L^n. Clearly, Θ is invariant under Γ so we have $\Theta \in \mathrm{Hom}_\Gamma(\mathscr{S}(V^n), \mathbf{C})$. Consequently if we define $\theta_\varphi(g', g)$ by

$$\theta_\varphi(g', g) = \Theta(\omega(g')\varphi(g^{-1}x)),$$

we find that $\theta_\varphi(g', g)$ is a right $\tilde{K}'$-finite map to $H^i(\Gamma, \mathbf{C})$. We then have the following theorem (see [11, Chapter II] for an elementary discussion).

THEOREM (WEIL [9]). *There exists an arithmetic subgroup $\Gamma' \subset G'$ and a diagram*

$$\begin{array}{ccc} & & \tilde{G}' \\ & \nearrow^{s} & \downarrow \\ \Gamma' & \rightarrow & G' \end{array}$$

such that Θ is invariant under $\omega|s(\Gamma')$.

COROLLARY. $\theta_\varphi(\gamma' g', g) = \theta_\varphi(g', g)$.

Here we identify Γ' with $s(\Gamma')$.

By taking inner products with θ_φ we get an antilinear map Λ' from a certain space of cuspidal automorphic forms on $M' = \Gamma' \backslash G'/K'$ to $H^i(M, \mathbf{C})$

regarded as classes of closed differential i-forms on M. Λ' is defined by the formula

$$\Lambda'(f) = ((\theta_\varphi, f)) = \theta_\varphi(f).$$

Here $((\, , \,))$ denotes the L^2 inner product on the hermitian vector bundle $\mathscr{E}$. We see that Λ' has kernel $\theta_\varphi(g', g)$ given by

$$\theta_\varphi(g', g) = \sum_{\xi \in L^n} \omega(g')\varphi(g^{-1}\xi).$$

θ_φ is a section of the bundle $\mathscr{E}$ over M' in the g' variable and a closed differential i-form in M in the g-variable. We can also use $\theta_\varphi(g', g)$ as the kernel of an integral transform Λ from cuspidal closed differential forms (or cohomology classes) on M to sections of $\mathscr{E}$ on M'. If η is a closed rapidly decreasing differential $(k - i)$-form (where $\dim M = k$), we define Λ by the formula

$$\Lambda(\eta) = \int_M \eta \wedge \theta_\varphi = \theta_\varphi(\eta).$$

We will prove Theorems 1 and 2 by computing the Fourier expansion of $\theta_\varphi(\eta)$ for a suitable φ (described in §4). We now explain what this means. Note that $G' = \mathrm{Sp}_n(\mathbf{R})$ (resp. $U(n, n)$) has a natural representation on R^{2n} (resp. $\mathbf{C}^{2n}$). We let N denote the abelian unipotent subgroup of G' consisting of those elements that leave fixed each of the first n standard basis vectors (recall that we are defining $U(n, n)$ to be the isometry group of the standard split skew-hermitian form). Then N is isomorphic to the space S of symmetric $n \times n$ real matrices (resp. $n \times n$ hermitian matrices). Then $N \cap \Gamma'$ is a lattice $\mathscr{L}_1$ in S^r. We let $\mathscr{L}_1^*$ be the dual lattice to $\mathscr{L}_1$ in S^r for the bilinear form B given by

$$B(X, Y) = \operatorname{tr} XY.$$

Then $\theta_\varphi(\eta)$ has a Fourier expansion indexed by the elements of $\mathscr{L}_1^*$:

$$\theta_\varphi(\eta)(u + iv) = \sum_{\beta \in \mathscr{L}_1^*} a_\beta(v) e_*(2\beta u).$$

Here $\tau = u + iv$ is the decomposition of τ into real and imaginary parts (resp. hermitian and skew-hermitian parts).

REMARK. We have shown in [8] that for any closed φ (not necessarily holomorphic) the nondegenerate (i.e., rank $\beta = n$) Fourier coefficients $a_\beta(v)$ are given by

$$a_\beta(v) = w_\beta(v) \int_{C_{2\beta}} \alpha \wedge \eta,$$

where w_β is a Whittaker function and α is a cohomology class on $C_{2\beta}$.

In the case φ is holomorphic, the calculation of the previous Fourier coefficients is greatly simplified by the following lemma.

LEMMA 3.1. *If* $[\varphi]$ *is holomorphic and* η *is closed and rapidly decreasing, then* $\theta_\varphi(\eta)$ *is a holomorphic section of* $\mathscr{E}$.

PROOF. Assume $\bar{\partial}\varphi = d\psi$ with $\psi \in C^{i-1,1}$. Then we have

$$\bar{\partial}\theta_\varphi = d\theta_\psi.$$

We then have

$$\bar{\partial}\theta_\varphi(\eta) = \bar{\partial}\left(\int_M \eta \wedge \theta_\varphi\right) = \int_M \eta \wedge \bar{\partial}\theta_\varphi$$
$$= \int_M \eta \wedge d\theta_\psi = \int_M d(\eta \wedge \theta_\psi) = 0.$$

The last equality holds [4] because $\eta \wedge \theta_\varphi$ and $\eta \wedge \theta_\psi$ are both easily seen to be L^1 since η is rapidly decreasing.

COROLLARY. *If* $n \geq 2$, *then* $a_\beta(v) = 0$ *unless* β *is positive semi-definite.*

PROOF. This is Koecher's Theorem [13, 4-04].

4. Construction of holomorphic Schwartz classes. In this section we will construct holomorphic cohomology classes (for $q \geq 1$)

$$\varphi^+_{nq} \in H^{nq}_{ct}(\mathrm{SO}_0(p,q), \mathscr{S}(V^n)),$$
$$\varphi^+_{nq,nq} \in H^{nq,nq}_{ct}(\mathrm{U}(p,q), \mathscr{S}(V^n)).$$

We recall the standard complex (in the orthogonal case):

$$(\Lambda^q \mathfrak{p}^* \otimes \mathscr{S}(V))^K \to (\Lambda^{q+1}\mathfrak{p}^* \otimes \mathscr{S}(V))^K \to,$$

where the differentials are described as follows. Let $\{\omega_{\alpha\mu}: 1 \leq \alpha \leq p; p+1 \leq \mu \leq p+q\}$ be a basis for the horizontal (i.e., in $\mathfrak{p}^*$) Maurer Cartan forms chosen as in [7, p. 368]. We let $A_{\alpha\mu}$ denote the operation of left multiplication by $\omega_{\alpha\mu}$. Then we have

$$d = \sum_{\alpha,\mu} A_{\alpha\mu} \otimes \left(x_\alpha \frac{\partial}{\partial x_\mu} + x_\mu \frac{\partial}{\partial x_\alpha}\right).$$

Here we choose coordinates $\{(x_i): 1 \leq i \leq m\}$ in V or complex coordinates $\{(z_i): 1 \leq i \leq m\}$ in the unitary case such that

$$(x,x) = \sum_{\alpha=1}^{p} x_\alpha^2 - \sum_{\mu=p+1}^{p+q} x_\mu^2$$

or (in the unitary case)

$$(x,x) = \sum_{\alpha=1}^{p} |z_\alpha|^2 - \sum_{\mu=p+1}^{p+q} |z_\mu|^2.$$

In the two cases we define φ_0, the Gaussian on V, by

$$\varphi_0(x) = e^{-\pi \sum_{i=1}^{m} x_i^2} \quad \text{or} \quad \varphi_0(x) = e^{-\pi \sum_{i=1}^{m} |z_i|^2}.$$

In the orthogonal case we define the Howe operator

$$D^+ \colon \Lambda^{\cdot} \mathfrak{p}^* \otimes \mathscr{S}(V) \to \Lambda^{\cdot+q} \mathfrak{p}^* \otimes \mathscr{S}(V)$$

by

$$D^+ = \frac{1}{2^q} \prod_{\mu=p+1}^{m} \left\{ \sum_{\alpha=1}^{p} \left[A_{\alpha\mu} \otimes \left(x_\alpha - \frac{1}{2\pi} \frac{\partial}{\partial x_\alpha} \right) \right] \right\}.$$

We define

$$\varphi^+ = D^+ \varphi_0.$$

In the unitary case we define

$$D^+ = \frac{1}{2^{2q}} \prod_{\mu=p+1}^{m} \left\{ \sum_{\alpha=1}^{p} \left[A_{\alpha\mu} \otimes \left(\bar{z}_\alpha - \frac{1}{\pi} \frac{\partial}{\partial z_\alpha} \right) \right] \right\}.$$

Here $\partial/\partial z_\alpha = \frac{1}{2}[\partial/\partial x_\alpha - i\partial/\partial y_\alpha]$ and $A_{\alpha\mu}$ denotes left multiplication by the element $\xi_{\alpha\mu}$ in $(\mathfrak{p}^*)^+$ defined in [7, p. 374]. In the unitary case we define

$$\varphi^+_{q,q} = D^+ \bar{D}^+ \varphi_0.$$

We have the following theorems; (i), (ii), and (iii) are proved in [7].

THEOREM 4.1. (i) φ^+_q and $\varphi^+_{q,q}$ are closed.
(ii) φ^+_q transforms under $\tilde{U}(1) \subset \tilde{\mathrm{Sp}}_1(\mathbf{R})$ according to $\mathrm{id}^{m/2}$.
(iii) $\varphi^+_{q,q}$ transforms under $U(1) \times U(1) \subset U(1,1)$ according to $\mathrm{id}^m \otimes \mathrm{id}^{-m}$.
(iv) φ^+_q and $\varphi^+_{q,q}$ are holomorphic.

Thus we obtain classes

$$\varphi^+_q \in H^q_{\mathrm{ct}}(\mathrm{SO}_0(p,q), \mathscr{S}(V)) \quad \text{and} \quad \varphi^+_{q,q} \in H^{q,q}_{\mathrm{ct}}(\mathrm{U}(p,q), \mathscr{S}(V)).$$

There are corresponding classes φ^-_p and $\varphi^-_{p,p}$ obtained by replacing D^+ by the operator D^- given by

$$D^- = \frac{1}{2^p} \prod_{\alpha=1}^{p} \left\{ \sum_{\mu=p+1}^{m} \left[A_{\alpha\mu} \otimes \left(x_\mu - \frac{1}{2\pi} \frac{\partial}{\partial x_\mu} \right) \right] \right\},$$

or

$$D^- = \frac{1}{2^{2p}} \prod_{\alpha=1}^{p} \left\{ \sum_{\mu=p+1}^{m} \left[A_{\alpha\mu} \otimes \left(\bar{z}_\alpha - \frac{1}{\pi} \frac{\partial}{\partial z_\mu} \right) \right] \right\}.$$

To construct more classes we note that we have an exterior product to be denoted $\wedge$

$$H^l_{\mathrm{ct}}(G, \mathscr{S}(V)) \otimes H^r_{\mathrm{ct}}(G, \mathscr{S}(V)) \to H^{l+r}_{\mathrm{ct}}(G, \mathscr{S}(V^2))$$

given by the usual formula for exterior product using the isomorphism from $\mathscr{S}(V) \otimes \mathscr{S}(V)$ to $\mathscr{S}(V^2)$.

We find the following results for the n-fold exterior power of the basic class. Recall $m = p + q$.

THEOREM 4.2. (i) $\varphi_{nq}^+ = \varphi_q^+ \wedge \cdots \wedge \varphi_q^+$ transforms under $\tilde{U}(n)$ according to $(\det)^{m/2}$.

(ii) $\varphi_{nq,nq}^+ = \varphi_{q,q}^+ \wedge \cdots \wedge \varphi_{q,q}^+$ transforms under $U(n) \times U(n)$ according to $\det_1^m \otimes \det_2^{-m}$.

(iii) φ_{nq}^+ and $\varphi_{nq,nq}^+$ are holomorphic.

Henceforth we will use a single symbol φ to denote one of the classes φ_{nq}^+ or $\varphi_{nq,nq}^+$.

We now let η be a closed rapidly decreasing $(p-n)q$ (resp. $2(p-q)q$) form on M and consider $\theta_\varphi(\eta)(\tau)$. Since $\theta_\varphi(\eta)(\tau)$ is holomorphic in τ, we obtain the following theorem.

THEOREM 4.3.

$$\theta_\varphi(\eta)(\tau) = \sum_{\substack{\beta \in \mathscr{L}_1^* \\ \beta \geq 0}} a_\beta e_*(2\beta\tau) \quad \text{with } a_\beta \in \mathbf{C}.$$

PROOF. In case $n \geq 2$ this is the corollary to Lemma 3.1. In case $n = 1$ it follows from a direct calculation.

It remains to calculate the positive semidefinite Fourier coefficients. We obtain the following theorems.

THEOREM 4.4. If we are in the orthogonal case, i.e., $G = SO_0(p,q)$, then $a_\beta = 0$ unless $2\beta \in \mathscr{L}$, and in this case

$$a_\beta = 2^{-nq/2} \int_{C_{2\beta}} \eta \wedge e_q^{n-t} \quad \text{for rank } \beta = t,$$

where e_q, the Euler form, is zero if q is odd and is given for $q = 2l$ by the formula

$$e_q = \left(-\frac{1}{4\pi}\right)^l \frac{1}{l!} \sum_{\sigma \in S_q} \text{sgn}(\sigma)\Omega_{\sigma(1),\sigma(2)} \cdots \Omega_{\sigma(2l-1),\sigma(2l)},$$

where

$$\Omega_{\mu\nu} = \sum_{\alpha=1}^p \omega_{\alpha\mu}\omega_{\alpha\nu}.$$

THEOREM 4.5. If we are in the unitary case, i.e., $G = U(p,q)$, then $a_\beta = 0$ unless $2\beta \in \mathscr{L}$, and in this case

$$a_\beta = i^{-nq} \int_{C_{2\beta}} \eta \wedge c_q^{n-t} \quad \text{for rank } \beta = t,$$

where c_q, the qth Chern form, is given by

$$c_q = \left(\frac{-i}{2\pi}\right)^q \frac{1}{q!} \sum_{\sigma,\bar{\sigma} \in S_q \times S_q} \text{sgn}(\sigma\bar{\sigma})\Omega_{\sigma(1),\bar{\sigma}(1)} \cdots \Omega_{\sigma(q),\bar{\sigma}(q)}.$$

In other words we alternate separately over holomorphic and anti-holomorphic indices. Here the curvature $\Omega_{\mu\nu}$ is given by

$$\Omega_{\mu\nu} = \sum_{\alpha=1}^{p} \bar{\xi}_{\alpha\mu} \wedge \xi_{\alpha\nu},$$

and $\{\xi_{\alpha\mu}\}$ is a basis for $(\mathfrak{p}^*)^+$ chosen as in [7, p. 374].

REFERENCES

1. A. Ash, *Non-square integrable cohomology of arithmetic groups*, Duke Math. J. **77** (1980), 435–449.

2. A. Borel and N. Wallach, *Continuous cohomology, discrete groups and representations of reductive groups*, Ann. of Math. Stud., vol. 94, Princeton Univ. Press, Princeton, N. J., 1980.

3. J. Cogdell, *The Weil representation and cycles on Picard modular surfaces*, Preprint.

4. M. Gaffney, *A special Stokes theorem for complete Riemannian manifolds*, Ann. of Math. (2) **60** (1954), 140–145.

5. P. Griffiths and J. Harris, *Principles of algebraic geometry*, Wiley, 1978.

6. F. Hirzebruch and D. Zagier, *Intersection numbers of curves on Hilbert modular surfaces and modular forms of Nebentypus*, Invent. Math. **36** (1976), 57–113.

7. S. Kudla and J. Millson, *The theta correspondence and harmonic forms*. I, Math. Ann. **274** (1986), 353–378.

8. ____, *The theta correspondence and harmonic forms*. II, Math. Ann. **277** (1987), 267–314.

9. ____, *Tubes, cohomology with growth conditions and an application to the theta correspondence*, Canad. J. Math. **XL** (1988), 1–37.

10. G. Lion and M. Vergne, *The Weil representation, Maslov index and theta series*, Progr. Math., vol. 6, Birkhäuser, 1980.

11. J. Millson, *Cycles and harmonic forms on locally symmetric spaces*, Canad. Math. Bull. **28** (1985), 3–38.

12. J. Millson and M. S. Raghunathan, *Geometric construction of cohomology for arithmetic groups*. I, Geometry and Analysis (Papers dedicated to the Memory of V. K. Patodi), The Indian Academy of Sciences, 1979, pp. 103–123.

13. Seminaire H. Cartan 10, *Fonctions Automorphes*, E. N. S., 1957/58.

14. Y. L. Tong and S. P. Wang, *Harmonic forms to geodesic cycles in quotients of* SU(p, q), Math. Ann. **258** (1982), 298–318.

15. ____, *Theta functions defined by geodesic cycles in quotients of* SU$(p, 1)$, Invent. Math. **71** (1983), 467–499.

16. ____, *Correspondence of Hermitian modular forms to cycles associated to* SU$(p, 2)$, J. Differential Geom. **18** (1983), 163–207.

17. ____, *Period integrals in non-compact quotients of* SU$(p, 1)$, Duke Math. J. **52** (1985), 649–688.

18. S. P. Wang, *Correspondence of modular forms to cycles associated to* O(p, q), J. Differential Geom. **22** (1985), 151–223.

19. A. Weil, *Sur certaines groupes d'operateurs unitaires*, Acta. Math. **111** (1964), 143–211.

UNIVERSITY OF MARYLAND

Proceedings of Symposia in Pure Mathematics
Volume **49** (1989), Part 2

Vector-Valued Modular Forms of Weight $\frac{g+j-1}{2}$

RICCARDO SALVATI MANNI

Introduction. The Siegel upper half-space $\mathbf{H}_g$ consists of all complex symmetric g by g matrices τ with positive definite imaginary part.

Let $[\Gamma_g, k]$ be the vector space of Siegel's modular forms of weight k with respect to Γ_g.

Siegel's Φ-operator gives a linear mapping from $[\Gamma_g, k]$ to $[\Gamma_{g-1}, k]$ in the following way: for every $\tau' \in \mathbf{H}_{g-1}$ we put

$$\Phi(\psi)(\tau') = \lim_{\lambda \to +\infty} \psi \begin{pmatrix} \tau' & 0 \\ 0 & i\lambda \end{pmatrix}.$$

For several reasons it is interesting to know when such a map is surjective. This can be done by lifting cusp forms of lower genus to modular forms of higher genus. This problem has been solved for even weights $k > 2g$ first by Maass using Poincaré series [6] and then by Klingen using Eisenstein series [5].

The restrictions on k come from two facts. First it is easy to prove that if g is odd then there are not modular forms of odd weight. Therefore for all of this paper we shall assume k even. The second restriction comes from the problems of the convergence of the above series. Recently, using the method of Hecke summation and studying the analytic continuation of the Eisenstein series, Weissauer has been able to show the surjectivity of the Φ-operator for $k \geq g + 2$ [9]. Another important related result obtained by Böcherer is the solution of the basis-problem. In fact in [1] he proved that $[\Gamma_{n,k}]$ is spanned by theta series for $k \equiv 0 \bmod 4$ and $k > 2n$. Moreover as reported in [2], using Weissauer's results there is the possibility that the last restriction can be omitted. Finally let us recall that if $k \leq (g-1)/2$ then $[\Gamma_g, k] \neq 0$ if and only if $k \equiv 0 \bmod 4$ [3]. Therefore the problem of the surjectivity of the Φ-operator for $k \equiv 2 \pmod 4$, $(g-1)/2 \leq h < k \leq g+1$, still remains open.

In this paper we disprove the surjectivity of the Φ-operator for almost all the remaining k when g is odd and greater or equal to 9. We do it by

---

1980 *Mathematics Subject Classification* (1985 *Revision*). Primary 14K24, 11F56.

generalizing the method we used in [8], where we proved the nonsurjectivity of the Φ-operator applied to $[\Gamma_g, g + 1]$ for $g \equiv 1 \bmod 4$, $g \neq 5, 13$.

1. We shall start by recalling some basic facts in the theory of characteristics. A characteristic m is a column vector with m' and m'' in $\mathbf{Z}^g$ as its first respectively second entry vectors. We put

$$e(m) = (-1)^{\,^t m' m''}.$$

We say that m is even or odd according as $e(m) = 1$ or -1. We recall the definition of the integral symplectic group Γ_g: an integral square matrix σ of degree $2g$ is in Γ_g if

$$^t\sigma \begin{pmatrix} 0 & 1 \\ -1 & 0 \end{pmatrix} \sigma = \begin{pmatrix} 0 & 1 \\ -1 & 0 \end{pmatrix}, \qquad \sigma = \begin{pmatrix} a & b \\ c & d \end{pmatrix},$$

in which 1 stands for the identity matrix 1_g of degree g.

Let $m_1, \ldots, m_r$ be a sequence of characteristics; we can think of it as an element M of $\mathrm{Mat}_{2g,r}(\mathbf{Z})$. We put

$$(1) \qquad \sigma \cdot M = \begin{pmatrix} d & -c \\ -b & a \end{pmatrix} \begin{pmatrix} M' \\ M'' \end{pmatrix} + \begin{pmatrix} \mathrm{diag}_r(c^t d) \\ \mathrm{diag}_r(a^t b) \end{pmatrix}.$$

We have denoted by "diag_r" the g by r matrix that has all the column vectors equal and each of them has as entries the diagonal coefficients of a square matrix. We recall that the above formula gives an action of Γ_g on $\mathrm{Mat}_{2g,r}(\mathbf{F}_2)$.

The following lemma proves the invariance of certain congruences under the action of Γ_g.

LEMMA 1. *Let M be in $\mathrm{Mat}_{2g,r}(\mathbf{F}_2)$, $r = 4k$, satisfying*

$$(2) \qquad M^t M \equiv 0 \quad \bmod 2 \qquad and \qquad \mathrm{diag}(M^t M) \equiv 0 \quad \bmod 4.$$

Then (2) *is preserved under the action of Γ_g.*

The proof of this lemma is similar to that of Lemma 2 in [4]. We are interested in finding solutions for these congruences using even and odd characteristics under the assumption that the odd one must be distinct.

In the following let F be the 9×11 matrix.

$$(3) \qquad F = \begin{pmatrix} 1 & 1 & 1 & 1 & 1 & 1 & 1 & 0 & 0 & 0 & 0 \\ 1 & 1 & 1 & 1 & 1 & 1 & 0 & 1 & 0 & 0 & 0 \\ 1 & 1 & 1 & 0 & 0 & 0 & 1 & 1 & 1 & 1 & 0 \\ 1 & 1 & 1 & 0 & 0 & 0 & 1 & 1 & 1 & 0 & 1 \\ 1 & 1 & 0 & 1 & 0 & 1 & 0 & 0 & 1 & 1 & 1 \\ 1 & 0 & 1 & 1 & 0 & 1 & 0 & 0 & 1 & 1 & 1 \\ 1 & 0 & 0 & 1 & 1 & 0 & 1 & 1 & 0 & 1 & 1 \\ 1 & 0 & 0 & 0 & 1 & 1 & 1 & 1 & 0 & 1 & 1 \\ 1 & 0 & 0 & 0 & 1 & 0 & 0 & 0 & 1 & 0 & 0 \end{pmatrix}$$

and $G = F + E_{2,7} - E_{2,8}$. We have denoted by $E_{i,j}$ the matrix with the (i, j)th entry equal to 1 and 0 elsewhere.

Let $e(r)$ be the square matrix of degree r with all its entries equal to 1. Then we put

$$M_8 = (1_4, e(4) - 1_4).$$

In our recent paper [8], we have considered the case in which all the m_i were odd characteristics. There we have proved that there exists at least a matrix satisfying (2) for $r \equiv 0 \bmod 4$, $r \neq 4, 12$. We point our attention to the cases in which in the matrix M occur even and odd characteristics. First of all it is easy to check that a necessary condition for M to satisfy (2) is the following: the number of the even (odd) characteristics must be even. In the proof of the next lemmas we shall use the following notation: Let A, B denote two matrices. Then $A \oplus B = \left(\begin{smallmatrix} A & 0 \\ 0 & B \end{smallmatrix}\right)$.

LEMMA 2. *Assume* $r = 4k$, $r \neq 4, 12$. *There exists at least a matrix* M, *belonging to* $\mathrm{Mat}_{2g,r}(\mathbf{F}_2)$ *with two even characteristics and all the others odd and distinct, satisfying* (2).

PROOF. We put
(4)
$$A = \begin{pmatrix} 1 & 1 & 1 & 0 & 0 & 1 & 0 & 0 \\ 1 & 1 & 0 & 1 & 0 & 0 & 1 & 0 \\ 1 & 1 & 0 & 0 & 1 & 0 & 0 & 1 \\ 1 & 0 & 1 & 1 & 1 & 0 & 0 & 0 \end{pmatrix} \quad \text{and} \quad B = \begin{pmatrix} 1 & 1 & 1 & 1 & 1 & 1 & 1 & 1 \\ 0 & 0 & 0 & 1 & 1 & 0 & 1 & 1 \\ 1 & 1 & 0 & 0 & 1 & 0 & 0 & 1 \\ 0 & 0 & 0 & 0 & 0 & 0 & 0 & 0 \end{pmatrix}.$$

Then a solution for $r = 8k$ is obtained, taking M' as the direct sum of A and $k - 1$ copies of M_8 and similarly M'' as the direct sum of B and $k - 1$ copies of M_8.

We put $M_{20} = (1_g, F)$ and $N_{20} = E_{21} - E_{22} + (1_g, G)$; then a solution for $r \equiv 4 \bmod 8$, $r \geq 20$, is obtained taking M' as the direct sum of M_{20} and copies of M_8 and M'' as the direct sum of N_{20} and copies of M_8. We observe that in this case m_2 and m_{17} are even.

LEMMA 3. *Assume* $r = 4k$. *There exists at least a matrix* M, *belonging to* $\mathrm{Mat}_{2g,r}(\mathbf{F}_2)$ *with* $2l$ $(l \geq 2)$ *even characteristics and all the others odd and distinct, satisfying* (2).

PROOF. It is similar to and even simpler than that of the lemma above. First of all let us remark that it is enough to find solutions for $l = 2$ and 3. In fact adding to them four times the vector 0 we get solutions for $l = 4$ and 5. Repeating this process we obtain solutions for all l. Let us discuss the case $l = 2$. Still it is enough to prove that there exists a solution for $r = 4$ and 8. Then for the other cases we obtain a solution considering the direct sum with copies of M_8. When $r = 4$ we have the solution $M' = M'' = 0$. When $r = 8$ we have the solution
(5)
$$M' = \begin{pmatrix} 0 & 1 & 0 & 0 & 1 & 1 & 0 & 1 \\ 0 & 0 & 1 & 0 & 1 & 0 & 1 & 1 \\ 0 & 0 & 0 & 1 & 0 & 1 & 1 & 1 \end{pmatrix} \quad \text{and} \quad M'' = \begin{pmatrix} 1 & 1 & 1 & 1 & 1 & 1 & 1 & 1 \\ 0 & 0 & 0 & 0 & 0 & 0 & 0 & 0 \\ 0 & 0 & 0 & 0 & 0 & 0 & 0 & 0 \end{pmatrix}.$$

Finally, we consider the case in which $l = 3$. Clearly we have to find a solution for $r = 8$ and 12.

In the first case a solution is the following

(6)
$$M' = \begin{pmatrix} 0 & 1 & 0 & 1 & 0 & 1 & 0 & 1 \\ 0 & 0 & 0 & 0 & 1 & 1 & 1 & 1 \end{pmatrix} \quad \text{and} \quad M'' = \begin{pmatrix} 0 & 0 & 1 & 1 & 0 & 0 & 1 & 1 \\ 0 & 0 & 0 & 0 & 0 & 0 & 0 & 0 \end{pmatrix}.$$

In the second case it is enough to add four times the vector 0 to A and B of (4).

We remark that in these two lemmas we didn't put any condition on g. Afterwards we shall consider only the cases in which g is greater or equal to the number of odd characteristics.

This doesn't give any problem; in fact, adding to a matrix M satisfying (2) as many 0 row vectors as we like, the new matrix still satisfies (2).

2. The real symplectic group $\text{Sp}(g, \mathbf{R})$ operates transitively on $\mathbf{H}_g$ as

$$\sigma \cdot \tau = (a\tau + b)(c\tau + d)^{-1}.$$

Let (V_ρ, ρ) be a finite-dimensional rational representation of $\text{GL}_g(\mathbf{C})$. Then we shall denote by $[\Gamma_g, \rho]$ the vector space of all homomorphic functions ψ on $\mathbf{H}_g$ with values in V_ρ that satisfy the functional equation

$$\psi(\sigma \cdot \tau) = \rho(c\tau + d)\psi(\tau)$$

for every σ in Γ_g (plus a condition at infinity for $g = 1$). Such functions are called vector-valued modular forms with respect to ρ and Γ_g. We shall write $[\Gamma_g, k]$ for $[\Gamma_g, \det^k]$. The weight $k(\rho)$ of ρ is, by definition, the greatest integer k such that $\rho \otimes \det^{-k}$ is a polynomial representation.

The Siegel Φ-operator can be defined also on vector-valued modular forms. In fact, we have that for every $\tau' \in \mathbf{H}_{g-r}$

$$\Phi'\psi(\tau') = \lim_{\lambda \to +\infty} \psi \begin{pmatrix} \tau & 0 \\ 0 & \lambda 1_r \end{pmatrix}$$

defines an element of $[\Gamma_{g-r}, \Phi_\rho]$. Here if $\rho = (\lambda_1, \dots, \lambda_g)$ is an irreducible representation, then $\Phi'\rho$ is the irreducible representation of $\text{GL}(g - r, \mathbf{C})$ associated to $(\lambda_1, \lambda, \dots, \lambda_{g-r})$. We shall say that ψ is a cusp form if $\Phi\psi = 0$. We shall denote by $[\Gamma_g, \rho]_0$ the space of cusp forms.

We have a decomposition

(7)
$$[\Gamma_g, \rho] = \bigoplus_{t=0}^{g} [\Gamma_g, \rho]_r,$$

where $[\Gamma_g, \rho]_0$ is the space of the cusp forms, $[\Gamma_g, \rho]'$ is its complement with respect to the Petersson's scalar product, and $[\Gamma_g, \rho]_r$ for $r \geq 1$ is defined inductively (cf. [9]) as

$$[\Gamma_g, \rho'] \cap \Phi^{-1}[\Gamma_{g-1}, \Phi\rho]_{r-1}.$$

We shall recall some of the main results of [9]. We don't state them in the more general form, but in a form that we shall use. From now on let $j < g$.

Given two irreducible representations ρ and ρ' of $GL(g, \mathbf{C})$ and $GL(j, \mathbf{C})$ respectively such that $\Phi^{g-j}\rho = \rho'$ and $\lambda_{j+1} = \lambda_{j+2} = \cdots = \lambda_g \equiv 0 \bmod 2$, then we say that ρ is a lifting of ρ'. Moreover if $\lambda_j = \lambda_n$ then we have a standard lifting.

Given a cusp form $\psi \in [\Gamma_j, \rho']_0$, then under certain assumptions we can lift ψ to an element of $[\Gamma_g, \rho]_{g-j}$.

Let us see how this works in the scalar case (i.e., $\rho = \det^k$). Let $\pi: \mathbf{H}_g \to \mathbf{H}_j$ be the projection map on the upper $j \times j$ submatrix. Then we put

$$(8) \quad E_k(\psi, \tau, s) = \sum_\sigma \psi(\pi(\sigma \cdot \tau)) \det(c\tau + d)^{-k} \left| \frac{\det(\operatorname{Im}(\pi(\sigma \cdot \tau)))}{\det(\operatorname{Im}(\sigma \cdot \tau))} \right|^{-s}.$$

Here we sum over the left coset in Γ_g with respect to the usual parabolic subgroup and s is a complex variable.

$E_k(\psi, \tau, s)$ has a meromorphic continuation as a function of s at the entire complex plane. We put

$$(9) \quad E_k(\psi, \tau) = \begin{cases} \lim_{s \to 0} E_k(\psi, \tau, s), & k \geq g + j + 1, \\ \operatorname{res}_{s=(g+j+1)/2-k} E_k(\psi, \tau, s), & k < g + j + 1. \end{cases}$$

Then we have

PROPOSITION 1. *For* $k \neq (g + j + 2)/2,\ (g + j + 3)/2,\ E_k(\psi, \tau) \in [\Gamma_g, \rho]$, *and for* $k > (n + j + 3)/2$ *we have*

$$\Phi^{n-j} E_k(\psi, \tau) = \psi.$$

When $k = g + j + 3$ we have the following

PROPOSITION 2. *Let* $k = (g + j + 3)/2$, *and let* ρ *and* $\tilde{\rho}$ *be two liftings of* ρ' *of weight* k *respectively* $k - 2$. *Let* $r = g - j$. *Then we have*

$$(10) \quad [\Gamma_j, \rho']_0 = \Phi^r[\Gamma_g, \rho]_r \oplus \Phi^r[\Gamma_g, \tilde{\rho}]_r.$$

This decomposition is orthogonal with respect to the Petersson scalar product.

From these two propositions it follows immediately that the Φ-operator is not surjective for scalar modular forms when $\Phi^r[\Gamma_g, \tilde{\rho}]_r \neq 0$. This is what we shall prove in the next section.

3. We have that $Sp(g, \mathbf{R})$ acts on $\mathbf{H}_g \times \mathbf{C}^g$ as

$$\sigma \cdot (\tau, z) = (\sigma \cdot \tau, {}^t(c\tau + d)^{-1} z).$$

We define the theta function $\vartheta_m(\tau, z)$ of characteristic m and modulus τ by

(11)

$$\vartheta_m(\tau, z) = \sum_{p \in \mathbf{Z}^g} \mathbf{C}\left(\frac{1}{2} {}^t\left(p + \frac{m'}{2}\right) \tau \left(p + \frac{m'}{2}\right) + {}^t\left(p + \frac{m'}{2}\right) \left(z + \frac{m''}{2}\right) \right)$$

in which z is a column vector of $\mathbf{C}^g$ and $\mathbf{C}(t) = \exp(2\pi i t)$. $\vartheta_m(\tau, z)$ is a holomorphic function on $\mathbf{H}_g \times \mathbf{C}^g$ and we have

$$\vartheta_{m+2n}(\tau, z) = (-1)^{{}^t m' n''} \vartheta_m(\tau, z), \qquad \vartheta_m(\tau, z) = e(m) \vartheta_m(\tau, -z)$$

for every m, n in $\mathbf{Z}^{2g}$. Therefore we can normalize the characteristics, considering m as a vector of $\mathbf{F}_2^{2g}$. We observe that the theta function $\vartheta_m(\tau, z)$ is even or odd according as the characteristic is even or odd. This parity is preserved under the action of Γ_g (i.e., $e(\sigma \cdot m) = e(m)$, for every $\sigma \in \Gamma_g$). We consider the action of Γ_g on the theta functions. We recall the transformation formula of theta functions: if $\sigma \in \Gamma_g$ we have

$$(12) \quad \vartheta_{\sigma \cdot m}(\sigma \cdot (\tau, z)) = k(\sigma) \det(c\tau + d)^{1/2} C(\Phi_m(\sigma) + \tfrac{1}{2} {}^t z (c\tau + d)^{-1} z) \vartheta_m(\tau, z)$$

in which

$$\Phi_m(\sigma) = (-\tfrac{1}{8})({}^t m'' b d m' - 2 {}^t m'' b c m'' + {}^t m''' a c m'' - 2 {}^t \operatorname{diag}(a^t b)(d m' - c m''))$$

and $k(\sigma)$ is an eighth root of the unity whose sign depends on the choice of the square root $\det(c\tau + d)^{1/2}$.

If m is an even characteristic, it is a well-known fact that the holomorphic function ϑ_m defined on $\mathbf{H}_g$ as $\vartheta_m(\tau) = \vartheta_m(\tau, 0)$ is not identically zero.

If m is an odd characteristic, the vector-valued holomorphic function ψ_m defined on $\mathbf{H}_g$ as $\psi_m(\tau) = \operatorname{grad}_z \vartheta_m(\tau, 0)$ is not identically zero.

From (12) it follows immediately that

$$(13) \qquad \vartheta_{\sigma \cdot m}(\sigma \cdot \tau) = k(\sigma) \det(c\tau + d)^{1/2} C(\phi_m(\sigma)) \vartheta_m(\tau)$$

and

$$(14) \qquad \psi_{\sigma \cdot m}(\sigma \cdot \tau) = k(\sigma) \det(c\tau + d)^{1/2} C(\phi_m(\sigma))(c\tau + d) \psi_m(\tau).$$

Let M be in $\operatorname{Mat}_{2g,g}(\mathbf{F}_2)$ with all m_i odd characteristics. We put

$$(15) \qquad D(M)(\tau) = \pi^{-g}(\psi_{m_1} \wedge \cdots \wedge \psi_{m_g})$$
$$= \det(\partial(\vartheta_{m_1} \cdots \vartheta_{m_g})/\partial(z_1, \ldots, z_g))(\tau, 0).$$

In [7] we proved that $D(M)(\tau)$ never vanishes identically provided that $m_i \neq m_j$ for $i \neq j$.

Let M be in $\operatorname{Mat}_{2g,r}(\mathbf{F}_2)$ with $m_1, \ldots, m_{2l}$ and $m_{2l+1}, \ldots, m_r$ even and odd distinct characteristics. Then it is a direct consequence of the above statement that

$$(16) \quad W(M)(\tau) = \pi^{-2r+4l}(\vartheta_{m_1} \cdots \vartheta_{m_{2l}})^{2t}(\psi_{m_{2l+1}} \wedge \cdots \wedge \psi_{m_r})(\psi_{m_{2l+1}} \wedge \cdots \wedge \psi_{m_r})$$

doesn't vanish identically provided that $g \geq r - 2l$.

$W(M)(\tau)$ has the following transformation formula

$$(17) \qquad W(\sigma \cdot M)(\sigma \cdot \tau) = k(\sigma)^{2r} \det(c\tau + d)^{r+2} C\left(2 \sum_{i=1}^{r} \phi_{m_i}(\sigma)\right)$$
$$\times \rho^{*[g-r+2l]}(c\tau + d) W(M)(\tau).$$

We have denoted by $\rho^{*[\mu]}$ the contragradient representation of the irreducible representation of $\operatorname{GL}(g, \mathbf{C})\rho^{[\mu]}$, having highest weight vector

$$(\underbrace{2, 2, \ldots, 2}_{\mu \text{ times}}, 0, \ldots, 0).$$

We shall denote by $\Gamma_g(2)$ the normal subgroup of Γ_g defined by $\sigma \equiv 1_{2g}$ mod 2. It can be verified as in [8] that if the matrix of the characteristics M satisfies (2) then $W(M)(\tau)$ is a vector-valued modular form with respect to $\Gamma_g(2)$. Moreover as we have done in [8] we can symmetrize such $W(M)(\tau)$. Before we state the theorem let us fix our notation. We put $\tilde{\rho} = \det^{r+2} \otimes \rho^{-[g-r+2l]}$. Then we have

THEOREM 1. *Let M be in* $\mathrm{Mat}_{2g,r}$ *with $2l$ even characteristics and all the others odd and distinct satisfying* (2). *Then the vector-valued modular form*

$$(18) \qquad \psi(\tau) = \sum_{\sigma \in \mathrm{Sp}(g,\mathbf{F}_2)} \tilde{\rho}(c\tau + d)^{-1} W(M)(\sigma \cdot \tau)$$

doesn't vanish identically and is in $[\Gamma_g, \tilde{\rho}]$. *Moreover $\Phi^{g-r+2l}\psi(\tau)$ is a nonvanishing element of* $[\Gamma_{r-2l}, r+2]_0$.

Let us derive some consequences from the above theorem.

COROLLARY 1. *Assume $g \equiv 3$ mod 4, $g \neq 7, 15$. Then*

$$\Phi: [\Gamma_g, g-1]_5 \to [\Gamma_{g-1}, g-1]_4$$

is not surjective.

PROOF. We take $l = 1$ and $g = r + 3$. Then $\tilde{\rho}$ becomes the irreducible representation with highest weight

$$(g-1, g-1, \ldots, g-1, \underbrace{g-3, \ldots, g-3}_{5 \text{ times}}).$$

Applying Proposition 2 we get

$$[\Gamma_{g-5}, g-1]_0 = \Phi^5 [\Gamma_g, g-1]_5 \oplus \Phi^5 [\Gamma_{g-1}, \tilde{\rho}]_5.$$

Since the second summand is different from 0, then Φ^5 from $[\Gamma_q, g-1]_5$ to $[\Gamma_{g-5}, g-1]_0$ is not surjective; but now applying Proposition 1 we see immediately that all the Φ^k from $[\Gamma_{g-5+k}, g-1]_k$ to $[\Gamma_{g-5}, g-1]_0$ are surjective for $k \leq 4$. Clearly the thesis follows from this.

Using the results of Lemma 3 and taking $r = g - 2l - 1$, we can prove, as above, the following:

COROLLARY 2. *Assume $g \equiv 1$ mod 4, $g \geq 9$, $l \geq 2$ even. Then*

$$\Phi: [\Gamma_g, g+1-2l]_{4l+1} \to [\Gamma_{g-1}, g+1-2l]_{4l}$$

is not surjective.

COROLLARY 3. *Assume $g \equiv 3$ mod 4, $g \geq 15$, $l \geq 3$ odd. Then*

$$\Phi: [\Gamma_g, g+1-2l]_{4l+1} \to [\Gamma_{g-1}, g+1-2l]_{4l}$$

is not surjective.

We conclude with some remarks. First let us recall that the case $l = 0$ has been already discussed in [8].

Second, when $g = 4l + 1$, the element of $[\Gamma_{g-1}, 2l + 2]_{g-1}$ is none other than the Eisenstein series (cf. [3]).

Finally when $g \equiv 3 \bmod 8$ our method does not work for analyzing $[\Gamma_g, (g + 1)/2]$.

REFERENCES

1. S. Böcherer, *Über die Fourier-Jacobi-Entwicklung Siegelscher Eisensteinreihen*, Math. Z. **183** (1983), 21–46.

2. ____, *Über die Funktionalgleichung automorpher L-Funktionen zur Siegelschen Modulgruppe*, J. Reine Angew. Math. **362** (1985), 146–168.

3. E. Freitag, *Stabile Modulformen*, Math. Ann. **230** (1977), 197–211.

4. J. I. Igusa, *On Jacobi's derivative formula and its generalizations*, Amer. J. Math. **102** (1980), 409–446.

5. H. Klingen, *Zum Darstellungssatz für Siegelsche Modulformen*, Math. Z. **102** (1967), 30–43 and **105** (1968), 399–400.

6. H. Maass, *Über die Darstellung der Modulformen n-ten Grades durch Poincaresche Reihen*, Math. Ann. **123** (1951), 125–151.

7. R. Salvati Manni, *On not identically zero Nullwerte of Jacobians of theta functions with odd characteristics*, Adv. in Math. **47** (1983), 88–104.

8. ____, *Holomorphic differential forms of degree N − 1 invariant under Γ_g*, Preprint.

9. R. Weissauer, *Stabile Modulformen und Eisensteinreihen*, Lecture Notes in Math., vol. 1219, Springer-Verlag, 1986.

UNIVERSITÀ LA SAPIENZA, ITALY

Proceedings of Symposia in Pure Mathematics
Volume **49** (1989), Part 2

Some Nonzero Harmonic Forms
and Their Geometric Duals

YUE LIN LAWRENCE TONG

Introduction. I shall discuss some recent joint work with S. P. Wang on harmonic forms and their dual geometric cycles in locally symmetric spaces, and state some possible generalizations as well as related open problems. The pioneering work in this subject is the paper of Hirzebruch and Zagier [**HZ**]. My involvement began with a question of Zagier about geometric interpretation of Fourier coefficients of modular forms of higher weight that can be lifted to Hilbert modular forms. It turned out to be interpretable in terms of an intersection theory with a coefficient bundle [**T**]. The idea was then liftings, or correspondences defined by theta functions, between reductive dual pairs should contain geometric information. More precisely certain lifted forms should be harmonic forms that are Poincaré dual to specific cycles. Kudla and Millson [**KM1**] succeeded in showing this for the pair $(O(p, 1), Sp(m, \mathbf{R}))$ and we found analogous results for $(U(p, 1), U(r, r))$ [**TW1**]. Somewhat earlier Oda [**Od**] had done related work on the pair $(O(p, 2), Sp(1, \mathbf{R}))$ although not formulated in geometric terms.

It was soon realized that the technicalities of directly generalizing those results to $O(p, q)$ or $U(p, q)$ are rather formidable. For example the construction of harmonic duals required analytic continuation techniques, which are complicated even for $U(p, 1)$. We found however the geometric interpretations of Fourier coefficients of modular forms can be achieved by using some invariant theory, and without constructing the harmonic duals [**TW2**]. Later we realized that the analytic continuation problem can be avoided by introducing coefficient bundles whose harmonic forms have good convergence properties. The appropriate generalization of this to higher rank groups was finally found in [**TW3**], after a computational approach generalizing [**TW2**] was carried out in [**W**]. The viewpoint of Rallis and Schiffmann [**RS**] on the construction of theta functions also helped to simplify some technical problems and give more direct representation-theoretic descriptions. Kudla and

---

1980 *Mathematics Subject Classification* (1985 *Revision*). Primary 11F75.

151

Millson on the other hand found another construction of the Poincaré duals [KM2] by making more intrinsic use of the reductive dual pairs [H1].

The above results for $O(p,q)$ or $U(p,q)$ have one serious limitation. The cycles must be of positive type (cf. §5.2), as technical problems got out of hand when one tried to consider mixed cycles. More or less similar difficulties also appear in [RSW, PR]. On the other hand we had known for some time that the unitary representations associated to the harmonic forms (of positive or mixed cycles) can be constructed in a uniform way by Flensted-Jensen's methods in discrete series for semisimple symmetric spaces [FJ1]. The idea became then to derive the harmonic forms and their geometric properties directly from the Flensted-Jensen functions. In fact this also suggests that geometric interpretations of harmonic forms are not restricted to dual pairs, but may be considered for any unitary representation with cohomology. The arguments needed to link Flensted-Jensen's construction to geometric cycles turned out to be mostly a translation of the arguments in [TW3] into the language of discrete series. The deeper results that we need have been supplied by the detailed work of Oshima and Matsuki [OM]. It was a pleasant surprise to find how readily prepared the entire theory of discrete series is for our application [TW4].

The lifting techniques also turned out to give results on nonvanishing of cohomology of discrete groups. In fact some versions of theta correspondence are implicit in the papers [K, BW, An, S] on nonvanishing of holomorphic differential forms. The paper [TW5] generalizes these results to type (k,l) cohomology where possibly $k \neq l$ and $kl \neq 0$. We make use of the results of [Ad, H2, KV] to find out the Zuckerman modules of $U(p,q)$ that correspond to the holomorphic discrete series of $U(r,r)$ in the pair $(U(p,q), U(r,r))$. The nonvanishing depends on a density argument of Kazhdan, Borel, and Wallach. The same argument can be adapted to harmonic forms constructed from discrete series for semisimple symmetric spaces [TW4].

1. Some problems about cohomology of discrete groups.

1.1. Let G be a real semisimple noncompact Lie group, K a maximal compact subgroup of G, and Γ a cofinite discrete subgroup of G. Let $D = G/K$ be the symmetric space, V a finite-dimensional irreducible G module, and E the associated locally constant vector bundle on $\Gamma\backslash D$. Then

$$(1) \qquad H^*(\Gamma, V) = H^*(\Gamma\backslash D, E) = H^*(\mathfrak{g}, K; C^\infty(\Gamma\backslash G) \otimes V),$$

where $\mathfrak{g}$ is the Lie algebra of G. If Γ is cocompact, then $L^2(\Gamma\backslash G)$ is a Hilbert space direct sum

$$L^2(\Gamma\backslash G) = \bigoplus_{\pi \in \hat{G}} m(\pi, \Gamma) H_\pi$$

with finite multiplicities $m(\pi, \Gamma)$. Then (1) can be expressed more precisely as

$$(2) \qquad H^*(\Gamma, V) = \bigoplus_{\pi \in \hat{G}} m(\pi, \Gamma) H^*(\mathfrak{g}, K; H_\pi \otimes V).$$

The representations $\pi \in \hat{G}$ such that $H^*(\mathfrak{g}, K; H_\pi \otimes V) \neq 0$ are referred to as representations with (nonzero) cohomology. These representations have been classified, and their Harish-Chandra modules can be constructed by Zuckerman's technique of derived functor modules [VZ,V]. A representation with cohomology π is uniquely determined by a single K type δ of π that lies in $\mathrm{Hom}_{\mathbb{C}}(V, \Lambda^* \mathfrak{p})$. In fact much representation-theoretic information of π can be calculated from δ. In particular the lowest degree $* = R$ for which $H^*(\mathfrak{g}, K; H_\pi \otimes V) \neq 0$ is determined.

1.2. We consider two questions which arise from (1) and (2).

I. If $\pi \in \hat{G}$ is a representation with cohomology, does it occur in the cohomology of a discrete group Γ?

This amounts to finding an imbedding

$$(1) \qquad H^*(\mathfrak{g}, K; H_\pi \otimes V) \to H^*(\mathfrak{g}, K; C^\infty(\Gamma \backslash G) \otimes V),$$

and it is referred to as a question about the nonvanishing of cohomology of Γ. If π belongs to the discrete series of G, then $H^R(\mathfrak{g}, K; H_\pi \otimes V) \neq 0$ only when $R = \frac{1}{2} \dim D$, and there are methods to estimate $m(\pi, \Gamma)$. Thus the interest here is to show examples where (1) is nontrivial for π not in the discrete series of G, e.g., R not in the middle dimension.

1.3. Our second question suggests that besides the representation information there may also be a considerable amount of geometric information coded into the representations with cohomology. We shall only formulate the question when Γ is cocompact in G so this is assumed here.

II. Given a representation with cohomology $\pi \in \hat{G}$, let $\alpha \in H^*(\Gamma \backslash D, E)$ be a cohomology class associated to π that is constructed from a suitable imbedding (3) independent of Γ. Then α has the K type δ (cf. 3.5(2)) that is the minimal K type of π. Is there a canonical way, *independent of* Γ, to find a geometric representative for α: namely, construct a geometric cycle C in $\Gamma \backslash D$ together with a locally constant section s of the bundle E that is supported on C. The pair (C, s) represents α in the sense of Poincaré (or Serre) duality: for all $\beta \in H^{\dim D - *}(\Gamma \backslash D, E^*)$ that has the K type δ^* contragredient to δ,

$$(1) \qquad \int_{\Gamma \backslash D} \alpha \wedge \beta = \int_C s(C^*(\beta)).$$

The supporting evidence for II is the paper [TW4]. The point is to find a suitable model of H_π that not only enables one to construct cohomology classes $\alpha \in H^*(\Gamma \backslash D, E)$, but also the cycles and sections.

If $\Gamma \backslash D$ is not compact, there will in general be correction terms in (1) coming from the cusps.

2. Some nonzero cohomology constructed from dual pair correspondences.

2.1. The first nonvanishing result we describe makes use of the theory of reductive dual pairs. Thus it applies only to the groups $\mathrm{U}(p, q)$, $\mathrm{O}(p, q)$, $\mathrm{Sp}(p, q)$. We will only describe the results for $G = \mathrm{U}(p, q)$ as in [TW5]. The other two cases can be formulated similarly and have the same proofs. We use the notations in [Ad, TW5] except that the Lie algebra of G is denoted by $\mathfrak{g}$ (rather than $\mathfrak{g}_0$) and its complexification is denoted by $\mathfrak{g}_\mathbb{C}$. Similarly for the subalgebras of $\mathfrak{g}$. Next let $G' = U(r,r)$, let $\mathfrak{g}'$ be its Lie algebra, etc. The groups G and G' are given by

$$G = \left\{ g \in GL(n, \mathbb{C}) | {}^t\bar{g} \begin{pmatrix} I_p & 0 \\ 0 & -I_q \end{pmatrix} g = \begin{pmatrix} I_p & 0 \\ 0 & -I_q \end{pmatrix} \right\},$$

$$G' = \left\{ g \in GL(2r, \mathbb{C}) | {}^t\bar{g} \begin{pmatrix} 0 & I_r \\ -I_r & 0 \end{pmatrix} g = \begin{pmatrix} 0 & I_r \\ -I_r & 0 \end{pmatrix} \right\}.$$

Let $(a_1, \ldots, a_p; b_1, \ldots, b_q)$ denote the matrix

$$\mathrm{diag}(\sqrt{-1}a_1, \ldots, \sqrt{-1}a_p, \sqrt{-1}b_1, \ldots, \sqrt{-1}b_q).$$

The elliptic coadjoint orbits of G in $\mathfrak{g}^*$ are parametrized by $\mathcal{O}_\lambda$ with

$$\lambda = (a_1, \ldots, a_p; b_1, \ldots, b_q)$$

satisfying $a_1 \geq \cdots \geq a_p$, $b_1 \geq \cdots \geq b_q$. Let $(c_1, \ldots, c_r; d_1, \ldots, d_r)$ denote the matrix

$$\alpha \, \mathrm{diag}(\sqrt{-1}c_1, \ldots, \sqrt{-1}c_r, \sqrt{-1}d_1, \ldots, \sqrt{-1}d_r)\alpha^{-1},$$

where

$$\alpha = \frac{1}{\sqrt{2}} \begin{pmatrix} I_r & \sqrt{-1}I_r \\ \sqrt{-1}I_r & I_r \end{pmatrix}.$$

The elliptic coadjoint orbits of G' in $\mathfrak{g}'$ are parametrized by $\mathcal{O}_{\lambda'}$ with $\lambda' = (c_1, \ldots, c_r; d_1, \ldots, d_r)$ satisfying $c_1 \geq \cdots \geq c_r$, $d_1 \geq \cdots \geq d_r$.

2.2. Let $\rho = \rho(u(\lambda))$, defined as in [TW5, §5]. Consider

(1) $\lambda = (a_1, \ldots, a_t, 0 \cdots 0, -b_r, \ldots, -b_1; 0 \cdots 0, -d_{r-t}, \ldots, -d_1),$

where $a_1 > \cdots > a_t > 0$, $b_1 > \cdots > b_r > 0$, $d_1 > \cdots > d_{r-t} > 0$ and $r > 0$, $0 \leq t \leq r$, such that $r - t \leq q$, and $2r \leq p$. Let

(2) $\lambda' = \det^{n/2} \bigotimes (-b_r, \ldots, -b_1; a_1, \ldots, a_t, -d_{r-t}, \ldots, -d_1).$

Now we have

(3) $\lambda - \rho = (A_1, \ldots, A_t, \overbrace{-(r-t)}^{p-t-r}, -B_r, \ldots, -B_1; \overbrace{-(r-t)}^{q-r+t}$

$$- (D_{r-t} + 2r), \ldots, -(D_1 + 2r)),$$

where $a_i = n/2 + A_i - (2i-1)/2$, $b_i = n/2 + B_i - (2i-1)/2$, and $d_i = n/2 + D_i + (2r - (2i-1))/2$, $n = p + q$. Assume $\lambda - \rho$ is the highest weight of a finite-dimensional irreducible representation of $\mathfrak{g}_\mathbb{C}$. Then

$$A_1 \geq \cdots \geq A_t, \quad B_1 \geq \cdots \geq B_r, \quad D_1 \geq \cdots \geq D_{r-t},$$

and all these numbers are integers. Let $A(\lambda)$ (resp. $A(\lambda')$) be the Zuckerman derived functor module associated to $A(\lambda')$ in the dual pair $(U(p,q), U(r,r))$. We assume $A(\lambda')$ belongs to the holomorphic discrete series: the condition is that $b_r > d_1$, or equivalently $B_r \geq D_1 + 2r$. The minimal K type μ_λ of $A(\lambda)$ is given by

$$(4) \quad \mu_\lambda = (q + A_1, \ldots, q + A_t, \overbrace{O, \ldots, O}^{p-t-r} - (q + B_r), \ldots, -(q + B_1);$$

$$\overbrace{O, \ldots, O}^{q+t-r}, -(p + D_{r-t}), \ldots, -(p + D_1)).$$

The minimal K type $\mu_{\lambda'}$ of $A(\lambda')$ is given by

$$(5) \quad \mu_{\lambda'} = (-B_r, \ldots, -B_1; n + A_1, \ldots, n + A_t, -D_{r-t}, \ldots, -D_1).$$

Let $n(\lambda)$ be the number defined in [**TW5**, 14.15(ii)].

2.3. THEOREM [**TW5**]. *Let $r > 0$, and $0 \leq t \leq r$ such that $r + t \leq p$ and $r - t \leq q$. Let $\nu = \lambda - \rho$ be given by 2.2(3) and let $V_{-\nu}$ be the finite-dimensional irreducible representation of $\mathfrak{g}_{\mathbf{C}}$ of lowest weight $-\nu$. Let μ_λ be given by 2.2(4).*

(i) If $4r \leq p + q$, there exists a compact discrete subgroup Γ of $SU(p,q)$ that has a nonzero cohomology class in

$$(1) \qquad\qquad H^{tq + (p-t-r)(r-t), rq}(\Gamma, V_{-\nu}).$$

Furthermore the nonzero class is represented by a harmonic form that is K isotypic and the K type is μ_λ.

(ii) If $n(\lambda) > nr$, $B_r - D_1 \geq 4r$, and $2r < p$, then there exists a cofinite discrete subgroup Γ of $U(p,q)$ such that exactly the same conclusions as in (i) hold for the cohomology of Γ. Furthermore the harmonic form is square integrable on $\Gamma \backslash D$.

2.4. The proof of Theorem 2.3 depends on the following ingredients.

A. Since $A(\lambda')$ is a holomorphic discrete series, the standard constructions of Poincaré series give automorphic forms φ on $\Gamma' \backslash G'$, for discrete subgroups $\Gamma' \subset G'$, whose right translates by G' give a copy of $A(\lambda')$.

B. Let W_k be a vector space over k, an imaginary quadratic extension of $\mathbf{Q}$, and assume W_k has a Hermitian form that has signature (p, q) on $W = W_k \otimes_{\mathbf{Q}} \mathbf{R}$. Let L be an $\mathcal{O}$ (ring of integers in k) lattice in W_k^r contained in its dual lattice L^*. A theta series θ is formed by summing a vector-valued Schwartz function Ψ of W^r over a coset of L in L^*. Ψ takes values in a vector space determined by the representations μ_λ and $\mu_{\lambda'}$. Ψ is not in the discrete spectrum of the oscillator representation restricted to $G \cdot G'$. However it is possible to calculate for suitable congruence subgroups $\Gamma' \subset U(r, r)(\mathcal{O})$ the pairing

$$(1) \qquad\qquad \mathcal{L}(\varphi) = \int_{\Gamma' \backslash G'} \langle \varphi, \theta \rangle \, dg'.$$

This is a basic calculation that has been used by various authors on liftings. Here we need some control over the growth of φ in fundamental domains of Γ'.

C. Since φ is of type $A(\lambda')$, one knows from 2.2 that $\mathscr{L}(\varphi)$ is of type $A(\lambda)$. The formula B.(1) allows one to adapt an argument of Kazhdan, Borel, and Wallach [**BW**] to conclude that there exists a lattice L and a coset of L in L^* for which $\mathscr{L}(\varphi) \neq 0$.

2.5. Simpler versions of the above argument had been used previously by Kazhdan, Shimura, Borel-Wallach, Anderson [**An, BW, K, S**] to get nonzero holomorphic differential forms on locally symmetric spaces.

3. Nonzero cohomology constructed from discrete series for semisimple symmetric spaces.

3.1. Let $\mathfrak{g}$ be a semisimple Lie algebra and σ an involution of $\mathfrak{g}$. Fix a Cartan involution θ such that $\sigma\theta = \theta\sigma$. Let $\mathfrak{g} = \mathfrak{h} + \mathfrak{q}$ (resp. $\mathfrak{g} = \mathfrak{k} + \mathfrak{p}$) be the decomposition of $\mathfrak{g}$ into $+1$ and -1 eigenspaces for σ (resp. θ). Let $\mathfrak{g}_{\mathbf{C}}$ be the complexification of $\mathfrak{g}$ and let $\mathfrak{g}^d$, $\mathfrak{k}^d$, $\mathfrak{h}^d$ be subalgebras in $\mathfrak{g}_{\mathbf{C}}$ defined by

$$\mathfrak{g}^d = \mathfrak{k} \cap \mathfrak{h} + \sqrt{-1}(\mathfrak{k} \cap \mathfrak{q}) + \sqrt{-1}(\mathfrak{p} \cap \mathfrak{h}) + \mathfrak{p} \cap \mathfrak{q},$$

$$\mathfrak{k}^d = \mathfrak{k} \cap \mathfrak{h} + \sqrt{-1}(\mathfrak{p} \cap \mathfrak{h}), \qquad \mathfrak{h}^d = \mathfrak{k} \cap \mathfrak{h} + \sqrt{-1}(\mathfrak{k} \cap \mathfrak{q}).$$

Let $G_{\mathbf{C}}$ be a connected complex Lie group with Lie algebra $\mathfrak{g}_{\mathbf{C}}$ and let G, K, H, G^d, K^d, H^d be the analytic subgroups of $G_{\mathbf{C}}$ corresponding to $\mathfrak{g}$, $\mathfrak{k}$, $\mathfrak{h}$, $\mathfrak{g}^d$, $\mathfrak{k}^d$, $\mathfrak{h}^d$ respectively. Let $\mathfrak{a}$ be a θ invariant Cartan subspace for G/H, i.e., $\mathfrak{a}$ is maximal abelian in $\mathfrak{q}$, and $\mathfrak{a} = \mathfrak{a} \cap \mathfrak{k} + \mathfrak{a} \cap \mathfrak{p}$. Then $\mathfrak{a}^d = \sqrt{-1}(\mathfrak{a} \cap \mathfrak{k}) + \mathfrak{a} \cap \mathfrak{p}$ is a Cartan subspace for G^d/K^d. Let $\Sigma(\mathfrak{a}_{\mathbf{C}}, \mathfrak{g}_{\mathbf{C}})$ be the restricted root system of $\mathfrak{a}_{\mathbf{C}}$ in $\mathfrak{g}_{\mathbf{C}}$ and Σ^+ a positive system. W denotes the Weyl group of Σ. For $\lambda \in \mathfrak{a}_{\mathbf{C}}^*$ the Harish-Chandra isomorphism determines algebra homomorphisms

$$\chi_\lambda : \mathbf{D}(G/H) \to \mathbf{C}, \quad \chi_\lambda^d : \mathbf{D}(G^d/K^d) \to \mathbf{C},$$

where $\mathbf{D}(G/H)$ (resp. $\mathbf{D}(G^d/K^d)$) are the invariant differential operators on G/H (resp. G^d/K^d). Let $\hat{K}$ (resp. $\hat{H}^d$) be the set of equivalence classes of finite-dimensional irreducible representations of $\hat{K}$ (resp. $\hat{H}^d$), and let $\hat{H}^d(K) \subset \hat{H}^d$ be those that are restrictions of holomorphic representations of $K_{\mathbf{C}}$. Then we may identify $\hat{K}$ and $\hat{H}^d(K)$. Let

$$\mathscr{A}_{\lambda,\delta}(G/H) = \{f \text{ is analytic on } G/H | f \text{ transforms according to}$$
$$\delta \text{ under } K \text{ action and } Df = \chi_\lambda(D)f \text{ for all } D \in \mathbf{D}(G/H)\}.$$

$\mathscr{A}_{\lambda,\delta}(G^d/K^d)$ is defined analogously where δ is an $\hat{H}^d(K)$ type. The duality principle of Flensted-Jensen gives an isomorphism

$$\eta : \mathscr{A}_{\lambda,\delta}(G/H) \to \mathscr{A}_{\lambda,\delta}(G^d/K^d).$$

3.2. Let $P^d = M^d A^d N^d$ be the minimal parabolic subgroup of G^d determined by $\mathfrak{a}^d$ and $\Sigma(\mathfrak{a}_{\mathbf{C}}, \mathfrak{g}_{\mathbf{C}})^+$. Let $\rho \in \mathfrak{a}_{\mathbf{C}}^{d*}$ be given by $\rho(Y) = \frac{1}{2}\mathrm{tr}(\mathrm{ad}(Y)|_{\mathfrak{n}^d})$ for $Y \in \mathfrak{a}^d$. For $\delta \in \hat{H}^d(K)$ and $\lambda \in \mathfrak{a}_{\mathbf{C}}^*$ put

$$\mathscr{B}_\delta(G^d/p^d; L_\lambda) = \{f \text{ is a hyperfunction on } G^d : f \text{ transforms according to}$$
$$\delta \text{ under } H^d, \text{ and } f(xman) = a^{\lambda-\rho}f(x),$$
$$x \in G^d, \ m \in M^d, \ a \in A^d, \ n \in N^d\}.$$

Then there is a G^d equivariant Poisson transform

$$\mathscr{P}_\lambda : \mathscr{B}_\delta(G^d/p^d; L_\lambda) \to \mathscr{A}_{\lambda,\delta}(G^d/K^d).$$

Now assume $\operatorname{rank}(G/H) = \operatorname{rank}(K/K \cap H)$. Then $\mathfrak{a}$ can be chosen in $\mathfrak{k}$. The Weyl group W is given by

$$W = N_{K^d}(\mathfrak{a}^d)/Z_{K^d}(\mathfrak{a}^d).$$

Let $W_{K \cap H} = N_{K \cap H}(\mathfrak{a}^d)/Z_{K \cap H}(\mathfrak{a}^d)$. The closed H^d orbits on G^d/P^d are given by [**Ma**]:

$$\mathscr{O}_{\overline{w}} = H^d w P^d / P^d, \qquad \overline{w} = W_{K \cap H} w \in W_{K \cap H} \backslash W.$$

To a closed orbit $\mathscr{O}_{\overline{w}}$ and $\lambda \in \mathfrak{a}_\mathbb{C}^*$ put

$$\mathscr{B}_{H^d}^{\overline{w}}(G^d/P^d; L_\lambda) = \{ f \in \mathscr{B}_{H^d}(G^d/P^d; L_\lambda) | \operatorname{supp} f \subset \mathscr{O}_{\overline{w}} \},$$

where $\mathscr{B}_{H^d}(G^d/P^d; L_\lambda) = \bigoplus_{\delta \in \hat{H}^d(K)} \mathscr{B}_\delta(G^d/P^d; L_\lambda)$.

$$\mathscr{V}_{w(\lambda)} = \mathscr{V}(\mathfrak{a}, w(\Sigma^+), w(\lambda)) = \eta^{-1} \mathscr{P}_\lambda(\mathscr{B}_{H^d}^{\overline{w}}(G^d/P^d; L_\lambda)).$$

In particular for the delta function $1_{\mathscr{O}_{\overline{w}}}$ supported on the orbit $\mathscr{O}_{\overline{w}}$ when it is H^d finite one denotes the Flensted-Jensen function by

$$\psi_{w,\lambda} = \eta^{-1} \mathscr{P}_\lambda(1_{\mathscr{O}_{\overline{w}}}).$$

3.3. Let Ω_X be the set of parameters for discrete series in $L^2(G/H)$ [**FJ2**, p. 62]. Assume $(w, \lambda) \in \Omega_X$ and furthermore that

$$(1) \qquad\qquad \langle \lambda, \alpha_i \rangle \geq 4 \langle \rho, \rho \rangle, \qquad \alpha_i \in \Sigma^+.$$

Then $\psi_{w,\lambda} \in L^1(G/H)$ [**TW4**, Proposition 2.4]. The condition (1) is by no means the best possible. We stated it for simplicity. Weaker conditions can be given but then more detailed discussions would be necessary.

Let ρ_c be half the sum of positive roots in $\Sigma_c = \Sigma(\mathfrak{a}_\mathbb{C}, \mathfrak{k}_\mathbb{C})$ where Σ_c^+ is chosen compatibly with $\Sigma(\mathfrak{a}_\mathbb{C}, \mathfrak{g}_\mathbb{C})^+$. Let $\mu_\lambda^w = w(\lambda + \rho) - 2\rho_c$. Then $-\mu_\lambda^w$ is the lowest weight of the unique minimal K type of $\mathscr{V}_{w(\lambda)}$.

3.4. For suitable choices of $k_i \in K$, let

$$f_i(g) = \psi_{w,\lambda}^v(g k_i)$$

be a basis of the representation space of $\tau_{-\mu_\lambda^w}(K)$, where $\psi \to \psi^v$ is the G equivariant isomorphism $\mathscr{A}_{\lambda,\delta}(G/H) \to \mathscr{A}_\lambda(H/G)_\delta$ given by $\psi^v(x) = \psi(x^{-1})$. Then $\overline{f_i(g)}$ is a basis of $\tau_{\mu_\lambda^w}$ with highest weight μ_λ^w. Let $R = \dim G/K - \dim H/H \cap K$, $\nu = \lambda - \rho$. Then

$$(1) \qquad\qquad (\Lambda^R \mathfrak{p}^* \otimes V_{-w\nu})_{-\mu_\lambda^w},$$

the isotypic component with lowest weight $-\mu_\lambda^w$, occurs with multiplicity one in $\Lambda^R \mathfrak{p}^* \otimes V_{-w\nu}$. Let $\{X_i\}$ be a basis of this component that is dual to $\{\overline{f_i}\}$. We define

$$\tilde{\omega}_\lambda^w(g) = \sum_i \overline{f_i(g)} X_i.$$

Then $\tilde{\omega}_\lambda^w(g)$ is the pull-back of a differential form

$$\omega_\lambda^w(g) \in C^\infty(D, \Lambda^R T^* \otimes E_{-w\nu}),$$

where $E_{-w\nu}$ is the vector bundle associated to $V_{-w\nu}$.

3.5. Let $\Gamma \subset G$ be a cofinite torsion free discrete subgroup such that $\Gamma \cap H/H$ also has finite volume. Let

$$(1) \qquad \hat{\omega}_\lambda^w = \sum_{\gamma \in \Gamma \cap H/\Gamma} \gamma^* \omega_\lambda^w.$$

The fact that $\psi_{w,\lambda} \in L^1(G/H)$ guarantees that $\hat{\omega}_\lambda^w$ converges absolutely and the convergence is uniform on compact subsets of D [**TW4**, §3]. This is similar to an argument of Borel and Godement [**Bo**].

From the infinitesimal character of $\mathcal{V}_{w(\lambda)}$, one concludes easily that ω_λ^w and thus $\hat{\omega}_\lambda^w$ is a harmonic form. Furthermore

$$(2) \qquad \hat{\omega}_\lambda^w \in C^\infty(\Gamma/D, \Lambda^R T^* \otimes E_{-w\nu})_{-\mu_\lambda^w},$$

where on $\Gamma\backslash D$ the K type condition is the requirement that for each $x \in \Gamma\backslash D$, $\hat{\omega}_\lambda^w(x) \in (\Lambda^R T_x^* \otimes (E_{-w\nu})_x)_{-\mu_\lambda^w}$. Here the isotropy group K_x acts on the left of the fibers at x, and there is a decomposition into irreducibles

$$(3) \qquad \Lambda^* T_x^* \otimes (E_{-w\nu})_x = \bigoplus_\delta (\Lambda^* T_x^* \otimes (E_{-w\nu})_x)_\delta.$$

We use the same notation to denote cohomology classes that are represented by harmonic forms that have a particular K type. Thus

$$[\hat{\omega}_\lambda^w] \in H^R(\Gamma\backslash D, E_{-w\nu})_{-\mu_\lambda^w}.$$

THEOREM [**TW4**]. *Let G be a semisimple algebraic group defined over* **Q**, *and assume the involution σ is also defined over* **Q**. *Let Γ be an arithmetic subgroup. Then there exist an element $\gamma_1 \in \Gamma$ and a subgroup $\Gamma_1 \subset \Gamma$ of finite index such that*

$$(4) \qquad \sum_{\gamma \in \Gamma \cap \gamma_1^{-1} H \gamma_1 \backslash \Gamma} \gamma^* \gamma_1^* \omega_\lambda^w \neq 0.$$

In particular if G is anisotropic over **Q**, *then*

$$H^R(\Gamma_1\backslash D, E_{-w\nu})_{-\mu_\lambda^w} \neq 0.$$

3.6. In case $\Gamma\backslash D$ has finite volume, to conclude the nonvanishing of cohomology from 3.5(4), one should either show the form is square integrable on $\Gamma\backslash D$ or calculate a nonzero period.

The nonvanishing results in Theorem 3.5 and Theorem 2.3 are different even when applied to the same group $SU(p,q)$. For example, let σ be the involution on $SU(p,q)$, which is conjugation by

$$\begin{pmatrix} I_{p-r} & & \\ & -I_{r+s} & \\ & & I_{q-s} \end{pmatrix}.$$

Then $H = S(U(p - r, q - s) \times U(r, s))$. Choose $w = e$ and $(e, \lambda) \in \Omega_X$. Then it is easy to see that μ_λ is of the form

$$
\mu_\lambda = \left(q + A_1, \ldots, q + A_r, \overbrace{0, \ldots, 0}^{p-2r}, -(q + B_r), \ldots, -(q + B_1) ; \right.
$$

$$
p - 2r + C_1 \ldots, p - 2r + C_s, \overbrace{0, \ldots, 0}^{q-2s},
$$

$$
\left. -(p - 2r + D_s), \ldots, -(p - 2r + D_1) \right).
$$

Comparing with 2.2(4) it is clear that if $rs \neq 0$, this K type is not covered in Theorem 2.3.

On the other hand for a general pair (p, q) where $p \neq q$, one sees from Berger's list [**Be**, p. 158] that the only candidates for H are

$$
S(U(p - r, q - s) \times U(r, s)) \quad \text{and} \quad SO(p, q).
$$

The cohomology given in Theorem 3.5 then appears in degree (k, k) where $k = s(p - r) + r(q - s)$ or in $(pq, 0)$. Thus the bidegree $(tq + (p - t - r)(r - t), rq)$ in Theorem 2.3 when $0 < t < r$ is not obtainable from discrete series for semisimple symmetric spaces.

3.7. In [**TW5**] we only considered the dual pairs $(U(p, q), U(r, r))$. Similar techniques should also apply to dual pairs $(U(p, q), U(r, s))$, and one can try to generate cohomology of $\Gamma \backslash U(p, q)$ from Poincaré series associated to holomorphic discrete series of $U(r, s)$. Besides holomorphic discrete series, the methods in §3.4 construct automorphic cohomology of $\Gamma' \backslash U(r, s)$, and these may be lifted to some nonzero classes in $\Gamma \backslash U(p, q)$. A key ingredient in these liftings is some control on the growth of these functions. This is now provided by the asymptotics of $\psi_{w, \lambda}$ based on hyperfunction theory. Other candidates for lifting are the harmonic forms realizing some highest weight representations of $U(r, s)$ constructed in [**RSW**].

Or one can use the boot-strapping method to vary r, s, p, q and go up. It seems that a combination of these methods should give rise to a much wider range of nonzero cohomology.

4. Geometric duals of harmonic forms.

4.1. From now on let $\Gamma \subset G$ be cocompact such that $\Gamma \cap H \subset H$ is also cocompact. Let D_H be the H orbit $H/H \cap K$ in G/K. Then there is a totally geodesic submanifold

$$
\Gamma \cap H \backslash D_H \subset \Gamma \cap H \backslash D.
$$

Let $\pi : \Gamma \cap H \backslash D \to \Gamma \backslash D$ be the intermediate covering projection. Then the restriction $\pi | \Gamma \cap H \backslash D_H$ is generically one to one. The image is a possibly singular totally geodesic cycle denoted by

$$
C_\Gamma^H = \pi(\Gamma \cap H \backslash D_H).
$$

Let $E_{-w\nu}$ be the locally constant vector bundle on $\Gamma\backslash D$ associated to the irreducible G module $V_{-w\nu}$. $V_{-w\nu}$ has a one-dimensional subspace that is fixed under H: $V^H_{-w\nu}$. Let $v_{-w\nu}$ be a nonzero vector in this subspace. We then define a section $s_{-w\nu,\Gamma}$ of the bundle $E_{-w\nu}$ restricted to the cycle C^H_Γ. At a point $h(H \cap K) \in \Gamma \cap H\backslash D_H$ it is given by the value

$$(h(H \cap K), v_{-w\nu}).$$

Note that $s_{-w\nu,\Gamma}$ is only determined up to a scalar multiple. $s_{-w\nu,\Gamma}$ can be chosen to be locally constant if λ satisfies an integrability condition [TW4].

4.2. Let $\dim D = n$. The duality pairing

$$H^*(\Gamma\backslash D, E)_\delta \times H^{n-*}(\Gamma\backslash D, E^*)_{\delta^*} \to \mathbf{C}$$

is given by

$$(\varphi, \psi) \to \int_{\Gamma\backslash D} \varphi \wedge \psi.$$

The section $s_{-w\nu,\Gamma}$ defines a linear functional on $\psi \in H^{n-R}(\Gamma\backslash D, E^*_{-w\nu})_{\mu^w_\lambda}$ by the formula

$$\int_{C^H_\Gamma} s_{-w\nu,\Gamma}(C^{H*}_\Gamma \psi)$$

where C^{H*}_Γ is the pull-back from $\Gamma\backslash D$ to C^H_Γ. Hence by duality we get a cohomology class

$$[s_{-w\nu,\Gamma}] \in H^R(\Gamma\backslash D, E_{-w\nu})_{-\mu^w_\lambda}.$$

4.3. Let $\alpha_1, \ldots, \alpha_m$ be a dual basis consisting of simple roots in Σ^+. Let $H_1, \ldots, H_m$ be a dual basis of $\mathfrak{a}^d$ and $Y_1, \ldots, Y_l$ determined from $H_1, \ldots, H_m$ as in [Sc, (7.14)].

$$F = \left\{ \lambda \in \mathfrak{a}^{d*} \;\middle|\; \begin{array}{c} 0 \le \langle w\lambda, Y_i \rangle \le \langle \rho, Y_i \rangle \\ \text{for some } w \in W, \; 1 \le i \le l \end{array} \right\}.$$

Note that F is the union of finitely many strips in $\mathfrak{a}^{d*}$ of width $2\langle \rho, Y_i \rangle$.

THEOREM. *Let $(w, \lambda) \in \Omega_X$, and assume $\lambda \notin F$, and λ satisfies 3.3(1). Then $s_{-w\nu,\Gamma}$ can be normalized, independently of Γ, such that in the cohomology group $H^R(\Gamma\backslash D, E_{-w\nu})_{-\mu^w_\lambda}$*

$$[\hat{\omega}^w_\lambda] = [s_{-w\nu,\Gamma}].$$

4.4 There are three ingredients in the proof of the above theorem.

A. By duality the theorem amounts to

$$(1) \qquad \int_{\Gamma\backslash D} \hat{\omega}^w_\lambda \wedge \psi = \int_{C^H_\Gamma} s_{-w\nu,\Gamma}(C^{H*}_\Gamma \psi)$$

for all $\psi \in H^{n-R}(\Gamma\backslash D, E^*_{-w\nu})_{\mu^w_\lambda}$. If one unfolds (pulls back) along the covering π: $\Gamma \cap H\backslash D \to \Gamma\backslash D$, (1) is equivalent to

$$(2) \qquad \int_{\Gamma\cap H\backslash D} \omega^w_\lambda \wedge \psi = \int_{\Gamma\cap H\backslash D_H} s_{-w\nu,\Gamma}(i^*\psi),$$

where i: $\Gamma \cap H\backslash D_H \to \Gamma \cap H\backslash D$.

B. Conditions on discrete series parameter λ imply that the representation of K with lowest weight $-\mu_\lambda^w$ has a one-dimensional subspace of $H \cap K$ fixed vectors. This fact implies that $\omega_\lambda^w|_{D_H}$ is a multiple of the projection of $\omega_N \otimes s_{-w\nu,\Gamma}$ to the subspace 3.4(1), where ω_N is the top exterior power of a basis of unit cotangent vectors normal to D_H. This is the crucial link from ω_λ^w to geometry.

C. The left-hand side of (2) can be pulled back to an integral over G and then transformed to

$$\int_{H\backslash G} \left(\tilde{\omega}_\lambda^w(g), \int_{\Gamma\cap H\backslash H} *\widetilde{\psi}(hg)\,dh \right) d(Hg),$$

where $*$ is the Hodge star operator defined by the invariant metric. We then must show that

(3)
$$\int_{\Gamma\cap H\backslash H} *\widetilde{\psi}(Hg)\,dh = c\tilde{\omega}_\lambda^w(g),$$

where c is a constant. This relies on the deep result of Oshima and Matsuki [OM] which asserts that the discrete series in $L^2(G/H) \cap \mathscr{A}_{\lambda,K}(G/H)$ are multiplicity free. Actually the left-hand side of (3) is only bounded, and the condition $\lambda \notin F$ is used to ensure that bounded K finite eigenfunctions are square integrable on G/H.

4.5. The condition $\lambda \notin F$ is somewhat awkward, but T. Oshima showed some rank one examples of $\lambda \in F$ where a space of bounded eigenfunctions is isomorphic to a $\mathscr{V}_{w(\lambda)}$ but nonetheless the bounded functions are not square integrable. Oshima also informed us that it can be shown if $\langle \lambda - \rho, \alpha_i \rangle > 0$, $i = 1, \ldots, m$, then

$$\dim_{\mathbb{C}} \mathrm{Hom}_{\mathfrak{g}}(\mathscr{V}_{w(\lambda)}, C^\infty(G/H)) = 1.$$

This is a better result. The proof depends on D-module theory.

5. Some remarks.

5.1. The first example of Theorem 4.3 appeared in [T] in the setting of Hilbert modular surfaces. Since I was familiar with intersection theory with coefficients in a bundle due to work on Riemann-Roch, it was my immediate reaction when Zagier posed the question to me. It is curious to note that such an intersection theory with Čech rather than harmonic representatives had answered some of Hirzebruch's questions about Riemann-Roch. The unfolding argument in 4.4A already appeared in [T] and was due to Zagier [Z]. Later B. Gordon [G] showed that in this case pairings of harmonic forms with coefficients in locally constant bundles can be interpreted as pairings of harmonic forms with *trivial coefficients* of middle dimensional degree in an appropriate Kuga fiber space over the Hilbert modular surface. It would be interesting to generalize this for the harmonic forms in Theorem 4.3.

5.2. An explicit and simple version of 4.4B, C is contained in [TW3] in the case $G = \mathrm{SU}(p,q)$, $H = S(\mathrm{U}(r) \times \mathrm{U}(p - r, q))$. In that approach it makes a great difference whether H contains one or two noncompact factors; the cycles determined by H are then called positive or mixed. It suffices to say that when H has two noncompact factors, its action on D is highly twisted and the normal bundle of D_H in D, for example, does not extend H equivariantly to a holomorphic bundle on D thus rendering differential geometric calculations of the type [TW1] very difficult. Similar difficulties have been encountered in [KM2] and in [RSW, PR]. Although we can now construct harmonic forms and prove Theorem 4.3 for mixed cycles, from the viewpoint of theta correspondence it is still not clear how to write down the Rallis-Schiffmann functions [RS].

5.3. Geometrically the interesting case is when the coefficients are trivial, i.e., $\lambda = \rho$. Of course then neither the L^1 condition for $\psi_{w,\lambda}$ nor the condition that guarantees multiplicity one holds. As remarked in the introduction [KM1] used an analytic continuation technique to obtain the harmonic form for the dual pair $(O(p, 1), Sp(r, \mathbf{R}))$. More complicated versions of analytic continuations were obtained for the pairs $(U(p, 1), U(r, r))$ [TW1] and $(U(p, 2), U(1, 1))$ [TW6].

For a fixed H, as one varies the choices of the different closed orbits on G^d/P^d parametrized by $W_{K \cap H} \backslash W$, the method in §4 suggests what are some of the K type components of the harmonic dual of C_Γ^H. What are all of the nontrivial K type components of a harmonic dual, including those which come from a cup product of an invariant form with a harmonic form of lower degree, is far from clear.

5.4. It is a very deep question whether our construction of harmonic forms actually exhausts the cohomology group $H^R(\Gamma \backslash D, E_{-w\nu})_{-\mu_\lambda^w}$ as H is varied among its conjugates compatible with Γ. The results of [MR, HLR] show that for Hilbert modular surfaces in general there will be $(1, 1)$ cohomology whose dual cycle cannot be so constructed. It appears that such cycles cannot be totally geodesic, and it seems to be difficult to construct them.

5.5. The discrete series for semisimple symmetric spaces give only a part of the Zuckerman modules, as the discussion in §3.6 showed. It is interesting to try to consider problem II in §1 for the other Zuckerman modules. Flensted-Jensen [FJ2, p. 70] has indicated some structure of representations associated to quasi-closed orbits. However since these functions are not square integrable in G/H, they are also not integrable on G/H, as Oshima remarked. Thus it is not clear how to utilize them to get harmonic forms on $\Gamma \backslash D$.

In connection with the Hodge conjecture one may ask if a Zuckerman module has nontrivial (k, k) cohomology, and if problem II can be answered, then is the cycle algebraic? As the remarks in §5.4 would suggest (and also discussed in §1) we are posing this for the (k, k) cohomology that can be constructed in some uniform manner independently of Γ. For this problem one may consider the case $G = SU(3, 2)$; then there is a Zuckerman module that has cohomology of bidegree $(2, 2)$ and K type $(1, 0, -1; 2, -2)$. This is not a member of discrete series. On the other hand, the only known complex codimension two cycles in this case come from the construction of §4 with $H = S(U(1) \times U(2, 2))$. This Zuckerman module therefore provides an interesting candidate for our problems I and II.

REFERENCES

[Ad] J. D. Adams, *Discrete spectrum of the reductive dual pair* $(O(p, q), Sp(2m))$, Invent. Math. **74** (1983), 449–475.

[An] G. W. Anderson, *Theta functions and holomorphic differential forms on compact quotients of bounded symmetric domains*, Duke Math. J. **50** (1983), 1137–1170.

[Be] M. Berger, *Les espaces symétriques noncompacts*, Ann. Sci. École Norm. Sup. (3) **74** (1957), 85–177.

[Bo] A. Borel, *Introduction to automorphic forms*, Proc. Sympos. Pure Math., vol. 9, Amer. Math. Soc., Providence, R.I., 1966, pp. 199–210.

[BW] A. Borel and N. Wallach, *Continuous cohomology, discrete subgroups and representations of reductive groups*, Ann. of Math. Stud., no. 94, Princeton Univ. Press, Princeton, N.J., 1980.

[FJ1] M. Flensted-Jensen, *Discrete series for semisimple symmetric spaces*, Ann. of Math. (2) **111** (1980), 253–311.

[FJ2] ____, *Analysis on non-Riemannian symmetric spaces*, CBMS Regional Conf. Ser. in Math., no. 61, Conf. Board Math. Sci., Washington, D.C., 1986.

[G] B. Gordon, *Intersections of higher weight cycles over quaternionic modular surfaces and modular forms of nebentypus*, Bull. Amer. Math. Soc. (N.S.) **14** (1986), 293–298.

[HLR] G. Harder, R. Langlands, and M. Rapoport, *Zyklen auf Hilbert-Blumenthal Flächen*, J. Reine Angew. Math. **366** (1986), 53–120.

[HZ] F. Hirzebruch and D. Zagier, *Intersection numbers of curves on Hilbert modular surfaces and modular forms of nebentypus*, Invent. Math. **36** (1976), 57–113.

[H1] R. Howe, *Remarks on classical invariant theory* (preprint).

[H2] ____, *On some results of Strichartz and Rallis and Schiffmann*, J. Funct. Anal. **32** (1979), 297–303.

[KV] M. Kashiwara and M. Vergne, *On the Segal-Shale-Weil representations and harmonic polynomials*, Invent. Math. **44** (1978), 1–47.

[K] D. Kazhdan, *Some applications of Weil representations*, J. Analyse Math. **32** (1977), 235–248.

[KM1] S. Kudla and J. Millson, *Geodesic cycles and the Weil representation*. I, Compositio Math. **45** (1982), 207–271.

[KM2] ____, *The theta correspondence and harmonic forms*. I, II, Math. Ann. **274** (1986), 353–378; **277** (1987), 267–314.

[Ma] T. Matsuki, *The orbits of affine symmetric spaces under the action of minimal parabolic subgroups*, J. Math. Soc. Japan **31** (1979), 331–357.

[MR] V. K. Murty and D. Ramakrishnan, *Period relations and the Tate conjecture for Hilbert modular surfaces*, Invent. Math. **89** (1987), 319–345.

[Od] T. Oda, *On modular forms associated with indefinite quadratic forms of signature* $(2, n - 2)$, Math. Ann. **231** (1977), 97–114.

[OM] T. Oshima and T. Matsuki, *A description of discrete series for semisimple symmetric spaces*, Adv. Stud. Pure Math., vol. 4, North-Holland, 1984, pp. 331–390.

[PR] C. Patton and H. Rossi, *Cohomology on complex homogeneous manifolds with compact subvarieties*, Contemp. Math., no. 58, part I, Amer. Math. Soc., Providence, R.I., 1986, pp. 199–211.

[RS] S. Rallis and G. Schiffmann, *Automorphic forms constructed from the Weil representation: Holomorphic case*, Amer. J. Math. **100** (1978), pp. 1049–1122.

[RSW] J. Rawnsley, W. Schmid and J. Wolf, *Singular unitary representations and indefinite harmonic theory*, J. Funct. Anal. **51** (1983), 1–114.

[S] G. Shimura, *Automorphic forms and the periods of abelian varieties*, J. Math. Soc. Japan **31** (1979), 561–592.

[Sc] H. Schlichtkrull, *Hyperfunctions and harmonic analysis on symmetric spaces*, Prog. Math., vol. 49, Birkhäuser, Boston, 1984.

[T] Y. L. Tong, *Weighted intersection numbers on Hilbert modular surfaces*, Compositio Math. **38** (1979), 299–310.

[TW1] Y. L. Tong and S. P. Wang, *Harmonic forms dual to geodesic cycles in quotients of SU(p, 1)*, Math. Ann. **258** (1982), 289–318; *Theta functions defined by geodesic cycles in quotients of SU(p, 1)*, Invent. Math. **71** (1983), 467–499.

[TW2] ____, *Period integrals in noncompact quotients of SU(p, 1)*, Duke Math. J. **52** (1985), pp. 649–688.

[TW3] ____, *Construction of cohomology of discrete groups*, Trans. Amer. Math. Soc. 306 (1988), 735–763.

[TW4] ____, *Geometric realization of discrete series for semisimple symmetric spaces* (to appear in Invent. Math.)

[TW5] ____, *Some nonzero cohomology of discrete groups*, J. Reine Angew. Math. **382** (1987), 85–144.

[TW6] ____, *Correspondence of Hermitian modular forms to cycles associated to SU(p, 2)*, J. Differential Geom. **18** (1983), 163–207.

[V] D. Vogan, *Representations of real reductive Lie groups*, Prog. Math., vol. 15, Birkhäuser, Boston, 1981.

[VZ] D. Vogan and G. Zuckerman, *Unitary representations with nonzero cohomology*, Compositio Math. **53** (1984), 51–90.

[W] S. P. Wang, *Correspondence of modular forms to cycles associated to O(p, q)*, J. Differential Geom. **22** (1985), 151–213.

[Z] D. Zagier, *Modular forms whose Fourier coefficients involve zeta functions of quadratic fields*, Lecture Notes in Math., vol. 627, Springer-Verlag, 1977, pp. 105–169.

PURDUE UNIVERSITY

Number Theory

Proceedings of Symposia in Pure Mathematics
Volume **49** (1989), Part 2

Transcendental Methods and Theta-Functions

D.V. CHUDNOVSKY AND G.V. CHUDNOVSKY

Introduction. The rich structure of theta-functions provides a variety of algebraic objects and numbers associated with them. These numbers include many classical constants of analysis and a multitude of periods of algebraic varieties. In algebraic geometry the main attention is devoted to algebraic numbers associated with theta-functions and their period structure (as an example, we can mention the complex multiplication). An attractive feature that helps in the study of these algebraic "factors" is the ability to consider specializations, p-adic and l-adic versions. At the same time there are transcendental numbers inherently associated with theta-functions, including periods of theta-functions and their values. These numbers are studied in transcendental number theory not totally divorced from algebraic geometry.

In these lectures we try to demonstrate the interrelationship between pure arithmetic arising from algebraic geometry and transcendental number theory that tries to analyze the diophantine properties of constants associated with theta-functions and period structures. Another dimension in this relationship is the use of experimental mathematics in the form of computer algebra and large scale numerical computations. In order to understand better the diophantine properties of classical constants such as π, we have to turn to the variation of the period (Hodge) structure of algebraic varieties, and, in general, investigate arithmetic properties of solutions of differential equations. A variety of auxiliary objects, needed in transcendental number theory, turn out to be associated with algebraic structures. These objects include Padé approximants to transcendental functions and their generating functions. The differential equations satisfied by these generating functions have to possess special arithmetic properties that are associated with the existence of the action of Frobenius. We are led therefore to the problem of determining which (linear) differential equations possess special arithmetic properties that include p-adic overconvergence and the existence of solutions

---

1980 *Mathematics Subject Classification* (1985 *Revision*). Primary 10F35, 10F37.

This research was supported in part by the National Science Foundation, U. S. Air Force, and Program OCREAE.

with nearly integral power series expansions. We examine these classes of
equations in Chapter 1, where we present what we hope are convincing evi-
dences for the conjecture ("Dwork-Siegel conjecture") that all such equations
with special arithmetic properties arise from geometry, as Picard-Fuchs equations on variations of periods. Interestingly enough, methods that we use
to study these, globally nilpotent, equations are methods of transcendental
number theory, including Padé-type approximation; see [14, 15, 19, 20, 27].
Another aspect of this interplay between algebraic and transcendental faces
of structures uniformized by theta-functions is a variety of algebraicity statements (see §§1.1–1.3). According to these statements (which include some
cases of the Tate conjecture) a function well uniformized in the archimedean
(complex) and p-adic domains is an algebraic one. In general Chapter 1 is
dedicated to arithmetic properties of (linear) differential equations (near integrality of power series expansion of solutions, solutions mod p, etc.) and
applications of these equations to irrationality and transcendence. A variety
of issues concerning uniformization problems and their analytic and numerical solutions is also investigated. We refer readers to the authors' papers [32,
48] for related computer oriented study.

We introduce a concept of "p-adic spectrum" to describe interesting numerical results observed in p-adic curvature of Lamé equations. Theoretical
and experimental motivations for these definitions are presented in §1.7.

In diophantine approximations a variety of algebraic structures can be associated with the study of a single number. Scattered examples are given in
Chapter 1 of special classes of linear differential equations (of Lamé type)
associated with Fuchsian arithmetic groups acting on H. These examples
are connected with such numbers as $\zeta(2), \zeta(3)$ and periods of elliptic curves.
We look at a possibility of a wider class of algebraic structures giving rise
to diophantine approximations of classical constants. In §§1.9–1.10 we introduce one such class generalizing the Stieltjes-Rogers continued fraction
expansions and associated with linear differential equations uniformized by
theta-functions (also known as "finite-band" potential equations). Rich in
arithmetic applications, this class of algebraic structures arises from Abelian
varieties of CM-type. In Chapter 2, §§2.1–2.3, we describe a Ramanujan-like
theory of period relations of CM-varieties interpreted from the point of view
of special values of solutions of Picard-Fuchs equations on periods specialized at CM-points. Interesting identities give the basis of better arithmetic
study of numbers arising from periods and quasiperiods of CM-varieties. π
is one of these numbers, and we use the machinery of Padé approximations
(to solutions of Picard-Fuchs equations, particularly generalized hypergeometric ones, see [43, 44, 48, 61]) to study the measure of irrationality of π
and its algebraic multiples. Those who can gain an understanding of π from
the first million of its decimal or continued fraction expansions can use presented identities for high speed computation. We leave aside the question of
complexity of these computations discussed in [48].

Explicit formulas of Padé approximations and their arithmetic/analytic study occupy §§2.4–2.6. We show how multiple-integral representations can be used to express various Padé-type approximations tailored to special constants. Integrals of Pochhammer type (Picard equations) are used as a case study in §2.5 for special cases of diophantine approximations to logarithms. We describe an iterative method of improvement of irrationality measures of such constants as $\ln 2$, $\pi/\sqrt{3}$, and π. This method leads to multidimensional integral representations in §2.6. These methods give a multiparameter scheme for construction of diophantine approximations using WKB in complex and archimedean domains. This method developed by us some years ago, see [44], can be compared with elementary methods in analytic number theory [57]. Applications of modular functions and equations will undoubtedly improve diophantine results. So far we see only a first glimpse of results and conjectures to come.

We thank R. Gunning and L. Ehrenpreis for their generous efforts in the organization and conduct of this conference, which benefited so many. We thank the AMS for their support of this school. Our work relied on computer algebra systems and foremost on IBM SCRATCHPAD. We thank the IBM Computer Algebra group and, particularly, R. Jenks. We are very thankful to the SYMBOLICS, Inc. for the use of a SYMBOLICS 3640 workstation. In the numerical realm, the use of IBM's IBM 3090-VF and of NSF's (sparsely sponsored) and Minnesota's CRAY-II are gratefully acknowledged. Conversations with R. Askey were most valuable and inspiring for us.

Chapter 1
Arithmetic and Geometry of Linear Differential Equations

In this chapter we focus on structures closely associated with θ-functions. These are periods of algebraic varieties and their deformations. We examine arithmetic and geometric properties of linear differential equations that distinguish the deformation equations (Picard-Fuchs equations) from other linear differential equations. Results of this chapter also serve as a basis for our further study of algebraic and transcendental values of functions (including θ-functions) defined by arithmetic conditions.

About 20 years ago it was discovered, due to works of Grothendieck, Dwork, Griffiths, Deligne, Katz, and many others, and well established that linear differential equations having a geometric sense, like the Picard-Fuchs equations satisfied by the variation of periods, possess strong arithmetic properties (global nilpotence, action of the Frobenius, Fuchsianity, etc). We refer to the review [1] and to exposition in [2].

Since then it was suspected that, in a certain sense, the converse is true too. We present the theoretic and experimental evidence intended to show that linear differential equations with arithmetic (or integrality) properties of their solutions arise from geometry, or, precisely, correspond to deformations of period (Hodge) structure of algebraic manifolds. A variety of problems from diophantine geometry arise here, including the irrationality and diophantine approximations to constants of classical analysis.

We start with defining our basic objects: integral or nearly integral classes of functions and related globally nilpotent differential equations.

1.1. Integrality properties and G-functions. In his seminal paper [3] on diophantine approximations, Siegel defined and targeted for future studies two classes of functions satisfying linear differential equations and given by power series expansions in x, for the values of which one can establish general theorems on irrationality, transcendence, and the measure of linear independence. These two classes are classes of E-functions and G-functions, which command attention of modern diophantine approximations research. The study of E-functions, started by Siegel [3, 4], has been significantly advanced since then by many researchers, see particularly [5, 6]. We would like to mention in this connection that only relatively recently have we proved results on the best possible measure of diophantine approximations of values of E-functions at rational points [7]. These results present an ultimate effective version of the Schmidt theorem [8] for the values of E-functions. The E-functions $f(x)$ are entire functions with power series expansion $f(x) = \sum_{n=0}^{\infty} a_n x^n / n!$, where $a_n \in \overline{\mathbf{Q}}$ and for every $\epsilon > 0$, we have $|a_n| < (n!)^{\epsilon}$, $\mathrm{denom}\{a_0, \ldots, a_n\} \leq (n!)^{\epsilon} : n \geq n_1(\epsilon)$. In algebraic geometry and analysis, however, most of the interesting functions are analytic only in the finite part of the complex plane and have much better p-adic convergence properties. Among these the G-functions play the crucial arithmetic role.

DEFINITION 1.1 A function $f(x)$ with the expansion at $x = 0$

$$f(x) = \sum_{n=0}^{\infty} a_n x^n$$

is called a G-function if $f(x)$ satisfies a linear differential equation over $\overline{\mathbf{Q}}(x)$, if coefficients a_n are algebraic numbers, and if there is a constant $C > 1$ such that for all $n \geq 0$ the sizes of coefficients a_n (i.e., the maximum of absolute values of a_n and all its conjugates) and the common denominators $\mathrm{denom}\{a_0, \ldots, a_n\}$ are bounded by C^n.

Note. In general, we say that a function with an algebraic number coefficient Taylor expansion has nearly integral expansion of denominators of nth coefficient grow significantly slower than factorial n!

Algebraic functions and their integrals are G-functions (by Eisenstein's theorem, see §1.2). Other subclasses of G-functions include generalized hypergeometric functions $_{p+1}F_p\binom{a_1,\ldots,a_{p+1}}{b_1,\ldots,b_p}|x)$ with rational parameters a_i, b_j, and solutions of Picard-Fuchs equations (periods of algebraic varieties depending on a parameter). As we will see later, we suspect that all G-functions arise from Picard-Fuchs equations and are closely connected with hypergeometric functions.

Siegel introduced this class of G-functions and put forward a program to prove the linear independence theorems for values of G-functions at algebraic points near the origin. Unfortunately, Seigel [3] never explicitly proved general theorems for G-functions, instead presenting examples of such theorems and presenting an outline of a theory that could be constructed similar to the theory of E-functions. Progress in the later study of G-functions becomes heavily dependent on additional very restrictive global "geometric" conditions, formulated for the first time by Galočhkin [9], that demand that the G-function property be shared by an expansion of any other solution of a differential equation satisfied by $f(x)$ at any algebraic point. We called these global conditions in [10, 11] the (G, C)-conditions. Previously known results on G-functions rely exclusively on these (G, C)-conditions, or on equivalent ones [9, 12, 13, 11]. These global conditions can also be reformulated [12] in terms of the p-adic "overconvergence at a generic point" of solutions of a linear differential equation satisfied by $f(x)$.

In [14] and [15] we had proved the general linear independence results for values of arbitrary G-functions at algebraic points (close to the origin), without any additional conditions. These results materialize Siegel's program after some 55 years. Also in [14, 15] proofs of results on the absence of algebraic relations are presented. We will present these results later in Chapter 2. It is more important, however, that we have proved the strong (G, C)-property for arbitrary G-functions [20]. This result, connected with our study of the Grothendieck conjecture, implies, e.g., that all previous results on G-function theory, proved under very restrictive conditions, are unconditionally valid for all G-functions. To describe our result and the (G, C)-function conditions, one needs the definition of the p-curvature (introduced and studied by Cartier, Grothendieck, Katz, Deligne, ...), to be studied in detail later in this chapter.

We consider a system of matrix first-order linear differential equations over $\mathbf{Q}(x)$, satisfied by functions $f_i(x)$, $i = 1, \ldots, n$,

$$(1.1.1) \qquad df_i(x)/dx = \sum_{j=1}^{n} A_{i,j}(x) f_j(x),$$

for $A_{i,j}(x) \in \mathbf{Q}(x)$, $i, j = 1, \ldots, n$. Rewriting the system (1.1) in the matrix form

$$df^t/dx = Af^t, \qquad A \in M_n(\mathbf{Q}(x)),$$

one can introduce the p-curvature operators Ψ_p, associated with the system (1.1) following [16, 17]. The p-curvature operators Ψ_p are defined for a prime p, as

$$\Psi_p = (d/dx - A)^p \quad (\mathrm{mod}\, p).$$

Then Ψ_p is a linear operator that can be represented as $\Psi_p = -A_p \pmod{p}$, where one defines for $m \geq 0$

$$(1.1.2) \qquad (d/dx)^m \equiv A_m \quad (\mathrm{mod}\, \mathbf{Q}(x)[d/dx](d/dx - A)).$$

Let $D(x)$ be a polynomial from $\mathbf{Z}[x]$ that is the denominator of A, i.e., $D(x)A_{i,j}(x)$ is a polynomial in $\mathbf{Z}[x]$ for $i, j = 1, \ldots, n$. The (G, C)-function condition [10, 11] of (1.1) means that (1.1) is satisfied by a system $(f_1(x), \ldots, f_n(x))$ of G-functions, and that there exists a constant $C_2 > 1$, such that for any N, the common denominator of all coefficients of all polynomial entries of matrices $D(x)^m A_m(x)/m!$, $m = 0, \ldots, N$, is growing not faster than C_2^N. With this condition is closely related a *global nilpotence* condition [15–18] stating that the matrices Ψ_p are nilpotent for almost all primes p. The (G, C)-condition implies the global nilpotence condition.

In [15] we proved the global nilpotence (and the (G, C)-function condition) of linear differential equations having a G-function solution. To prove this result we used Padé approximants of the second kind (Germanic polynomials) to construct, starting from a single solution of a linear differential equation, the basis of approximate solutions of this equation. Using these methods of Padé approximations we proved [15]

THEOREM 1.2. *Let* $f_1(x), \ldots, f_n(x)$ *be a system of G-functions, satisfying a system of first-order linear differential equations* (1.1) *over* $\overline{\mathbf{Q}}(x)$. *If* $f_1(x), \ldots, f_n(x)$ *are linearly independent over* $\overline{\mathbf{Q}}(x)$, *then the system* (1.1) *satisfies a* (G, C)-*function condition and is globally nilpotent. Any solution of* (1.1) *with algebraic coefficients in Taylor expansions is a G-function.*

Let $Ly = 0$ be a linear differential equation of order n over $\overline{\mathbf{Q}}(x)$ satisfied by a G-function $y(x)$, and $y(x)$ does not satisfy a linear differential equation over $\overline{\mathbf{Q}}(x)$ of order $< n$. Then the equation $Ly = 0$ is globally nilpotent, and all solutions of the equation $Ly = 0$ with algebraic initial conditions at an algebraic point $x = x_0$ have G-function expansions at $x = x_0$.

1.2. Algebraicity principle for functions overconvergent in the complex and non-archimedean domains. An important part of applications of our new approximation technique used in [14, 15] was the establishment of the connection between the complex analytic (convergence) properties of Padé approximations to power series and their arithmetic properties (p-adic convergences). This connection is of major importance by itself, for it sheds light on the inverse to the famous Eisenstein theorem on the near-integrality of coefficients in the power series expansion of algebraic functions. According to this Eisenstein theorem, if a power series expansion $f(x) = \sum_{n=0}^{\infty} c_n x^n$ satisfies an algebraic equation $Q(x, f) = 0$ with a polynomial $Q(x, y)$ having integral coefficients, then there exists an integer D, such that all numbers $D^n c_n$ are integral for $n \geq 1$.

The trivial converse to this theorem is obviously false, even if a function $f(x)$ satisfies a Fuchsian linear differential equation (i.e., a globally nilpotent hypergeometric equation can have a nonalgebraic function solution, e.g., $f(x) = {}_2F_1(\frac{1}{2}, \frac{1}{2}; 1; x)$).

The only examples of converses to Eisenstein's theorem were some rationality results of Borel, Fatou, Polya (and related ones by Selberg, Gelfand,

Bieberbach, Straus, and others). A typical early example of such a statement is the Borel-Polya theorem according to which a power series expansion with rational integer coefficients, meromorphic in the domain of conformal radius > 1, represents a rational function.

We report briefly on our work on application of Padé approximation technique to establishment of the converse to the Eisenstein theorem for functions uniformized by meromorphic or automorphic functions.

Our main tool in the study of the Grothendieck conjecture, and in the current study of globally nilpotent equations, is the analytic method of Padé-(rational) and more general algebraic approximations to functions satisfying nontrivial complex analytic and arithmetic (p-adic) conditions. The corresponding group of results can be considered as a certain "local-global" principle. According to this principle, algebraicity of a function occurs whenever one has a near integrality of coefficients of power series expansion—*local conditions*—coupled with the assumptions of the analytic continuation (controlled growth) of an expanded function in the complex plane (or its Riemann surface)—a global, *archimedean* condition.

Our plan for proof of algebraicity follows pretty much the general idea in applications of transcendence methods: one constructs Padé (algebraic) approximations to a given function (and its powers) and then compares the complex analytic and arithmetic (p-adic) properties of the remainder function in the approximation problem to show that the remainder function is identically zero. Namely, one tries to show that the approximation is too good to be nontrivial and, as a consequence, the approximation by an algebraic function is, in fact, the approximated function. In doing so, the arithmetic properties of the approximated function govern the arithmetic properties (sizes of the coefficients) of approximants, while analytic properties of the original function determine the convergence rate of the approximation. If the approximation is very good in a complex (real) plane, then its expansion cannot have nearly integral coefficients that are nonzero. □

To prove the algebraicity of an integral expansion of an analytic function, assumptions on a uniformization of this function have to be made.

Our results from [19] and [20] were proved in the multidimensional case as well, to include the class of functions, uniformized by Jacobi's theta-functions (e.g., integrals of the third kind on an arbitrary Riemann surface). Moreover, our result includes "the nearly-integral" expansions, when the denominators grow slower than a typical factorial $n!$ denominator. We present one of our results [19–20], according to which $g+1$ functions in g variables having nearly integral power series expansions at $\overline{x} = \overline{0}$ and uniformized near $\overline{x} = \overline{0}$ by meromorphic functions of finite order of growth *are algebraically dependent*.

To present this result, let $f_1(\overline{x}), \ldots, f_n(\overline{x})$ be functions of x in $\mathbf{C}^g$ analytic at $\overline{x} = 0$ and uniformized by n meromorphic functions $U_1, \ldots, U_n$ of u in $\mathbf{C}^g$ of order of growth at most ρ, with the Jacobian $D(\overline{u})/D(\overline{x})$ nonzero at $\overline{x} = 0$. Then we have

THEOREM 2.1. *Let* $f_1(\overline{x}),\ldots,f_n(\overline{x})$ *be* $n \geq g + 1$ *functions of* $\overline{x} = (x_1,\ldots,x_g)$ *as above, uniformized by a meromorphic functions* $U_1(\overline{u}),\ldots,$ $U_n(\overline{u})$ *in* $\mathbf{C}^g$ *of finite order* $\leq \rho$ *of growth. Let* $f_i(\overline{x}) = \sum_{\overline{m} \in \mathbf{Z}} g a_{\overline{m},i} \cdot \overline{x}^{\overline{m}}$ *be expansions at* $\overline{x} = \overline{0}$ *with coefficients from an algebraic number field* $\mathbf{K}$, *whose sizes grow not faster than geometric progression (i.e.,* $|\overline{a_{m,i}}| \leq C^{\|\overline{m}\|}$, *and* $\| \overline{m} \| = m_1 + \cdots + m_g$ *for* $\overline{m} = (m_1,\ldots,m_g)$).

For expansion of monomials

$$f_1(\overline{x})^{k_1} \cdots f_n(\overline{x})^{k_n} = \sum_{\overline{m}} a_{\overline{m},}\, {}_{\overline{k}} \overline{x}^{\overline{m}}$$

$(k_i \geq 0)$, *we denote by* Δ_M *the common denominator of*

$$a_{\overline{m};\overline{k}}, \qquad \| \overline{m} \| < M, \| \overline{k} \| < M.$$

If

$$\xi \overset{\text{def}}{=} \limsup_{M \to \infty} \frac{\ln \Delta_M}{M \ln M} < \frac{1 - g/n}{\rho[\mathbf{K} : \mathbf{Q}]},$$

then the functions $f_1(\overline{x}),\ldots,f_n(\overline{x})$ *are algebraically dependent over* $\mathbf{K}$.

The proof of this and other similar results is given in [19, §5, Theorems 5.2, 5.9, 5.12] and in [20, Theorem 1.1]. Among the generalizations of this result is the version of Theorem 2.1 when the functions $f_i(x)$ are expanded in several algebraic points. In this case the theorem generalizes the Schneider-Lang and Bombieri-Lang theorems on transcendence of values of meromorphic functions, see, e.g., [6]. Also, our results are effective in the sense that one has to examine only finitely many $M \leq M_0$ whenever $f_1,\ldots,f_n$ satisfy a system of g algebraic differential equations in $x_1,\ldots,x_g$.

Among the applications of this result are some partial but effective results on the Tate conjecture on the bijectivity of a map $\text{Hom}(A, B) \otimes \mathbf{Z}_l \to \text{hom}(T_l(A), T_l(B))$ for Abelian varieties A and B over algebraic number fields. The Tate conjecture for elliptic curves was proved by Serre [21] for all cases but the one in which A and B have integral invariants but no complex multiplication. Finally Faltings [22] proved (ineffectively) the finiteness of the isogeny classes for arbitrary Abelian varieties, solving the Tate, Schafarevich, and Mordell conjectures. We proved the effective version of the Tate conjecture for elliptic curves using only Theorem 2.1 and the Honda [23] criterion of isogeny of elliptic curves in terms of the logarithms of their formal groups over $\mathbf{F}_p$:

COROLLARY 2.2. *If two elliptic curves* $E_1/\mathbf{Q}$ *and* $E_2/\mathbf{Q}$ *have the same number of points* mod p *for almost all* p *(or even for almost all* p *below an effective bound) then* $E_1/\mathbf{Q}$ *and* $E_2/\mathbf{Q}$ *are isogenous over* $\mathbf{Q}$.

To see why this is true, one looks at the functions $f_1(x) = x$ and $f_2(x) = L_{E_2}^{-1}(L_{E_1}(x))$, where $u = L_{E_i}(x)$ $(= \int \omega_i = \int(dx/\sqrt{P_i(x)}))$ is an elliptic logarithm corresponding to E_i, uniformized by the Weierstrass elliptic function: $x = \mathscr{P}_i(u_i)$. Honda's isomorphism theorem [23] for 1-dimensional formal groups over $\mathbf{F}_p$ shows that $f_2(x)$ has nearly integral expansions. On the other hand, $f_1(x)$ and $f_2(x)$ are uniformized by meromorphic functions:

$$f_1(x) = \mathscr{P}_1(u), \qquad f_2(x) = \mathscr{P}_2(u).$$

Thus we apply Theorem 2.1 in the case $g = 1$, and $\mathscr{P}_1$ and $\mathscr{P}_2$ are algebraically dependent over $\mathbf{Q}$ and E_1, E_2 are isogeneous over $\mathbf{Q}$. In [20, §2] we give an effective version of Corollary 2.2, where one has a bound on p that is to be tested to ensure the isogeny of E_1 and E_2. Also our methods are applied in [20, §3] to determine effectively all isomorphism classes of elliptic curves isogeneous to a given elliptic curves over $\mathbf{Q}$. Methods of formal groups lead to some further progress in the effectivization of the Tate conjecture. E.g., Honda's results [23] and Theorem 2.1 can be used to get an effective answer to the Tate conjecture for Abelian varieties with real multiplications (generalizations of Serre and Ribert's results).

The demand to have a meromorphic uniformization of Theorem 2.1 is a pretty strict one, and, combined with the near integrality condition, limits us pretty much to the class of functions uniformized by θ-functions. This leaves other classes of functions out of this approach, including, particularly, solutions of linear differential equations uniformized by Fuchsian groups. Apparently, results similar to Theorem 2.1 can be proved in this case, at least in the case of special Fuchsian groups. Namely, for functions with nearly integral coefficients that satisfy linear differential equations, whose monodromy group is up to the conjugation a subgroup of $GL_n(\overline{\mathbf{Q}})$ (in particular, an arithmetic subgroup) an analog of Theorem 2.1 holds.

Let us look at applications of these local-global results to the Grothendieck conjecture and other similar problems.

1.3. The Grothendieck conjecture and the monodromy group of a linear differential equation. We have applied the Padé approximations methods, which were successful in diophantine approximations and G-functions, to the Grothendieck conjecture [16, 17] that determines the global (monodromy) properties of a linear differential equation in terms of reductions (mod p) of this differential equation.

The Grothendieck conjecture was geometrically formulated by Katz in terms of the p-curvature operators Ψ_p, associated with the system (1.1) of linear differential equations over $\mathbf{Q}(x)$ [16, 17]. The p-curvature operators are linear operators that can be represented as

$$\Psi_p = -A_p \quad (\mathrm{mod}\, p),$$

according to (1.2).

THE GROTHENDIECK CONJECTURE. *If a scalar linear differential equation of order n over $\mathbf{Q}(x)$ has n solutions (mod p) in $\mathbf{F}_p(x)$, linearly independent over $\mathbf{F}_p(x^p)$, for almost all prime numbers p, then this linear differential equation has only algebraic function solutions.*

Equivalently, if a matrix system (1.1) of differential equations over $\overline{\mathbf{Q}}(x)$ has a zero p-curvature $\Psi_p = 0$ for almost all p, then this system (1.1) has algebraic function solutions only.

According to this conjecture, strong integrality (Eisenstein-like) properties of *all* power series expansions of solutions of a given linear differential equation imply that all these solutions are algebraic functions.

Katz [16] proved the Grothendieck conjecture for Picard-Fuchs's equations (including Gauss's hypergeometric functions). However, the Grothendieck conjecture was left open in many important cases, particularly for Lamé equations [18] and linear differential equations of rank one over algebraic curves [16, 17].

Methods of Padé approximations allowed us to solve the Grothendieck conjecture in the important cases mentioned above [24, 15, 20]. Our first results solved the case of the Lamé equation, left open in [18] for all integral n. We will present this result from [25] later in this chapter.

This result was proved in [25], where it was shown that the Grothendieck conjecture is true for any linear differential equation all solutions of which can be parametrized by the meromorphic functions. The result was considerably generalized in [20] in connection with the inverse to the Eisenstein theorem of §1.2. Among the generalizations ([20, §§5–8]) there was a solution of the Grothendieck conjecture for equations, solutions of which can be parametrized by means of multidimensional theta-functions. To the class of these equations belong equations of rank one over arbitrary (finite) Riemann surfaces [20]:

THEOREM 3.1. *Any rank one linear differential equation over an algebraic curve, i.e., a first-order equation with algebraic function coefficients, satisfies the Grothendieck conjecture. Namely, if* Γ *is an algebraic curve (given by the equation* $Q(z, w) = 0$) *over* $\overline{\mathbf{Q}}$, *and if the rank one equation*

$$(1.3.1) \qquad\qquad \frac{dF}{F} = \omega(z, w)\,dz$$

over $\overline{\mathbf{Q}}(\Gamma)$ *(for an Abelian differential* $\omega\,dz$ *on* Γ) *is globally nilpotent, then all solutions of* (3.1) *are algebraic functions.*

This result implies, according to Katz [26], that for any equation of rank 2 (over an algebraic curve), and the Lie algebra $\mathbf{G}$, generated by the p-curvature operators Ψ_p for almost all p, if $\mathbf{G} \neq 0$, then the Lie algebra of the Galois group of the equation coincides with $\mathbf{G}$.

The relationship of the p-curvature operators with the monodromy (Galois) group of a differential equation is extremely interesting. Our methods, involving various generalizations of Padé approximations, allow us to prove the Grothendieck conjecture for a larger class of differential equations, when additional information on a monodromy group is available. For example, if a second-order linear differential equation has a commutative monodromy group, then this equation satisfies the Grothendieck conjecture (the Lamé equation with an *integral* parameter n belongs to this class). A technique from [27] using a random walk method (in the free group, corresponding to the representation of a full modular group $SL_2(\mathbf{Z})$) allowed us to treat this crucially important class of equations. The random walk technique used in [27] is the extension of the earlier work [28]. Among the results of [27] is

THEOREM 3.2. *Let $Ly = 0$ be an nth-order linear differential equation over $\overline{\mathbf{Q}}(x)$ satisfying the assumptions of the Grothendieck conjecture. If the monodromy group of $Ly = 0$ is up to a conjugation a subgroup of $GL_n(\overline{\mathbf{Q}})$, then all solutions of $Ly = 0$ are algebraic functions.*

We will discuss the p-curvature operators for Lamé equations with non-integral n later in this chapter in connection with the uniformization theory and the irrationality proofs.

Our results on the Grothendieck conjecture are effective (i.e., one has to examine only a finite set of primes p), and they can be used in various algorithms, including algorithms that determine the reduction of Abelian integrals to elementary functions, see [24].

Various applications of our new methods to the proof of transcendence of numbers appearing as elements in the monodromy matrices of linear differential equations are possible. One of our results is the following [28]:

THEOREM 3.3. *Let us look at a G-function solution $(f_1(x), \ldots, f_n(x))$ of (1.1) with algebraic initial conditions at a nonsingular point of (1.1). Then its analytic continuations along (the basis of) all possible paths leads to at least one transcendental number.*

Another application is the Matthews problem [29] on indefinite integration of algebraic functions. Let Γ be a curve over $\mathbf{Q}$ and D be any derivation of a function field $\mathbf{Q}(\Gamma)$. Let $f \in \mathbf{Q}(\Gamma)$ be such that for almost all p, we can find g in $\mathbf{F}_p(\Gamma \bmod p)$ such that, mod p, $f = Dg$. Is it true then that

$$f = Dg$$

for some g in $\mathbf{Q}(\Gamma)$? (I.e., if an integral is locally algebraic at the same field, is it globally algebraic?) The answer is yes.

1.4. P-curvature operator and overconvergence in a nonarchimedean domain. Let us specialize the p-curvature operator from §1.1 in the case of a scalar linear differential equation.

Let $\mathbf{K}$ be an algebraic number field (for all practical purposes $\mathbf{K}$ is always $\mathbf{Q}$) and let L be a linear differential operator with coefficients that are rational functions over $\mathbf{K}$:

(1.4.1) $$L = \sum_{i=0}^{n} a_i \left(\frac{d}{dx} \right)^i$$

for $a_i = a_i(x) \in \mathbf{K}[x]$, $i = 0, \ldots, n$. All derivatives $(\frac{d}{dx})^j$ are expressed as linear combinations of $(\frac{d}{dx})^i$, $i = 0, \ldots, n-1$, with coefficients from $\mathbf{K}(x)$ mod $\mathbf{K}(x)[\frac{d}{dx}] \cdot L$. For $m \geq 0$ we denote

$$(1.4.2) \qquad \left(\frac{d}{dx}\right)^m \equiv \sum_{i=0}^{n-1} H_{m,i} \cdot \left(\frac{d}{dx}\right)^i \quad \text{mod } \mathbf{K}(x) \left[\frac{d}{dx}\right] \cdot L$$

where $H_{m,i} \in \mathbf{K}(x)$ belong to a differential ring generated by a_i/a_n, $i = 0, \ldots, n-1$, over $\mathbf{Z}$.

These rational functions satisfy the following recurrence:

$$H_{m+1,i} = H'_{m,i} + H_{m,i-1} - \frac{a_i}{a_n} H_{m,n-1},$$

$i = 0, \ldots, n-1$. The rational functions $H_{m,i} = H_{m,i}(x)$ determine the Taylor expansion of the solution $y = y(x)$ of a linear differential equation:

$$(1.4.3) \qquad\qquad\qquad Ly = 0$$

with the initial conditions $y^{(i)}|_{x=t} = c_i$, $i = 0, \ldots, n-1$, at a nonsingular point $x = t$ (where $a_n|_{x=t} \neq 0$). Namely for y in (4.3) we have

$$(1.4.4) \qquad\qquad y(x) = \sum_{i=0}^{n-1} \frac{c_i}{i!} \cdot \sum_{m=0}^{\infty} H_{m,i}(t)(x-t)^m/m!.$$

REMARK. As a "generic" point t in the p-adic domain Dwork (see, e.g., [18]) suggests a t such that t generates a transcendental extension of $\mathbf{Q}_p$, the residue class of t is transcendental over $\mathbf{F}_p$, and $|t|_p = 1$.

As (4.4) shows, the convergence of a solution $y(x)$ of (4.3) near the generic point $x = t$ in the p-adic domain is, at worst, that of the (p-adic) exponent (namely the only source of p-adic divergence of $y(x)$ are factorials in the denominator of (4.4)). This means that any function $y(x)$ in (4.4) converges near the generic point $x = t$ in the disc

$$\operatorname{ord}_p(x-t) > 1/(p-1).$$

The overconvergence of solutions $y(x)$ (i.e., their convergence in a larger disc) is equivalent to p-divisibility properties of $H_{m,i}$ and is equivalent to the nilpotence of the p-curvature operator of L. The precise statement to this effect was proved by Katz, see [2] (there exists an elementary treatment of this subject by Honda [17]—those are posthumous notes of Honda prepared in 1972). Hereafter we assume $\mathbf{K} = \mathbf{Q}$.

THEOREM 4.1 (Katz). *If all solutions of an equation*

$$Ly = 0$$

at a generic point $x = t$ converge in a nontrivial disc

$$\operatorname{ord}_p(x-t) > \frac{1}{p-1} - \epsilon$$

for some $\epsilon > 0$, then and only then the p-curvature of L is nilpotent. The nilpotence of the p-curvature of L means that for some $l \geq 1$ we have

(1.4.5) $$\left(\frac{d}{dx}\right)^{lp} \equiv 0 \mod \mathbf{F}_p(x)\left[\frac{d}{dx}\right] \cdot (L \mod p)$$

(i.e., $H_{lp,i} \equiv 0 \mod p$ for $i = 0, \ldots, n-1$).

If (4.5) is true for some $l \geq 1$, then (4.5) is true for $l = n$.

The annihilation of the p-curvature of L means that (4.5) is true for $l = 1$. This is equivalent to the existence of n linearly independent solutions of $Ly = 0$ in $\mathbf{F}_p[[x]]$, see the formulation of Grothendieck conjecture in §1.3.

REMARK. For $x = t$ to be a generic point one has to have $|t|_p = 1, a_n(x)$ has no zeros in the unit circle $D(t, 1^-)$ around t, and the Gauss norm of $a_i(x)/a_n(x)$ does not exceed 1, $i = 0, \ldots, n-1$ (see, e.g., [18]).

The overconvergence of all solutions of $Ly = 0$ at a (nontrivial) point is equivalent to the nilpotence of the p-curvature of L. Often a stronger convergence property holds, when there are n linearly independent solutions that converge in the unit circle, and their Wronskian does not have zeros in that circle. In this case Dwork and Robba observed, see [18] and [30], that there holds "the law of logarithmic grown" of solutions of $Ly = 0$, which means that for any solution $y = y(x)$ with an expansion

$$y(x) = \sum_{N=0}^{\infty} c_N x^N$$

one has for some $v > 0$,

$$|c_N|_p = O(N^v)$$

for $N \geq N_0$. Moreover,

$$|c_N|_p \leq \text{const.} \cdot \frac{1}{\inf|j_1 \cdots j_{n-1}|_p}$$

where inf is taken over all *distinct* $j_k \leq N$.

This "law of logarithmic growth," if it holds for almost all p ($p \geq p_0$), implies a very definitive structure of denominators of any solution $y(x)$ of $Ly = 0$. Namely, let $y(x) = \sum_{N=0}^{\infty} c_N x^N$ be a solution of $Ly = 0$ with *rational c_N* and $Ly = 0$ has, for almost all p, n linearly independent solutions convergent in the p-adic unit circle. Then there are two integers A_0 and A_1 such that

the denominator of c_N divides

(1.4.6) $$A_0 \cdot A_1^N \cdot (\text{lcm}\{1, \ldots, N\})^{n-1}.$$

Thus y is a G-function!

REMARK. This structure of denominators suggests several things. First, if one puts above $n = 1$, then the corresponding condition means that the function $A_0 \cdot y(A, x)$ has integral coefficients—the statement of Eisenstein's theorem(see §1.2) for algebraic functions $y(x)$. Eisenstein's theorem also implies that the law (4.6) holds for functions that can be represented as $(n-1)$-dimensional integrals of algebraic functions. This brings us immediately to the next point.

The overconvergence and the "law of logarithmic growth" of solutions hold for linear differential equations with a natural action of Frobenius. A class of equations where the action of Frobenius was studied by Dwork, Katz, and others is the class of Picard-Fuchs differential equations (for variation of periods or homologies of smooth and singular varieties). In these cases, the law (4.6) holds (cf. our remark above about multiple integrals); moreover, Deligne's estimates show that the exponent $n - 1$ in the estimate (4.6) can be substituted by the dimension of the corresponding (smooth) variety.

It is natural to assume that the "law of logarithmic growth" for almost all p, and the law (4.6) for the denominators, hold for any linear differential equation that is globally nilpotent (i.e., its p-curvature is nilpotent for almost all p). Unfortunately, we know currently only that the nilpotence of the p-curvature implies the overconvergence.

Next, all evidence points towards the conjecture that the globally nilpotent equations are only those equations that are reducible to Picard-Fuchs equations (i.e., equations satisfied by Abelian integrals and their periods depending on a parameter).

As Dwork puts this conjecture, all globally nilpotent equations come from geometry.

Our results on G-functions (see §1.1) allow us to represent this conjecture even in a more fascinating form:

DWORK-SIEGEL CONJECTURE. *Let* $y(x) = \sum_{N=0}^{\infty} c_N x^N$ *be a G-function (i.e., the sizes of* c_N *and the common denominators of* $\{c_0, \ldots, c_N\}$ *grow not faster than the geometric progression in* N*).*

If $y(x)$ *satisfies a linear differential equation over* $\overline{\mathbf{Q}}(x)$ *of order* n *(but not of order* $n - 1$*), then the corresponding equation is reducible to Picard-Fuchs equations.*

In this case $y(x)$ *can be expressed in terms of multiple integrals of algebraic functions.*

Siegel, in fact, put forward a conjecture that is, in a sense, stronger than the one given above. To formulate Siegel's conjecture we have to look again at his E-functions defined in the same paper [3] of 1929, where G-functions were defined, and where Siegel's theorem on integral points was proved. We remind the reader (cf. §1.1) that a function $f(x) = \sum_{N=0}^{\infty} (c_N/N!)x^N$ is called an E-function if c_N are algebraic numbers and for any $\epsilon > 0$, the size $|\overline{c_N}|$ of c_N and the common denominator of $\{c_0, \ldots, c_N\}$ is bounded by $O(N!^\epsilon)$; one also assumes that $f(x)$ satisfies a linear differential equation over $\overline{\mathbf{Q}}(x)$. Siegel showed that the class of E-functions is a ring closed under differentiation and integration. Siegel also studied the hypergeometric functions

$$_mF_n\binom{a_1,\ldots,a_m}{b_1,\ldots,b_n}|\lambda x)$$

for algebraic $\lambda \neq 0$, *rational* parameters $a_1,\ldots,a_m$ and $b_1,\ldots,b_n$, and $m \leq n$. These functions he called hypergeometric E-functions and suggested in [4] all E-functions can be constructed from hypergeometric E-functions.

Looking at the (inverse) Laplace transform of $f(x)$, we see that Siegel's conjecture translates into a conjecture on G-function structure stronger than the Dwork-Siegel conjecture given above. Indeed, it would seem that all Picard-Fuchs equations might be expressed in terms of generalized hypergeometric functions. This conjecture should be very flattering for the generalized hypergeometric functions (with rational parameters!), but is it true?

This stronger conjecture is not entirely without merit; e.g., one can reduce linear differential equations over $\overline{\mathbf{Q}}(x)$ satisfied by G-functions to higher order equations over $\overline{\mathbf{Q}}(x)$ with regular singularities at $x = 0, 1, \infty$ only—like the generalized hypergeometric ones; cf. [27]. This problem of hypergeometric functions modulo the usual Dwork-Siegel conjecture is, in fact, a statement of deformations of algebraic varieties and has little to do with transcendental methods (unlike the Grothendieck and Dwork-Siegel conjecture). In such a formulation one has to remember that the original equation has to be defined over $\overline{\mathbf{Q}}(x)$; otherwise it is definitely not true.

We are unable so far to give a positive answer to this Dwork-Siegel conjecture that all arithmetically interesting $(G$-)functions are solutions of Picard-Fuchs equations. Nevertheless, in some cases we can prove that this conjecture is correct. For now our efforts are limited to the second-order equations (which provide an extremely rich class of functions).

PROPOSITION 4.2. *Let a second-order equation over* $\mathbf{Q}(x)$: $Ly = 0$ *be a globally nilpotent one and it has zero p-curvature* $\Psi_p = 0$ *for primes p lying in the set of destiny* $1/2$. *Then the corresponding linear differential equation either has all of its solutions as algebraic functions, or is reducible to a Picard-Fuchs equation (corresponding to the deformation of the curve), or has at least one transcendent element in a monodromy matrix for any representation of the monodromy group.*

It is very likely that, at least for the second-order equations, the Dwork-Siegel conjecture can be proved *modulo* the full Grothendieck conjecture.

Our methods are based on Padé approximations and, unfortunately, are conditional on the algebraic representation of the monodromy group as in Theorems 3.2–3.3. That is why we obtained only partial results in the problem of algebraicity and explicit determination of all globally nilpotent equations.

1.5. Lamé equations and their analytic applications. Let us look in detail at a very interesting case of the second-order linear differential equation with a particular attention paid to the first highly nontrivial case of equations with

4 regular singularities. Among these equations, a special role is played by the general Lamé equation.

We are looking at linear differential equations

$$(1.5.1) \qquad y'' + ay' + by = 0$$

for $a, b \in \mathbf{Q}(x)$ that are globally nilpotent. These equations according to Katz's result (see [2] and [1]) are Fuchsian with rational exponents at regular singularities. The first nontrivial class of such equations, depending on a number of undetermined (accessory) parameters, is the class of Heun's equation [31]:

$$(1.5.2) \qquad y'' + \left\{ \frac{\gamma}{x} + \frac{\delta}{x-1} + \frac{\epsilon}{x-a} \right\} y' + \frac{\alpha\beta x - q}{x(x-1)(x-a)} y = 0,$$

with regular singularities at $x = 0, 1, a$, and ∞ (here $\alpha + \beta - \gamma - \delta - \epsilon + 1 = 0$).

The Heun equation (5.2) with $\gamma = \delta = \epsilon = 1/2$ is called a general Lamé equation; it is represented in the form [31–33]:

$$(1.5.3) \qquad y'' + \frac{1}{2} \left\{ \frac{1}{x} + \frac{1}{x-1} + \frac{1}{x-a} \right\} y' + \frac{B - n(n+1)x}{4x(x-1)(x-a)} y = 0$$

(depending on n and on accessory parameters B).

A more familiar form of the Lamé equation is the transcendental one (with the change of variables: $a = k^{-2}$, $x = (sn(u,k))^2$) [31]:

$$(1.5.4) \qquad \frac{d^2 y}{du^2} + k^2 \cdot \{B - n(n+1)sn^2(u,k)\} y = 0$$

in terms of the Jacobi sn-function. An alternative form of (5.3–5.4) is in terms of Weierstrass's elliptic function:

$$(1.5.5) \qquad \frac{d^2 y}{du^2} + \{H - n(n+1)\mathcal{P}(u)\} y = 0.$$

Lamé equations are considered usually for integral values of the parameter n in (5.3–5.5). This is the only case when solutions of (5.5) (or (5.4)) are meromorphic functions in the u-plane. In the case of integral n the following facts are known [31–33]:

(i) There exist $2n + 1$ values of an accessory parameter (B in (5.3–5.4) or H in (5.5)) for which the algebraic form of the Lamé equation (5.3) has algebraic function solutions. These numbers B_n^m, $m = 1, \dots, 2n + 1$, are the ends of lacunae of the spectrum of an equation (5.4) considered as the spectral problem for the Lamé potential.

(ii) All solutions of (5.4) and (5.5) are meromorphic functions of u of order of growth 2.

Moreover, for every $B \neq B_n^m$, two linearly independent solutions of (5.5) have the form

$$F_{\pm} = \prod_{i=1}^{n} \frac{\sigma(a_i \pm u)}{\sigma(u)\sigma(a_i)} \cdot \exp\left\{ \mp u \sum_{i=1}^{n} \zeta(a_i) \right\}$$

for parameters a_i determined from B—all $\mathscr{P}(a_i)$ are algebraic in terms of B.

If the Lamé equation (5.3) is defined over $\overline{\mathbf{Q}}$ (i.e., $a \in \overline{\mathbf{Q}}$ and $B \in \overline{\mathbf{Q}}$) our local-global principle of algebraicity can easily solve the Grothendieck conjecture for Lamé equations with integral n. We have proved in [25]

THEOREM 5.1. *For integer $n \geq 0$ the Lamé equation has zero p-curvature for almost all p if and only if all its solutions are algebraic functions. The Lamé equation with integral n is globally nilpotent for $2n + 1$ values of B, $B = B_n^m$-ends of lacunae of spectrum of* (5.5).

For all other values of B, the global nilpotence of the Lamé equation with integral n over $\overline{\mathbf{Q}}$ is equivalent to the algebraicity of all solutions of (5.5).

The possibility of all solutions of (2.3) with $B \neq B_n^m$ was shown by Baldassari and kindly communicated to us by B. Dwork. Such a possibility corresponds only to special torsion points $H = \mathscr{P}(u_0)$ for special elliptic curves. All such cases for any given integral n and Lamé equations defined over $\mathbf{Q}$ can be explicitly determined.

For nonintegral n no simple uniformization of solutions of the Lamé equation exists. Moreover, Lamé equations themselves provide the key to several interesting uniformization problems. An outstanding Lamé equation is that with $n = -1/2$. This equation (and some of its equivalents to be seen later) determines the uniformization of the punctured torus. This leads to the classical Poincaré (Poincaré-Klein [34–35]) problem of the accessory parameter, which in the case of (5.3) with $n = -1/2$ means the determination for any $a \neq 0, 1, \infty$ a unique value of B, for which the monodromy group of (5.3) is represented by real 2×2 matrices. This complex-analytic investigation of the complex-analytic structure of the Lamé (and of the more general) equation and the accessory parameter had been actively pursued by Klein, Poincaré, Hilbert, Hilb [39], V. I. Smirnov [36], and recently Bers [37] and Keen [38]. Recently the accessory parameter problem was studied in connection with conformally invariant field theories by Polyakov, Takhtajan, Zograf, and others, cf. [40].

The uniformization problem for the punctured torus case is particularly easy to formulate, and our efforts towards the examination of the arithmetic nature of Fuchsian groups uniformizing algebraic curves were initially focused on this case. The punctured torus case can be easily described in terms of the Lamé equation with $n = -1/2$. If one starts with a torus corresponding to an elliptic curve $y^2 = P_3(x)$, then the function inverse to the automorphic function, uniformizing the torus arises from the ratio of two solutions of the Lamé equation with $n = -1/2$ [38, 32, 37]:

(1.5.6)
$$P_3(x)y'' + \frac{1}{2}P_3'(x)y' + \frac{x+C}{16}y = 0,$$

or

$$\frac{d^2y}{du^2} + \left[H + \frac{1}{4}\mathscr{P}(u)\right]y = 0.$$

If $P_3(x) = x(x-1)(x-a)$ (i.e., the singularities are at $x = 0$, a, and ∞), then the monodromy group of (5.6) is determined by 3 traces $x = \text{tr}(M_0M_1)$, $y = \text{tr}(M_0M_a)$, $z = \text{tr}(M_1M_a)$. Here M is a monodromy matrix in a fixed basis corresponding to a simple loop around the singularity a. These traces satisfy a single Fricke identity [38]:

$$x^2 + y^2 + z^2 - xyz = 0.$$

The (Poincaré) uniformization case is that, when in (5.6) the monodromy group can be represented by real 2×2 matrices.

There exists a single value of the accessory parameter C for which the uniformization takes place. Equivalently, C is determined by conditions of reality of x, y, z.

One can always assume that

$$2 < x \leq y \leq z \leq xy/2,$$

and these conditions remove Markov's famous triplets that satisfy the Fricke identity studied by H. Cohn and others in not unrelated circumstances.

ALGEBRAICITY PROBLEM [32]. *Let an elliptic curve be defined over $\overline{\mathbf{Q}}$ (i.e., $a \in \overline{\mathbf{Q}}$). Is it true that the corresponding (uniformizing) accessory parameter C is algebraic? Is the corresponding Fuchsian group a subgroup of $GL_2(\overline{\mathbf{Q}})$ (i.e., x, y, and z are algebraic)? Conversely, if x, y, and z are algebraic corresponding to a Fuchsian subgroup of $SL_2(\overline{\mathbf{Q}})$, is it true that the uniformized torus is defined over $\overline{\mathbf{Q}}$?*

Extensive multiprecision computations, we first reported in [32], of accessory parameters showed rather bleak prospects for algebraicity in the accessory parameter problems.

Namely, as it emerged, there are only 4 (classes of isomorphisms of) elliptic curves defined over $\overline{\mathbf{Q}}$, for which the values of uniformizing accessory parameters are algebraic. These 4 classes of algebraic curves are displayed below in the next section in view of their arithmetic importance.

There is absolutely no evidence *for* the existence of the fifth elliptic curve over $\overline{\mathbf{Q}}$ having algebraic accessory parameter C *or* algebraic x, y, and z. For some particularly interesting curves (e.g., $a = 1/3$ or curves with complex multiplication by $\sqrt{-2}$ or $2i$) our multiprecision computations found no nontrivial algebraic relations on C, x, y, or z with degree up 200 and sizes of integral coefficients up to 10^{100}. (Of course, there is a possibility that there is a relation of degree say, 300, and you are welcome to find it.)

1.6. Lamé equations and their arithmetic applications. What is left of that arithmetic sense of the uniformization problem for the punctured torus and why are we interested in algebraicity (rationality) of the accessory parameters anyway?

It seems that attention to the arithmetic properties of the Lamé equation with $n = -1/2$ arose shortly after Apéry's proof of the irrationality of $\zeta(2)$

and $\zeta(3)$. His proof (1978), see [42], was soon translated into assertions of integrality of power series expansions of certain linear differential equations.

To look at these differential equations we will make use of the classical equivalence between the punctured torus problem and that of 4 punctures on the Riemann sphere. For differential equations this means Halphen's algebraic transformation from [31, 41] between the Lamé equation with $n = -1/2$:

$$(1.6.1) \qquad P(x)y'' + \frac{1}{2}P'(x)y' + \frac{x+C}{16}y = 0,$$

for $P(x) = x(x-1)(x-a)$, and the Heun equation (see(5.1)) with zero-differences of exponents at all singularities:

$$(1.6.2) \qquad P(x)y'' + P'(x)y' + (x+H)y = 0.$$

The relation between two accessory parameters is the following:

$$C = 4H + (1+a).$$

Let us denote the equation (6.2) by $Ly = 0$.

We have already stated in §1.5 that there are 4 Lamé equations with $n = -1/2$ (up to Möbius transformations) for which the value of the accessory parameter is known explicitly and is algebraic. These are 4 cases when the Fricke equation

$$x^2 + y^2 + z^2 = xyz,$$

with $0 \le x \le y \le z \le xy/2$, has solutions whose squares are integers. It is in these 4 cases when the corresponding Fuchsian group uniformizing the punctured torus is the congruence (arithmetic) subgroup, see references in [32, 27].

Let us look at these 4 cases, writing down the corresponding equation (6.2):
(1) $x(x^2 - 1)y'' + (3x^2 - 1)y' + xy = 0$.
Here $H = 0$, $C = 0$, and the modular invariant J of the corresponding curve with $\lambda = -1$ is $J = 1728$.
(2) $x(x^2 + 3x + 3)y'' + (3x^2 + 6x + 3)y' + (x+1)y = 0$.
Here $H = 1$, $C = 8$, and the modular invariant J of the corresponding curve with $\lambda = \rho$ is $J = 0$.
(3) $x(x - 1)(x + 8)y'' + (3x^2 - 14x - 8)y' + (x+2)y = 0$.
Here $H = 2$, $C = 1$, and J of the corresponding elliptic curves is $J = 2^2 \cdot 73^3/3^4$.
(4) $x(x^2 + 11x - 1)y'' + (3x^2 + 22x - 1)y' + (x+3)y = 0$.
Here $H = 3$, $C = 12$, and J of the corresponding elliptic curve is $J = 2^{14}31^3/5^3$.

Each of the equations (1)–(4) is a pull-back of a hypergeometric function by a rational map. Each of the equations (1)–(4) is a Picard-Fuchs equation for an appropriate family of elliptic curves parametrized by the corresponding congruence subgroup, and, as a consequence, each of the equations is globally nilpotent.

In fact, for each of the equations (1)–(4) there exists an integral power series $y(x) = \sum_{N=0}^{\infty} c_N x^N$ satisfying $Ly = 0$ and regular at $x = 0$. This solution is unique, up to normalization, because the second regular solution of $Ly = 0$ at $x = 0$ has logarithmic terms.

It is here where Apéry's dense rational approximations to $\zeta(2)$ and $\zeta(3)$ appear. Apéry's example for $\zeta(2)$ arises from the equation (4). In this case the solution $y(x)$ of (4), regular at $x = 0$, has the form

$$y(x) = \sum_{N=0}^{\infty} c_N x^N$$

with integral c_N, whose explicit expression was given by Apéry:

$$c_N = \sum_{j=0}^{N} \binom{N}{j}^2 \binom{N+j}{j}.$$

It is easy to see that a nonhomogeneous equation $Ly = \text{const} \neq 0$ has a solution $z(x) = \sum_{N=0}^{\infty} d_N x^N$ regular at $x = 0$ with nearly integral d_N (this is according to the global nilpotence of the corresponding L). In fact, the denominator of d_N always divides $\mathrm{lcm}\{1, \ldots, N\}^2$. An explicit expression of d_N was again given by Apéry; see [42].

The only (possible) singularities of $y(x)$ and $z(x)$ in the finite part of the plane are $x = (-11 \pm 5\sqrt{5})/2$, where all local exponents are zero. Thus we can always find a constant ζ such that

$$\zeta \cdot y(x) + z(x)$$

is regular at $x = \frac{-11+5\sqrt{5}}{2} = (\frac{\sqrt{5}-1}{2})^5$.

Apéry determined the constant ζ; if one takes $z(x) = \sum_{N=0}^{\infty} d_N x^N$ as $Lz = 5$, then $\zeta = \zeta(2) = \pi^2/6$.

Consequently, one has $c_N \in \mathbf{Z}$, $d_N \cdot \mathrm{lcm}\{1, \ldots, N\}^2 \in \mathbf{Z}$, and

$$0 < |c_N - d_N \zeta(2)| < \left(\frac{\sqrt{5}-1}{2}\right)^{5N},$$

since $\zeta \cdot y(x) + z(x)$ converges in the disc $|x| < (\frac{\sqrt{5}+1}{2})^5$.

Thus one obtains the irrationality of $\zeta(2)$ and, because c_N/d_N are dense approximations, a nontrivial measure of the irrationality of π^2 is derived too.

Apéry did not prove his results via differential equations; differential equations appeared in reformulations of Apéry's proofs by H. Cohen and A. Van der Poorten [42]. Later Dwork realized the global nilpotence of the corresponding equations and explicitly computed their solutions as periods in the Picard-Fuchs problem for a family of elliptic curves

$$Y^2 = X^3 + \tfrac{1}{4}(1 + 6x + x^2)X^2 + \tfrac{1}{2}x(x+1)X + \tfrac{1}{4}x^2$$

parametrizing the 5-torsion elliptic curves, (i.e., corresponding to $\Gamma_1(5)$).

Other equations (1)–(3) can be used in a similar way, because of their global nilpotence, and the existence of only 4 regular singularities. No irrationality results arise from them directly. However, Apéry approximations to $\zeta(3)$ arise from the equation (3). To see this, one turns to the Lamé form (6.1) of equation (3) and takes a symmetric square of the corresponding operator L; i.e., we are looking at the third-order linear differential equation satisfied by all products of solutions of (6.1). Since differences between local exponents of (6.1) are always $1/2$ (in the finite part of the plane), the corresponding symmetric square has, as above, integral exponent differences at regular points (see Hermite's transformations in [31, 42]). After simple transformations we arrive at Apérys third-order equation:

$$Ly = x^2(x^2 - 34x + 1)y''' + x(6x^2 - 153x + 3)y''$$
$$+ (7x^2 - 112x + 1)y' + (x - 5)y = 0.$$

Again there exist a unique solution $y(x) = \sum_{N=0}^{\infty} c_N x^N$ of $Ly = 0$ and a nearly integral solution $z(x) = \sum_{N=0}^{\infty} d_N x^N$ of $Lz = 1$ such that

$$|c_N - \zeta(3)d_N| < (\sqrt{2} - 1)^{4N}.$$

This shows that $\zeta(3)$ is irrational.

These examples lead to a method of the construction of sequences of dense approximation to numbers using nearly integral solutions of globally nilpotent equations. Often the corresponding equations are Picard-Fuchs equations satisfied by generating functions of Padé approximants to solutions of special linear differential equations; see examples in [43, 44]. This is, particularly, the case of Padé approximations to generalized hypergeometric functions, though the corresponding Picard-Fuchs equations (already for $_3F_2$ case) are very complicated (cf. [43]). As we shall see, Apéry's approach is not the best for the purpose of improving measures of irrationality. Nevertheless it gives a good explanation why one needs globally nilpotent equations of this particular kind.

Since we have motivated ourselves towards the search for globally nilpotent, or simply p-adically nilpotent (for fixed p), equations, particularly, the Lamé equations, let us try to establish the goals, bearing in mind an archimedean (complex) situation.

Which values of the accessory parameter H in (6.2) (call it a spectral parameter; cf. the transcendental form of the Lamé equation (5.3–5.4)) are we interested in?

First of all, if $H \in \mathbf{Q}$ (or $\overline{\mathbf{Q}}$), we want to determine all cases of global nilpotence.

Our intensive numerical experiments reveal an unpleasant, but predictable, phenomenon: it seems that, with the exception of equations (1)–(4) (and all equations equivalent to them via Möbius transformations), there are *no* Lamé equations with $n = -1/2$ over $\overline{\mathbf{Q}}$ that are globally nilpotent. We put these observations as a

CONJECTURE 6.1. *Lamé equations with $n = -1/2$ defined over $\overline{\mathbf{Q}}$ are not globally nilpotent except for 4 classes of equations corresponding to the congruence subgroups, with representatives of each class given by (1)–(4).*

What are our grounds for this conjecture?

First of all, the Padé approximation technique related to the Dwork-Siegel conjecture allows us to prove one positive result in the direction of this conjecture for the $n = -1/2$ case of the Lamé equation.

THEOREM 6.2. *For fixed $a \in \overline{\mathbf{Q}}$ $(a \neq 0, 1)$, there are only finitely many algebraic numbers C of bounded degree d such that the Lamé equation with $n = -1/2$ is globally nilpotent.*

The similar result holds for any Lamé equation with rational n.

Of course, one wants a more specific answer for any n (e.g., for $n = -1/2$, there are only 4 classes of a and C given above with the global nilpotence). However for half-integral n, there are always $n + 1/2$ trivial cases of global nilpotence, where solutions are expressed in terms of elliptic integrals, see [**31, 41**].

We have started the study of globally nilpotent Lamé equations (6.1) or (6.2) with numerical experiments. This ultimately led to Conjecture 6.1.

We checked possible equations of the form (6.2) with

$$P(x) = x(x^2 - a_1 x + a),$$

i.e., 4 singularities at $x = 0, \infty$, and 2 other points, for values of $a_1, a \in \mathbf{Z}$ in the box: $|a|, |a_1| \leq 200$.

For all these equations (6.2) we checked their p-curvature for the first 500 primes (see below in §1.7). Our results clearly show that with the exception of 4 classes of equations equivalent to (1)–(4), any other equation has a large proportion of primes p such that the p-curvature is not nilpotent for any $H \in \mathbf{Q}$!

An additional argument for Conjecture 6.1 is the Dwork-Siegel conjecture. If an equation were globally nilpotent, it would be a Picard-Fuchs equation or, at least, would give rise to the uniformization of the corresponding punctured torus. Our numerical evidence against the algebraicity of the accessory parameter in the (complex) uniformization problem for punctured torus, see §1.5, is even more impressive.

There is, unfortunately, always a possibility that there exists somewhere a fifth elliptic curve over $\overline{\mathbf{Q}}$ (or even $\mathbf{Q}$) for which the Lamé equation (6.2) is globally nilpotent with some $H \in \overline{\mathbf{Q}}$ of high degree and height.

1.7. P-adic spectrum of second-order equations. From globally nilpotent equations let us now turn to local integrality conditions; i.e., instead of the integrality of power series expansions of solutions we look at p-integrality of power series expansions for a given p. This means that we look at the nilpotence condition of the p-curvature for a fixed p.

Trying to keep the analogy with the archimedean real case, we will try to define the (arithmetic) spectrum of the Lamé equation in the p-adic or mod p case. It is instructive to look at a special case of the p-adic analysis of the simplest Lamé equation with $n = 0$, which was performed in 1958 by John Tate. We refer to Dwork's review [45] for the exposition of these results.

In the case $n = 0$ the transcendental form of the Lamé equation is very simple:

$$(1.7.1) \qquad \frac{d^2}{du^2}y - C^2 \cdot y = 0$$

with $x = \mathscr{P}(u)$ corresponding to the elliptic curve E, given by the general elliptic curve equation $Y^2 + a_1 XY + a_3 Y = X^3 + a_2 X^2 + a_4 X + a_6$.

Let the curve E be defined over p-adic number field k. If the reduced curve $\overline{E}$ over the residue class field $\overline{k}$ is nonsingular (good reduction case), *and if* $\overline{E}$ has nonzero Hasse invariant, then there exists a unit C in the maximal unramified extension K of k such that the functions $\exp(\pm Cu)-$ solutions of (7.1) have, as power series in x-plane (see equation (5.3)), integral coefficients in K. The parameter C is closely connected with eigenvalues of the Frobenius:

$$C^{\sigma-1} = \alpha,$$

where α is the unit root of the ζ-function of $\overline{E}$ (an eigenvalue of the Frobenius), and σ is the Frobenius of K over k. Here $\alpha \cdot \overline{\alpha} = q$ for $\overline{k} = \mathbf{F}_q$.

This result of Tate, and its wide generalizations due to Dwork [45] and Katz, was a starting point of p-curvature study. It was also used some ten years later by Deligne, who noticed that for a curve E defined over $\mathbf{Q}$ and for a nonsupersingular p, the algebraic form of Lamé equation (5.3) with $n = 0$ has a nilpotent p-curvature *only* when $C = 0$. See [2] and [18] for the presentation of this result.

This immediately provides one with an example of an equation, defined over $\mathbf{Q}(x)$, that is Fuchsian, has rational exponents at all singularities, but for infinitely many prime p has a nonnilpotent p curvature. Of course, now we know due to Theorem 5.1 that many Lamé equations with integral n provide one with such an equation for most values of accessory parameter B.

This example of Tate suggests the following definition of the (arithmetic) spectrum of a Lamé equation, which we study in the most interesting case of $n = -1/2$ (we take the form (7.2)):

$$(1.7.2) \qquad P(x)y'' + P'(x)y' + (x + H)y = 0,$$

for

$$P(x) = x \cdot (x^2 - a_1 x + a) \quad (= x(x - 1)(x - a)).$$

We are interested in those H mod p for which the p-curvature of (7.2) is nilpotent, and particularly in those p-adic $H \in \mathbf{Z}_p$ for which there exists a solution $y = y(x)$ of (7.2) whose expansion has p-integral coefficients. We call those $H \in \mathbf{Z}_p$, for which such $y(x)$ exists, eigenvalues of (7.2) in the "p-adic domain," and their set we call "an integral p-adic spectrum." The problem of studying the arithmetic nature of the Lamé equation was proposed by I.M. Gelfand. Of course, if a p-curvature is not zero, there is no second solution $y(x)$ with the same property.

REMARK. As a little derivation, we want to note the relationship between the usual spectrum of the Lamé equation for $n = -1/2$ and the uniformization problems. The unique value of the (Fuchsian) accessory parameter does not have a simple spectral sense. On the other hand, some values of the accessory parameter that correspond to boundaries of quasi-Fuchsian domain and different Kleinian cases do have spectral interpretation.

We had conducted extensive p-adic computations to determine "an integral p-adic spectrum" of the Lamé equations. Some interesting phenomena emerged, which again point to the special role played by the 4 congruence subgroup equations (1)–(4).

We summarize some of these observations below.

The main computational tool to check the nilpotence of the p-curvature of the equation (7.2) $Ly = 0$ is to check directly in the identity (4.2)

$$\left(\frac{d}{dx}\right)^m \equiv a_m \left(\frac{d}{dx}\right) + b_m \mod \mathbf{F}_p(x) \left[\frac{d}{dx}\right] \cdot L \quad (\mathrm{mod}\, p)$$

whether one has $a_{2p} \equiv 0 \mod p$, and $b_{2p} \equiv 0 \mod p$. An easier task is to look at the formal power series expansion of a single solution of (7.2) regular at $x = 0$

$$y(x) = \sum_{N=0}^{\infty} \frac{A_N}{A^N \cdot N!^2} x^N,$$

$A_0 = 1$. The condition of nilpotence of the p-curvature of (7.2) is equivalent to

$$A_p \equiv 0 \quad \mod p.$$

To check these conditions and the conditions of p-integrality of coefficients in the expansion of $y(x)$, we used the SCRATCHPAD (IBM) and MACSYMA (Symbolics Inc.) systems.

We start with the observations of the "mod p" spectrum as p varies.

I. For noncongruence equations (7.2) with rational $a \neq 0, 1$ (i.e., for an elliptic curve defined over $\mathbf{Q}$ with a point of order 2) there seem always to be infinitely many primes p for which no value of the accessory parameter H mod p gives a nilpotent p-curvature (thus the mod p spectrum is empty).

On the basis of our observations (for $p \leq 10,000$) we conjecture that for every noncongruence equation (7.2) the number of such primes p, with null mod p spectrum, is infinite.

One can go further and predict the distribution of such p; one can compare their distribution to that of anomalous primes p (i.e., those p for which $a_p = 1$ for the trace a_p of the Frobenius of an elliptic curve mod p), and to the distribution of primes in quadratic progressions. We cannot state with definitiveness any detailed distribution conjecture.

Sometimes the first prime p for which the mod p spectrum of (7.2) is null occurs quite far. Here are a few statistics for noncongruence equations with rational integers a:

For $a = 3$ the first p's with the null spectrum mod p are $p = 61, 311, 677,$ 1699, 1783, 1811, 2579, 2659, 3253,... .

For $a = 5$ the first p's with the null spectrum mod p are $p = 659, 709,$ 1109, 1171, 1429, 2539, 2953, 2969, 3019, 3499, 3533, 3803, 3863, 4273, 4493, 4703, 4903, 5279, 5477, 5591, 6011, 7193, 7457, 7583,... .

For $a = 4$ the corresponding p's with the null spectrum are $p = 101, 823,$ 1583, 2003, 3499,... .

For $a = 13$ the corresponding list starts at $p = 1451, 1487, 2381,...$.

Observation I above was checked by us for all noncongruent $P(x) = x(x^2 - a_1 x + a)$ with integral $a_1 a$ not exceeding 250 in absolute value.

II. An integral p-adic spectrum of equations (7.2) with (p-integral) a has a complicated structure depending on the curve.

A p-adic spectrum can be null, finite (typically a single element), or infinite, resembling the Cantor set. Additional information about the p-adic spectrum of (7.2) can be derived from the p-adic spectrum of the Lamé equations with integral $n = (p^k - 1)/2$. Unfortunately, our lack of knowledge of algebraic properties of the (classical/complex/real) spectrum of the Lamé equation with integral $n > 2$ does not allow a nontrivial analytic description.

Numerical analysis is not easy either. To study p-adic expansions of p-adic numbers from p-adic spectra up to the order $\mod p^k$, one has to carry all the computations of rational functions and coefficients of power series expansions in the modular arithmetic $\mod p^N$ for $N = 2(p^k - 1)/(p - 1)$.

For example, in order to determine the 3-adic spectrum with 14 digits of precision (in the 3-adic expansion), one has to carry out computations with numbers over 2,000,000 decimal digits long!

For congruence equations (1)–(4) the p-adic spectrum seems to be an infinite one with a Cantor-like structure. For noncongruence equations, the p-adic spectrum is possibly a finite one, often null, and in many other cases having a single element. Here the computations were not carried out too far for reasons of exponential increase in complexity. We can present some amusing data. We present the picture of the 3-adic integral spectrum of the congruence equation (we took $\lambda = a = 2$), where the spectrum exhibits a typical Cantor form: for an increase of two digits in the 3-adic expansion three possibilities out of nine are allowed.

To represent graphically 3-adic numbers we divide an interval into 3 parts according to whether the digit is 0, 1 or 2, and then continue the division of subintervals in the same manner. In the next picture we represent in this way elements of $\mathbf{Z}_3$ of the 3-adic spectrum of the congruence curve. We present 2,187 elements of the spectrum, which gives correctly 12 leading digits in the 3-adic expansion of all spectrum elements.

FIGURE 1

The first part of this Figure 1 shows the display of the whole spectrum, and the second one magnifies an underlying portion of the spectrum.

As with the complex-analytic uniformization problem, one can point to an obvious relationship with the p-adic uniformization of Mumford, generalizing Tate's construction. In our case this should correspond to the p-adic spectrum for curves having multiplicative reduction at p.

1.8. Arithmetic subgroups and second-order equations. It might appear from the discussion above that one cannot construct new examples of an irrationality proof á la Apéry, because of the lack of globally nilpotent equations. This is wrong; there is an abundance of globally nilpotent equations that can be used for various irrationality and transcendence proofs. Picard-Fuchs equations, for example, provide generating functions for Padé approximants in the Padé approximation problem to generalized hypergeometric functions, Pochhammer integrals, and other classes of solutions of linear differential equations (with appropriate integral representation); see Chapter 2. In these cases, however, orders of Picard-Fuchs equations are high and the number of apparent and regular singularities is rather large. That is why it is often more convenient to substitute the analysis of a linear differential equation by the direct examination of its solution, when represented by multiple integrals as we describe below in Chapter 2. One always needs, though, a starting point—an initial Picard-Fuchs equation and the integral formula for its solutions, which one can then improve upon. We suggest, as a starting equation, when it is of the second-order, an equation corresponding to an arithmetic Fuchsian subgroup. Congruence subgroups of $\Gamma(1)$ and quaternion groups provide interesting families of globally nilpotent equations. As the conjecture in §1.4 claims, these are the only globally nilpotent equations.

One can start with equations uniformizing punctured torus with more than one puncture, to add to the list in §1.4. The complete description of arithmetic Fuchsian groups of signature $(1; e)$ had been provided by Maclachlan and Rosenberger [46] and Takeuchi [47]. These groups of signature $(1; e)$ are defined according to the following representation:

$$\Gamma = \langle \alpha, \beta, \gamma | \alpha\beta\alpha^{-1}\beta^{-1}\gamma = -1_2, \gamma^e = -1_2 \rangle,$$

where α and β are hyperbolic elements of $SL_2(\mathbf{R})$ and γ is an elliptic (respectively a parabolic) element such that $\mathrm{tr}(\gamma) = 2\cos(\pi/e)$.

For all $(1;e)$ arithmetic subgroups there exists a corresponding Lamé equation with a rational n, uniformized by the corresponding arithmetic subgroup. This way we obtain *73* Lamé equations, all defined over $\overline{\mathbf{Q}}$ (i.e., the corresponding elliptic curves and accessory parameter C are defined over $\overline{\mathbf{Q}}$). Some of these equations give rise to nearly integral sequences satisfying three-term linear recurrences with coefficients that are quadratic polynomials in n, and have the growth of their denominators and the convergence rate sufficient to prove the irrationality of numbers arising in this situation in a way similar to that of Apéry. Unfortunately, we cannot yet report the identification of these numbers with classical constants, because they are represented as integrals of combinations of hypergeometric functions (i.e., as double integrals), which we cannot yet explicitly integrate. Among Lamé equations arising this way we can list the following series.

Groups of the signature $(1;e)$ correspond to the Lamé equations (see (5.3)):

$$P(x)y'' + \frac{1}{2}P'(x)y' + \left\{C - \frac{n(n+1)}{4}x\right\}y = 0$$

with $n + \frac{1}{2} = \frac{1}{2e}$.

In the arithmetic case one looks at totally real solutions of the modified Fricke's identity, which now takes the form:

$$x^2 + y^2 + z^2 - xyz = 2(1 - \cos(\pi/e)).$$

Using the numerical solution of the (inverse) uniformization problem, we determined the values of the accessory parameters. Among the interesting cases are the following:

Here $P(x) = x(x - 1)(x - A)$ and

$(1;2)$-case:
(1) $A = 1/2$, $C = -3/128$ $(x = y = (1 + \sqrt{2})^{1/2} \cdot 2^{3/4}, z = 2 + \sqrt{2})$;
(2) $A = 1/4$, $C = -1/64$;
(3) $A = 3/128$, $C = -13/2^{11}$;
(4) $A = (2 - \sqrt{5})^2$, $C = \sqrt{5} \cdot (2 - \sqrt{5})/64$;
(5) $A = (2 - \sqrt{3})^4$, $C = -(2 - \sqrt{3})^2/2^4$;
(6) $A = (21\sqrt{33} - 27)/256$.

$(1;3)$-case:
(1) $A = 1/2$, $C = -1/36$;
(2) $A = 32/81$, $C = -31/2^4 \cdot 3^4$;
(3) $A = 5/32$, $C = -67/2^9 \cdot 3^2$;
(4) $A = 1/81$, $C = -1/2 \cdot 3^4$;
(5) $A = (8 - 3\sqrt{7})/2^4$.

(1;4)-case:
(1) $A = -11 + 8\sqrt{2}$, C-cubic;
(2) $A = (3 - \sqrt{8})/4$, C-cubic;
(1;5)-case:
(1) $A = 3/128$, $C = -397/2^{11} \cdot 5^2$.

In all cases above, A is real (as well as C) and $0 < A \le 1/2$.

Not all elliptic curves corresponding to $(1; e)$-groups are defined over $\mathbf{Q}$— there is a nontrivial action of the Galois group (cf. a different situation in [47]).

1.9. Nearly integral solutions of linear recurrences and arithmetic continued fractions. The problem of explicit determination of all linear differential equations that have arithmetic sense (i.e., an overconvergence property or the existence of nontrivial solutions mod p) can be easily translated into a classical problem of nearly integral solutions to linear recurrences. This problem arose in works of Euler, Lambert, Lagrange, Hermite, Hurwitz, Stieltjes, and others in connection with irrational continued fraction expansions of classical functions and constants.

Let us start with a reformulation of the arithmetic problems á la Dwork-Siegel of §1.4:

PROBLEM 9.1. Let u_n be a solution of a linear recurrence of rank r with coefficients that are rational (polynomial) in n:

$$u_{n+r} = \sum_{k=0}^{r-1} A_k(n) \cdot u_{n+k}$$

for $A_k(n) \in \overline{\mathbf{Q}}(n)$, $k = 0, \ldots, r - 1$, and such that u_n are "nearly integral." Then the generating function of u_n is a geometric object.

In Problem 9.1 a "geometric object" means a function whose local expansion represents either an integral of an algebraic function or a period of an algebraic integral, i.e., a solution of a Picard-Fuchs-like equation. The "near integrality" of u_n means that the u_n are algebraic numbers whose sizes grow slower than factorials (i.e., for any $\epsilon > 0$, the sizes of u_n are bounded by $(n!)^{\epsilon}$) and whose common denominator also grows slower than a factorial (i.e., for any $\epsilon > 0$ the common denominator of $\{u_0, \ldots, u_n\}$ is bounded by $(n!)^{\epsilon}$).

REMARK. As mentioned above, this condition of near integrality is close to the overconvergence condition imposed on the generating function of u_n. The growth of the size of a denominator of u_n as a factorial in n is typical for a generic solution of a generic linear recurrence.

Problem 9.1 concerning near integral solutions of linear recurrences initially arose for three-term linear recurrences, where this problem is crucial for the proof of irrationality of a number defined by a continued fraction expansion. In this particular case (rank r is two, in the recurrence above) Problem 9.1 can be reformulated as

PROBLEM 9.1′. Let us look at an explicit continued fraction expansion with partial fractions being rational functions of indices:

$$\alpha = [a_0; a_1, \ldots, A(n), A(n+1), \ldots],$$

for $A(n) \in \mathbf{Q}(n)$. Let us look then at the approximations P_n/Q_n to α defined by this continued fraction expansion

$$\frac{P_n}{Q_n} = [a_0; a_1, \ldots, A(n-1), A(n)],$$

$n \geq 1$, where $P_n, Q_n \in \mathbf{Z}$.

If the continued fraction representing α is convergent *and* for *some* $\epsilon > 0$

$$\left| \alpha - \frac{P_n}{Q_n} \right| < |Q_n|^{-1-\epsilon},$$

$n \geq n_0(\epsilon)$, i.e., if α is *irrational*, then the sequences P_n and Q_n of numerators and denominators in the approximations to α are arithmetically defined sequences; their generating functions represent solutions of Picard-Fuchs and generalized Picard-Fuchs equations.

REMARK. The later equations correspond to deformations with possible irregular singularities, arising from Laplace and Borel transforms of solutions of ordinary Picard-Fuchs equations.

REMARK. While partial fractions $a_n = A(n)$ are rational functions of n, the sequences P_n and Q_n are *not* rational or algebraic functions of n except in very special cases, when α is reducible to a rational number. We can ask, in Problem 9.1′, whether P_n and Q_n have sizes and denominators that grow slower than a factorial of n, say, as a geometric progression in n. This problem is slightly weaker than that of the full Problem 9.1′, because the requirement of convergence of a continued fraction to α is dropped. We conjecture in this case that, as in Problem 9.1′, the generating functions of P_n and Q_n are solutions of Picard-Fuchs equations.

It is incorrect to assume that for a generic $A(n)$, or even for a large class of rational functions $A(n)$, the approximations P_n/Q_n to α have *any* arithmetic sense. Appropriate numbers α in general represent certain ratios of elements of monodromy matrices of a linear differential equation associated with a linear recurrence. For the discussion of the transcendental nature of α in the solution of the monodromy problem, see [32]. The constant α and P_n/Q_n acquire an arithmetic sense only when the corresponding differential equation is globally nilpotent. Our conjectures say that in this case it is also a Picard-Fuch equation or is reducible to one. There are only a few such equations! Even when α is itself a period of an algebraic variety (curve) defined over $\mathbf{Q}$, this by no means guarantees the existence of the convergent explicit continued fraction expansion of α with nearly integral P_n and Q_n.

How often do such continued fraction expansions occur, apart from classical cases known to Euler (Hermite in the multidimensional case)?

One of the main purposes of our investigation was an attempt to establish, first empirically, that there are only finitely many classes of such continued fraction expansions, all of which can be determined explicitly. One has to distinguish three types of numbers/functions α and Picard-Fuchs-like equations that can occur when such a continued fraction expansion of α exists:

A. Θ-function parametrization. This is the case when a linear differential equation can be parametrized by Abelian or θ-functions. This is the case of linear differential equations reducible to the so-called finite band/isospectral deformation equations. Such equations include the Lamé equations with integral n. In general, the continued fraction expansions representing appropriate α do not have an arithmetic sense (are not convergent, do not give irrationality, etc.). Here α depends on the spectral parameter (uniformizing parameter of the curve) and on the curve moduli. For special values of spectral parameter ("ends of lacunae"), α is represented as a convergent continued fraction expansion with an arithmetic sense. In all these cases we had completely determined all the cases of global nilpotence in our work on the Grothendieck conjecture; see [19]. We return to the class A of continued fractions in connection with Stieltjes-Rogers continued fraction expansions.

B. In this case the monodromy group of a linear differential equation associated with a linear recurrence of any rank is connected with one of the triangle groups. These groups do not have to be arithmetic. The cases of finite Schwarz groups and elliptic groups are relatively easy to describe (cf. Klein's description of linear differential equations having algebraic function solutions only). The hyperbolic (Fuchsian) cases provide a large class of equations of high rank that are the blowups of hypergeometric equations using a rational function map. This is the case of Apéry's recurrences and continued fractions. However, for any *given* rank r there are only *finitely many* linear differential equations that occur this way.

C. Not all arithmetic Fuchsian groups are directly related to triangle ones, though Jacquet-Langland correspondence suggests some relationship at least on the level of representations and underlying algebraic varieties in the SL_2 case. In any case, to class C belong those α's and continued fraction expansions for which the corresponding differential equation has an arithmetic monodromy group. Again, for a fixed rank and order, there are only finitely many such equations, if one is restricted to one-dimensional arithmetic Fuchsian groups. Multidimensional arithmetic groups, particularly Picard groups and associated Pochhamer differential equations, provide *classes* of continued (more precisely, multidimensional continued) fractions corresponding to periods on algebraic surfaces and varieties.

We conclude this section with new examples of arithmetic three-term linear recurrences of particularly attractive form (cf. §6).

In applications to diophantine approximations, particular attention is devoted to three-term linear recurrences such as

$$n^d \cdot u_n = P_d(n) \cdot u_{n-1} - Q_d(n) \cdot u_{n-2}, \qquad n \geq 2,$$

for $d \geq 2$. Apart from trivial cases (reducible to generalized hypergeometric functions), our conjectures claim that for every $d \geq 1$, there are only finitely many classes of such recurrences that correspond to deformations of algebraic varieties.

For $d = 2$ (second-order equations) we have classified nontrivial three-term recurrences whose solutions are always nearly integral, assuming our integrality conjectures. Most of these recurrences are useless in arithmetic applications (apart from two known cases in §6). There are a few new ones that give some nontrivial results. Among these recurrences are the following:

(i) $2n^2 u_n = 2(-15n^2 + 20n - 7) \cdot u_{n-1} + (3n - 4)^2 \cdot u_{n-2}$;

(ii) $3n^2 u_n = (-12n^2 + 18n - 7) \cdot u_{n-1} + (2n - 3)^2 \cdot u_{n-2}$;

(iii) $n^2 u_n = (-12n^2 + 18n - 7) \cdot u_{n-1} + (2n - 3)^2 \cdot u_{n-2}$;

(iv) $n^2 \cdot u_n = (56n^2 - 70n + 23) \cdot u_{n-1} - (4n - 5)^2 \cdot u_{n-2}$.

There is a larger class of rank $r > 2$ linear recurrences of the form

$$n^2 \cdot u_n = \sum_{k=1}^{r} A_k(n) \cdot u_{n-k},$$

all solutions of which are nearly integral. Many of these recurrences (like iii) above) give rise to new irrationalities. E.g., we present the following new globally nilpotent equation ($r = 3$):

$$4x(x^3 + 16x^2 + 77x - 2)y''$$
$$+ 8(2x^3 + 24x^2 + 77x - 1)y'$$
$$+ (9x^2 + 70x + 84)y = 0.$$

1.10. Explicit continued function expansions associated with Lamé equations and elliptic functions. The correspondence between the power series expansion and the continued fraction expansion of the same analytic function is a nontrivial and almost always an implicit one. A few classes of cases, when the function is known in the "closed" form and its continued fraction expansion can be "explicitly" determined, are rare and treasured instances. In function theory these cases are usually associated with classical orthogonal polynomials. Askey and his coworkers compiled a full list of neoclassical orthogonal polynomials, which include a variety of q-basic generalized hypergeometric polynomials; see [49]. These classes of polynomials include virtually all cases when the function is known explicitly, as well as its continued fraction expansions (i.e., with an explicit expression for three-term linear recurrence) and the corresponding orthogonal polynomials (i.e., the generating function of these polynomials is also known in a closed form). This list is perhaps close to being complete, at least when one deals with generalized (q-basic) hypergeometric polynomials.

It appears, however, that there are other classes of explicit continued fraction expansions of known analytic functions, not immediately related to classical hypergeometric functions. These continued fraction expansions include many of Rogers's continued fractions and are related to elliptic theta-functions. They are not a part of the famous Rogers-Ramanujan identities, which are interpreted as a part of generalized hypergeometric function relations.

The study of these peculiar continued fractions was initiated by Stieltjes [50–51], and we have recently found a whole class of similar continued fraction expansions that can be easily interpreted from the point of view of Lamé equations, providing even a few quite interesting arithmetic applications.

But first a little bit of the classical theory of Stieltjes continued fraction expansions. The theory of J-continued fraction expansions describes the relationship between the power series expansion at ∞,

$$f = \sum_{n=0}^{\infty} \frac{(-1)^n c_n}{x^{n+1}},$$

and the corresponding continued fraction expansion (J-fraction)

$$\cfrac{1}{b_1 + x - \cfrac{a_1}{b_2 + x - \cfrac{a_2}{b_3 + x -}}} \,.$$

In this theory the orthogonal polynomials associated with f satisfy the three-term linear recurrence:

$$Q_n(x) = (b_n + x)Q_{n-1}(x) - a_{n-1}Q_{n-2}(x),$$

$$n = 2,\dots (Q_0 = 1, \qquad Q_1 = x + b_1).$$

There are classical algorithms that determine a_n, b_n from c_n and vice versa. They all, as shown by Stieltjes [50, 51] (Rogers [52] and Schur [53]), are reduced to the question of representing an infinite-variable quadratic form as a sum of squares:

$$\sum_{n,m=0}^{\infty} c_{n+m} x_n x_m = (x_0 + k_{01}x_1 + k_{02}x_2 + \cdots)^2$$

$$+ a_1(x_1 + k_{12}x_2 + k_{13}x_3 + \cdots)^2$$
$$+ a_1 a_2(x_2 + k_{23}x_3 + k_{24}x_4 + \cdots)^2 + \cdots$$

$$[b_1 = k_{01}, \qquad b_{n+1} = k_{n,n+1} - k_{n-1,n}, \quad n \geq 2].$$

A nice reformulation of this identity by Rogers [52] gives rise to the following "addition" law:

$$G(x + y) = G(x) \cdot G(y) + a_1 G_1(x) \cdot G_1(y) + a_1 a_2 G_2(x) \cdot G_3(y) + \cdots,$$

where

$$G(x) \overset{\text{def}}{=} \sum_{n=0}^{\infty} c_n \frac{x^n}{n!}$$

and

$$G_m(x) = \sum_{n=m}^{\infty} k_{m,n} \frac{x^n}{n!} = O(x^m).$$

From these formal identities one can inherit explicit continued fraction expansions, whenever $G(x)$ satisfies algebraic laws of addition of special form. The corresponding classes of continued fraction expansions were determined by Stieltjes [50, 51], Rogers [52], Schur [53], and Carlitz [54]. To see them, one needs only a book on classical analysis. An easy example is the following:

$$G(x) = \text{sech}^k(x), \qquad \text{sech}^k(x + y) = (\cosh x \cdot \cosh y + \sinh x \cdot \sinh y)^{-1}.$$

Then the J-fraction is the following one:

$$\int_0^\infty \text{sech}^k u \cdot e^{-xu} du = \cfrac{1}{x + \cfrac{1 \cdot k}{x + \cfrac{2 \cdot (k+1)}{x + \cfrac{3 \cdot (k+2)}{x + \ddots}}}}.$$

One generalizes this continued fraction expansion to elliptic functions, particularly to $sn(u,k)$, $cn(u,k)$, $dn(u,k)$, and their transformations. The most representative is

$$\int_0^\infty cn(u, k) \cdot e^{-xu} du = \cfrac{1}{x + \cfrac{1^2}{x + \cfrac{2^2 \cdot k^2}{x + \cfrac{3^2}{x + \cfrac{4^2 \cdot k^2}{x + \cfrac{5^2}{\ddots}}}}}}.$$

In the notations of J-fractions among other examples are the following (1.10.1)

$$\int_0^\infty sn(u, k^2) e^{-ux} du = \frac{1}{z^2 + a-} \frac{1 \cdot 2^2 \cdot 3k^2}{z^2 + 3^2 a-} \frac{3 \cdot 4^2 \cdot 5k^2}{z^2 + 5^2 a-} \frac{5 \cdot 6^2 \cdot 7k^2}{z^2 + 7a^2 - \cdots},$$

$$z \int_0^\infty sn^2(u, k^2) e^{-uz} du = \frac{2}{z^2 + 2^2 a-} \frac{2 \cdot 3^2 \cdot 4k^2}{z^2 + 4^2 a-} \frac{4 \cdot 5^2 \cdot 6k^2}{z^2 + 6^2 a-} \frac{6 \cdot 7^2 \cdot 8k^2}{z^2 + 8^2 a - \cdots},$$

$a = k^2 + 1$.

In the case of expansion (10.1) the approximations P_m/Q_m to the integral in the left-hand side of (10.1) are determined from a three-term linear recurrence satisfied by P_m and Q_m:

$$(2m + 1)(2m + 2)\phi_{m+1}(z) = (z + (2m + 1)^2 a)\phi_m(z) - 2m(2m + 1)k^2 \phi_{m-1}(z).$$

Here $\phi_m = P_m$ or Q_m, and Q_m are orthogonal polynomials. The generating function of Q_n satisfies a Lamé equation in the algebraic form with a parameter $n = 0$. Here z plays a role of the accessory or spectral parameter in the Lamé equation, and the corresponding solution is

$$y(x) = \sum_{m=0}^{\infty} Q_m(z) \cdot x^m,$$

the only solution regular at $x = 0$. The generating function of P_m is a regular at $x = 0$ solution of the nonhomogeneous Lamé equation.

These special continued fraction expansions can be generalized to continued fraction expansions associated with any Lamé equation with an arbitrary parameter n. The theory of such continued fraction expansions is presented in [32]. According to this theory, the ratios of the elements of monodromy matrices of the general Lamé equation have explicit continued fraction representations associated with the three-term linear recurrences on two classes of orthogonal polynomials [32, §13].

For an arbitrary n, the monodromy matrices of the Lamé equation are not known to be reducible to any classical transcendences (see the discussion above for $n = -1/2$), and the continued fraction expansions arising this way have no closed form expressions. However, when n is integral, we have closed form expressions for monodromy matrices (as for the solutions themselves) that can be expressed in terms of integrals of elliptic functions. For $n = 0$ these closed form expressions represent the Stieltjes-Rogers expansions. For $n = 1$ two classes of continued fractions from [32, §13] have arithmetic applications, because for three values of the accessory parameter H (corresponding to e_i-nontrivial second-order points) the Lamé equation is a globally nilpotent one and we have p-adic as well as archimedean convergence of continued fraction expansions. This way we obtain the irrationality and bounds on the measure of irrationality of some values of complete elliptic integrals of the third kind, expressed through traces of the Floquet matrices. Similarly, for an arbitrary integral $n \geq 1$, among continued fraction expansions, expressed as integrals of elliptic θ-functions, there are $2n + 1$ cases of global nilpotence, when continued fractions have arithmetic sense and orthogonal polynomials have nearly integral coefficients.

Among new explicit continued fraction expansions is the expansion of the following function generalizing Stieltjes-Rogers:

$$\int_0^{\infty} \frac{\sigma(u - u_0)}{\sigma(u)\sigma(u_0)} e^{\zeta(u_0)u} du,$$

or

$$\int_{\omega}^{\omega+\omega'} \frac{\sigma(u - u_0)}{\sigma(u)\sigma(u_0)} e^{\zeta(u_0)u} du,$$

as a function of $x = \mathscr{P}(u_0)$. In Jacobi's notations this function can be presented as

$$\int_0^\infty \frac{H(u + u_0)}{\Theta(u)} e^{-uZ(u_0)} du,$$

where Θ and H are Jacobi's notations for functions.

The three-term linear recurrence determining the J-fraction for the corresponding orthogonal polynomials has the following form:

$$Q_N(x) = Q_{N-1}(x) \cdot \{(l + k^2) \cdot (N - 1)^2 + x\}$$
$$+ Q_{N-2}(x) \cdot k^4 \cdot (N - 1)^2 \cdot N \cdot (N - \tfrac{1}{2}) \cdot (N - \tfrac{3}{2}) \cdot (N - \tfrac{5}{2}).$$

Here $x = sn^2(u_0; k^2)$.

The more general J-fraction of the form

$$\cdots b_{n-1} + x - \cfrac{a_{n-1}}{b_n + x - \cfrac{a_n}{b_{n+1} + x - \cdots}},$$

with

$$a_n = k^4 \cdot n(n + 1) \cdot (n + \tfrac{1}{2})(n - \tfrac{1}{2}) \cdot \{(n - 1) \cdot (n - \tfrac{1}{2}) - m \cdot (m + 1)/4\};$$
$$b_n = (1 + k^2) \cdot (n - 1)^2, \qquad n \geq 2,$$

is convergent to the integral of the form

$$\int_0^\infty \prod_{i=1}^m \frac{H(u - u_i)}{\Theta(u)\theta(u_i)} e^{-Z(u_i)u} du.$$

The generating function of the corresponding orthogonal polynomials is expressed in terms of solutions of a Lamé equation with parameter $m \geq 1$.

There is also a nontrivial relationship between these continued fraction expansions and that of Rogers-Ramanujan. The relationship is based on identification of solutions of generalized Lamé equations with eigenfunctions of quantum models that are continuous versions of XYZ, 8-vertex, and hard hexagon models.

Chapter 2
Diophantine Approximations and Period Relations

In Chapter 1 we presented a few examples where the arithmetic properties of classical constants were investigated by means of sequences of very good approximations arising from linear differential equations uniformized by arithmetic subgroups. In this chapter we pursue this subject further, describing in detail methods of construction of very good approximations to functions and their values. These are the methods of Padé approximations, and the functions values that we are approximating are G-functions. As explained in Chapter 1, G-functions are connected with periods of algebraic varieties. In this chapter we study period relations in the complex multiplication case (for elliptic curves and Abelian varieties) and show how the existence of period relations allows one to represent classical constants (such as

π and other periods) in terms of rapidly convergent nearly integral G-series. We then describe methods of Padé approximations and demonstrate how archimedean/p-adic convergence influences irrationality proofs and bounds on measures of diophantine approximations. To get nontrivial diophantine results, one relies first on underlying arithmetic properties of linear differential equations (start with Chapter 1 and Picard-Fuchs equations) and then improves approximations. In these improvements, multiple integrals representing deformations of period structures associated with a given G-function play a crucial role. These multiple integrals in arithmetical cases, when the G-function itself is a period, give an explicit (closed form) representation of Padé-type approximations to G-functions. Earlier [55, 56, 43] we studied the deformation structures associated with Padé approximations by means of Bäcklund transformations and addition of apparent singularities. In the cases under consideration now these Bäcklund transformations can be seen explicitly as multiple integrals, and a variety of parameters that can be controlled allow for greater freedom in constructing approximations. Extremal problems arising here can be reduced to the balance of archimedean/p-adic convergence in the style of [57, 44].

2.1. Ramanujan's period relations and rapidly convergent hypergeometric representations of logarithms.

In this chapter we briefly describe one aspect of complex multiplication theory for elliptic curves associated with Ramanujan's name. In our exposition we follow closely our own review paper [48], to which the readers are referred for details and complete references.

Complex multiplications of arbitrary Abelian varieties (including elliptic curves) usually mean nontrivial endomorphisms of Abelian varieties manifesting themselves as nontrivial algebraic relations between $2g \times g$ elements of the Riemann period matrix. In the one-dimensional case ($g = 1$) a complex multiplication is a single algebraic relation between a pair of periods, whose ratio is an imaginary quadratic number, known as a singular module. An interesting phenomenon discovered by Ramanujan in this classical field was the existence of new quasiperiod relations. In the Weierstrass notation, period and quasiperiod relations in the elliptic curve case can be described as follows. One starts with an elliptic curve over $\overline{\mathbf{Q}}$ with a Weierstrass equation $y^2 = 4x^3 - g_2 x - g_3$ ($g_2, g_3 \in \overline{\mathbf{Q}}$), having the fundamental periods ω_1, ω_2 (with $\mathrm{Im}(\omega_2/\omega_1) > 0$) and the corresponding quasiperiods η_1, η_2. Then ω_i and η_j are related by the Legendre identity:

$$\eta_1 \omega_2 - \eta_2 \omega_1 = 2\pi i.$$

Thus π belongs to the field generated by periods and quasiperiods over $\overline{\mathbf{Q}}$.

In the complex multiplication case $\tau = \omega_2/\omega_1$ is an imaginary quadratic number. Whenever $\tau \in \mathbf{Q}(\sqrt{-d})$ for $d > 0$, invariants g_2 and g_3 can be chosen from the Hilbert class field of $\mathbf{Q}(\sqrt{-d})$, and this field is the minimal extension with this property. A priori complex multiplication means only a single relation between ω_i.

It seems that until Ramanujan's paper [58] nobody explicitly stated the existence of the second relation between periods and quasiperiods. This relation is the following one:

Whenever τ is a quadratic number, the four numbers ω_1, ω_2, η_1, η_2 are linearly dependent over $\mathbf{Q}$ only on two of them.

Explicitly, if $A\tau^2 + B\tau + C = 0$ for $A, B, C \in \mathbf{Z}$, and $\omega_2 = \tau\omega_1$, then $A\tau\eta_2 - C\eta_1 + \alpha\omega_1 = 0$ for $\alpha \in \overline{\mathbf{Q}}$ ($\alpha \in \mathbf{Q}(\tau, g_2, g_3)$).

Invariant α—a nontrivial part of the Ramanujan quasiperiod relation—can be identified with the value of "nonholomorphic Eisenstein series," described in Weil's treatise [59].

The usual Eisenstein series is defined as

$$E_k(\tau) = 1 - \frac{2k}{B_k} \cdot \sum_{n=1}^{\infty} \sigma_{k-1}(n) \cdot q^n$$

for $\sigma_{k-1}(n) = \sum_{d|n} d^{k-1}$, and $q = e^{2\pi i \tau}$.

In the $E_k(\tau)$ notations, the quasiperiod relation is expressed by means of the function

(2.1.1) $$s_2(\tau) \stackrel{\text{def}}{=} \frac{E_4(\tau)}{E_6(\tau)} \cdot \left(E_2(\tau) - \frac{3}{\pi \text{Im}(\tau)} \right),$$

which is nonholomorphic but invariant under the action of $\Gamma(1)$.

It is this function that Ramanujan studied in connection with α. Ramanujan proved in [58] that this function admits algebraic values whenever τ is imaginary quadratic. Moreover, Ramanujan [58] presented a variety of algebraic expressions for this function, differentiating modular equations. His work, or Weil's, shows that the function in (1.1) has values from the Hilbert class field $\mathbf{Q}(\tau, j(\tau))$ of $\mathbf{Q}(\tau)$ for quadratic τ. The relation of (1.1) to α is simple: for $\beta = s_2(\tau)$ from (1), $\alpha = (B + 2A\tau)\beta \cdot g_3/g_2$. These relations were rediscovered many times; see references in [48]. On can combine (1.1) with the Legendre relation to arrive to a "quadratic relation" derived by Ramanujan that expresses π in terms of squares of ω_1 and η_1 only (no ω_2 and η_2!). Moreover, Ramanujan transformed these quadratic relations into a rapidly convergent generalized hypergeometric representation of simple algebraic multiples of $1/\pi$. To do this he used only modular functions and hypergeometric function identities. Let us start with Ramanujan's own favorite:

$$\frac{9801}{2\sqrt{2}\pi} = \sum_{n=0}^{\infty} \{1103 + 26390n\} \frac{(4n)!}{n!^4 \cdot (4 \cdot 99)^{4n}}.$$

The reason for this representation of $1/\pi$ lies in the representation of $(K(k)/\pi)^2$ as a $_3F_2$-hypergeometric function. Apparently there are four classes of such representations, all of which were determined by Ramanujan. All are based on four special cases of the Clausen identity of a hypergeometric function (and all presented by Ramanujan):

$$F(a, b; a + b + \tfrac{1}{2}; z)^2 = {}_3F_2 \left(\begin{matrix} 2a, a+b, 2b \\ a+b+\tfrac{1}{2}, 2a+2b \end{matrix} \Big| z \right).$$

Unfortunately, the Clausen identity is a unique one—no other nontrivial relation between parameters makes a product of hypergeometric functions a generalized hypergeometric function.

All Ramanujan's quadratic period relations (four types) can be deduced from one series by modular transformations, and we choose the series as the one associated with the modular invariant $J = J(\tau)$. We take, as in (1.1):

$$s_2(\tau) = \frac{E_4(\tau)}{E_6(\tau)} \left(E_2(\tau) - \frac{3}{\pi \mathrm{Im}(\tau)} \right).$$

Then the Clausen identity gives the following $_3F_2$- representation for an algebraic multiple of $1/\pi$:

$$(2.1.2) \qquad \sum_{n=0}^{\infty} \left\{ \frac{1}{6}(1 - s_2(\tau)) + n \right\} \cdot \frac{(6n)!}{(3n)!n!^3} \cdot \frac{1}{J(\tau)^n}$$

$$= \frac{(-J(\tau))^{1/2}}{\pi} \cdot \frac{1}{(d(1728 - J(\tau))^{1/2}}.$$

Here $\tau = (1 + \sqrt{-d})/2$. If $h(-d) = 1$, then the second factor in the right-hand side is a rational number. The largest one-class discriminant $-d = -163$ gives the most rapidly convergent series among those series where all numbers in the left side are *rational*:

$$(2.1.3) \quad \sum_{n=0}^{\infty} \{c_1 + n\} \cdot \frac{(6n)!}{(3n)!n!^3} \frac{(-1)^n}{(640,320)^n} = \frac{(640,320)^{3/2}}{163 \cdot 8 \cdot 27 \cdot 7 \cdot 11 \cdot 19 \cdot 127} \cdot \frac{1}{\pi}.$$

Here

$$c_1 = \frac{13,591,409}{163 \cdot 2 \cdot 9 \cdot 7 \cdot 11 \cdot 19 \cdot 127}$$

and $J(\frac{1+\sqrt{-163}}{2}) = -(640,320)^3$.

Ramanujan provided instead of this a variety of other formulas connected mainly with the three other triangle groups commensurable with $\Gamma(1)$. All four classes of $_3F_2$ representations of algebraic multiples of $1/\pi$ correspond to four $_3F_2$ hypergeometric functions (which are squares of $_2F_1$-representations of complete elliptic integrals via the Clausen identity). These are

$$_3F_2(^{1/2,1/6,5/6}_{1,1}|x) = \sum_{n=0}^{\infty} \frac{(6n)!}{(3n)!n!^3} \left(\frac{x}{12^3} \right)^n,$$

$$_3F_2(^{1/4,3/4,1/2}_{1,1}|x) = \sum_{n=0}^{\infty} \frac{(4n)!}{n!^4} \left(\frac{x}{4^4} \right)^n,$$

$$_3F_2(^{1/2,1/2,1/2}_{1,1}|x) = \sum_{n=0}^{\infty} \frac{(2n)!^3}{n!^6} \left(\frac{x}{2^6} \right)^n,$$

$$_3F_2(^{1/3,2/3,1/2}_{1,1}|x) = \sum_{n=0}^{\infty} \frac{(3n)!}{n!^3} \cdot \frac{(2n)!}{n!^2} \left(\frac{x}{3^3 \cdot 2^2} \right)^n.$$

Representations similar to (1.3) can be derived for any of these series for any singular moduli $\tau \in \mathbf{Q}(\sqrt{-d})$ and for any class number $h(-d)$, thus extending the Ramanujan list ad infinitum. For a simple recipe to generate these new identities, see [48].

Even before Ramanujan's remarkable approximations to π, singular moduli evaluations were used to approximate multiples of π by logarithms of algebraic numbers (usually the values of modular invariants); see references in [48]. These approximations are not technically approximations to π, but rather to a linear form in π and in another logarithm. All of them are natural consequences of Schwarz theory and the representation of the function inverse to the automorphic one (say $J(\tau)$) as a ratio of two solutions of a hypergeometric equation. One such formula is

$$(2.1.4) \qquad \pi i \cdot \tau = \ln(k^2) - \ln(16) + \frac{G(\frac{1}{2}, \frac{1}{2}; 1; k^2)}{F(\frac{1}{2}, -\frac{1}{2}; 1; k^2)},$$

and another is Fricke's

$$(2.1.5) \qquad 2\pi i \cdot \tau = \ln J + \frac{G(\frac{1}{12}, \frac{5}{12}; 1-; \frac{12^3}{J})}{F(\frac{1}{12}, \frac{5}{12}; 1; \frac{12^3}{J})}.$$

Here

$$G(a, b; c; x) = \sum_{n=0}^{\infty} \frac{(a)_n (b)_n}{(c)_n n!} \cdot \left\{ \sum_{j=0}^{n-1} \left(\frac{1}{a+j} + \frac{1}{b+j} - \frac{2}{c+j} \right) \right\}$$

is the hypergeometric function (of the second kind) in the exceptional case, when there are logarithmic terms.

Perhaps the most popular approximations to linear forms in π and in another logarithm are associated with the Stark-Baker solution to one-class and two-class problems; see [6].

Ramanujan's algebraic approximations to $1/\pi$ can be extended to the analysis of linear forms in logarithms arising from class number problems. In fact, each term in these linear forms can be separately represented by series in $1/J$ with nearly integral coefficients. For this, one takes an automorphic function $\phi(\tau)$ with respect to one of the congruence subgroups of $\Gamma(1)$ and expands functions like $F(\frac{1}{12}, \frac{5}{12}; 1; 12^3/J)$, $G(\frac{1}{12}, \frac{5}{12}; 1, ; 12^3/J)$ in powers of $\phi(\tau)$ instead of $J(\tau)$. Whenever $\phi(\tau)$ is automorphic with respect to a classical triangle group, we arrive at the corresponding usual hypergeometric functions.

Logarithms of quadratic units, like π, can be represented as values of convergent series satisfying Fuchsian linear differential equations. This holds for $\ln \epsilon_{\sqrt{k}}$ of a fundamental unit $\epsilon_{\sqrt{k}}$ of a real quadratic field $\mathbf{Q}(\sqrt{k})$. To represent this number as a convergent series (in, say, $1/J(\tau)$) one uses Kronecker's limit formula expressing this logarithm $\ln \epsilon_{\sqrt{k}}$ in terms of products of values of Dedekind 's Δ-function ("Jugendtraum", see [59]). Such an expression of $\ln \epsilon_{\sqrt{k}}$ in terms of power series in $1/J(\tau)$ for $\tau = (1 + \sqrt{-d})/2$ for any $d \equiv$

3 (8), depends on k, because k is related to the level of the appropriate modular from $\phi = \phi_k(\tau)$.

For $k = 5$ Siegel [60] made an explicit computation that expresses $J(\tau)$ in terms of the resolvent $\phi_5(\tau)$ of the fifth-degree modular equation known from the classical theory of fifth-degree equations. His relations [60] were

$$(\phi - \epsilon^3)((\phi - \epsilon)(\phi^2 + \epsilon^{-1}\phi + \epsilon^{-2}))^3 + (\phi/\sqrt{5})^5 J = 0$$

and

$$\phi(\tau)(= \phi_5(\tau)) = \epsilon^{h(-5p)/2}$$

for $\tau = (1 + \sqrt{-p})/2$ and $\epsilon = \epsilon_{\sqrt{5}}$. Here one has $p \equiv 3$ (5) if $p \equiv 2$ (5) and replaces ϵ by ϵ^{-1}.

This, in combination with Ramanujan's approximation to π, allows one to express the multiple of $\ln \epsilon$ as a convergent series in $1/J$ or in $1/\phi$.

2.2. Period relations and diophantine approximations. The Ramanujan identities of §2.1 are not just good numerical schemes for evaluation of π (there are better ones). In diophantine approximations it is important that there exists a representation of the number that we study, a multiple of $1/\pi$ in this case, as a rapidly convergent series with nearly integral coefficients, i.e., as a value of a G-function at a point very close to the origin. This allows us to unleash the machinery of Padé approximations, see §2.4, to G-functions and to use the global nilpotence properties of the corresponding linear differential equations (in this case hypergeometric ones).

We already mentioned in §1.1 the basic linear independence results for values of G-functions. We present now the corresponding results together with the measure of linear independence for values of G-functions at rational and algebraic points. Proofs of these results use Padé-type approximations of the second kind.

THEOREM 2.1 [14]. *Let $f_1(x), \ldots, f_n(x)$ be G-functions, satisfying a system of equations*

$$(2.2.1) \qquad df_i(x)/dx = \sum_{j=1}^{n} A_{ij}(x) f_j(x),$$

for $A_{ij}(x) \in \mathbf{Q}(x)$, $i, j = 1, \ldots, n$, and such that the functions $1, f_1(x), \ldots, f_n(x)$ are linearly independent over $\mathbf{Q}(x)$. If $\epsilon > 0$ and $r = a/b \neq 0$ for integers a and b with $|b|^\epsilon > c_1 |a|^{(n+1)(n+\epsilon)}$, then for arbitrary rational integers $h_0, h_1, \ldots, h_n$ we have

$$|h_0 + h_1 f_1(r) + \cdots + h_n f_n(r)| > H^{-n-\epsilon}$$

for $H = \max(|h_0|, |h_1|, \ldots, |h_n|)$, provided that $H \geq h_0$. Here $c_1 = c_1(f_1, \ldots, f_n, \epsilon) > 0$ and $h_0 = h_0(f_1, \ldots, f_n, \epsilon, r) > 0$ are effective constants. In particular $1, f_1(r), \ldots, f_n(r)$ are linearly independent over $\mathbf{Q}$.

Another simple example of a G-function result deals with the absence of algebraic relations between values of G-functions at algebraic points.

PROPOSITION 2.2 [14]. *Let $f_1(x), \ldots, f_n(x)$ be G-functions, satisfying a system of equations (1) and algebraically independent with 1 over $\mathbf{Q}(x)$. Then for any $d \geq 1$, there exists an effective constant $c_2 = c_2(f_1, \ldots, f_n, d) > 0$ such that for an arbitrary algebraic number $\xi \neq 0$ of degree at most d, the numbers $f_1(\xi), \ldots, f_n(\xi)$ are not related by an algebraic relation of degree d over $\mathbf{Q}(\xi)$, provided that*

$$|\xi| > \exp(-c_2\{\ln H(\xi)\}^{2n/(2n+1)}).$$

These results can be considerably improved using the methods of graded Padé approximations; see [14, 15, 7].

Ramanujan's and other similar identities that express π and other similar numbers in terms of values of G-functions very close to the origin give us the basis of applications for Theorems 2.1–2.2 and other similar results. In these applications the exponent in the measure of diophantine approximations strongly depends on the proximity of an evaluation point to the point of expansion of a G-function. To make this dependence explicit, we quote the following result [15]:

THEOREM 2.3 [15]. *Let $f_1(x), \ldots, f_n(x)$ be G-functions satisfying linear differential equations over $\mathbf{Q}(x)$. Let $r = a/b$, with integers a and b, be very close to the origin. Then we get the following lower bound for linear forms in $f_1(r), \ldots, f_n(r)$.*

For arbitrary nonzero rational integers $H_1, \ldots, H_n$ and $H = \max\{|H_1|, \ldots, |H_n|\}$, if $H_1 f_1(r) + \cdots + H_n f_n(r) \neq 0$,

$$|H_1 f_1(r) + \cdots + H_n f_n(r)| > |H_1 \cdots H_n|^{-1} \cdot H^{1-\epsilon}$$

provided that r is very close to 0,

$$|b| \geq c_1 \cdot |a|^{n(n-1+\epsilon)},$$

and $H \geq c_2$ with effective constants $c_i = c_i(f_1, \ldots, f_n, r, \epsilon)$. If r is not as close to 0, we get only

$$|H_1 f_1(r) + \cdots + H_n f_n(r)| > H^{\lambda-\epsilon}$$

for $\lambda = -(n-1)\ln|b|/\ln|b/a^n|$ (< 0).

We can apply these G-functions results, particularly Theorem 2.3, to the $_3F_2$-series of Ramanujan-type representing algebraic multiplies of $1/\pi$ as G-function series of $J(\tau)$. These results *might* have shown that numbers connected with π all have the exponent of irrationality $-2 - \epsilon$ for any $\epsilon > 0$. For example, if there are infinitely many negative quadratic discriminants $-d$ with class number one (similarly, for any *fixed* class number h), then *all* elements of the field $\mathbf{Q}(\pi^2)$, irrational over $\mathbf{Q}$, have measures of irrationality with the exponent $-2 - \epsilon$ for all $\epsilon > 0$. Unfortunately, degrees of number $J(\tau)$ for $\tau = \frac{1+\sqrt{-d}}{2}$ grow to infinity as $d \to \infty$, so only finitely many cases can be treated. The exponent then stays away from 2, and to obtain nontrivial results we are using instead explicit Padé approximants of the second kind to generalized hypergeometric functions $_3F_2$ and multidimensional integrals of Pochhammer type.

We will present some of the results for numbers connected with π based on effective Padé approximations with schemes described in [44, 61]. The first two bounds are connected with Ramanujan-like series:

$$|\pi\sqrt{2} - p/q| > |q|^{-16.67\ldots}$$

for rational integers p, q, $|q| \geq q_0$;

$$|\pi\sqrt{640320} - p/q| > |q|^{-12.11\ldots}$$

for $|q| \geq q_1$.

For $\pi\sqrt{3}$ we use a different system of Padé-type approximations. Below we present the corresponding integral representations, leading to the bound

$$|\pi\sqrt{3} - p/q| > |q|^{-5.791\ldots}$$

for $|q| \geq q_2$.

PROPOSITION 2.4. *For arbitrary rational integers p, q with $|q| \geq 2$ we have*

$$|\pi - p/q| > |q|^{-15.0\ldots}.$$

For π^2, in connection with multidimensional integrals arising from $\zeta(2)$, we get

$$|\pi^2 - p/q| > |q|^{-7.51\ldots}$$

for $|q| \geq q_3$.

What is the class of numbers, in addition to π, that can be an object of similar study? One looks for this at generalizations of Ramanujan period relations.

2.3. Periods of CM-varieties. The extension of Ramanujan theory to Abelian varieties with complex multiplication (of CM-type) was finalized in works of Shimura [62], who initiated studies of algebraic values of automorphic forms of **Q**-rational algebraic groups, and Deligne [63]. We refer the readers to references in [62–64] and to the exposition in [48].

In general, it is difficult to translate period relations into power series identities, because one needs for this explicit representation of deformations of periods in terms of solutions of explicit Picard-Fuchs equations. This is relatively easy in the case of elliptic curves, and we look, following [48], at other cases, when periods can be expressed in terms of hypergeometric functions.

Thus we look at arithmetic subgroups acting on the upper half-plane, whose automorphic functions can be expressed in terms of hypergeometric functions. This means that we look at Schwarz's triangle groups, whose signature is

$$(0, 3; l_1, l_2, l_3),$$

where $2 \leq l_i \leq \infty$, and the non-euclidean condition is satisfied:

$$(2.3.1) \qquad\qquad \sum_{i=1}^{3} \frac{1}{l_i} < 1.$$

Among triangle groups we are looking only at arithmetic ones. Among these arithmetic triangle groups there are those that are commensurable with the whole modular group. There are 4 representatives of this class, corresponding to 4 classes of hypergeometric functions studied by Ramanujan in connection with elliptic period relations. If one puts $l_1 = 2$, and this is necessary in order to take advantage of the Clausen identity, then the 4 (elliptic) triangle subgroups referred to in §2.1 are

$$(0, 3; 2, 3, \infty), \quad (0, 3; 2, 4, \infty), \quad (0, 3; 2, 6, \infty), \quad (0, 3; 2, \infty, \infty).$$

The list of all arithmetic triangle subgroups is rather large, and its compilation was initiated by Fricke and Klein in [35] in connection with ternary quadratic forms. Later at the turn of the twentieth century an extensive investigation of triangle groups was conducted by a group of American mathematicians, Hutchinson, Young, and Morris, following earlier work by Hurwitz and Burkhardt. In these works detailed investigations of special classes of arithmetic triangle groups were carried out; see references in R. Morris [65]. Recently a complete classification of all arithmetic triangle groups was presented by Takuchi [66]. Of special interest to us are 17 classes [66] of commensurability of triangle subgroups corresponding to 12 totally real fields with class number 1 and quaternion algebras over them. In each of these cases there is a rich theory of arithmetic values of functions automorphic with respect to the arithmetic triangle groups Γ acting on H with a compact H/Γ.

One of these arithmetic subgroups is a Hurwitz group $(0, 3; 2, 3, 7)$. This is a group with the minimal volume of H/Γ among all Fuchsian groups of the first kind. Factors of Hurwitz's group are known to attain the maximal order $84(g - 1)$ of an automorphism group of a Riemann surface of genus $g > 1$.

Little is known about the arithmetic properties of values of automorphic functions corresponding to Hurwitz's and other arithmetic triangle compact groups, though Shimura in his papers since the 1950s (see the review in [64]) has built a theory of complex multiplication in most of these cases.

A single commensurability class of triangle subgroups corresponds to quaternion algebras over $\mathbf{Q}$ with discriminant $D = 2 \cdot 3$. This class contains

$$(0, 3; 2, 4, 6) \quad \text{and} \quad (0, 3; 2, 6, 6).$$

Let us look now at the solution of a Picard-Fuchs equation corresponding to periods of hypergeometric curves, i.e., curves with 4 critical points only. In these cases all periods of differentials of the first kind are expressed in terms of hypergeometric functions only.

The simplest case of $(0; 3; \infty, \infty, \infty)$ corresponds to an elliptic function field. Already other triangle subgroups, commensurable with $\Gamma(1)$ give rise to nontrivial Riemann surfaces of hypergeometric type. E.g., two triangle groups in (2) lead to Riemann surfaces of genera 2. These are

$$y^3 = (x - a_0)(x - a_1)(x - a_2)^2(x - a_3)^2$$

for $(0, 3; 2, 6, \infty)$, and

$$y^4 = (x - a_0)(x - a_1)^2(x - a_2)^2(x - a_3)^3$$

for $(0, 3; 2, 4, \infty)$, respectively.

In both cases $g = 2$, and the Jacobian of each of the curves is isogenous to the product of two elliptic curves, which are isogenous to each other. Explicit modular equations express integrals of the first kind on these curves in terms of elliptic integrals.

For two triangle subgroups associated with the quaternion algebra over $\mathbf{Q}$ with the discriminant $D = 2 \cdot 3$, the corresponding hypergeometric curves are even more complicated. For the triangle group $(0, 3; 2, 6, 6)$ we get a genus $g = 3$ equation

$$y^6 = (x - a_0)^2(x - a_1)^2(x - a_2)^3(x - a_3)^5,$$

and for $(0, 3; 2, 4, 6)$ we arrive at genus $g = 23$ curve

$$y^{24} = (x - a_0)(x - a_1)^{11}(x - a_2)^{17}(x - a_3)^{19}.$$

The cases of $(0, 3; 2, 4, 6)$ and $(0, 3; 2, 6, 6)$ and other quaternion arithmetic triangle groups immediately lend themselves to the generalization of the Ramanujan period relations. For each of these cases we can look at the function $z = \phi(\tau)$ automorphic in H with respect to the corresponding arithmetic triangle group, cf. [64]. We can normalize this function by its values in the vertices of the fundamental triangle, following [64]

$$\phi(e_2) = 1, \quad \phi(e_m) = 0, \quad \phi(e_n) = \infty$$

for vertices e_i (say, $m = n = 6$ or $m = 4, n = 6$) that are fixed points of elliptic elements γ_i of orders i in the triangle group: $\prod \gamma_i = -I_2$.)

The theory of complex multiplication in the quaternion case [64] shows that for τ in H that is imaginary quadratic, the field $\mathbf{Q}(\tau, \phi(\tau))$ is an explicit Abelian extension of $\mathbf{Q}(\tau)$.

For example, whenever $\mathbf{Q}(\tau)$ has the class number 1, the values of $\phi(\tau)$ have a structure similar to that of $J(\tau)$. For the numbers $z = \phi(\tau)$ one obtains Ramanujan's period relations as in the elliptic case. This way one gets 2 new algebraic relations between values of

$$F(\tfrac{1}{24}, \tfrac{5}{24}; \tfrac{3}{4}; z) \quad \text{and} \quad F'(\tfrac{1}{24}, \tfrac{5}{24}; \tfrac{3}{4}; z)$$

for $(0, 3; 2, 4, 6)$ case, and between values of

$$F(\tfrac{1}{12}, \tfrac{1}{4}; \tfrac{5}{6}; z) \quad \text{and} \quad F'(\tfrac{1}{12}, \tfrac{1}{4}; \tfrac{5}{6}; z)$$

for $(0, 3; 2, 6, 6)$, where $z = \phi(\tau)$.

When $\tau \in \mathbf{Q}(\sqrt{-d})$ and $-d$ is a one-class discriminant, we arrive at new Ramanujan-like period identities. There are 3 classes of hypergeometric functions for these two triangle subgroups, where the Clausen identity is satisfied, and a product of the periods can be expressed in terms of $_3F_2$ function.

We can develop a theory [48] of Ramanujan-like relations for arbitrary arithmetic groups Γ. In addition to the theory of complex multiplication for these groups (see [62–64, 67]) one needs an analog of Ramanujan's nonholomorphic function $s_2(\tau)$. We look at linear differential equations corresponding to Γ. We look at the derivatives of the automorphic function $\phi(\tau)$ for the arithmetic group Γ normalized by its values at vertices. The function itself satisfies the third-order nonlinear differential equation over $\overline{\mathbf{Q}}$ that follows from the determination of the Schwarzian $\{\phi, \tau\}$ in terms of ϕ. An analog of $s_2(\tau)$ in §2.1 that is a nonholomorphic automorphic form for Γ is

$$(2.3.2) \qquad -\frac{1}{\phi'(\tau)} \cdot \left\{ \frac{\phi''(\tau)}{\phi'(\tau)} - \frac{i}{\operatorname{Im}(\tau)} \right\}.$$

For $\phi(\tau) = J(\tau)$ one gets $s_2(\tau)$ in (1) of §2.1.

For example, let us look at a quaternion triangle group $(0; 3; 2, 6, 6)$. In this case, instead of an elliptic Schwarz formula (4–5) of §2.1 one has the following representation of the normalized automorphic function $\phi = \phi(\tau)$ in H in terms of hypergeometric functions:

$$\frac{\tau + i(\sqrt{2} + \sqrt{3})}{\tau - i(\sqrt{2} + \sqrt{3})} = -\frac{3^{1/2}}{2^2 \cdot 2^{1/6}} \cdot \left\{ \frac{\Gamma(1/3)}{\sqrt{\pi}} \right\}^6 \cdot \frac{F(\frac{1}{12}, \frac{1}{4}; \frac{5}{6}; \phi)}{\phi^{1/6} \cdot F(\frac{1}{4}, \frac{5}{12}; \frac{7}{6}; \phi)}.$$

Thus the role of π in Ramanujan's period relations is occupied in $(0, 3; 2, 6, 6)$-case by the transcendence $\{ \frac{\Gamma(1/3)}{\pi} \}^6$.

In the case $(0, 3; 2, 4, 6)$-group the representation of $\phi = \phi(\tau)$ is [35, 48]:

$$\frac{(\sqrt{3} - 1)\tau - i\sqrt{2}}{(\sqrt{3} - 1)\tau + i\sqrt{2}} = -2(\sqrt{3} - \sqrt{2}) \frac{\Gamma(-\frac{1}{24})\Gamma(-\frac{5}{24})}{\Gamma(-\frac{13}{24})\Gamma(-\frac{17}{24})} \cdot \phi^{1/2} \cdot \frac{F(\frac{13}{24}, \frac{17}{24}; \frac{3}{2}; \phi)}{F(\frac{1}{24}, \frac{5}{24}; \frac{1}{2}; \phi)}.$$

This leads to a new transcendence:

$$\frac{\Gamma(\frac{1}{24})^4}{\{\Gamma(\frac{1}{3})\Gamma(\frac{1}{4})\}^2}.$$

Thus, generalizations of Ramanujan identities allow us to express constants, such as π and other Γ-factors, as values of rapidly convergent series with nearly integral coefficients in a variety of ways, with convergence improving as the discriminant of the corresponding singular moduli increases. One can ask: what kind of constants allow these representations? Can one express this way all combinations of periods and quasiperiods? This problem has an obvious relationship with another "period" problem—the Dwork-Siegel conjecture from Chapter 1. One hopes that period identities will give an interpretation of the best rational (diophantine) approximations to such constants as π in terms of values of some modular or automorphic functions, in the fashion similar to interpretation of the best rational approximations to cubic irrationalities and solutions of Fermat-like equations via the Tanniyama-Weil conjecture.

Let us look now briefly at the transcendence results for periods and quasi-periods of hypergeometric curves, and the corresponding statements for the transcendence of values of hypergeometric functions at algebraic points.

We look first at Legendre's function $F(\frac{1}{2}, \frac{1}{2}; 1; z)$. Then the next theorem follows from our 1977 results [68, 69]:

THEOREM 3.1. *For an arbitrary algebraic* $z \neq 0, 1$ *two numbers* $F(\frac{1}{2}, \frac{1}{2}; 1; z)$ *and* $F(-\frac{1}{2}, \frac{1}{2}; 1; z)$ *are algebraically independent over* **Q**.

Moreover, we proved the measure of algebraic independence of two numbers $\alpha = F(\frac{1}{2}, \frac{1}{2}; 1; z)$, $\beta = F(-\frac{1}{2}, \frac{1}{2}; 1; z)$ that is very close to the best possible. Namely, if $P(x, y) \in \mathbf{Z}[x, y]$, $P \neq 0$, P has the (total) degree d and height—the maximum of absolute values of coefficients—$H > 1$, then [69]

$$|P(\alpha, \beta)| > H^{-c_0 d^2 \ln^2(d+1)}$$

with $c_0 = c_0(\alpha, \beta, z) > 0$—an effective constant.

Theorem 3.1 can be supplemented with a stronger statement [68] that whenever $z \neq 0, 1, \infty$ two of three numbers

$$z, F(\frac{1}{2}, \frac{1}{2}, 1; z) \quad \text{and} \quad F(-\frac{1}{2}, \frac{1}{2}; 1; z)$$

are algebraically independent over **Q**.

We can prove the analog of Theorem 3.1, if all hypergeometric functions giving periods of hypergeometric curves are considered.

In particular, for all hypergeometric functions $F(z)$ algebraically reducible to the triangle case $(0, 3; 2, 4, 6)$, the values $F(z)$ and $F'(z)$ are algebraically independent over **Q**, whenever z is algebraic $\neq 0, 1$.

We also can look at the algebraic independence results for a general hypergeometric function when all branches of the function are involved. Methods of [27] based on the uniformization of general hypergeometric functions $F(x)$ lead to

THEOREM 3.2. *Let*

$$F_1(x)(= F(a, b; c; x)), \qquad F_2(x)(= x^{1-c} \cdot F(a - c + 1, b - c + 1; 2 - c; x))$$

be two algebraically independent (over **Q**(x)*) solutions of hypergeometric equation with rational* a, b, c. *If* F_1, F_2 *do not correspond to genus* $= 0, 1$ *cases, then for every algebraic* $x \neq 0, 1$ *at least two of the numbers* $F_1(x), F_1'(x), F_2(x)$, $F_2'(x)$ *are algebraically independent.*

In general, there are at most 3 algebraically independent numbers among these and in singular moduli case there are exactly 2 algebraically independent ones.

2.4. Methods of Padé approximations in diophantine approximations. We have referred often to methods of Padé approximations. It is time to examine the Padé approximation methods themselves and present some examples in the context of diophantine approximations to special numbers.

Analytic methods in the theory of diophantine approximations to the values of a function with "good" arithmetic properties are based on the construction of systems of approximations to functions themselves. Results on the arithmetic nature of values of these functions (their irrationality, transcendence, and measures of rational approximations) are deduced by specialization of these approximating systems. These approximation systems applied to diophantine approximations of specializations of functions are Padé-type approximations. By Padé-type approximations one understands rational or algebraic approximations to functions that satisfy algebraic differential or functional equations that are determined by local conditions.

Padé-type approximations are determined by matching the orders (degrees) of approximations at given points with the degrees of approximants. As an example we can point at the multipoint Hermite-Padé approximation, which is determined (often uniquely) by matching the maximal allowable orders of approximations at given points with the number of undetermined coefficients in the approximants.

In order to be specific we define the schemes of Padé approximations.

We start with a scheme of Padé approximations referred to above as a scheme of Padé approximations of the second kind.

DEFINITION 4.1. Let $f_1(x), \ldots, f_n(x)$ be functions analytic at $x = 0$. For any (weight) $D \geq 0$, there exist polynomials $q(x), p_1(x), \ldots, p_n(x)$ of degree at most D such that

$$\mathrm{ord}_{x=0}\{q(x) \cdot f_i(x) - p_i(x)\} \geq D + 1 + [D/n]$$

for all $i = 1, \ldots, n$.

This means that all functions $f_i(x)$ are simultaneously approximated by $(p_i(x)/q(x))$–rational functions with the common denominator.

The counterpart of the scheme 4.1 is a more popular scheme of Hermite-Padé approximations to a system of functions, defined as follows:

DEFINITION 4.2. Let $f_1(x), \ldots, f_n(x)$ be analytic at $x = 0$ (or, at least, having formal power series expansions at $x = 0$). Then for a given nonnegative D there exist polynomials (Padé approximants) $p_0(x), p_1(x), \ldots, p_n(x)$ of degrees at most D such that the remainder function

$$R(x) \overset{\text{def}}{=} p_0(x) + p_1(x)f_1(x) + \cdots + p_n(x)f_n(x)$$

has a zero at $x = 0$ of order at least $(n + 1)(D + 1) - 1$.

These Padé approximants are called approximants of the first kind. The Khintchine transference principle establishes the duality between Padé approximants of the first and the second kind, which allows us to express a sequence of one kind of approximants in terms of the other—this duality principle had been made explicit by Mahler [**70**] in 1938.

Padé approximations always exist, but in applications to arithmetic problems, one requires good convergence both in the archimedean (complex) and nonarchimedean domains in the neighborhood of a sequence of points of the approximation.

The arithmetic (nonarchimedean) properties of Padé approximants are virtually unknown except in the cases when there exists a closed expression of Padé approximations. It is not a closed expression that is needed, but the control of the growth of the coefficients (local-nonarchimedean conditions) versus the order of convergence of approximations (an archimedean condition). Of course, to assume good arithmetic properties of Padé approximations we start with functions having good arithmetic properties. Thus we look at Padé approximations to functions "coming from Arithmetic or Geometry," i.e., at functions satisfying globally nilpotent equations—G- functions. These functions have nearly integral power series expansions that are *not* immediately transferable to local/global conditions on Padé approximations.

This local/global requirements on Padé approximations can be stated as follows:

If $f_i(x) \in \mathbf{Q}[[x]]$ $(i = 1,\dots,n)$, can one find Padé approximants of the second kind (Definition 4.1), $q(x), p_1(x),\dots,p_n(x)$ of weight (degree) D from $\mathbf{Z}[x]$ such that heights of all polynomials are bounded by C^D for a constant $C > 1$ (independent of D)?

These arithmetic conditions are often violated even for the simplest $f_i(x)$. This is the case when $n = 1$ and $f_1(x)$ is an algebraic function with more than 3 critical points, e.g., $f_1(x) = \sqrt{(x - a_1) \cdots (x - a_{2k})}$. In this and other generic cases the heights of Padé approximants from $\mathbf{Z}[x]$ grow as fast as C^{D^2} for $D \to \infty$. Such growth makes Padé approximations often unsuited for applications in diophantine approximations.

This poor (archimedean) convergence leads to the need of improving upon approximations given by Padé approximation schemes. One way to do it is to investigate more complicated schemes of Padé approximations involving multivariable generalizations, and the construction of "graded Padé approximations." Also to prove many general results one does not need Padé approximations but rather Padé-type approximations, when the number of approximation conditions (in terms of degrees and orders of Padé approximants) is significantly less than the maximal allowable. In this approach one uses "ineffective" constructions based on counting arguments, when Padé-type approximations are not constructed but proved to exist, with upper bounds on sizes of their coefficients. Unfortunately, in most interesting cases, such as for the numbers $\pi, \ln 2, \sqrt[3]{5}, \dots$, most of the "ineffective" proofs do not give very good results in view of poor convergence rates for these approximations and involvement of extremely large constants—a typical consequence of simple counting arguments.

There are, fortunately, large classes of arithmetically important functions for which closed form expressions of Padé approximations can be determined *and* arithmetic conditions on local-global behavior stated above can be verified. Most of these classes are related to a variety of generalizations of hypergeometric functions.

There are two circumstances that help to deduce these explicit expressions: (a) an explicit expression for the monodromy group and contiguous relations; (b) integral representations. In all cases when explicit construction is possible, considerations (a) and (b) play important roles.

We describe some of the important cases when there are closed form expressions for Padé approximations and when local-global convergence conditions are satisfied:

These cases are the following (see also in [**44, 10, 56**]):

A.

$$f_i(x) =_2 F_1(1, b_i; c_i; x) \quad \text{or} \quad f_i(x) =_2 F_1(1, b; c; \omega_i x);$$

e.g., $f_i(x) = (1 - x)^{\nu_i}$ or $f_i(x) = (1 - \omega_i x)^\nu$, $i = 1, \ldots, n$.

B. When

$$f(x) = {}_{p+1}F_p\binom{a_1, \ldots, a_{p+1}}{b_1, \ldots, b_p} | x)$$

and one looks at Padé approximations of the second kind to p functions $f_1(x), \ldots, f_p(x)$ defined as

$$(1 : f_1 : \ldots : f_p) = (f : \delta_x f : \ldots : \delta_x^p f_p)$$

for $\delta_x = x \frac{d}{dx}$, [**44, 61**]. This is a generalization of Euler-Gauss continued fraction expansions of $\frac{d}{dx} \ln_2 F_1 (a, b; c; x)$.

An important counterpart to B is given by Hermite-Padé approximations of the first kind to contiguous functions:

$$_{p+1}F_p\binom{a_1 + i, a_2, \ldots, a_{p+1}}{b_1, \ldots, b_p} | \omega_j x), \qquad i = 0, \ldots, p.$$

This system of Padé approximations for $p = 0$ was used by us in applications of Ramanujan identities; see [**44, 61**].

In cases A and B we determined the asymptotic expansions of Padé approximants and of the corresponding remainder functions. Explicit expressions for the asymptotics of the denominators of coefficients of Padé approximants turned out to be complicated arithmetic functions of rational parameters a_i, b_i, c_i. See examples in [**71, 44**].

C. Padé approximations can be also explicitly determined for Picard generalizations of hypergeometric functions

$$F(\mu_0, \ldots, \mu_{d+1}) = \int_1^\infty t^{-\mu_0}(t - 1)^{\mu_1} \prod_{i=2}^{d+1}(t - x_i)^{-\mu_i} dt.$$

Here the μ_i are rational numbers and the singularities x_i are linear functions of a single variable x.

The generating functions of Padé approximants to Picard integrals are themselves expressed as periods of algebraic varieties.

D. There are also explicit Padé approximations to multidimensional generalizations of generalized hypergeometric functions, which are expressed as integrals over powers of polynomials in complex variables taken over polytops. Some of these integrals naturally arise from Feynman diagrams, and the monodromy theory of the corresponding multidimensional Fuchsian differential equations is an interesting multidimensional object.

We present one example of a new multidimensional integral arising in our work as a generalization of the Hermite-Lindemann proof of the transcendence of e and π. This formula can be considered as a multidimensional analog of the operations calculus formula for Laplace transforms.

In this formula:

$$I_\Delta = \int \cdots \int_\Delta e^{-\sum_{i=1}^n x_i y_i} \cdot P(y_1, \ldots, y_n) \prod_{i=1}^n dy_i,$$

where Δ is a polytop in n-dimensional space and P is a polynomial vanishing up to high orders at vertices of Δ. The integral I_Δ can be evaluated through values of P and its derivatives at vertices of Δ only. This is a generalization of the so-called Hermite identity (see [4, 6, 48]):

$$I_\Delta = \sum_{\bar{e} \in V(\Delta)} e^{-(\bar{x},\bar{e})} \cdot \prod_{i=1}^n \left(l_i \left(\frac{\partial}{\partial y_1}, \ldots, \frac{\partial}{\partial y_n} \right) - l(x_1, \ldots, x_n) \right)^{-1} \cdot P|_{\bar{y}=\bar{e}}$$

where $l_i(y_1, \ldots, y_n) = 0$ are hyperplane equations of sides of Δ intersecting at the vertex $\bar{e} \in V(\Delta)$.

The choice of the polytop and the polynomials vanishing up to high order in all its vertices determines a variety of Padé approximations to linear combinations of exponents. Hermite's simultaneous Padé approximations correspond to $n = 1$ and give the Padé approximations to $e^{\omega_i x}$. Whenever Δ is an n-polytop (tetrahedron in $\mathbf{R}^n$), one obtains the Padé approximations used by Hermite, Mahler, and Siegel to estimate diophantine approximations to e^α and π, see [4]. An interesting intermediate case corresponds to the so-called graded Padé approximations [7] that provide sharp measures of diophantine approximations to such numbers as $\sin 1$, $\cos 1$, etc.

In the next section we discuss applications of integral representations of Padé approximants and the remainder function.

2.5. Explicit expressions of Padé-type approximations to classical functions and constants. In this section we describe an effective procedure to fine-tune explicit formulas for Padé approximations to such classical functions as logarithms, to deduce the best measures of diophantine approximations of special values of such functions, including $\ln 2$, $\pi/\sqrt{3}$, and π. This exposition follows [44] and can serve as an introduction into this mixture of numerical and p-adic techniques.

In general, one would like to see an effective realization of Padé-type approximations for a large class of functions including solutions of arbitrary Fuchsian linear differential equations. One approach to an effective construction of the Padé-type approximation was studied by the authors from the point of view of Bäcklund transformations and contiguous relations arising from the isomonodromy deformation of linear differential equations in [10, 55, 56, 43]. In this approach ordinary Padé approximations were determined from the local multiplicities at regular singularities, while a variety of new Padé-type approximations were determined from isomonodromy deformations with apparent singularities. We presented in our papers, see [43], several explicit examples of such an approach, together with recurrences connecting consecutive approximants, for particular values of logarithmic and inverse trigonometric functions such as $\ln 2$, $\pi/\sqrt{3}$, and π.

Within the framework of this general approach one can often express solutions of appropriate auxiliary linear differential equations in a closed form as integrals of rational or algebraic functions. In fact, many linear differential equations that are satisfied by the remainder function and the Padé-type approximants are Pochhammer equations solvable in quadratures. This led to the investigation of an optimal form of the integrand in a Pochhammer-type integral representation that provides the best Padé-type approximations for values of logarithmic, binomial, and inverse trigonometric functions. An optimal choice of the integrand is equivalent to the extremal problem of the best rational (polynomial) approximation on a given interval (continuum) of a rational function. This is a classical problem in approximation theory, with new arithmetic extremality conditions including the integrality (or near integrality with a fixed set of possible bad primes and controllable orders at these primes) of polynomials involved. This problem of mixed analytic-arithmetic extremality of polynomials also appeared in connection with a closely related problem [57] of elementary methods in the prime number theorem. Our recipes [57] are based on integral equations and are important in our study of diophantine approximations. Another important component is the WKB method that can reduce the asymptotic (estimate) of the integrals $\int_{\gamma} Q(\overline{x})^N d\overline{x}$ as $N \to \infty$ to the study of critical points of integrands $Q(\overline{x})$. We present below a variety of applications of these integral representations. A key element in this approach is the ability to determine with high precision critical points of integrands of Pochhammer type. This was achieved by means of new high-precision packages for polynomial root finding developed by us, reported in [77].

The necessary analytic tool in the construction of a variety of approximations to arithmetically important functions is the method of (multiple) integral representation of periods of algebraic varieties (solutions of Picard-Fuchs equations). We look in detail now at the one-dimensional case, when our main object is the study of periods of Abelian integrals of the form

$$\int_\gamma \prod_{i=1}^n (\zeta - a_i)^{\alpha_i - 1} d\zeta,$$

where γ is a fixed path on the Riemann surface of the integrand. The class of such integrals is known as Pochhammer integrals. In his study [73] Pochhammer chose one of the a_i as a variable, say z and investigated the dependence of the integral on z. Namely, let us denote

$$(2.5.1) \qquad W(\alpha_1, \ldots, \alpha_m, \mu; z) \overset{\text{def}}{=} \int_C (\zeta - z)^{\mu + n - 1} \prod_{i=1}^m (\zeta - a_i)^{\alpha_i - 1} d\zeta,$$

where the contour C is one of two types: (i) C is a closed path not containing any of the a_i or z; or (ii) C is a path beginning and ending at points where the integrand vanishes. The function in (5.1) is the Euler representation of a solution of a linear differential equation with polynomial coefficients. Namely, if one defines two polynomials $Q(z)$ and $P(z)$ as

$$Q(z) = \prod_{i=1}^m (z - a_i),$$

and

$$\frac{P(z)}{Q(z)} = \sum_{i=1}^m \frac{\alpha_i}{z - a_i}.$$

Then $W(z) = W(a_1, \ldots, a_m, \mu; z)$ (for all contours C in (i,ii) above) satisfies an equation:

$$(2.5.2) \qquad \begin{aligned} &\sum_{i=0}^m (-1)^{m-i} \binom{\mu}{i} Q^{(m-i)}(z) \frac{d^i}{dz^i} W \\ &= \sum_{i=0}^{m-1} (-1)^{m-1-i} \binom{\mu+1}{i} P^{(m-1-i)}(z) \frac{d^i}{dz^i} W. \end{aligned}$$

The Pochhammer equation (5.2) together with some other classes of equations whose solutions can be determined explicitly as integrals belong to the few cases when the monodromy group is known explicitly. The knowledge of the monodromy group allows one to construct, following our methods [10, 55, 56], Padé approximations and Padé-type approximations to solutions of Pochhammer-type equations (5.2). All Padé approximations have again the form (5.1) with exponents α_i, differing from the original ones by integers, and with, possibly, additional apparent singularities a_j. These Padé and Padé-type approximants are related by contiguity formulas that represent linear relations between functions (5.1), when exponents α_i are changing in integral increments. In the case of functions (5.1) all contiguity relations can be obtained from the following few basic relations.

$$(5.3.i) \qquad \frac{dW(\alpha_1, \dots, \alpha_m, \mu; z)}{dz} = -(\mu + n - 1)W(\alpha_1, \dots, \alpha_m, \mu - 1; z).$$

Similar to $(5.i)$ one gets other identities replacing d/dz by d/da_i, $\mu + n$ by α_i. In addition to $(5.i)$ two more classes of basic contiguous relations exist:

$$W(\alpha_1 + 1, \alpha_2, \dots, \alpha_m, \mu; z)$$

$$(2.5.3) \qquad = W(\alpha_1, \alpha_2, \dots, \alpha_m, \mu + 1; z) + W(\alpha_1, \dots, \alpha_m, \mu; z) \cdot (z - a_1);$$

$$(\mu + n - 1)\frac{\partial W}{\partial a_1} - (\alpha_1 - 1)\frac{\partial W}{\partial z} = (z - a_1)\frac{\partial^2 W}{\partial a_1 \partial z}.$$

This allows us to express the functions $W(\alpha_1 + n_1, \dots, \alpha_m + n_m, \mu + n; z)$ for integers n_i in terms of any m linearly independent functions of the same form (contiguous functions) with coefficients that are rational in a_i, α_i, z, and μ. The integral representation (5.1), differential equation (5.2), and contiguous relations (5.3) are major ingredients in the application of Pochhammer integrals to explicit construction of Padé-type approximations.

Further topics on Pochhammer integrals include the study of Picard's group of monodromy of these integrals in $(m+1)$-dimensional complex space of parameters $a_1, \dots, a_m, z$, developed by Deligne-Mostow [74].

We present applications of Pochhammer-type integrals to diophantine approximations of values of the logarithmic and inverse trigonometric functions. One studies these integrals in the most general form

$$(2.5.4) \qquad I = \int_C \prod_\xi (\zeta - \alpha_\xi)^{v_\xi} d\zeta,$$

where α_ξ are algebraic numbers such that exponents are equal: $v_\xi = v_{\xi'}$, whenever α_ξ and $\alpha_{\xi'}$ are algebraically conjugate. Varying paths C we obtain all analytic continuations of integrals in (5.4). Looking at the monodromy of (5.4) one sees that integrals in (5.4) are linear combinations of logarithmic functions only if all exponents v_ξ in (5.4) are *integers*. In fact, one can see that under these conditions and the assumptions above on v_ξ, any integral I of the form (5.4) with $C = \overline{\alpha_{\xi_1} \alpha_{\xi_2}}$—being a path connecting two α_{ξ_1} and α_{ξ_2}, for which v_{ξ_1} and v_{ξ_2} are positive—is a linear combination of logarithms. Namely, we have an explicit evaluation:

$$(2.5.5) \qquad I\left(= \int_{\alpha_{\xi_1}}^{\alpha_{\xi_2}} \prod_\xi (\zeta - \alpha_\xi)^{v_\xi} d\zeta\right) = \sum_{v_\xi < 0} \ln\left(\frac{\alpha_{\xi_2} - \alpha_\xi}{\alpha_{\xi_1} - \alpha_\xi}\right) \cdot D_\xi + D_0,$$

where D_ξ are algebraic numbers and such that $D_{\xi'}$ and D_ξ are algebraically conjugate whenever $\alpha_{\xi'}$ and α_ξ are algebraically conjugate. This simple observation is sufficient to deduce a variety of pure Padé approximations to logarithmic functions. Hermite's original construction of Padé approximations to exponential, binomial, and logarithmic functions follows exactly the same outline (see the multiple integral above in §2.4 and [75]). We present only examples:

EXAMPLE 5.1. In the Padé approximation problem of the first kind to the system of functions $\ln(1 - \omega_i x)$, $i = 1, \ldots, m$ (with distinct ω_i), at $x = 0$ one considers the integrals of the form:

$$(2.5.6) \qquad I = \int_C \zeta^n (\zeta - 1)^n \prod_{i=1}^m \left(\zeta - \frac{1}{\omega_i x} \right)^{-n-1} d\zeta.$$

If one takes a path $C = \overline{01}$ in (5.6), I is the remainder function in the Padé approximation problem to $\ln(1 - \omega_i x)$, $i = 1, \ldots, m$, at $x = 0$ with the weight n. The Padé approximants in this problem also have the form (5.6) for $C = \gamma_i$—a loop around $\alpha_i = 1/\omega_i x$ for $i = 1, \ldots, m$.

The Padé approximation problem of the second kind adjoint to Example 5.1 (in the sense of the transference principle of [70]) consists of simultaneous approximation to 1, $\ln(1 - \omega_i x)$, $i = 1, \ldots, m$, at $x = 0$, i.e., of systems of polynomials $Q_i(x)$, $i = 0, \ldots, m$, such that the remainder functions $R_i(x) \overset{\text{def}}{=} Q_i(x) - Q_0(x) \ln(1 - \omega_i x)$ at $x = 0$ have zeros of orders $n + [\frac{n}{m}] + 1$, $i = 0, \ldots, m$, where degrees of polynomials $Q_i(x)$, $i = 0, \ldots, m$, are bounded by n. These are the Padé approximations of the second kind, and n is the weight of Padé approximations. As in Example 5.1, these Padé approximations arise as Pochhammer integrals:

EXAMPLE 5.2. One looks at integrals

$$(2.5.7) \qquad I = \int_C \prod_{i=1}^m \left(\zeta - 1 - \frac{1}{\omega_i x} \right)^n \cdot \frac{(\zeta - 1)^n}{\zeta^{n+1}} d\zeta.$$

If $C = \overline{1\alpha_i}$ for $\alpha_i = 1/\omega_i x$, $i = 1, \ldots, m$, then one obtains the remainder functions $R_i(x)$ above, (after multiplication by appropriate powers of x). The appropriate Padé approximants with the weight n correspond to different choices of C.

Examples 5.1 and 5.2 were used by Hermite, and later they and their multidimensional generalizations (cf. the expression of §2.1) were used by Mahler, Baker and others (referenced in [43, 71]) to obtain measures of irrationality of logarithms of rational and algebraic numbers. Particularly successful were ordinary Padé approximations to a single $\ln(1 - x)$ corresponding to $m = 1$ in (5.6) or (5.7) with $\omega_1 = 1$, with Padé approximants and the remainder function expressed in terms of Legendre polynomials and Legendre functions of the first and second kind. Later in 1979–83 we developed Padé-type approximations to logarithmic function giving better measures of diophantine approximations; see [43]. These methods of new Padé-type approximations can now be reformulated in terms of appropriate Pochhammer integrals.

New methods of better rational approximations to a number $\ln h$ for a rational or algebraic h are based on the study of integrals

$$(2.5.8) \qquad I = \int_1^h \frac{\prod_\xi (\zeta - \alpha_\xi)^{v_\xi}}{\zeta^{n+1}} d\zeta,$$

where $\alpha_0 = 1$, $\alpha_1 = h$; v_ξ are (nonnegative) integers (and, as above, $v_\xi = v_{\xi'}$ whenever α_ξ and $\alpha_{\xi'}$ are algebraically conjugate). The integrals (5.8) thus can be represented in the form

$$(2.5.8') \qquad I = \int_C \frac{(\zeta - 1)^{v_0}(\zeta - h)^{v_1} \prod_{k>2} P_k(\zeta)^{v_k}}{\zeta^{n+1}} d\zeta,$$

where $P_k(\zeta) \in \mathbf{Z}[\zeta]$. The case of $v_k = 0$ for $k \geq 2$ corresponds to the standard Padé approximations $I = P_n(h) \ln h - Q_n(h)$ to $\ln h$ at $h = 1$ for I in (5.8).

Extremality conditions that determine which integrals (5.8) give the best rational approximations to $\ln h$ are determined by the following criteria: (a) how close I in (5.8) is to zero; (b) how far from 0 are Padé-type approximants in (5.8'); and (c) how small is the common denominator of the coefficients of the polynomial Padé-type approximants.

The answer to questions (a)–(c) for a *given* integrand in (5.8–5.8') can be obtained effectively from the WKB-method (in complex domain for (a) and (b) and in the p-adic domain for (c)). To make the formulation simple, let us assume that exponents v_ξ depend linearly on a single parameter n in (5.8–8'), i.e., we can assume that $v_\xi = [w_\xi \cdot n]$ for fixed (rational) w_ξ, and that n is a sufficiently large integer.

As follows from the monodromy discussion above (cf. Example 5.2), the integral I has the form

$$(2.5.9) \qquad I = P \cdot \ln h + Q,$$

where P and Q are rational functions in h (and polynomials in α_ξ) with rational number coefficients. Then the "Padé-type approximants" P and Q in (5.9) also can be represented as linear combinations of integrals (5.8') for various paths C. The expression of P in (5.9) is very simple:

$$(2.5.10) \qquad P = \operatorname*{res}_{\zeta=1} \left\{ \frac{(\zeta - 1)^{v_0}(\zeta - h)^{v_1} \prod_{k>2} P_k(\zeta)^{v_k}}{\zeta^{n+1}} \right\}.$$

Thus P is a polynomial in h and in coefficients of $P_k(\zeta)$ for $k \geq 2$.

We can give a complete answer to a 3-part question (a)–(c) of the analytic and arithmetic asymptotics of the remainder function I and "Padé-type approximants" P and Q in (5.9) as $n \to \infty$ for $v_\xi = [w_\xi \cdot n]$, as above.

PROPOSITION 5.3. *Let h, α_ξ be all algebraic numbers from the field* $\mathbf{K}$. *Let S be the set of all non-archimedean valuations v of $\mathbf{K}$ such that $v(h) \neq 0$ or $v(\alpha_\xi) \neq 0$ for some ξ. If $M = \max\{n, \sum v_\xi - n\}$, and $\Delta_n = \operatorname{lcm}\{1, \ldots, M\}$, then numbers $\Delta_n \cdot P$ and $\Delta_n \cdot Q$ for P and Q from (5.9) are S-integral numbers from* $\mathbf{K}$.

Moreover, for any $v \in S$, $\min\{v(P), v(Q)\} = \min\{\mu_1, \mu_2\}$, where

$$\mu_1 (= \mu_1(v)) = \sum_{\xi, v(\alpha_\xi) < 0} v_\xi \cdot v(\alpha_\xi),$$

$$\mu_2 (= \mu_2(v)) = \sum_{\xi, v(\alpha_\xi) < v(h)} (v(\alpha_\xi) - v(h)) \times v_\xi + v(h) \cdot \left[\sum_\xi v_\xi - n \right].$$

In particular, if $\mathscr{P}$ are prime ideals of $\mathbf{K}$ corresponding to v, then

$$\mathscr{D} = \Delta_n \cdot \prod_{\mathscr{P}, v_{\mathscr{P}}(\alpha_\xi \cdot h) \neq 0} \mathscr{P}^{- \min\{\mu_1(v_{\mathscr{P}}), \mu_2(v_{\mathscr{P}})\}}$$

is the common denominator of $P, Q \in K$ from (5.9).

The archimedean part of the height of P and Q together with the asymptotics of $|I|$ in (5.9) is determined from the steepest descent method applied to (5.9) and (5.8'). Laplace's method allows us to apply the classical form of the steepest descent method used by Riemann [76] to integrals (5.8–5.8') with $v_k = 0$ for $k \geq 2$:

PROPOSITION 5.4. *As above, let* $v_\xi = [w_\xi \cdot n], \xi = 0, 1, \ldots,$ *for fixed* w_ξ *and sufficiently large* n, *where* α_ξ *in* (5.8–5.8') *are independent of* n. *We put* $R(\zeta) = \prod_\xi (\zeta - \alpha_\xi)^{w_\xi} \cdot \zeta^{-1}$, *and denote by* z_j *all critical points of* $R(\zeta)$, *i.e., roots of the equation* $\sum_\xi w_\xi / (\zeta - \alpha_\xi) = 1/\zeta$. *Then the asymptotics of integrals* (5.8') *is determined by the contribution of critical points* z_j *lying on the steepest descent paths homological to* C. *In particular, whenever* I *in* (5.8') *is not identically zero as* $n \to \infty$,

$$(2.5.11) \qquad \frac{1}{n} \ln |I| \leq \max_{z_j} \ln |R(z_j)|$$

where max *is taken over all critical points* z *of* $R(\zeta)$. *If* h *is real* > 1 *and* $w_0 > 0$, $w_1 > 0$, *then*

$$(2.5.12) \qquad \frac{1}{n} \ln |I| \leq \max_{z_j \in (1,h)} \ln |R(z_j)|,$$

where max *is taken only over those critical points* z_j *that lie in the interval* $(1, h)$.

The asymptotics of (5.12) holds, in fact, for arbitrary complex h not lying on the cut from $-\infty$ to 1. In this case one chooses as a path between 1 and h the steepest descent/ascent path.

REMARK 5.5. One can explicitly determine conditions under which equalities in (5.11) and (5.12) hold. Moreover, if all critical points z_j are determined, all steepest descent/ascent paths determining I, P, and Q in (5.9) can be explicitly determined from solutions of one-particle dynamical systems governed by the integrand $R(\zeta)^n$ in the complex ζ-plane.

The key problem becomes a problem of determination of all critical points of $R(\zeta)$, i.e., the solution of a polynomial equation of degree $\sum_{\xi, v_\xi > 0} 1$ with high accuracy. This is achieved by our high-precision version of a polynomial root finding package [77]. The ability to determine all critical points of $R(\zeta)$ and the asymptotics of integrals in (5.11–5.12) allows us to solve the extremality problem and to find the best choice of exponents v_ξ and polynomials $P_k(\zeta) \in \mathbf{Z}[\zeta]$ in (5.8'). The figure of merit, which determines the extremality condition, is expressed in terms of the ratio of arithmetic and analytic asymptotics of $|I|$ and heights of P and Q in (5.9). The corresponding measure of irrationality is described in the following well-known simple lemma on dense approximations [71]:

LEMMA 5.6. *Let us assume that there exist sequences of rational integers* P_n, Q_n *such that*

$$\left. \begin{array}{c} \ln|P_n| \\ \ln|Q_n| \end{array} \right\} \sim an \text{ as } n \to \infty$$

and $\ln|P_n\theta - Q_n| \sim bn$ *as* $n \to \infty$ *for* $b < 0$. *Then the number* θ *is irrational and for any* $\epsilon > 0$ *and for all rational integers* p, q *with* $|q| \geq q_0(\epsilon)$, *we have*

(2.5.13) $$|\theta - p/q| > |q|^{a/b - 1 - \epsilon}.$$

Thus one tries to find α_ξ and v_ξ such that for the sequence of approximations $\mathscr{D} \cdot I_n = (\mathscr{D} \cdot P_n) \cdot \ln h + (\mathscr{D} \cdot Q_n)$ from (5.9), and Proposition 5.3 satisfies the conditions of Lemma 5.6 with the minimal $|a/b|$.

One sees that the extremality problem formulated this way does not have a unique solution if the number of distinct zeros α_ξ of the integrand is unbounded. When one bounds the number of α_ξ's, then the best form of the integrand can be (numerically) determined. One can see that, by increasing the number of zeros α_ξ, any bound on the exponent $a/b - 1$ as in Lemma 5.6 for $\theta = \ln h$ can be improved. To improve the exponent it is enough to introduce a new factor $P_{k+1}(\zeta)^{v_{k+1}}$ into the integrand, whose roots are exactly the critical points of the previous integrand $R(\zeta)^n$. Consequently, any bound on the exponent of measure of irrationality of $\ln h$ that can be obtained using integrals (5.8–5.8′) can be always further slightly improved, with the only change in the constant $q_0(\epsilon)$ in the bound (5.13).

This process of iterative improvement can be best illustrated by starting from the case of $v_k = 0$ for $k \geq 2$ corresponding to the usual Padé approximations $I_n = P_n(h) \ln h - Q_n(h)$ to $\ln h$ at $h = 1$. The equation of the critical points of the integrand in this case is $\zeta^2 - h = 0$ giving rise to asymptotics of $\frac{1}{n} \ln|I_n|$, and to $\frac{1}{n} \ln|P_n(h)|$ and $\frac{1}{n} \ln|Q_n(h)|$ given by two numbers: $\ln|1 - \sqrt{h}|^2$, $\ln|1 + \sqrt{h}|^2$ respectively. Thus the next choice for improved integrals in (5.8–5.8′) is

(2.5.14) $$I = \int_1^h \frac{\{(\zeta - 1)(\zeta - h)\}^m \cdot (\zeta^2 - h)^{n-m}}{\zeta^{n+1}} d\zeta.$$

These new Padé-type approximants to logarithmic functions were presented by us in [43] and identified with the integral representation (5.14). The integral I in (5.14) is a linear combination of 1 and $\ln h$:

(2.5.15) $$I = P_n(h) \cdot \ln h - Q_n(h),$$

where

(2.5.16) $$P_n(h) = \sum_{i=0}^{m} \sum_{\substack{j=0, n-m \leq i+2j \leq n}}^{n-m} \binom{m}{i}\binom{m}{n-i-2j}\binom{n-m}{j}(-1)^j h^{i+j},$$

and the other approximant $Q_n(h)$ is determined as

$$\int_1^h \frac{P_n(h) - P_n(\zeta)}{h - \zeta} d\zeta.$$

Consequently $P_n(h)$ has integral coefficients and $Q_n(h)$ has a common denominator of $\mathrm{lcm}\{1, \ldots, n\}$. For various h and ratios m/n, various Padé-type approximations to $\ln h$ are obtained. We refer to [43] for recurrence relations and for applications to the measure of irrationality of $\ln 2$, i.e., for $h = 2$.

For imaginary quadratic numbers h, this approach gives a new rational approximation to inverse trigonometric functions. It is far better to use a different form of integrals in (5.4) with two distinct poles instead of one in (5.8). These integrals have the form

$$(2.5.17) \qquad I = \int_0^1 \frac{\{\zeta(\zeta - 1)\}^{v_0} \prod_{k \geq 2} P_k(\zeta)^{v_k}}{(\zeta - z)^{n+1} (\zeta - (z - 1))^{n+1}} d\zeta.$$

These integrals (5.17) give the rational approximations to

$$\frac{1}{2z - 1} \ln \left(\left(1 - \frac{1}{z} \right)^2 \right)$$

when $(2z - 1)^2$ is positive. When $(2z - 1)^2 = -\Delta$ is negative (i.e., when $z = \frac{1}{2} + it$), the integral (5.17) gives the rational approximations to

$$\frac{1}{\sqrt{\Delta}} \arctan \frac{1}{\sqrt{\Delta}}.$$

Instead of the form (5.17) of integrals we consider the following integral representation, as in [57]:

$$(2.5.18) \qquad I' = \int_{u-1/4}^u \frac{\prod_\xi (Z - \beta_\xi)^{v_\xi} dZ}{Z^{n+1} \sqrt{4Z + (2\zeta - 1)^2}},$$

where $u = -\zeta(\zeta - 1)$, and $\beta_0 = u$, $\beta_1 = u - 1/4 = -1/4(2\zeta - 1)^2$.

It follows from (5.18) that any integral of the form I' is a linear combination

$$(2.5.19) \qquad I' = P' \frac{4}{\sqrt{\Delta}} \arctan \left(\frac{1}{\sqrt{\Delta}} \right) + Q'$$

for $\Delta = -(2\zeta - 1)^2 = 4u - 1$, where P', Q' are rational functions in u, polynomials in β_ξ, and have the rational number coefficients.

The derivation of asymptotics of $|I'|$ in (5.18) and $|P'|$, $|Q'|$ using the steepest descent method is identical to that of integrals (5.8–5.8') because the main term $R(Z)^n = \prod_\xi (Z - \beta_\xi)^{v_\xi} / Z^n$ in the integrand of (5.18) has the form identical to that of (5.8). In particular, the statement of asymptotics in Proposition 5.4 holds for (5.18) with the natural substitution of $R(\zeta)$ by $R(Z)$. The asymptotics of (5.11) and (5.12) as $n \to \infty$ also holds if one replaces the interval of integration $[1, h]$ by $[u - 1/4, u]$ in (5.18).

The only difference between the integrals of the form (5.8) and (5.18) is in their arithmetic asymptotics, particularly in the term corresponding to $\mathscr{D}$ and Δ_n of Proposition 5.3; see [44].

One can improve the figure of merit provided by Lemma 5.6—given by a ratio a/b, by taking v_0, v_1 in (5.18) to be less than n. An optimal value of $v_0 + v_1$ is around $3n$ for those ζ for which $\frac{\zeta-1}{\zeta}$ is a unit (or $\frac{1}{\sqrt{\Delta}}$ arctan $\frac{1}{\sqrt{\Delta}}$ is commensurable with π). The case of $\pi/\sqrt{3}$ is the most revealing. This corresponds to $\zeta = (1 + \sqrt{-3})/2$. Padé approximations of the form (5.19) corresponding to (5.18) for this ζ with $v_0 = v_1 = [3n/4]$, $v_k = 0$ for $k \geq 2$ (and n divisible by 4) were analyzed in detail in 1980; see §5 of [43]. Let us present in this case, following [43], a three-term recurrence relating consecutive P', Q' that have coefficients polynomial in m. The derivation of these and other properties had been achieved in IBM SCRATCHPAD environment.

For this we look at sequences of dense approximations to $\pi/\sqrt{3}$ arising from (5.18–5.19) with $v_0 = v_1 = [3n/4]$, $v_k = 0$ for $k \geq 2$, and $4|n$. We arrive thus at a sequence X_n/Y_n approximating $4\pi/3\sqrt{3}$, where Y_n are rational integers and X_n are rational numbers with denominators dividing $\mathrm{lcm}\{1,\ldots,4n\}$. The expression of Y_n is the following:
(2.5.20)

$$Y_n = \sum_{i_1=0}^{3n} \sum_{i_2=0, i_1+i_2 \leq 4n}^{3n} \binom{3n}{i_1}\binom{3n}{i_2}\binom{2(4n - i_1 - i_2)}{4n - i_1 - i_2} \times (-1)^{i_1+i_2} \cdot 4^{i_2} \cdot 3^{i_1}.$$

The contiguity relations (5.3) for the integral representations (5.18–5.19) show the existence of the scalar three-term linear recurrence relation satisfied by each of the sequences X_n and Y_n that has coefficients polynomial in n. It has the form

(2.5.21)
$$A_2(n)Y_{n+2} + A_1(n)Y_{n+1} + A_0(n)Y_n = 0;$$
$$A_2(n)X_{n+2} + A_1(n)X_{n+1} + A_0(n)X_n = 0.$$

Here $A_0(n)$, $A_1(n)$, and $A_2(n)$ are polynomials in n of degree 9 with integer coefficients and are determined explicitly as follows:

$$A_2(n) = -2^3 \cdot (4n + 7) \cdot (4n + 5) \cdot (4n + 3) \cdot (4n + 1) \cdot (2n + 3) \cdot (n + 2)$$
$$\times (27279n^3 + 52164n^2 + 31511n + 6046);$$

$$A_1(n) = 3 \cdot (4n + 3) \cdot (4n + 1) \cdot (15484624281n^7 + 12251806648n^6$$
$$+ 401859218160n^5 + 706125904254n^4$$
$$+ 715282318379n^3 + 415975459648n^2$$
$$+ 128021157420n + 16022087856)$$

$$A_0(n) = 2 \cdot 3^3 \cdot (6n + 5) \cdot (6n + 1) \cdot (3n + 2) \cdot (3n + 1)$$
$$\times (2n + 1) \cdot (n + 1) \cdot (27279n^3 + 134001n^2 + 217676n + 117000).$$

The initial conditions and the first few terms X_n and Y_n from (5.20–5.21) are the following:

$$Y_0 = 1, \quad Y_1 = 1250, \quad Y_2 = 5915250, \quad Y_3 = 32189537978\ldots,$$

$$X_0 = 0, \quad X_1 = 3023,\ldots, \quad \text{and} \quad X_2/Y_2 = 111264499/46007500,\ldots.$$

The limit quadratic equation that determines the asymptotics of Y_n and X_n follows from (5.21): $2^8 x^2 - 3^3 \cdot 59 \cdot 1069x - 3^5 = 0$.

Thus according to Lemma 5.6 one deduces the following measure of irrationality of $\pi/\sqrt{3}$:

$$\left| q\frac{\pi}{\sqrt{3}} - p \right| > |q|^{-4.8174417\ldots}$$

for arbitrary rational integers p, q, with $|q| \geq q_1$ [43]. A slightly different choice of v_0 and v_1 can improve this exponent. For example, if one puts $v_0 = [0.8 \cdot n]$, $v_1 = [0.7n]$ (and $v_k = 0$ for $k \geq 2$) in (5.18–5.19), then the following measure of rational approximations is achieved:

$$\left| q\frac{\pi}{\sqrt{3}} - p \right| > |q|^{-4.792613804\ldots}$$

for $|q| \geq q_2$.

As we have remarked, the process of improvement of the exponent by means of adding terms to the integrand in (5.18) can be continued indefinitely. E.g., by taking $v_0 = [0.8005 \cdot n]$, $v_1 = [0.6995 \cdot n]$ one improves the exponent to $-4.7926098\ldots$, etc.

Another interesting application of integrals (5.18) is to π. Though several good measures of irrationality of π are known, see references in [78] and [43], all these measures are achieved by means of multidimensional integral representations. This means that all those measures are based on approximation of π and its high powers, involving π^2, etc. Not a single system of dense approximations to π of the type demanded by Lemma 5.6 was known. Integrals of the form (5.18) with $\zeta = (1 + i)/2$ and $\Delta = 1$ provide such sequences of dense approximations to $\pi = 4\arctan 1$. A search for appropriate integrands of the form (5.18) leads to the following simple class of integrals:

$$(2.5.22) \qquad I' = \int_{1/4}^{1/2} \frac{(2Z - 1)^{v_0} \cdot (4Z - 1)^{v_1} \cdot (3Z - 1)^{v_2} dZ}{Z^{n+1}\sqrt{4Z - 1}}.$$

A choice of v_0, v_1, v_2 as $v_0 = [0.435 \cdot n]$, $v_1 = [0.305 \cdot n]$, $v_2 = [0.755 \cdot n]$ leads to the dense sequence of approximations $\mathscr{D}_n \cdot I' = P'_n \cdot \pi - Q'_n$ to π with integral P'_n, Q'_n satisfying the condition of Lemma 5.6. This choice of parameters in (5.22) leads to an exponent in the measure of irrationality of π of $-55.9829\ldots$. This exponent is far from the best that we obtained, but a simple system of rational approximations Q'_n/P'_n to π is important for various problems in approximations to π. We note that from contiguous relations (5.3) it follows that sequences P'_n and Q'_n satisfy a four-term linear recurrence with coefficients polynomial in n.

2.6. Multidimensional integrals in diophantine approximations. Following techniques explained in §2.5, one can use complex and nonarchimedean WKB-like methods for multidimensional integrals representing a variety of Padé-type approximations. In fact, nontrivial applications of diophantine approximations can be obtained only this way.

These multidimensional integrals of the Pochhammer type typically have the form

$$\int_\gamma f_1^{\alpha_1} \cdots f_m^{\alpha_m} \prod_{k=m+1}^{l} P_k^{N_k} \prod_{i=1}^{n} dx_i,$$

where $f_1, \ldots f_m$ are linear functions in $x_1, \ldots, x_n$, and $\alpha_1, \ldots, \alpha_m$ are rational numbers, while P_k are polynomials in $x_1, \ldots, x_n$ and N_k are nonnegative integers. A variety of Padé-type approximations is represented in this form. We saw one example of such formula in §2.4 in connection with Hermite's identities. To be more specific, the (Hermite-) Padé approximation problem to functions $(1 - x)^{\omega_i}$, $i = 1, \ldots, m$, at $x = 0$ with weight D has a solution in terms of hypergeometric-like integrals (G-functions in Meyer's, not Siegel's sense).

The remainder function in this problem has the following integral representation:

(2.6.1)
$$R_D(x) = \int_0^x dt_1 \int_0^{t_1} dt_2 \cdots \int_0^{t_{m-2}} dt_{m-1}(x - t_1)^{D-1}(t_1 - t_2)^{D-1} \cdots t_{m-1}^{D-1}$$
$$\times (1 - t_1)^{\omega_2 - \omega_1 - D} \cdots (1 - t_m)^{\omega_m - \omega_{m-1} - D}.$$

The asymptotics of this remainder function and the corresponding Padé approximants (as $D \to \infty$) are easy to determine following §2.5; see [71] and [78]. It is harder to determine the asymptotics of the nonarchimedean part— the growth of the denominator of the coefficients of the corresponding Padé approximants. For rational exponents $\omega_i = \frac{k_i}{n}$, $i = 1, \ldots, m$, the denominator $\mathrm{Den}_D(\omega_1, \ldots, \omega_m)$ of the coefficients of the Padé approximants to $(1 - x)^{\omega_i}$ with the weight D has the asymptotics

$$\overline{\lim_{D \to \infty}} \frac{1}{D} \ln \mathrm{Den}_D(\omega_1, \ldots, \omega_m) = \mathscr{C}_n^m,$$

where the constant $\mathscr{C}_n^m$ has an extremely complicated expression in terms of values of $\psi(z) = \Gamma'(z)/\Gamma(z)$ at rational points with the denominator n (depending on $\omega_i = k_i/n$) that reflect the distribution of primes in arithmetic progressions mod n in various intervals. See computations in [71] and in [44].

One application of this approach is the effective improvement over the Liouville theorem for algebraic numbers from Kummer extensions $\mathbf{Q}(\sqrt[n]{\alpha})$. These results are not as strong as an ineffective Thue-Siegel-Roth [6] theorem, giving the exponent $-2 - \epsilon$ for any $\epsilon > 0$ in the measure of irrationality for an arbitrary algebraic number. Nevertheless these effective bounds are better than any other effective bounds. Baker's nontrivial effective bounds based on linear forms in logarithms are extremely close to the (trivial) Liouville one (see [6]).

We present a little table of the best effective exponents λ in the bound

$$\left| \alpha - \frac{p}{q} \right| > |q|^{-\lambda},$$

for all rational integers p, q with $|q| \geq q_0(\alpha)$ and effective $q_0(\alpha)$ for all algebraic (irrational) α in $\mathbf{Q}(\sqrt[3]{D})$:

D	λ
2	$2.427\ldots$
3	$2.690\ldots$
5	$2.998\ldots$
6	$2.320\ldots$
7	$2.725\ldots$
10	$2.413\ldots$
11	$2.997\ldots$
12	$2.907\ldots$
13	$2.824\ldots$

Also, if $n \geq 89$, every algebraic number from the field $\mathbf{Q}(\sqrt[n]{2})$ satisfies the effective version of the Thue theorem:

$$\left| \alpha - \frac{p}{q} \right| > |q|^{-n/2-1}$$

for $|q| \geq q_0(\alpha)$ and effective constant $q_0(\alpha)$ depending on $\alpha \in \mathbf{Q}(\sqrt[n]{2})$.

The complete effectivization of Roth's theorem for cubic irrationalities, as it stands now, is conditional upon the Tanniyama-Weil conjecture of functional equations satisfied by all L-functions of elliptic curves over $\mathbf{Q}$, and an additional conjecture on the bound of the Patterson norm of the modular forms of weight 2 of congruence subgroups. To the same category belongs the still-open question of the unboundedness of the partial fractions in the continued fraction expansion of cubic irrationalities. This group of questions is closely connected with Hall's conjecture, which is very easy to formulate:

HALL'S CONJECTURE. *For integers x, y whenever $x^3 \neq y^2$*

$$|x^3 - y^2| \geq C \cdot \sqrt{|x|}$$

for an absolute constant C.

From algebraic numbers we return to transcendental and to multiple integrals to obtain good bounds on measures of irrationality and transcendence of logarithms $\ln h$ for algebraic h. For this, as above, one uses Padé approximations to

$$\ln(1-x), \ldots, \ln(1-x)^m$$

at $x = 0$. These Padé approximations can be deduced from the integral representation (6.1) in the binomial function case if one puts $\omega_i = 0$, $i = 1,\ldots,m$.

We can express this integral, representing the remainder function in the following transparent form:

$$R_D(h) = \int dx_1 \cdots \int\limits_{\substack{1\le x_1,\ldots,1\le x_m \\ x_1\ldots x_m\le h}} dx_m \frac{(h - x_1\cdots x_m)^D \prod_{i=1}^m (x_i - 1)^D}{x_1^D \cdots x_m^D},$$

taken over an m-polytop, and modified appropriately for complex h.

This integral is, as required, an approximation to powers of $\ln h$:

$$R_D(h) = P_0(h) + P_1(h)\ln h + \cdots + P_m(h)\ln^m h,$$

vanishing at $h = 1$ up to order $(h-1)^{(m+1)D+m}$. This integral can be efficiently used for h even quite far from $h = 1$ if m is large. See, e.g., [71] and [78]. E.g., for $h = e^{\pi i/2}$ one takes $m = 5$ and gets the exponent in the measure of irrationality of $\pi : -19.88999444\ldots$. We got in this way [78]

$$\left| \pi - \frac{p}{q} \right| > |q|^{-19.88999444\ldots}$$

for $|q| \ge 2$.

Efficient new methods of rational approximations to powers of logarithms, proposed in §2.5, occur when one modifies the integral $R_D(h)$ to improve the asymptotics in the neighborhood of a particular h, by adding more factors to the integrand to dampen the contribution of critical points of the integrand. We studied, in particular, the following forms of the integrand:

$$(2.6.2)\quad R'_D(h) = \int \cdots \int\limits_{1\le x_i, x_1\cdots x_m\le h} \prod_{i=1}^m dx_i$$

$$\times \frac{(h - x_1\cdots x_m)^{[\nu_0 D]} \prod_{i=1}^m (x_i - 1)^{[\nu_1 D]} \prod_{i<j}(x_i - x_j)^{[\nu_2 D]} \prod_{i=1}^m (h - x_i^2)^{[\nu_3 D]}}{\prod_{i=1}^m x_i^D}$$

with various rational exponents ν_α. This gives new measures of irrationality of π and π^2, see §2.2.

Unfortunately, the complex structure of the critical points and the need to determine the contributions from all of them into the remainder function and, particularly, into all Padé approximants makes it impossible at this point to analyze (6.2) with $m \ge 5$.

We conclude with a problem on multiple integrals. Let

$$I = \int_\gamma f_1^{N_1} \cdots f_m^{N_m} \prod_{i=1}^n dx_i,$$

where f_i are polynomials in $x_1,\ldots,x_n$ (or even linear functions) and N_i are integers. When is I a combination of powers of logarithmic and rational functions of coefficients of f_i?

REFERENCES

1. P. Griffiths, *Periods of integrals on algebraic manifolds: summary of main results and discussion of open problems*, Bull. Amer. Math. Soc., **75** (1970), 228–296.

2. N. Katz, *Nilpotent connections and the monodromy theorem: applications of a result of Turrittin*, Inst. Hautes Études Sci. Publ. Math., **32** (1970) 232–355.

3. C.L. Siegel, *Über einige Anwendungen diophantischer Approximationen*, Abh. Preuss. Akad. Wiss. Phys. Math. Kl. **1** (1929).

4. ____, *Transcendental numbers*, Princeton Univ. Press, Princeton, N.J., 1949.

5. A.B. Shidlovsky, *The arithmetic properties of the values of analytic functions*, Trudy Mat. Inst. Steklov, **132** (1973), 169–202.

6. A. Baker, *Transcendental number theory*, Cambridge Univ. Press, 1975.

7. G. V. Chudnovsky, *On some applications of diophantine approximations*, Proc. Nat. Acad. Sci. U.S.A., **81** (1984), 1926–1930.

8. W. M. Schmidt, *Diophantine approximations*, Lecture Notes in Math., vol. 785, Springer-Verlag, 1980.

9. A.L. Galočhkin, *Lower bounds of polynomials in the values of a certain class of analytic functions*, Mat. Sb. (N.S.), **95** (1974), 396–417.

10. G. V. Chudnovsky, *Padé approximations and the Riemann monodromy problem*. Bifurcation Phenomena in Mathematical Physics and Related Topics, D. Reidel, Boston, 1980, pp. 448–510.

11. ____, *Measures of irrationality, transcendence and algebraic independence. Recent progress*, Journees Arithmetiques 1980 (J.V. Armitage, ed.), Cambridge Univ. Press, 1982, pp. 11–82.

12. E. Bombieri, *On G-functions*, Recent Progress in Analytic Number Theory (H. Halberstram and C. Hooly, editors), Academic Press, 1981, pp. 1–67.

13. K. Väänänen, *On linear forms of certain class of G-functions and p-adic G-functions*, Acta Arith. **36** (1980), 273–295.

14. G. V. Chudnovsky, *On applications of diophantine approximations*, Proc. Nat. Acad. Sci. U.S.A. **81** (1984), 7261–7265.

15. D.V. Chudnovsky and G.V. Chudnovsky, *Applications of Padé approximations to diophantine inequalities in values of G-functions*, Lecture Notes in Math., vol. 1135, Springer-Verlag, 1985, pp. 9–51.

16. N. Katz, *Algebraic solutions of differential equations*, Invent. Math. **18** (1972), 1–118.

17. T. Honda, *Algebraic differential equations*, Sympos. Math., vol. 24, Academic Press, 1981, pp. 169–204.

18. B. Dwork, *Arithmetic theory of differential equations*, Sympos. Math., vol. 24, Academic Press, 1981, pp. 225–243.

19. D. V. Chudnovsky and G.V. Chudnovsky, *Applications of Padé approximations to the Grothendieck conjecture on linear differential equations*, Lecture Notes in Math., vol. 1135, Springer-Verlag, 1985, pp. 52–100.

20. ____, *Padé approximations and diophantine geometry*, Proc. Nat. Acad. Sci. U.S.A. **82** (1985), 2212–2216.

21. J.-P. Serre, *Quelques applications du théoreme de densité de Chebotarev*, Inst. Hautes Études Sci. Publ. Math. **54** (1981), 323–401.

22. G. Faltings, *Eudichkeitssätze für abelsche varietäten über zahlkörpern*, Invent. Math. **73** (1983), 349–366.

23. T. Honda, *On the theory of commutative formal groups*, J. Math. Soc. Japan **22** (1970), 213–246.

24. D.V. Chudnovsky and G.V. Chudnovsky, *p-adic properties of linear differential equations and Abelian integrals*, IBM Research Report RC 10645, 7/26/84.

25. ____, *The Grothendieck conjecture and Padé approximations*, Proc. Japan Acad. Sci. Ser. A Math. Sci. **61** (1985), 87–90.

26. N. Katz, *A conjecture in the arithmetic theory of differential equations*, Bull. Soc. Math. France **110** (1982), 203–239; corr., 347–348.

27. D. V. Chudnovsky and G.V. Chudnovsky, *A random walk in higher arithmetic*, Adv. Appl. Math. **7** (1986), 101–122.

28. G.V. Chudnovsky, *A new method for the investigation of arithmetic properties of analytic functions*, Ann. of Math. (N.S.) **109** (1979), 353–377.

29. C. Matthews, *Some arithmetic problems on automorphisms of algebraic varieties*, Number Theory Related to Fermat's Last Theorem, Birkhäuser, 1982, pp. 309–320.

30. B. Dwork and P. Robba, *Effective p-adic bounds for solutions of homogeneous linear differential equations*, Trans. Amer. Math. Soc. **259** (1980), 559–577.

31. E. Whittaker and G. Watson, *Modern analysis*, Cambridge Univ. Press, 1927.

32. D.V. Chudnovsky and G.V. Chudnovsky, *Computer assisted number theory with applications*, Lecture Notes in Math., vol. 1240, Springer-Verlag, 1987, pp. 1–68.

33. ____, *Remark on the nature of the spectrum of Lamé equation. Problem from transcendence theory.*, Lett. Nuovo Cimento **29** (1980), 545–550.

34. H. Poincaré, *Sur les groupes les équations linéaires*, Acta Math. **5** (1884), 240–278.

35. R. Fricke and F. Klein, *Vorlesungen über die theorie der Automorphen Functionen*, bd. 1., Teubner, Leipzig, 1925.

36. V. I. Smirnov, *Sur les équations différentielles linéaires du second ordre et la théorie des fonctions automorphes*, Bull. Soc. Math. France **45**(2) (1921), 93–120, 126–135.

37. L. Bers, *Quasiconformal mappings, with applications to differential equations, function theory and topology*, Bull. Amer. Math. Soc. **83** (1977), 1083–1100.

38. L. Keen, H.E. Rauch, and A.T. Vasques, *Moduli of punctured tori and the accessory parameter of Lamé's equation*, Trans. Amer. Math. Soc. **255** (1979), 201–229.

39. E. Hilb, *Lineare Differentialgleichungen im komplexen Gebiet*, Enzyklopädie Math. Wissensch. II, Band 6, Teubner, Leipzig, 1917, 471–562.

40. L. A. Takhtadjan and P.G. Zograf, *The Liouville equation action—the generating function for accessory parameter*, Funct. Anal. **19** (1975), 67–68.

41. E. G. C. Poole, *Introduction to the theory of linear differential equations*, Oxford Univ. Press, 1936.

42. A. J. Van der Poorten, *A proof that Euler missed ... Apéry's proof of the irrationality* $\zeta(3)$. Math. Intelligencer **1** (1978/79), 195–203.

43. D.V. Chudnovsky and G.V. Chudnovsky, *Padé and rational approximations to systems of functions and their arithmetic applications*, Lecture Notes in Math., vol. 1052, Springer-Verlag, 1984, pp. 37–84.

44. ____, *The use of computer algebra for diophantine and differential equations*, Computer Algebra as a Tool for Research in Mathematics and Physics (Proc. New York Conf., 1984), M. Dekker (to appear).

45. B. Dwork, *A deformation theory for the zeta function of a hypersurface*, Proc. Internat. Congr. Math. (Stockholm, 1962), Inst. Mittag-Leffler Djursholm, 1963, pp. 247–259.

46. C. Maclachlan and G. Rosenberg, *Two-generator arithmetic Fuchsian groups*, Math. Proc. Cambridge Philos. Soc. **93** (1983), 383–391.

47. K. Takeuchi, *Arithmetic Fuchsian groups with signature* $(1; e)$, J. Math. Soc. Japan **35** (1983), 381–407.

48. D.V. Chudnovsky and G.V. Chudnovsky, *Approximations and complex multiplication according to Ramanujan*, Proc. Ramanujan Centenary Conf., Academic Press, 1988, pp. 375–472.

49. R. Askey, *Orthogonal polynomials and theta functions*, this volume.

50. T.J. Stieltjes, *Sur la réduction en fraction continue d'une série procédant suivant les puissances descendantes d'une variable*, Ann. Fac. Sci. Toulouse **3** (1889), 1–17; reprinted in Oeuvres, t. II, Groningen, 1918, pp. 184–200.

51. ____, *Recherches sur lest fractions continue*, Ann. Fac. Sci. Toulouse **8** (1894), 1–22; **9** (1895), 1–47; reprinted in Ouevres, t. II, Groningen, 1918, pp. 402–566.

52. L. J. Rogers, *On the representation of certain asymptotic series as convergent continued fractions*, Proc. London Math. Soc. (2) **4** (1907), 72–89.

53. I.J. Schur, *Ueber Potenzreihen die im Invern des Einheitskreises beschränkt sind*, J. Reine Angew. Math. **147** (1916), 205–232; **148** (1917), 122–145.

54. L. Carlitz, *Some orthogonal polynomials related to elliptic functions*, Duke Math. J. **27** (1960), 443–460.

55. G.V. Chudnovsky, *The inverse scattering problem and its applications to arithmetic, algebra and transcendental numbers*, Lecture Notes in Physics, vol. 120, Springer-Verlag, 1980, pp. 150–198.

56. _____, *Rational and Padé approximations to solutions of linear differential equations and the monodromy theory*, Proc. Houches Internat. Colloq. Complex Analysis and Relativistic Quantum Field Theory, Lecture Notes in Physics, vol. 126, Springer-Verlag, 1980, pp. 136–169.

57. _____, *Number theoretical applications of polynomials with rational coefficients defined by extremality conditions*, Arithmetic and Geometry, Progr. Math., vol. 35, Birkhäuser, 1983, pp. 67–107.

58. S. Ramanujan, *Modular equations and approximations to π*, Collected Papers, Cambridge Univ. Press, 1927, pp. 23–39.

59. A. Weil, *Elliptic functions according to Eisenstein and Kronecker*, Springer-Verlag, 1976.

60. C.L. Siegel, *Zum Beneise des Starkschen Satzes*, Invent. Math. **5** (1968), 180–191.

61. G.V. Chudnovsky, *Padé approximations to the generalized hypergeometric functions. I*, J. Math. Pures Appl. **58** (1979), 445–476.

62. G. Shimura, *Automorphic forms and the periods of Abelian varieties*, J. Math. Soc. Japan **31** (1979), 561–579.

63. P. Deligne, *Cycles de Hodge absolus et périodes des intégrales des varietés abéiennes*, Mém. Soc. Math. France (N.S.) no. 2, 1980, pp. 23–33.

64. G. Shimura, *Introduction to the arithmetic theory of automorphic forms*, Princeton Univ. Press, Princeton, N.J., 1971.

65. R. Morris, *On the automorphic functions of the group* $(0, 3; l_1, l_2, l_3)$, Trans. Amer. Math. Soc. **7** (1906), 425–448.

66. K. Takeuchi, *Arithmetic triangle groups*, J. Math. Soc. Japan **29** (1977), 91–106.

67. H.P.F. Swinnerton-Dyer, *Arithmetic groups*, Discrete Groups and Automorphic Functions, Academic Press, 1977, pp. 377–401.

68. G.V. Chudnovsky, *Algebraic independence of values of exponential and elliptic functions*, Proc.Internat. Congr. Math. (Helsinki, 1979), Vol. 1, Acad. Sci. Fenn., Helsinki, 1980, pp. 339–350.

69. _____, *Contributions to the theory of transcendental numbers*, Math. Surveys Monogr., vol. 19, Amer. Math. Soc., Providence, R.I., 1984.

70. K. Mahler, *Perfect systems*, Compositio Math. **19** (1968), 95–166.

71. G.V. Chudnovsky, *On the method of Thue-Siegel*, Ann. of Math. (2) **117** (1983), 325–382.

72. _____, *Bäcklund transformations and deformations of linear differential equations with applications to diophantine approximations*, Problems of Mathematical Physics, Festschrift in honor of F. Gürsey, I. Bars, A. Chodos, and C.-N. Tze, editors, Plenum Press, New York, 1984, pp. 201–220.

73. L. Pochhammer, *Ueber ein Integral mit doppeltem Umlauf; Veber eine Classe von Integralen mit geschlossener Integrationscurve*, Math. Ann. **35** (1889) 470–494; **37** (1890), 500–511.

74. P. Deligne and G.D. Mostow, *Monodromy of hypergeometric functions and non-lattice integral monodromy*, Inst. Hautes Études Sci. Publ. Math. **63** (1986), 5–89.

75. Ch. Hermite, *Sur la fonction exponentielle*, C.R. Acad. Sci. Paris **77** (1873), 18–24, 74–79, 226–233, 285–293; reprinted in Oeuvres, v. III, Gauthier-Villars, 1912, pp. 150–179.

76. B. Riemann, *Oeuvres mathématiques*, Blanchard, Paris, 1968.

77. D.V. Chudnovsky and G.V. Chudnovsky, *Computer algebra in the service of mathematical physics and number theory*, Proc. Internat. Conf. Computers and Mathematics (Stanford University, 1986), M. Dekker, (to appear).

78. G.V. Chudnovsky, *Hermite-Padé approximations to exponential functions and elementary estimates of the measure of irrationality of π*, Lecture Notes in Math., vol. 925, Springer-Verlag, 1978, pp. 299–322.

COLUMBIA UNIVERSITY

Proceedings of Symposia in Pure Mathematics
Volume **49** (1989), Part 2

Singular Moduli and Modular Equations
For Fricke's Cases

HARVEY COHN

1. Introduction. It could be argued that a century ago a seemingly specialized development in mathematics had become "central" by bestowing a high degree of unity and motivation on seemingly diverse fields. Today this development is recognized as "ring class field theory" largely through work by H. Weber [**12**].

As Parkinson's Laws would mandate, by the time that ring class field theory was properly defined and understood in perspective it had lost its eminent position (even within what became "class field theory," see [**2**]). It is nevertheless appropriate to reexamine some parts of it that were closely tied to modular forms, and in particular to theta-functions. Indeed, some of the early numerical aims are only now being realized through the use of computers.

Deferring definitions to later sections of this talk, we first consider the chain of requirements involving ring class field theory:

(a) ring class field theory requires singular moduli;

(b) singular moduli require modular equations;

(c) modular equations require modular forms with the attendant theory of theta-functions, differentials, Riemann surfaces, etc.

These connections (roughly speaking, between "algebra" and "analysis") motivated the work of R. Fricke in almost the opposite order. He began as "half of Fricke-Klein [**6**]", preoccupied with transformations of Riemann surfaces, suitably illustrated by applications to modular functions. His later work on elliptic functions [**7**] became clarified as "algebra" [**8**] through his concentration on the search for modular equations of order b. His search was narrowed to 37 special values,

$$(1.1) \quad b = 2, \ldots, 21, 23, 24, 25, 26, 27, 29, 31, 32,$$

$$35, 36, 39, 41, 47, 49, 50, 59, 71,$$

1980 *Mathematics Subject Classification* (1985 *Revision*). Primary 11F11.
Research supported in part by National Science Foundation Grant DMS 8602077.

which we refer to as "Fricke's cases" (see §5 below). These values of b are of unexpected interest today, since the prime divisors of these b are precisely the prime divisors of the order of the Monster group [5].

Fricke showed that, in principle, these special values of b permit an explicit modular equation of a special type to be computed (namely the "two-valued type" explained in §5). He did not complete a single calculation of this type, perhaps because such numerical computations are so formidable (see, for example, some recent work for comparison [10]). Yet he did provide enough information for all but six cases, and the remaining six ($b = 39, 41, 47, 50, 59, 71$) follow largely from his methods [3]. We use $b = 59$ as an illustration because the role of theta-functions is more complex there. (This case was curiously absent in Fricke's original list, of only 36 cases.)

Once the modular equations are constructed, we go on to the computation of singular moduli by a general algorithm of Fricke, which we illustrate for $b = 21$ (see §10).

Actually Fricke's cases are so diverse that a very reasonable technique for proving a theorem is to verify it for all 37 cases individually. In a closely related context, W. Magnus [11] cited a quotation of C. L. Siegel: "**The mathematical universe is inhabited not only by important species but also by interesting individuals.**"

2. Ring class field theory. We consider the binary quadratic forms in $\mathbb{Z}$ with discriminant $d < 0$, but not necessarily fundamental. The *principal* quadratic form is defined as

$$(2.1) \qquad f_d(x,y) = \begin{cases} x^2 - dy^2/4, & d \equiv 0 \bmod 4, \\ x^2 + xy - (d-1)y^2/4, & d \equiv 1 \bmod 4. \end{cases}$$

The *class number* $h(d)$ is defined as the number of inequivalent quadratic forms in $\mathbb{Z}$ with discriminant d under unimodular transformations. This is the number of classes of forms

$$(2.2a) \qquad \{R, S, T\} = Rx^2 + Sxy + Ty^2$$

with

$$(2.2b) \qquad \gcd(R, S, T) = 1, \quad S^2 - 4RT = d, \quad R > 0,$$

subject to equivalence under the *modular group* $\Gamma = SL(2, \mathbb{Z})$,

$$(2.3) \quad x' = Ax + By, \quad y' = Cx + Dy, \qquad A, B, C, D \in Z, \ AD - BC = 1.$$

The problem is to represent a (positive) prime p by

$$(2.4) \qquad\qquad p = f_d(x, y) \qquad (x, y \in \mathbb{Z}).$$

A necessary condition is found by taking (2.4) modulo p,

$$(2.5) \qquad\qquad (d/p) = 1$$

(if we exclude $p|2d$). This condition is sufficient if $h(d) = 1$. When $h(d) > 1$, we look for conditions of the following type: We ask for a so-called "class polynomial" $F_d(t)$ of degree $h(d)$, monic in $\mathbb{Z}[t]$, such that (2.4) holds if and only if (2.5) is valid together with the splitting of the equation

$$(2.6) \qquad\qquad F_d(t) = 0$$

into $h(d)$ distinct linear factors when taken modulo p. (As before, a finite set of p are excluded, but the theory shows [2] that such a polynomial is necessarily of the stated degree.) Also, the splitting field of (2.6) is normal over $\mathbb{Q}(\sqrt{d})$ and defines by adjunction a field $K\{d\}$, the *ring class field* for discriminant d. In terms of ideal theory, we are saying that the representation (2.4) holds for large p if and only if (p) splits in $K\{d\}/\mathbb{Q}$.

The fountainhead of this theory is the theorem of Fermat that for p odd, then $p = x^2 + y^2$ if and only if $p \equiv 1 \bmod 4$ (i.e., $(-4/p) = 1$). Of course for $d = -4$, $h(d) = 1$ so $K\{d\} = \mathbb{Q}(\sqrt{d})$, or $F_d(t)$ is any polynomial of degree 1.

3. The singular moduli. At this point the classes of quadratic forms can be understood by use of the *modular invariant* $j(z)$. We begin with the upper half z-plane

$$(3.1a) \qquad\qquad H = \{z: \operatorname{Im} z > 0\}$$

on which the modular group of (2.3) acts in its inhomogeneous form $\mathrm{PSL}(2, \mathbb{Z})$, also denoted by Γ, i.e.,

$$(3.1b) \qquad\qquad \Gamma: z' = (Az + B)/(Cz + D).$$

The group is generated by the transformations

$$(3.1c) \qquad\qquad z' = z + 1, \qquad z' = -1/z.$$

The quotient space H/Γ is seen to be the set

$$(3.1d) \qquad\qquad R(1): \{0 \le |\operatorname{Re} z| \le 1/2, \ |z| \ge 1\},$$

together with a suitable compactification at ∞ and identification of boundaries according to (3.1c). Then H/Γ has a complex structure [9] of a Riemann surface. The modular function $j(z)$ is a uniform parameter on this surface that maps $R(1)$ onto the entire j-sphere in one-to-one fashion. An important fact is that the locus of equalities of (3.1d) is the real j-axis. Furthermore,

$$(3.2) \qquad j(z) = 1/q + 744 + 196884q + \cdots \qquad (q = \exp 2\pi i z),$$

in a power series with integral coefficients.

For given d, we consider a collection of $h(d)$ forms $\{R_i, S_i, T_i\}$, $i = 1, \ldots, h(d)$. Then each form has roots $(z = x/y)$ such that

$$(3.3a) \qquad\qquad z_i = (-S_i + \sqrt{d})/R_i \qquad (i = 1, \ldots, h(d)).$$

Thus, the *singular moduli* for d are defined as follows:

$$(3.3b) \qquad\qquad \{j(z_1), j(z_2), \ldots, j(z_{h(d)})\}.$$

This set of numbers is the set of roots of a monic polynomial, which can be taken as the definition of $F_d(t)$, so for any i,

$$(3.4) \qquad\qquad K\{d\} = \mathbb{Q}(\sqrt{d}, j(z_i)).$$

(This rather concrete construction is Weber's theorem [2].) Actually, for any imaginary quadratic surd μ, with positive imaginary part, it can be shown that a d exists so that $j(\mu)$ belongs to the set (3.3b).

Every class of forms $\{R, S, T\}$ is represented by a root in $R(1)$ (see (3.1d)); thus

$$(3.5a) \qquad\qquad 0 \le |S| \le R \le T.$$

We define an *ambiguous* class of forms as one for which one or more of the inequalities in (3.5a) hold, i.e., z lies on the boundary of $R(1)$. More important for our algorithm (in §10 below) is the fact that ambiguous forms are precisely those for which the singular moduli are real. There are $h^*(d)$ ($\le h(d)$) of them, and actually, $h^*(d)$ is a power of 2. The principal form is one, say

$$(3.5b) \qquad\qquad z_1 = \begin{cases} \sqrt{d}/2, & d \equiv 0 \bmod 4, \\ (-1 + \sqrt{d})/2, & d \equiv 1 \bmod 4, \end{cases}$$

with $j(z_1) > 0$ or ≥ 0 respectively. With $|z| = 1$, $j(z) \ge 0$. Where $h^*(d) = h(d)$, it is possible to construct $K\{d\}$ wholly by quadratic radicals (over $\mathbb{Z}$), but this equality fails as early as $d = -32$ (see [2]).

4. The modular equation. The *modular equation* of order b is the polynomial connecting $j(z)$ with $j(z/b)$. It can be shown that under Γ, $j(z/b)$ is one of m conjugates

$$(4.1a) \qquad\qquad z_i = (Uz + V)/W, \qquad i = 1, \ldots, m,$$

$$(4.1b) \qquad\qquad UV = b, \qquad 0 \le W < V, \gcd(U, V, W) = 1.$$

Here the number of conjugates is

$$(4.1c) \qquad\qquad m = b \prod (1 + 1/p), \qquad \text{(prime)} p | b.$$

Thus $j(z/b)$ is defined on an m-sheeted Riemann surface

$$(4.2a) \qquad\qquad R(b) = H/\Gamma^0(b),$$

where, referring to (3.1b),

$$(4.2b) \qquad\qquad \Gamma^0(b) = \{\Gamma: B \equiv 0 \bmod b\}.$$

Thus $|\Gamma/\Gamma^0(b)| = m$. Note also that $\Gamma^0(b)$ possesses the transformation $z' = z + b$, so a uniform parameter for $j(z/b)$ at $b = \infty$ is not $q\ (= \exp 2\pi i z)$ but

$$(4.3) \qquad\qquad r = q^{1/b} = \exp 2\pi i z/b.$$

We now have the modular equation of degree m and order b

$$(4.4) \qquad\qquad \Phi_b(X, Y) = 0, \qquad X = j(z/b), \ Y = j(z).$$

This is a polynomial that is symmetric in X and Y and of degree m in each since $j(z/b)$ is one of the conjugates (4.1a). If $X = Y$,

(4.5) $\Phi_b(X, X) = 0$

is an equation whose roots are given by values of X for which

(4.6) $j(z) = j(z/b) = X$.

This means that for some z and element of Γ (see (3.1b)),

(4.7) $(Az + B)/(Cz + D) = z/b$.

We have seemingly many choices to satisfy (4.7), and it is an elementary matter to see that for any form $\{R, S, T\}$ (in (2.2a)) infinitely many b exist such that $j((-S + \sqrt{d})/R)$ is a root of (4.5), or, equivalently, the class polynomial $F_d(t)|\Phi_b(t, t)$.

The Riemann surface $R(b)$ is of genus g, which is asymptotically $b/12$, so not many cases of genus 0 exist (i.e., $b = 2, \ldots, 10, 12, 13, 16, 18, 25$; see [8]). For these cases Klein and Fricke [6] wrote out a parametrized form of the modular equation as

(4.8)
$$\begin{cases} j(z) = F_b(x), \\ j(z/b) = F_b(1/x) \end{cases}$$

for $F_b(x)$ rational.

To illustrate take $b = 2$. Here [6]

(4.9) $F_2(x) = 64(x + 3)^3/x^2$.

Thus (4.8) is rather easy to comprehend when compared with
(4.10a)
$$\Phi_2(X, Y) = X^3 + Y^3 - X^2 Y^2 + 2^4 \cdot 3 \cdot 31(X + Y)XY - 2^4 \cdot 3^4 \cdot 5^3(X^2 + Y^2)$$
$$+ 3^4 \cdot 5^3 \cdot 4027XY + 2^8 \cdot 3^7 \cdot 5^6(X + Y) - 2^{12} \cdot 3^9 \cdot 5^9$$

We set $Y = X$ and factor (4.10a):

(4.10b) $\Phi_2(X, X) = -(X - 12^3)(X + 15^3)^2(X - 20^3)$.

Now we obtain singular moduli for $d = -4, -7, -8$, namely

(4.10c) $j(i) = 12^3, \quad j((1 + \sqrt{7})/2) = -15^3, \quad j(\sqrt{2}) = 20^3$.

In general we require the algorithm of §10 (below) to coordinate values of z and X as we have just done for $b = 2$.

5. The two-valued modular equation. Actually, Fricke was able to cope with genus $g > 0$ by using the involution (later generalized in [1])

(5.1) $W: z' = -b/z$.

This transformation commutes with $\Gamma^0(b)$ so that

(5.2) $\Gamma^*(b) = \Gamma^0(b) + W\Gamma^0(b)$

is a group extension of second order, which generates a Riemann surface which is a double covering of $R(b)$, namely

(5.3) $$R^*(b) = H/\Gamma^*(b).$$

This is a Riemann surface of genus g^* ($< g$) that is asymptotically $b/24$. We defer details of the genus computation to [4], but we note that when b is odd (> 3), then the condition for $g^* = 0$ is

(5.4) $$2g + 2 = \begin{cases} h(-b) + h(-4b), & b \equiv -1 \bmod 4, \\ h(-4b), & b \equiv -3 \bmod 4. \end{cases}$$

(In effect the roots of the classes of forms determine the fixed points that reduce the genus on projection from $R(b)$ to $R^*(b)$.)

The cases of higher genus embraced by Fricke's method include

$$g = 1: \quad b = 11, 14, 15, 17, 19, 20, 21, 24, 27, 32, 36, 49;$$
$$g = 2: \quad b = 23, 26, 29, 31, 50; \qquad g = 3: \quad b = 35, 39, 41;$$
$$g = 4: \quad b = 47; \qquad g = 5: \quad b = 59; \qquad g = 6: \quad b = 71.$$

If we let t denote a suitable global uniformizing parameter for $R^*(b)$ then since $R(b)$ is two-valued for $R^*(b)$, it is determined by the parametrization (t, s) where

(5.5) $$s^2 = P_b(t),$$

where $P_b(t)$ denotes a polynomial of degree $2g + 2$. In other words, the extension is hyperelliptic for $g > 1$. Indeed, from (5.1),

(5.6) $$W : \{z \to -b/z\} \equiv \{(t, s) \to (t, -s)\}.$$

The two-valued modular equation develops from the pair of rational functions of $R(b)$:

(5.7a) $$j(z) = F_b(t, s),$$
(5.7b) $$j(z/b) = F_b(t, -s).$$

These functions were given by Fricke for all but six of his cases (see §1 above). A more useful rational form is

(5.8a) $$j(z)j(z/b) = N_b(t),$$
(5.8b) $$j(z) + j(z/b) = S_b(t),$$
(5.8c) $$j(z) - j(z/b) = D_b(t, s).$$

Here $N_b(t)$, the *norm*, and $S_b(t)$, the *trace*, are both rational in t, while $D_b(t, s)$, the *different*, is s times a rational function in t. Of course since we are dealing with two unknowns, $j(z)$ and $j(z/b)$, we do not need both the trace and the norm; indeed

(5.9) $$D_b(t, s)^2 + 4N_b(t) = S_b(t)^2.$$

We prefer the norm since, as the results show, the norm has a large cubic component, which serves to keep the coefficients small.

Fricke did not find these equations for any specific b, but in principle, he used this method to find the singular moduli arising from $b = 11$ (see [8]). He solved simultaneously $F_b(t, s) = F_b(t, -s)$ and (5.5). We use MACSYMA to perform the simplifications of the norm and trace as well as the factorization of the different (see Tables I and III below).

To illustrate a case of genus 0, take $b = 2$. We make the substitutions into (4.9)

$$(5.10a) \qquad x = (1 + w)/(1 - w),$$
$$(5.10b) \qquad w^2 = (t - 128)/(t + 128).$$

Then $z' = -2/z$ engenders $x' = 1/x$ and $w' = -w$. Actually, here $s = w(t + 128)$, so that the two-valued modular equations are

$$(5.11a) \qquad s^2 = t^2 - 128^2,$$
$$(5.11b) \qquad j(z)j(z/2) = (t + 272)^3,$$
$$(5.11c) \qquad j(z) + j(z/2) = t^2 + 49t - 6656,$$
$$(5.11d) \qquad j(z) - j(z/2) = (t + 47)s.$$

The values (4.10c) can now be found by substituting the roots of $D_2(t, s)$ (namely $t = -128, -47, 128$) from (5.11d) into either (5.11b) or (5.11c), since $j(z) = j(z/2)$ for these values. To correlate j and t is not trivial and (as we noted) requires the algorithm of §10.

6. Use of theta functions. For cases with b prime and $g > 0$, the two-valued modular equation starts with theta-functions.

When $b \equiv 3 \mod 4$, we define a theta-function for the form of discriminant $-b$ (see (2.2a))

$$(6.1a) \qquad f(x, y) = Rx^2 + Sxy + Ty^2, \qquad S^2 - 4RT = -b,$$
$$(6.1b) \qquad \theta(z) = \theta\{R, S, T\} = \sum r^{f(m,n)}$$

for all $m, n \in \mathbb{Z}$. Here r is the parameter for $j(z/b)$, namely

$$(6.1c) \qquad r = \exp 2\pi i z/b.$$

By Poisson's summation formula, it is easy to prove [7]

$$(6.2) \qquad \theta(z) = i\sqrt{b}\,\theta(-b/z)/z.$$

A more difficult result is that for a transformation in $\Gamma^0(b)$

$$(6.3) \qquad \theta((Az + B)/(Cz + D)) = (Cz + D)\theta(z)(A/b)$$

for the Jacobi symbol (A/b) (see [7]).

We usually can construct several such theta functions (when $h(d) > 1$); in fact, some might vanish identically. Otherwise assume we have enough theta-functions to construct quadratic forms

$$(6.4a) \qquad\qquad T_0(z) = \sum a_{ij}\theta_i\theta_j,$$

$$(6.4b) \qquad\qquad T_1(z) = \sum b_{ij}\theta_i\theta_j.$$

These forms lead to an invariant ration in $R(b)$, single-valued in t,

$$(6.5) \qquad\qquad T_0(z)/T_1(z) = g(z).$$

With a suitable choice of the forms in (6.4ab), we take $t = g(z)$. It then follows that

$$(6.6) \qquad\qquad g'(z)/T_1(z) = h(z)$$

has the property that $h(z)/s$ is single-valued in t. Again, by the same choice of forms in (6.4ab), we take $s = h(z)$ (also see [3]).

We need even more theta-functions in general. When $b \equiv 3 \bmod 4$, we can also define (in the notation of (6.1abc))

$$(6.7) \qquad \phi(z) = \phi\{R, S, T\} = \sum(-1)^n r^{f(m,n)/2} \quad (m \text{ odd}).$$

This function has similar invariance properties, since

$$(6.8) \qquad \begin{aligned} \phi\{R, S, T\} = &-\theta\{R/2, S/2, T/2\} + 2\theta\{R/2, S, 2T\} \\ &+ \theta\{2R, S, T/2\} - 2\theta\{2R, 2S, 2T\}. \end{aligned}$$

For the case where $b \equiv 1 \bmod 4$, only the value $-4b$ (not $-b$) can be the discriminant of the form $f(x, y)$ in (6.1a). Adjusting the notation, we define a third type of theta-function

$$(6.9) \qquad \sigma(z) = \sigma\{R, S, T\} = \sum(-1)^n r^{f(m,n)/4} \quad (m \text{ odd}).$$

It is necessary to form the quadratics (6.4ab) from forms of the same type (with R in the same congruence class mod 4) since the multipliers are more complicated than Jacobi symbols (see [3]).

This might be the place to point out that by a variety of methods, all 37 cases lead to two-valued modular equations (5.8a–d) with functions involving monic polynomials in $\mathbb{Z}[t]$. There surely must be a theoretical explanation, but the easiest way to see this result is to see a collection (à la Siegel) of 37 individuals!

7. Illustration of theta-functions for $b = 59$.
With $b = 59$, $g = 5$, $h(-59) = 3$, $h(-4.59) = 9$ (see (5.4)), we construct four theta-functions of two types:

$$(7.1a) \qquad \begin{aligned} t_1 &= \theta\{1, 1, 15\} = 1 + 2r + \cdots, \\ t_2 &= \theta\{3, 1, 5\} = 1 + 2r^3 + \cdots; \end{aligned}$$

$$(7.1b) \qquad \begin{aligned} t_3 &= \phi\{1, 1, 15\} = 2\sqrt{r} + 2\sqrt{r}^9 + \cdots, \\ t_4 &= \phi\{3, 1, 5\} = 2\sqrt{r}^3 - 2\sqrt{r}^7. \end{aligned}$$

We need quadratic polynomials of integral order in r,

(7.2a)
$$t_{11} = (t_1 + t_2)^2/4, \qquad t_{12} = (t_1^2 - t_2^2)/4,$$
$$t_{22} = (t_1 - t_2)^2/4;$$

(7.2b)
$$t_{33} = t_3^2/4, \quad t_{34} = t_3 t_4/4, \quad t_{44} = t_4^2/4.$$

We form combinations of maximal order in r

(7.3a) $\qquad T_0 = (t_{34} - 2t_{22} + t_{12} - t_{33} + t_{44})/4 = r^4 + \cdots ,$

(7.3b) $\qquad T_1 = (-3t_{34} + 2t_{22} + t_{12} - t_{33} + t_{44})/4 = r^5 + \cdots .$

Then we define

(7.4)
$$t = T_0/T_1 = 1/r + r + r^2 + 2r^3 + \cdots$$

and

(7.5)
$$s = -r(dt/dr)/T_1 = 1/r^6 + 2/r^5 + \cdots .$$

It is then a fairly routine problem to expand s^2 as a Laurent series starting with $1/r^{12}$ and to transform this Laurent series into a polynomial in t (see (7.4)) of degree 12. We obtain the product of two factors whose degrees come from (5.4):

(7.6a) $\qquad\qquad\qquad P_{59}(t) = p(t)q(t),$

(7.6b) $\qquad\qquad\qquad p(t) = t^3 + 2t^2 + 1,$

(7.6c) $\qquad q(t) = t^9 + 2t^8 - 4t^7 - 21t^6 - 44t^5 - 60t^4$
$$-61t^3 - 46t^2 - 24t - 11.$$

We finally are ready to calculate the two-valued modular equations for this case $b = 59$. The finer techniques (see [3] and [4]) are seldom of uniform use, so perhaps we might just use a fairly crude and quick method applicable to any prime b. Here the choice of t gives us one pole in $R^*(b)$, namely at the cusps $z = 0$ and $i\infty$ (identified by $z' = -b/z$). Thus $D_b(t, s)/s$ is a polynomial of degree $b - g - 1$ in t, which is the principal part of the series

(7.7) $\qquad D_b(t, s)/s = (1/r^b - 1/r + \cdots)/s = 1/r^{b-g-1} + \cdots .$

We use the known expansion for s in r, but no other information is needed from the coefficients of $j(z)$ or $j(z/b)$. The right-hand side of (7.7) is an exact polynomial in t. The trace $S_b(t)$ is found by a similar approximation, and the norm $N_b(t)$ is found by (5.9). For $b = 59$, the answers are shown in Table I (compare [3]).

$$N_{21}(t) = (t^{20} + 248t^{19} + 4104t^{18} + 30020t^{17} + 134312t^{16} + 421040t^{15}$$
$$+ 999366t^{14} + 1890200t^{13} + 2944888t^{12}$$
$$+ 3861732t^{11} + 4327848t^{10} + 4183680t^{9} + 3501793t^{8}$$
$$+ 2541584t^{7} + 1594800t^{6} + 856640t^{5} + 388832t^{4}$$
$$+ 144896t^{4} + 42432t^{2} + 8960t + 1024)^{3}$$

$$S_{21}(t) = t^{59} - 59t^{57} - 59t^{56} + 1534t^{55} + 3127t^{54} - 21417t^{53} - 72098t^{52}$$
$$+ 143193t^{51} + 916447t^{50} + 174286t^{49} - 6516727t^{48} - 11515738t^{47}$$
$$+ 19221846t^{46} + 92182957t^{45} + 65239368t^{44} - 284253976t^{43} - 754619145t^{42}$$
$$- 308671598t^{41} + 1977259920t^{40} + 4550403910t^{39} + 2575381270t^{38}$$
$$- 7632475764t^{37} - 20657216762t^{36} - 19963085715t^{35} + 9507261180t^{34}$$
$$+ 59805257547t^{33} + 91253130157t^{32} + 55180018686t^{31} - 60666698493t^{30}$$
$$- 201693103797t^{29} - 266853603370t^{28} - 176572038603t^{27} + 59881108419t^{26}$$
$$+ 333403380486t^{25} + 495749968855t^{24} + 451258867420t^{23} + 214958799438t^{22}$$
$$- 100904630525t^{21} - 356956184970t^{20} - 462413784548t^{19} - 410304447353t^{18}$$
$$- 260679395796t^{17} - 94424809972t^{16} + 28887189420t^{15} + 8847227848t^{14}$$
$$+ 95279288800t^{13} + 73656872866t^{12} + 45446452976t^{11} + 22881418704t^{10}$$
$$+ 9210419200t^{9} + 2701208800t^{8} + 347512832t^{7} - 190061184t^{6} - 170524160t^{5}$$
$$- 78057472t^{4} - 24891392t^{3} - 5679104t^{2} - 860160t - 65536$$

$$D_{21}(t) = (t - 1)t(t + 1)(t + 2)(t^{2} - 2t - 4)(t^{2} - t - 7)(t^{2} - t - 4)$$
$$(t^{2} - t - 1)(t^{2} + t - 1)(t^{3} - t - 1)(t^{3} - 3t^{2} + t - 2)(t^{3} - 2t^{2} - 2)$$
$$(t^{3} + 2t^{2} + t + 1)(t^{4} - 3t^{2} - 2)(t^{4} - t^{3} - 5t^{2} - 3t - 1)$$
$$(t^{4} + t^{3} - 3t^{2} - 6t - 4)(t^{4} + 3t^{3} + 3t^{2} + 3t + 1)$$
$$(t^{5} - t^{4} - 4t^{3} - 7t^{2} - 3t - 2)(t^{6} + t^{5} - 5t^{4} - 10t^{3} - 10t^{2} - 4t - 4)$$
$$((t^{3} + 2t^{2} + 1)(t^{9} + 2t^{8} - 4t^{7} - 21t^{6} - 44t^{5} - 60t^{4} - 61t^{3} - 46t^{2} - 24t - 11))^{1/2}$$

TABLE I. Output data for the modular equation with $b = 59$.

8. Geometry of the fundamental domain. We now construct the Riemann surface for $H/\Gamma^{0}(b)$ as a polygon bounding $R(b)$, by matching the arcs. We have a special, simple construction valid when b has at most two prime factors (hence valid in all of Fricke's cases). This construction is vital to the identification of singular moduli later on.

We first of all use the translation $z' = z + b$ to place $R(b)$ within the vertical lines

$$(8.1) \qquad\qquad\qquad |\operatorname{Re} z| \leq b/2.$$

We now have to construct the lower boundary by considering each interval $(N, N + 1/2)$ and $(N, N - 1/2)$ for $N \in \mathbf{Z}$ within the limits (8.1). To start, N is a cusp (where the boundary touches $z = N$) exactly when $(N, b) > 1$.

When $(N, b) = 1$, we fill the interval $(N - 1/2, N + 1/2)$ with the arc from (the fixed-points of Γ)$N - 1/2 + i\sqrt{3}/2$ to $N + 1/2 + i\sqrt{3}/2$, which has unit radius:

$$(8.2) \qquad |z - N| = 1, \qquad |\text{Re } z - N| \leq 1/2.$$

When $(N, b) > 1$, we consider two cases. If

$$(8.3a) \qquad (N + 1, b) = 1,$$

then we fill the interval $(N, N + 1/2)$ with the arc of unit radius

$$(8.3b) \qquad |z - N - 1| = 1, \qquad 0 \leq \text{Re } z - N - 1 \leq 1/2.$$

(We make a symmetrical construction if $(N - 1, b) = 1$.)

If, however, $(N, b) > 1$ and $(N + 1, b) > 1$, we fill the interval $(N, N + 1)$ with the arc of radius $1/2$ joining the cusps N and $N + 1$,

$$(8.3c) \qquad |z - N - 1/2| = 1/2, \qquad N \leq \text{Re } z \leq N + 1.$$

Thus if b is a prime, only $N = 0$ is a cusp on the interval in (8.1), so the boundary consists entirely of arcs (8.2) except for the arcs of $|z - 1|$ and $|z + 1| = 1$ with $|\text{Re } z| \leq 1/2$. The arcs of type (8.3c) (of radius $1/2$) occur only if b has two distinct prime factors.

The identifications of the boundary arcs of $R(b)$ are as follows: If $(N, b) = 1$ and $(M, b) = 1$ where

$$(8.4a) \qquad NM \equiv -1 \quad \mod b,$$

then the arc containing $N + i$ is mapped into the arc containing $M + i$ by

$$(8.4b) \qquad (z' - M)(z - N) = -1.$$

If $(N, b) > 1$ and $(N + 1, b) > 1$, then the arc (8.3c) between N and $N + 1$ is mapped into the arc connecting M and $M + 1$ by

$$(8.5a) \qquad (1 - 1/(z' - M))(1 - 1/(z - N)) = -1$$

provided M and N satisfy

$$(8.5b) \qquad (2N + 1)M + N + 1 \equiv 0 \quad \mod b.$$

Of course both transformations (8.4ab) and (8.5ab) lie in $\Gamma^0(b)$.

For $b = 21$, Figure 1 shows the arcs on the positive real half of $R(21)$. From (8.4ab), the following are matching pairs of (M, N): $(1, -1), (2, 10), (4, 5), (8, -8)$. From (8.5ab), $N = 6$ and 7 match $M = -7$ and -6 (respectively). It is an amusing exercise to verify the genus in each case ($g = 1$ for $b = 21$).

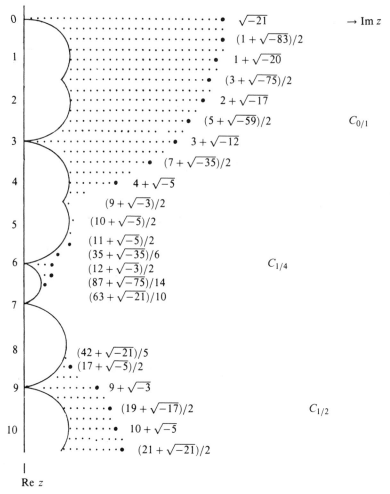

FIGURE 1. Locus of real t-axis in $R(21)$. Real roots of $D_{21}(t)$ are shown (as listed in Table IV) on the portion of the locus which intersects $C_{0/1}$, $C_{1/4}$, and $C_{1/2}$

9. The symmetry circles. We do not explicitly construct the fundamental domain of $H/\Gamma^*(b)$, which essentially would consist of bisecting $R(b)$ in accordance with the symmetry of $z' = -b/z$ in (5.1) (see [8]). We do something much simpler: we trace the locus of the real t-axis, which essentially follows circles of symmetry. We take only the right quadrant (Re $z \geq 0$, Im $z \geq 0$). It contains the image of the positive and negative t-axis.

Consider the relations on the real t locus

(9.1a) $$j(z) - j(z/b) = D_b(t,s),$$

(9.1b) $$j(z) + j(z/b) = S_b(t).$$

Recall that $S_b(t)$ and $D_b(t,s)/s$ are real functions of t, while s is a radical that can be either real or purely imaginary.

When s is real, $j(z)$ and $j(z/b)$ are both real, so z must lie on an image of the boundary of $R(1)$, the fundamental domain of Γ. In effect, since s is of even degree, this happens for t large,

$$(9.2a) \qquad t \to +\infty, \qquad z \to +i\infty,$$
$$(9.2b) \qquad t \to -\infty, \qquad z \to +i\infty + b/2.$$

When s becomes purely imaginary (as t falls below the largest real root of s or lies between other real roots), then $j(z)$ and $j(z/b)$ must be complex conjugates (which we shall denote by "$\char`^$"). Thus $j(z)\char`^ = j(-z\char`^)$ (by symmetry with the imaginary z-axis). Thus

$$(9.3a) \qquad j(-z\char`^) = j(z/b) = j(-b/z).$$

This relation is easily true on the circle $zz\char`^ = b$, or

$$(9.3b) \qquad C_{0/1}: |z|^2 = b.$$

We use analytic continuation (or translation by $\Gamma^0(b)$) to change this circle, by (3.1b) and (4.2b), to

$$(9.3c) \qquad C_{p/q}: |z - pb/q|^2 = b/q^2$$

for certain special values of p and q,

$$(9.4a) \qquad p = (AB/b - CD),$$
$$(9.4b) \qquad q = A^2 - bC^2.$$

A more useful result is that under the mapping (8.4b) as $z \to z'$, $C_{p/q} \to C_{p'/q'}$ where

$$(9.5a) \qquad p' = [(pb - qN)^2 - b]/q,$$
$$(9.5b) \qquad q' = [(pb - qN)[Mp - q(NM + 1)/b] - M]/q.$$

In summary, the real t-axis in $R(b)$ consists of a patchwork of Moebius images of the boundary of $R(1)$ when $j(z)$ and $j(z/b)$ are both real, and special arcs $C_{p/q}$ when $j(z)$ and $j(z/b)$ are complex conjugates. Aside from the above heuristics, a formal proof will again consist of verifying the fact for all of Fricke's 37 cases!

The case $b = 21$ (in Figure 1) illustrates how the real t-axis consists of seven arcs in $R(21)$, which correspond to seven intervals of the real t-axis

b (odd)	$\{p/q\}$	b (even)	$\{p/q\}$
13	$\{0/1, 1/3, 1/2\}$	10	$\{0/1, 1/3\}$
19	$\{0/1, 1/3, 1/2\}$	14	$\{0/1, 1/5\}$
21	$\{0/1, 1/4\}\{1/2\}$	16	$\{0/1, 1/3\}$
25	$\{0/1, 1/4, 1/3, 1/2\}$	18	$\{0/1, 1/7\}$
29	$\{0/1, 1/4, 2/5, 1/2\}$	24	$\{0/1\}\{2/5\}$
31	$\{0/1, 1/5, 1/3, 1/2\}$	26	$\{0/1, 1/5, 5/11\}$
39	$\{0/1, 3/10\}\{2/5, 1/2\}$	32	$\{0/1, 3/7\}$
41	$\{0/1, 1/5, 1/4, 1/2\}$	36	$\{0/1, 1/5\}$
49	$\{0/1, 1/6, 1/4, 1/3, 2/5, 1/2\}$	50	$\{0/1, 1/7, 4/17, 11/23\}$
59	$\{0/1, 1/5, 1/2\}$		
71	$\{0/1, 1/14, 1/7, 1/5, 1/2\}$		

TABLE II. Occurrences of circles $C_{p/q}$ for $p/q \neq 0/1, 1/2$. The positive values of p/q are given for which $C_{p/q}$ follows the real t-axis in $R(b)$. Grouping in braces shows interruption of the circles $C_{p/q}$ by boundary segments of $R(b)$.

(which we now enumerate in terms of z):

$$\{8.169\cdots \leq t < \infty\} \Leftrightarrow \text{line from } z = i\infty \text{ to } z = i\sqrt{21},$$

$$\{1 \leq t \leq 8.169\ldots\} \Leftrightarrow \text{arc of } C_{0/1} \text{ to } z = (9 + \sqrt{-3})/2,$$

$$\{.122\cdots \leq t \leq 1\} \Leftrightarrow \text{arc of } C_{1/4} \text{ to } z = (63 + \sqrt{-21})/10,$$

$$\{0 \leq t \leq .122\ldots\} \Leftrightarrow \text{arc of boundary } |z - 13/2| = 1/2 \text{ to } z = 7,$$

$$\{-.586\cdots \leq t \leq 0\} \Leftrightarrow \text{arc of boundary } |z - 8| = 1 \text{ to}$$
$$z = (42 + \sqrt{-21})/5,$$

$$\{-1.705\cdots \leq t \leq -.586\ldots\} \Leftrightarrow \text{arc of } C_{1/2} \text{ to } \{21 + \sqrt{-21}\}/2,$$

$$\{-\infty < t \leq -1.705\ldots\} \Leftrightarrow \text{line continuing to } i\infty + 21/2.$$

It is verified that for b odd (and > 3) the real t-axis will have as locus $C_{0/1}$ and $C_{1/2}$ (among other arcs). Sometimes, these circles are in analytic continuation, but this requires (see (9.4b)) that 2 or -2 be represented by $A^2 - bC^2$ (which might be excluded on number-theoretic grounds). When b is even, only $C_{0/1}$ is necessarily present in the real t-locus. The extra circles $C_{p/q}$ that occur are listed in Table II.

10. The algorithm for singular moduli. We consider a given b together with $R(b)$ (as given in §8) and the locus of the t-axis (as given in §9). We then have an ordered listing of the values of z that produce singular moduli (from the intersections of the parts of the real t-axis with the circles $C_{p/q}$). This is shown in Figure 1 for $b = 21$.

The next thing we must do is find the class polynomials $F_d(t)$ that correspond to the $D_b(t,s)$ as factors of its numerator. For this purpose we need the two-valued modular equation. For the case $b = 21$, the data in Fricke

$$s^2 = t^4 - 6t^3 - 17t^2 - 6t + 1$$

$$F_{21}(t,s) = \frac{(w^2 + 5w + 1)^3(w^2 + 13w + 49)}{w}$$

$$w = (t^2 - 3t + 1 - s)^2/4t$$

$$N_{21}(t) = (t-1)^2(t^2 + t + 1)^3(t^2 + 5t + 1)^3$$
$$\times (t^6 + 228t^5 - 180t^4 - 34t^3 + 60t^2 - 12t + 1)^3/t^{10}$$

$$S_{21}(t) = (t-1)(t^{27} - 41t^{26} + 694t^{25} - 6124t^{24}$$
$$+ 28253t^{23} - 47263t^{22} - 123507t^{21} + 598257t^{20}$$
$$- 245187t^{19} - 1969735t^{18} + 1817111t^{17} + 2974127t^{16}$$
$$- 2762933t^{15} - 2213993t^{14} + 1465267t^{13}$$
$$+ 728811t^{12} - 256635t^{11} - 77967t^{10} + 9841t^9$$
$$- 659t^8 - 1435t^7 + 49t^6 + 95t^5$$
$$+ 103t^4 - 50t^3 - 36t^2 + 13t - 1)/t^7$$

$$D_{21}(t) = (t-1)(t+1)(t^2 - 7t + 1)(t^2 - 4t + 1)(t^2 - 3t + 1)$$
$$(t^2 - t - 1)(t^2 + t - 1)(t^3 - 8t^2 - 1)$$
$$(t^3 - 4t^2 - 4t - 1)(t^4 - 8t^3 + 3t^2 + 2t + 1)$$
$$(t^4 - 5t^3 - 6t^2 + 3t - 1)s/t^7$$

TABLE III. Data for the modular equation with $b = 21$.

[8] (see Table III) are sufficient (in fact in this case, no theta-functions were needed).

Actually, we discover class polynomials of varying degree as factors of $D_b(t,s)$, but we are not able to immediately see which value of d belongs to each factor (of degree $h(d)$). We are able to find the real roots of each (which are $h^*(d)$ in number), and we can list them in decreasing order. *The algorithm arising from the previous results is that the ordered listing of the values of z will correspond precisely to the ordered listing of the roots t.* This is shown in Table IV in the first two columns. Each root t corresponds to one class polynomial, which is now identified by the discriminant of the surd in the first column.

We can illustrate this method for $b = 21$ where $h(-84) = h^*(-84) = 4$. The roots of

(10.1) $$F_{-84}(t) = t^4 - 6t^3 - 17t^2 - 6t + 1$$

correspond to the four real singular moduli
(10.2)
$$j(\sqrt{-21}), \quad j((63 + \sqrt{-21})/10), \quad j((42 + \sqrt{-21})/5), \quad j((21 + \sqrt{-21})/2).$$

We can therefore state that the condition that a prime p (with $(-84/p) = 1$), satisfies

(10.3) $$p = x^2 + 21y^2 \qquad (x, y \in \mathbb{Z})$$

z	t	Class polynomial
$\sqrt{-21}$	$8.169\ldots$	$t^4 - 6t^3 - 17t^2 - 6t + 1$
$(1 + \sqrt{-83})/2$	$8.015\ldots$	$t^3 - 8t^2 - 1$
$1 + \sqrt{-20}$	$7.566\ldots$	$t^4 - 8t^3 + 3t^2 + 2t + 1$
$(3 + \sqrt{-75})/2$	$6.854\ldots$	$t^2 - 7t + 1$
$2 + \sqrt{-17}$	$5.930\ldots$	$t^4 - 5t^3 - 6t^2 + 3t - 1$
$(5 + \sqrt{-59})/2$	$4.864\ldots$	$t^3 - 4t^2 - 4t - 1$
$3 + \sqrt{-12}$	$3.732\ldots$	$t^2 - 4t + 1$
$(7 + \sqrt{-35})/2$	$2.618\ldots$	$t^2 - 3t + 1$
$4 + \sqrt{-5}$	$1.618\ldots$	$t^2 - t - 1$
$(9 + \sqrt{-3})/2$	1	$t - 1$
$(10 + \sqrt{-5})/2$	$.903\ldots$	$t^4 - 8t^3 + 3t^2 + 2t + 1$
$(11 + \sqrt{-5})/2$	$.618\ldots$	$t^2 + t - 1$
$(35 + \sqrt{-35})/6$	$.381\ldots$	$t^2 - 3t + 1$
$(12 + \sqrt{-3})/2$	$.267\ldots$	$t^2 - 4t + 1$
$(87 + \sqrt{-75})/14$	$.145\ldots$	$t^2 - 7t + 1$
$(63 + \sqrt{-21})/10$	$.122\ldots$	$t^4 - 6t^3 - 17t^2 - 6t + 1$
7	0	(pole)
$(42 + \sqrt{-21})/5$	$-.586\ldots$	$t^4 - 6t^3 - 17t^2 - 6t + 1$
$(17 + \sqrt{-5})/2$	$-.618\ldots$	$t^2 - t - 1$
$9 + \sqrt{-3}$	-1	$t + 1$
$(19 + \sqrt{-17})/2$	$-1.369\ldots$	$t^4 - 5t^3 - 6t^2 + 3t - 1$
$10 + \sqrt{-5}$	$-1.618\ldots$	$t^2 + t - 1$
$(21 + \sqrt{-21})/2$	$-1.705\ldots$	$t^4 - 6t^3 - 17t^2 - 6t + 1$

TABLE IV. Matching of real roots of $D_{21}(t)$ with class polynomials.

is that the polynomial in (10.1) splits into four factors modulo p. We can continue down the list of z in Table IV and obtain a similar result for each. For instance the second entry belongs to $d = -83$, $h(-83) = 3$, $h^*(-83) = 1$ (one real root). Thus, for an odd prime p, (with $(-83/p) = 1$),

$$(10.4) \qquad\qquad p = x^2 + xy + 21y^2 \qquad (x, y \in \mathbb{Z})$$

precisely when the polynomial

$$(10.5) \qquad\qquad F_{-83}(t) = t^3 - 8t^2 - 1$$

splits modulo p, etc.

In conclusion, the algorithm might seem so "natural" that one could expect it to work in other cases where the geometry of §8 holds (only two primes divide b) if some special set of symmetry circles locates the arguments z of the singular moduli. From preliminary attempts, this seems impossible. Fricke's cases are (conjecturally) the only ones where the present algorithm works.

REFERENCES

1. A. O. L. Atkin and J. Lehner, *Hecke operators on* $\Gamma^0(m)$, Math. Ann. **185** (1970), 134–140.

2. H. Cohn, *Introduction to the construction of class fields*, Cambridge Univ. Press, 1985.

3. _____, *Fricke's two-valued modular equation*, Math. Comp. **51** (1988), 787–807.

4. _____, *Iteration of two-valued modular equation* (manuscript).

5. J. H. Conway and S. P. Norton, *Monstrous moonshine*, Bull. London Math. Soc. **11** (1979), 308–339.

6. R. Fricke and F. Klein, *Vorlesungen ueber die Theorie der Elliptischen Modulfunctionen.* II, Leipzig, 1892, Chapter 5.

7. R. Fricke, *Die Elliptische Funktionen und ihre Anwendungen.* II, Leipzig, 1922.

8. ____, *Lehrbuch der Algebra.* III (Algebraische Zahlen), Braunschweig, 1928.

9. R. C. Gunning, *Lectures on modular forms*, Princeton, 1962.

10. E. Kaltofen and N. Yui, *Explicit construction of the Hilbert class fields of imaginary quadratic fields with class numbers 7 and 11*, Eurosam 84, Lecture Notes in Comp. Sci., vol. 174, Springer-Verlag, 1984, pp. 310–320.

11. W. Magnus, *Noneuclidean tesselations and their groups*, Academic Press, 1974, p. ix.

12. H. Weber, *Elliptische Funktionen und algebraischen Zahlen*, Braunschweig, 1891.

CITY COLLEGE (CUNY)

Proceedings of Symposia in Pure Mathematics
Volume **49** (1989), Part 2

The Sums of the Kloosterman Sums

N. V. KUZNETSOV

1. The domain of mathematics that will be discussed was called by M. Huxley "Kloostermania". This name outlined (but not too sharply) the boundary field between number theory, the theory of the modular and automorphic functions, spectral theory, and geometry.

The beginning was due to Poincaré. The contributions that defined the base of this theory were given by Petersson, Hecke, Rankin, and Maass. More recently the development was stimulated by the famous talk of A. Selberg at the Tata Institute and by L. Faddeev's works, which cleared the spectral expansion.

It was ten years ago when I found the explicit form of the connection between sums of the Kloosterman sums ("known quantities") and the Fourier coefficients of cusp forms (unknown quantities that are very mysterious up to this day). In the next year R. Bruggeman rediscovered (independently) part of these results. From that moment the number of publications has increased rapidly in this domain.

The goal of my talk is to make more popular this dynamic branch of mathematics.

So it happened that the Kloosterman sums (these sums will be defined below) were arising first for improving the Hardy-Littlewood "circle method".

But these sums would have arisen earlier if Poincaré had wanted to calculate the Fourier coefficients of the series that are called today "the Poincaré series".

2. The Lobatčevskii plane. This plane will be considered as the upper half-plane **H** of the complex variable $z = x + iy$, $x, y \in \mathbf{R}$, $y > 0$, with the metric

$$(1) \qquad ds^2 = y^{-2}(dx^2 + dy^2),$$

measure

$$(2) \qquad d\mu(z) = y^{-2} dx\, dy$$

1980 *Mathematics Subject Classification* (1985 *Revision*). Primary 11F30.

and with the corresponding Laplace operator

$$(3) \qquad L = -y^2 \left(\frac{\partial^2}{\partial y^2} + \frac{\partial^2}{\partial y^2} \right).$$

The full modular group acts on this plane in the natural way

$$(4) \qquad z \mapsto \gamma z = \frac{az + b}{cz + d}, \qquad a, b, c, d \in \mathbf{Z}, \ ad - bc = 1.$$

Most of the results may be developed for certain Fuchsian groups, but there are no essentially new ideas for these cases, so I restrict myself to the full modular group Γ only.

3. The appearance of the Kloosterman sums. This appearance is inescapable if one can calculate the Fourier coefficients of the automorphic function that is defined as a sum over a group. For example let us define the classical Poincaré series by the equality

$$(5) \qquad P_n(z; k) = \frac{(4\pi n)^{k-1}}{\Gamma(k-1)} \sum_{\gamma \in \Gamma_\infty \backslash \Gamma} j^{-k}(\gamma, z) e(n\gamma z), \qquad n \geq 1$$

(here Γ_∞ is generated by the shift $z \mapsto z + 1$, $j(\gamma, z) = cz + d$ if the transformation γ is defined by a matrix $\left(\begin{smallmatrix} * & * \\ c & d \end{smallmatrix} \right)$; we assume that k is an even integer and $k \geq 4$). Then for the mth Fourier coefficients of this series we have an almost obvious equality (the so-called "Petersson formula"):

$$(6) \quad p_{n,m}(k) \equiv \int_0^1 P_n(z; k) e(-mx) \, dx \cdot e^{2\pi m y}$$

$$= \frac{(4\pi n)^{k-1}}{\Gamma(k-1)} \delta_{n,m} + 2 \frac{(4\pi \sqrt{nm})^{k-1}}{\Gamma(k-1)}$$

$$\times i^{-k} \sum_{c \geq 1} \frac{1}{c} S(n, m; c) J_{k-1} \left(\frac{4\pi \sqrt{nm}}{c} \right), \qquad n, m \geq 1,$$

where J_{k-1} is the Bessel function of the order $k - 1$ and S is the Kloosterman sum,

$$(7) \qquad S(n, m; c) = \sum_{\substack{1 \leq d \leq c, (d,c)=1 \\ dd' \equiv 1 \ (\mathrm{mod}\ c)}} e \left(\frac{nd + md'}{c} \right).$$

We have a similar (but more complicated) representation for the Fourier coefficients of the nonholomorphic Poincaré-Selberg series, which for $\mathrm{Re}\, s > 1$ is defined by the equality

$$(8) \qquad U_n(z, s) = \sum_{\gamma \in \Gamma_\infty \backslash \Gamma} (\mathrm{Im}\, \gamma z)^s e(n\gamma z), \qquad n \geq 1$$

(for $n = 0$ it is the Eisenstein series).

For the Kloosterman sums we have the famous estimate due to A. Weil:

$$(9) \qquad |S(n, m; c)| \leq (2n, 2m; c)^{1/2} d^2(c).$$

But for applications we are lacking estimates for the average of these sums. Yu. V. Linnik was the first to conjecture that the Kloosterman sums are regular oscillating; his conjecture is

$$(10) \qquad \left| \sum_{c \leq X} \frac{1}{c} S(n, m; c) \right| \ll_{n,m} X^{\varepsilon}$$

for every $\varepsilon > 0$ as $X \to +\infty$.

It is obvious that the A. Weil estimate gives only $O(X^{1/2+\varepsilon})$ on the right side, and A. Selberg destroyed hopes of near progress in this conjecture when he constructed the counterexamples of the groups for which the Linnik conjecture is not valid (1963).

As background my first paper on this subject contained a nice result (1977): for every fixed $\varepsilon > 0$ we have

$$(11) \qquad \left| \sum_{1 \leq c \leq T} \frac{1}{c} S(n, m; c) \right| \ll_{n,m} T^{1/6+\varepsilon}.$$

At the same time for the "smoothing" average we have a stronger estimate; if $\varphi \in C^{\infty}(0, \infty)$, $\varphi \equiv 0$ outside of the interval $(a, 2a)$, and for every fixed integer $r \geq 0$ we have

$$\left| \frac{\partial^r}{\partial x^r} \varphi(x) \right| \ll a^{-r},$$

then for every fixed $A > 0$ the following estimate is valid:

$$(12) \qquad \left| \sum_{a \leq c \leq 2a} S(n, m; c) \varphi(c) \right| \leq a^{-A}$$

as $a \to +\infty$.

Thus is a confirmation of the Linnik conjecture.

4. The eigenfunctions of the automorphic Laplacian. As the generalization of the classical cusp forms of the even integral weight k (these are regular functions on the upper half-plane with conditions $f(\gamma z) = j^k(\gamma, z) f(z)$ for any $\gamma \in \Gamma$ and $y^{k/2} |f(z)|$ is bounded for $y > 0$; the Poincaré series $P_n(z; k)$ is an example of cusp form of weight k with respect to full modular group) Maass introduced the nonholomorphic cusp forms (so-called Maas's waves).

The Laplace operator in $L^2(\Gamma \backslash \mathbf{H})$ has the continuous spectrum on the half axis $\lambda \geq \frac{1}{4}$ and the discrete spectrum $\lambda_0 = 0$, $0 < \lambda_1 < \lambda_2 \leq \lambda_3 \leq \cdots$ with the limit point at ∞ (note that $\lambda_1 \simeq 91.07\ldots$). For the case of the full modular group there are no exceptional eigenvalues in the interval $(0, \frac{1}{4})$ (and Huxley proved that the same is true for the congruence subgroup with the level ≤ 19). So $L^2(\Gamma \backslash \mathbf{H})$ is decomposed in $L^2_{\mathrm{con}}(\Gamma \backslash \mathbf{H}) \oplus L^2_{\mathrm{cusp}}(\Gamma \backslash H)$, where L^2_{con} is the continuous direct sum of $E(z, \frac{1}{2} + it)$, $t \in \mathbf{R}$ (E is the Eisenstein series), and L^2_{cusp} is spanned by the eigenfunctions $u_j(z)$,

$$(13) \quad \begin{cases} Lu_j(z) \equiv -y^2\left(\frac{\partial^2}{\partial x^2} + \frac{\partial^2}{\partial y^2}\right)u_j = \lambda_j u_j, \quad u_j(\gamma z) = u_j(z), \\ (u_j, u_j) \equiv \int_{\Gamma\backslash H} |u_j|^2 \, d\mu(z) < \infty. \end{cases}$$

Any $f \in L^2(\Gamma\backslash H)$ that is smooth enough can be expanded into eigenfunctions of L and we have

$$(14) \quad f(z) = \sum_{j\geq 0}(f, u_j)u_j(z) + \frac{1}{4\pi}\int_{-\infty}^{\infty}\left(f, E\left(\cdot, \frac{1}{2} + ir\right)\right)E\left(z, \frac{1}{2} + ir\right)dr$$

if we choose u_j so that we have an orthonormal basis $\{u_j\}_{j\geq 0}$.

Note that the Eisenstein series has the Fourier expansion

$$(15) \quad E(z, s) = y^s + \frac{\xi(1-s)}{\xi(s)}y^{1-s} + \frac{2\sqrt{y}}{\xi(s)}\sum_{n\neq 0}\tau_s(n)e(nx)K_{s-1/2}(2\pi|n|y),$$

where $K_{s-1/2}(\cdot)$ is the modified Bessel function of order $s - \frac{1}{2}$ and

$$(16) \quad \tau_s(n) = n^{s-1/2}\sum_{\substack{d|n \\ d>0}}d^{1-2s}.$$

The eigenfunctions of a point λ_j of the discrete spectrum have a similar Fourier expansion

$$(17) \quad u_j = \sum_{n\neq 0}\rho_j(n)e(nx)\sqrt{y}K_{i\kappa_j}(2\pi|n|y)$$

with $\kappa_j = \sqrt{\lambda_j - \frac{1}{4}}$, $\lambda_j > \frac{1}{4}$.

5. Hecke operators. The ideas behind Hecke operators go back to Poincaré, and Mordell used them to prove that Ramanujan's τ-function was multiplicative. The main observation is a simple fact that if H is a subgroup of Γ of finite index, so that Γ is a finite coset union $\bigcup H\gamma_j$, and f is automorphic on H, then $\sum_j f(\gamma_j z)$ is automorphic on Γ. By appropriately choosing the set of the representatives, we can define the nth Hecke operator as the average

$$(18) \quad (T_n f)(z) = \frac{1}{\sqrt{n}}\sum_{ad=n}\sum_{b \,(\text{mod } d)}f\left(\frac{az+b}{d}\right).$$

For this normalization we have

$$(19) \quad T_n T_m = \sum_{d|(n,m)}T_{\frac{nm}{d^2}}$$

and all these operators are commutative.

We have the set of the Hermitian commutative operators with the same set of the eigenfunctions that arose for the Laplace operator. Thus we can choose the eigenfunctions of the Laplace operator so that in the basis which is constructed from these ones each Hecke operator has a diagonal form. Then these eigenfunctions will be called "Maass waves". Now we have

(20) $T_n u_j = t_j(n) u_j, \qquad n \geq 1, \ j \geq 0,$

(21) $T_n E(\cdot, s) = \tau_s(n) E(\cdot, s).$

The eigenvalues $t_j(n)$ of the discrete spectrum of nth Hecke operator T_n are connected with the Fourier coefficients of u_j by the equalities

(22) $\rho_j(1) t_j(n) = \rho_j(n), \qquad n \geq 1, \ j \geq 1.$

It is convenient to choose the eigenfunctions so that ones will be an eigenfunction of the operator T_{-1}, $(T_{-1} f)(z) = f(-\bar{z})$. Then $T_{-1} u_j = \varepsilon_j u_j$ with $\varepsilon_j = \pm 1$, and we have

(22a) $\rho_j(-n) = \varepsilon_j \rho_j(1) t_j(n).$

6. The sum formulae for the Kloosterman sums. The natural generalization of the classical Petersson formula

(23) $(f, P_n) \equiv \displaystyle\int_{\Gamma \backslash \mathbf{H}} f(z) \overline{P_n(z; k)} y^k \, d\mu(z)$

$$= a_f(n) \quad (= \ n\text{th Fourier coefficient of } f)$$

for any f from the space $\mathfrak{M}_k$ of the regular cusp forms of the weight k is the same formula for the inner product $(f, V_n(\cdot, \bar{s}))$ if f is an automorphic function from the space of cusp forms $\mathfrak{M}_0$ of the weight 0. It is possible to show that the nonholomorphic Poincaré series $U_n(z; s)$ may be continued analytically (with its Fourier expansion) in the half-plane $\mathrm{Re}\, s > \frac{3}{4}$ (it is based on the A. Weil estimate for the Kloosterman sum). So for $\mathrm{Re}\, s_1, \mathrm{Re}\, s_2 > \frac{3}{4}$ the inner product $(U_n(\cdot, s_1), U_m(\cdot, \bar{s}_2))$ is well defined. Because U_n may be expressed as a sum over a group, this inner product is a sum of the Kloosterman sums. On the other hand, the inner product (u_j, U_n) may be calculated explicitly in terms of Γ-functions and the nth Fourier coefficient of the jth eigenfunction u_j of the automorphic Laplacian. Hence the bilinear form of nth and mth Fourier coefficients of the eigenfunctions

$$\sum_{j \geq 0} \rho_j(n) \overline{\rho_j(m)} h(\kappa_j)$$

for the certain taste function h may be expressed as a sum of Kloosterman sums. Of course we have for the case $(U_n(\cdot, s_1), U_m(\cdot, s_2))$ two free variables s_1, s_2, and we can try to construct an arbitrary taste function in our bilinear form by the integration over these variables.

This plan was fulfilled in my first paper, and thus we have the following sum formula (some authors called it the "Kuznetsov trace formula").

THEOREM 1. *Let us assume that function $h(r)$ of the complex variable r is regular in the strip $|\operatorname{Im} r| \leq \delta$ with some $\delta > \frac{1}{2}$ and*

$$|h(r)| \ll |r|^{-B}$$

for some $B > 2$ when $r \to \infty$ in this strip. Then for any integers $n, m \geq 1$

(24)
$$\zeta \sum_{j=1}^{\infty} \alpha_j t_j(n) t_j(m) h(\kappa_j) + \frac{1}{4\pi} \int_{-\infty}^{\infty} \tau_{1/2+ir}(n) \tau_{1/2+ir}(m) h(r) \frac{dr}{|\zeta(1+2ir)|^2}$$

$$= \frac{\delta_{n,m}}{\pi^2} \int_{-\infty}^{\infty} r \tanh(\pi r) h(r)\, dr + \sum_{c \geq 1} \frac{1}{c} S(n,m;c) \varphi\left(\frac{4\pi\sqrt{nm}}{c}\right),$$

where

(25)
$$\alpha_j = \frac{|\rho_j(1)|^2}{\cosh(\pi \kappa_j)},$$

ζ *is the Riemann zeta-function, and for $x > 0$ the function $\varphi(x)$ is defined in terms of h by the integral transform*

(26)
$$\varphi(x) = \frac{2i}{\pi} \int_{-\infty}^{\infty} J_{2ir}(x) \frac{r h(r)}{\cosh(\pi r)}\, dr.$$

The identity (24) is modified in the following manner if the integers n, m on the right-hand side have different signs.

THEOREM 2. *Assume that the function h satisfies the conditions of the preceding theorem. Then for any integers $n, m \geq 1$ we have*

(27) $$\sum_{j \geq 1} \alpha_j \varepsilon_j t_j(n) t_j(m) h(\kappa_j) + \frac{1}{4\pi} \int_{-\infty}^{\infty} \tau_{1/2+ir}(n) \tau_{1/2+ir}(m) \frac{h(r)\, dr}{|\zeta(1+2ir)|^2}$$

$$= \sum_{c \geq 1} \frac{1}{c} S(n,-m;c) \psi\left(\frac{4\pi\sqrt{nm}}{c}\right),$$

where $\psi(x)$ for $x > 0$ is defined in terms of h by the integral

(28)
$$\psi(x) = \frac{4}{\pi^2} \int_{-\infty}^{\infty} K_{2ir}(x) h(r) r \sinh(\pi r)\, dr.$$

We can invert the identities (24) and (27), and we shall assume that the sum of the Kloosterman sums is given rather than the bilinear form of the Fourier coefficients.

THEOREM 3. *Assume that to the function $\psi: [0, \infty) \to \mathbf{C}$ the integral transform*

(29)
$$h(r) = 2\cosh(\pi r) \int_{0}^{\infty} K_{2ir}(x) \psi(x) \frac{dx}{x}$$

associates the function $h(r)$, satisfying the conditions of Theorem 1. Then for this ψ and for integers $n, m \geq 1$ we have identity (27), where the function h is defined by the integral (29).

THEOREM 4. *Let $\varphi \in C^3(0, \infty)$, $\varphi(0) = \varphi'(0) = 0$, and assume that $\varphi(x)$ together with derivatives up to the third order is estimated by the quantity $O(x^{-B})$ for some $B > 2$ as $x \to +\infty$. Then for integers $n, m \geq 1$ we have*

$$\sum_{c \geq 1} \frac{1}{c} S(n, m; c) \varphi_H \left(\frac{4\pi\sqrt{nm}}{c} \right)$$

(30)
$$= -\frac{\delta_{n,m}}{2\pi} \int_0^\infty J_0(x)\varphi(x)\, dx + \sum_{j \geq 1} \alpha_j t_j(n) t_j(m) h(\kappa_j)$$

$$+ \frac{1}{4\pi} \int_{-\infty}^\infty \tau_{1/2+ir}(n) \tau_{1/2+ir}(m) \frac{h(r)\, dr}{|\zeta(1 + 2ir)|^2}$$

where the functions $\varphi_H(x)$ and $h(r)$ are defined in terms of φ by the integral transforms

(31)
$$\varphi_H(x) = \varphi(x) - 2\sum_{k=1}^\infty (2k-1) J_{2k-1}(x) \int_0^\infty J_{2k-1}(y)\varphi(y) \frac{dy}{y},$$

(32)
$$h(r) = \frac{i\pi}{2\sinh(\pi r)} \int_0^\infty (J_{2ir}(x) - J_{-2ir}(x))\varphi(x) \frac{dx}{x}.$$

It should be useful to note that the transformation $\varphi \mapsto \varphi_H$ in (31) is a projection by which to a given φ one associates its component orthogonal on the semiaxis $x \geq 0$ (with respect to measure $x^{-1}\, dx$) to all the Bessel functions of odd integer order.

Together with (31), this projection can be defined by the equality

(33)
$$\varphi_H(x) = \varphi(x) - x \int_0^\infty \varphi(u) \left(\int_0^1 \xi J_0(x\xi) J_0(u\xi)\, d\xi \right) du$$

$$= \varphi(x) - x \int_0^\infty \varphi(u) \frac{x J_0(u) J_1(x) - u J_0(x) J_1(u)}{x^2 - u^2}\, du$$

and any sufficiently smooth φ admits a decomposition

(34)
$$\varphi = \varphi_H + (\varphi - \varphi_H),$$

where $\varphi - \varphi_H$ is a combination of Bessel functions, defined by (31), while φ_H is equal to integral (26), in which by h one means the integral transform (32) of the function φ.

The Petersson classical formula

(35)
$$\sum_{j=1}^{\nu_k} \|f_{j,k}\|^{-2} t_{j,k}(n) t_{j,k}(m) = i^k \delta_{n,m} + 2\pi \sum_{c \geq 1} \frac{1}{c} S(n, m; c) J_{k-1} \left(\frac{4\pi\sqrt{nm}}{c} \right)$$

(here $f_{j,k}$ are the orthonormal basis in the space $\mathfrak{M}_k$ of the cusp forms of the weight k, $\|f_{j,k}\|^2 = (f_{j,k}, f_{j,k})$, and $\nu_k = \dim \mathfrak{M}_k$) allow us to represent as a bilinear form the eigenvalues of the Hecke operators of the sum

$$(36) \qquad \sum_{c \geq 1} \frac{1}{c} S(n, m; c) \varphi \left(\frac{4\pi \sqrt{nm}}{c} \right)$$

for the case when φ is represented by a Neumann series of the Bessel functions of odd order. Together with (30), this gives a representation of such bilinear forms of the sums (36) for an arbitrary, sufficiently smooth, and sufficiently rapidly decreasing function φ for $x \to 0$ and $x \to \infty$.

7. Some relations with Bessel functions. The special case of the following expansion over Bessel functions is the crucial key to proving the identities of the preceding theorems from the initial identity, which is a result of the comparison of two different expressions for $(U_n(\cdot, 1 + it), U_m(\cdot, 1 - it))$, $t \in \mathbf{R}$.

THEOREM 5. *Let* $f \in C^2(0, \infty)$, $f(0) = 0$, *and*

$$\sum_{r=0}^{2} |f^{(r)}(x)| \ll x^{-B}$$

for some $B > 2$ *as* $x \to +\infty$. *Let* $\alpha \in \mathbf{R}$ *and* $F(x, t; \alpha)$ *is defined by the equality*

$$(37) \qquad F(x, t; \alpha) = J_{it}(x) \cos \tfrac{\pi}{2}(\alpha - it) - J_{-it}(x) \cos \tfrac{\pi}{2}(\alpha + it).$$

Let

$$(38) \qquad \hat{f}(t; \alpha) = \int_0^\infty F(x, t; \alpha) f(x) \frac{dx}{x}$$

and

$$(39) \qquad h_n(f) = 2(2n + 1 - \alpha) \int_0^\infty J_{2n+1-\alpha}(x) f(x) \frac{dx}{x}.$$

Then we have the representation with a parameter $\alpha \in \mathbf{R}$

$$(40) \qquad f(x) = - \int_0^\infty F(x, t; \alpha) \hat{f}(t; \alpha) \frac{t \, dt}{\sinh(\pi t)(\cosh(\pi t) + \cos(\pi \alpha))}$$
$$+ \sum_{n > (\alpha - 1)/2} h_n(f) J_{2n+1-\alpha}(x)$$

8. Some consequences. Because we have the explicit form of the connection between quantities $\rho_j(n)$ and the sum of the Kloosterman sums, we can transform the information about the Kloosterman sums into information about the Fourier coefficients of the eigenfunctions and vice versa.

The first example is the confirmation of the Linnik conjecture. The second is

THEOREM 6. *For any* $n \geq 1$ *as* $T \to +\infty$

$$(41) \qquad \sum_{\kappa_j \leq T} \alpha_j t_j^2(n) = \frac{T^2}{\pi^2} + O(T(\log T + d^2(n))) + O\left(\sqrt{nd_3(n)} \log^2 n \right)$$

where

$$\alpha_j = \frac{|\rho_j(1)|^2}{\cosh \pi \kappa_j}, \qquad d_3(n) = \sum_{d_1 d_2 d_3 = n} 1 = \sum_{d|n} d(n/d).$$

The following (nondirect) consequence is due to V. Bykovskij:

(42) $$\sum_{n \leq T} d(n^2 - D) = T(c_1(D) \log T + c_0(D)) + O_D((T \log T)^{2/3}),$$

where a fixed D is not a full square and c_1, c_0 are constants.

H. Iwaniec proved the excellent estimate for the number $\pi_\Gamma(X)$ of the conjugate primitive hyperbolic classes $\{P_0\}$ with $NP_0 < X$:

(43) $$\pi_\Gamma(X) = liX + O(X^{35/48+\varepsilon})$$

for any $\varepsilon > 0$. The proof is based essentially on the sum formulae for the Kloosterman sums.

We have made progress in the additive divisors problem (N. Kuznetsov, H. Iwaniec and J.-M. Deshouillers):

(44) $$\sum_{n \leq T} d(n)d(n + N) = TP_2(\log T, N) + O_N(T^{2/3}(\log T)^{2/3}),$$

where $P_2(z, N)$ is a polynomial in z of degree 2.

9. The Hecke series. To each eigenfunction of the ring of Hecke operators (in $\mathfrak{M}_k$ with $k > 0$ it is regular and in $\mathfrak{M}_0$ it is real analytic) we associate a Dirichlet series whose nth coefficient is the eigenvalue corresponding to this eigenfunction of the nth Hecke operator.

As we have the relations connecting the spectra of the Hecke operators with the Fourier coefficients of the eigenfunctions, these series differ only by normalization from the series associated by Hecke to regular parabolic cusps by means of the Mellin transform.

We set

(45) $$\mathscr{H}_{j,k}(s, x) = \sum_{n \geq 1} e(nx)n^{-s} t_{j,k}(n), \qquad \mathscr{H}_j(s) = \sum_{n \geq 1} e(nx)n^{-s} t_j(n),$$

and we denote by $\mathscr{L}_\nu(s, x)$ the Hecke series associated with the Eisenstein-Maass series $E(z, \nu)$,

(46) $$\mathscr{L}_\nu(s, x) = \sum_{n \geq 1} e(nx)n^{-s} \tau_\nu(n).$$

For $x = 0$ these series are denoted by $\mathscr{H}_{j,k}(s)$, $\mathscr{H}_j(s)$, $\mathscr{L}_\nu(s)$ respectively.

THEOREM 7. *Let x be rational, $x = d/c$ with $(d, c) = 1$, $c \geq 1$. Then*
(1) $\mathscr{H}_{j,k}(s, d/c)$, $\mathscr{H}_j(s, d/c)$ are entire functions of s.
(2) For $\nu \neq \frac{1}{2}$ the unique singularities of $\mathscr{L}_\nu(s, d/c)$ are the simple poles at the points $s_1 = \nu + \frac{1}{2}$ and $s_2 = \frac{3}{2} - \nu$ with residues

$$c^{-2\nu} \zeta(2\nu) \quad and \quad c^{2\nu-2} \zeta(2 - 2\nu);$$

the function $((s-1)^2 - (v - \frac{1}{2})^2)\mathscr{L}_v(s, d/c)$ *is an entire function of s.*

For what follows it is convenient to set

(47) $$\gamma(u, v) = \frac{2^{2u-1}}{\pi}\Gamma\left(u + v - \frac{1}{2}\right)\Gamma\left(u - v + \frac{1}{2}\right);$$

as a consequence of the functional equation for the gamma function, this function (for any $u, v \in \mathbf{C}$) satisfies the relation

(48) $$\gamma(u, v)\gamma(1 - u, v) = -\frac{1}{\cos^2 \pi u - \sin^2 \pi v}.$$

THEOREM 8. *The Hecke series have functional equations of the Riemann type; moreover,*

(1) *for even integers* $k \geq 12$ *and for* $(d, c) = 1, c \geq 1$, *we have*

(49) $$\mathscr{H}_{j,k}\left(s, \frac{d}{c}\right) = -\left(\frac{4\pi}{c}\right)^{2s-1}\gamma\left(1 - s, \frac{k}{2}\right)\cos(\pi s)\mathscr{H}_{j,k}\left(1 - s, -\frac{d'}{c}\right),$$

where d' *is defined by the congruence* $dd' \equiv 1 \pmod{c}$;

(50) $$\mathscr{L}_v\left(s, \frac{d}{c}\right) = \left(\frac{4\pi}{c}\right)^{2s-1}\gamma(1 - s, v)\left(-\cos(\pi s)\mathscr{L}_v\left(1 - s, -\frac{d'}{c}\right)\right.$$
$$\left. + \sin(\pi v)\mathscr{L}_v\left(1 - s, \frac{d'}{c}\right)\right);$$

(51) $$\mathscr{H}_j\left(s, \frac{d}{c}\right) = \left(\frac{4\pi}{c}\right)^{2s-1}\gamma\left(1 - s, \frac{1}{2} + i\kappa_j\right)\left(-\cos(\pi s)\mathscr{H}_j\left(1 - s, -\frac{d'}{c}\right)\right.$$
$$\left. + \varepsilon_j \cosh(\pi\kappa_j)\mathscr{H}_j\left(1 - s, \frac{d'}{c}\right)\right).$$

We conclude by the simple but important consequence of the multiplicative relations (19) for the Hecke operators: for $\operatorname{Re} s > 1 + |\operatorname{Re} v - \frac{1}{2}|$

(52) $$\sum_{n=1}^{\infty} \frac{\tau_v(n)t_j(n)}{n^s} = \frac{1}{\zeta(2s)}\mathscr{H}_j\left(s + v - \frac{1}{2}\right)\mathscr{H}_j\left(s - v + \frac{1}{2}\right)$$

(if we shall replace $t_j(n)$ by $\tau_\mu(n)$ what corresponds to the continuous spectrum of the Hecke operators, then the known Ramanujan identity will arise instead of (52)).

10. The spectral mean of Hecke series. Let $N \geq 1$ be an integer, and let s, v be complex variables. We set

(53) $$Z_N^{(d)}(s, v; h) = \sum_{j \geq 1}\alpha_j t_j(N)\mathscr{H}_j\left(s + v - \frac{1}{2}\right)\mathscr{H}_j\left(s - v + \frac{1}{2}\right)h(\kappa_j),$$

$$\alpha_j = \frac{|\rho_j(1)|^2}{\cosh(\pi\kappa_j)},$$

(54) $\quad Z_{-N}^{(d)}(s,\nu;h) = \sum_{j\geq 1} \alpha_j \varepsilon_j t_j(N) \mathscr{H}_j\left(s + \nu - \frac{1}{2}\right) \mathscr{H}_j\left(s - \nu + \frac{1}{2}\right) h(\kappa_j).$

Here the summation is over the positive discrete spectrum of the automorphic Laplacian and one assumes that its eigenfunctions have been selected in such a manner that they are at the same time the eigenfunctions of the ring of Hecke operators and of the reflection operator T_{-1} ($\varepsilon_j = \pm 1$ are the eigenvalues of T_{-1}).

Further we define the square mean of the Hecke series over the continuous spectrum by the equality

(55)

$Z_N^{(c)}(s,\nu;h)$

$= \frac{1}{\pi} \int_{-\infty}^{\infty} \frac{\zeta\left(s+\nu-\frac{1}{2}+ir\right)\zeta\left(s+\nu-\frac{1}{2}-ir\right)\zeta\left(s-\nu+\frac{1}{2}+ir\right)\zeta\left(s-\nu+\frac{1}{2}-ir\right)}{\zeta(1+2ir)\zeta(1-2ir)}$

$\times \tau_{1/2+ir}(N)h(r)\,dr$

with the stipulation that by integral (55) the function $Z_N^{(c)}(s,\nu;h)$ is defined under the conditions

$$\operatorname{Re}\left(s+\nu-\tfrac{1}{2}\right) < 1, \qquad \operatorname{Re}\left(s-\nu+\tfrac{1}{2}\right) < 1.$$

If the points $s \pm (\nu - \frac{1}{2})$ (or one of them) lie to the right-hand side of the unit line, then integral (55) defines another function, connected with $Z_N^{(c)}$ by the Sokhotskii formulae. For example, if by $\check{Z}_N^{(c)}$ we denote the function that is defined by (55) for $\operatorname{Re} s > 1$, $\operatorname{Re}\nu = \frac{1}{2}$, then a simple computation gives

$$\check{Z}_N^{(c)}(s,\nu;h) = Z_N^{(c)}(s,\nu;h) + 4\xi_N(s,\nu)h\left(i\left(s - \nu - \tfrac{1}{2}\right)\right)$$
$$+ 4\xi_N(s,1-\nu)h\left(i\left(s+\nu-\tfrac{3}{2}\right)\right),$$

where we have introduced the notation

$$\xi_N(s,\nu) = \frac{\zeta(2s-1)\zeta(2\nu)}{\zeta(2-2s+2\nu)}\tau_{s-\nu}(N)$$

and the regularity strip of h is assumed to be sufficiently wide so that the right-hand side should make sense.

Now we need the mean with respect to the weights of the Hecke series associated with regular cusp forms. For integer $k \geq 1$ we set

(56)

$Z_{N,k}(s,\nu) = 2(-1)^k \dfrac{\Gamma(2k-1)}{(4\pi)^{2k}}$

$\times \displaystyle\sum_{1\leq j\leq\nu_{2k}} |a_{j,2k}(1)|^2 t_{j,2k}(N)\mathscr{H}_{j,2k}\left(s+\nu-\frac{1}{2}\right)\mathscr{H}_{j,2k}\left(s-\nu+\frac{1}{2}\right),$

where $t_{j,2k}(N)$ is the eigenvalue of the Nth Hecke operator in the space $\mathfrak{M}_{2k}$ of regular cusp forms of weight $2k$, $\nu_{2k} = \dim\mathfrak{M}_{2k}$; the empty sum for $1 \leq k \leq 5$ and $k = 7$ is assumed to be equal to zero. Assume now that

$h^* = \{h_{2k-1}\}_{k=1}^\infty$ is a sufficiently rapidly decreasing sequence; we define the mean of the Hecke series with respect to weights by the equality

$$(57) \qquad Z_N^{(p)}(s, \nu; h^*) = \sum_{k \geq 6} h_{2k-1} Z_{N,k}(s, \nu).$$

11. The convolution formula. Some of the consequences of the algebra of modular forms are the so-called "exact formulae", an example of which is the identity

$$\sum_{n=1}^{N-1} \sigma_3(n)\sigma_3(N - n) = \frac{1}{120}(\sigma_7(N) - \sigma_3(N))$$

$(\sigma_a(n) = \sum_{d|n} d^a)$. A source of similar identities is the obvious assertion that the product of modular forms of weights k and l is a modular form of weight $k + l$.

There are analogues of these identities for the real analytic Eisenstein series of weight zero. For an integer $N \geq 1$ we associate to a pair of series $E(z, s)$ and $E(z, \nu)$ the expression of convolution type

$$(58)$$

$$W_N(s, \nu; w_0, w_1)$$

$$= N^{s-1} \sum_{n=1}^\infty \tau_\nu(n)\left(\sigma_{1-2s}(n - N)w_0\left(\sqrt{\frac{n}{N}}\right) + \sigma_{1-2s}(n + N)w_1\left(\sqrt{\frac{n}{N}}\right)\right),$$

where $\sigma_{1-2s}(0)$ means $\zeta(2s - 1)$ and w_0, w_1 are assumed to be sufficiently smooth and sufficiently rapidly decreasing for $x \to +\infty$.

THEOREM 9. *Assume that the functions w_0, w_1 are continuous on the semi-axis $x \geq 0$ together with the derivatives up to the fourth order, $w_j(0) = w_j'(0) = 0$ for $j = 0, 1$, and for $x \to +\infty$ the functions $w_j(x)$ themselves as well as their derivatives up to third order are estimated by the quantity $O(x^{-B})$ with some $B > 4$. Then for any integer $N \geq 1$ and $s, \nu \in \mathbf{C}$ satisfying $\mathrm{Re}\,\nu = \frac{1}{2}, \frac{1}{2} < \mathrm{Re}\,s < 1$ we have*

$$(59)$$

$$W_N(s, \nu; w_0, w_1)$$

$$= Z_N^{(d)}(s, \nu; h_0) + Z_{-N}^{(d)}(s, \nu; h_1) + Z_N^{(c)}(s, \nu; h_0 + h_1) + Z_N^{(p)}(s, \nu; h^*)$$

$$+ \zeta_N(s, \nu)V(\tfrac{1}{2}, \nu) + \zeta_N(s, 1 - \nu)V(\tfrac{1}{2}, 1 - \nu) + \zeta_N(1 - s, \nu)V(s, \nu)$$

$$+ \zeta_N(1 - s, 1 - \nu)V(s, 1 - \nu),$$

where

$$(60) \qquad \zeta_N(s, \nu) = \frac{\zeta(2s)\zeta(2\nu)}{\zeta(2s + 2\nu)}\tau_{s+\nu}(N),$$

$$(61) \quad V(s, \nu) = 2\int_0^\infty (|1 - x^2|^{1-2s}w_0(x) + (1 + x^2)^{1-2s}w_1(x))x^{2\nu}\,dx,$$

and the column vector $h(r; s, \nu) = \begin{pmatrix} h_0 \\ h_1 \end{pmatrix}$ *is defined in terms of* $w = \begin{pmatrix} w_0 \\ w_1 \end{pmatrix}$ *by the integral transform*

$$(62) \quad h(r; s, \nu) = \pi \int_0^\infty \begin{pmatrix} R_0(x, \tfrac{1}{2} + ir) & 0 \\ 0 & R_1(x, \tfrac{1}{2} + ir) \end{pmatrix} \left(\frac{x}{4\pi} \right)^{2s-1}$$

$$\times \int_0^\infty \begin{pmatrix} R_0(xy, \nu) & R_1(xy, \nu) \\ R_1(xy, \nu) & R_0(xy, \nu) \end{pmatrix} \begin{pmatrix} w_0(y) \\ w_1(y) \end{pmatrix} y \, dy \, dx$$

with the kernels

$$(63) \quad R_0(x, \nu) = \frac{J_{2\nu-1}(x) - J_{1-2\nu}(x)}{2 \cos(\pi\nu)}, \qquad R_1(x, \nu) = \frac{2}{\pi} \sin(\pi\nu) K_{2\nu-1}(x).$$

12. Some consequences of the convolution formula. The first example using (59) is the additive divisors problem; if we choose $s = \nu = \tfrac{1}{2}$, $w_1 = 0$, and w_0 so that it is near to 1 in the interval $(0, \sqrt{T/N})$ (so $w_0(\sqrt{n/N})$ will be near to 1 for $n \leq T$), then the left-hand side of (59) gives the sum on the left side (44). Terms with the integral (61) are leading terms and all other terms give the remainder term.

Of course the asymptotic formula for the additive divisors problem is crucial for the investigation of the fourth power moment of the Riemann zeta-function. The consequence of (59) in this direction is the following

THEOREM (N. ZAVOROTNYI, KHABAROVSK, 1987). *Let* $T \to +\infty$; *then for any* $\varepsilon > 0$ *we have*

$$(64) \qquad \int_0^T |\zeta(\tfrac{1}{2} + it)|^4 \, dt = TP_4(\log T) + O(T^{2/3+\varepsilon}),$$

where $P_4(z)$ *is the polynomial in* z *of the fourth degree with constant coefficients.*

We can consider the functions h_0 and h_1 in (59) as given; the following unusual integral transform is useful to invert (59). Let us define the matrix kernel $\mathbf{K}(x, \nu)$ by the equality

$$(65) \qquad \mathbf{K}(x, \nu) = \begin{pmatrix} R_0(x, \nu) & R_1(x, \nu) \\ R_1(x, \nu) & R_0(x, \nu) \end{pmatrix}$$

with R_j from (63). Now we shall consider the matrix equation

$$(66) \qquad w(x) = \int_0^\infty \mathbf{K}(xy, \nu) v(y) \sqrt{xy} \, dy,$$

where $w = \begin{pmatrix} w_0 \\ w_1 \end{pmatrix}(x), v = \begin{pmatrix} v_0 \\ v_1 \end{pmatrix}(x).$

THEOREM 10. *Let* $\operatorname{Re} \nu = \tfrac{1}{2}$ *and* $w \in L^2(0, \infty)$ *in the sense that* $w_0, w_1 \in L^2(0, \infty)$. *Then there exists a unique solution* $v \in L^2(0, \infty)$ *of equation* (66), *and this solution is given by the formula*

$$(67) \qquad v(x) = \int_0^\infty \mathbf{K}(xy, \nu) w(y) \sqrt{xy} \, dy,$$

where the integral is understood in the square-mean sense.

Now, as a special case of the convolution formula (59), we have the following asymptotic formulae.

THEOREM 11. *Let $T \to +\infty$. Then for fixed σ and $t \in \mathbf{R}$ for $\frac{1}{2} < \sigma < 1$ we have*

$$(68) \quad \sum_{\kappa_j \leq T} \alpha_j |\mathscr{H}_j(\sigma + it)|^2 = \frac{T^2}{\pi^2}\left(\zeta(2\sigma) + \frac{\zeta(2 - 2\sigma)}{2(1 - \sigma)}\left(\frac{T^2}{2\pi}\right)^{1-2\sigma}\right) + O(T \log T)$$

while for $\sigma = \frac{1}{2}$ the right-hand side has to be replaced by

$$(69) \quad \frac{2T^2}{\pi^2}(\log T + 2\gamma - 1 + 2\log(2\pi)) + O(T \log T),$$

where γ is the Euler constant.

13. The fourth spectral moment of the Hecke series. The next step is the investigation of the fourth power moment over the spectrum of the Laplacian.

We define the fourth spectral mean of the Hecke series over a discrete spectrum by the equality

$$(70) \quad Z^{\mathrm{dis}}\left(\begin{matrix} s, & \nu \\ \rho, & \mu \end{matrix}\middle| h\right)$$

$$= \sum_{j \geq 1} \alpha_j \frac{\mathscr{H}_j(s + \nu - \frac{1}{2})\mathscr{H}_j(s - \nu + \frac{1}{2})\mathscr{H}_j(\rho + \mu - \frac{1}{2})\mathscr{H}_j(\rho - \mu + \frac{1}{2})}{\zeta(2s)\zeta(2\rho)} h(\kappa_j)$$

with $\alpha_j = (\cosh(\pi\kappa_j))^{-1}|\rho_j(1)|^2$. We have the same definition for the sum

$$Z^{\mathrm{cusp}}\left(\begin{matrix} s, & \nu \\ \rho, & \mu \end{matrix}\middle| h^*\right)$$

over regular cusp forms. The sum (70) with α_j replaced by $\varepsilon_j \alpha_j$ we shall denote by

$$Z_{-1}^{\mathrm{dis}}\left(\begin{matrix} s, & \nu \\ \rho, & \mu \end{matrix}\middle| h\right).$$

The corresponding sum over a continuous spectrum is defined by the integral

$$(71) \quad Z^{\mathrm{con}}\left(\begin{matrix} s, & \nu \\ \rho, & \mu \end{matrix}\middle| h\right)$$

$$= \frac{1}{4\pi}\int_{-\infty}^{\infty} z\left(s; \nu, \frac{1}{2} + ir\right) z\left(\rho;, \mu, \frac{1}{2} + ir\right) \frac{h(r)}{\zeta(1 + 2ir)\zeta(1 - 2ir)} dr$$

with

$$(72) \quad z(s; \nu, \mu) \equiv \sum_{n=1}^{\infty} \frac{\tau_\nu(n)\tau_\mu(n)}{n^s}$$

$$= \frac{\zeta(s + \mu - \nu)\zeta(s + \nu - \mu)\zeta(s + \nu + \mu - 1)\zeta(s - \nu - \mu + 1)}{\zeta(2s)}$$

and the same symbol

$$Z^{\mathrm{con}}\left(\begin{matrix} s, & \nu \\ \rho, & \mu \end{matrix}\middle| h\right)$$

will be used for an analytical continuation of this integral. Then we have the following functional equation for $\mathrm{Re}\,\nu = \mathrm{Re}\,\mu = \frac{1}{2}$, $\frac{1}{2} < \mathrm{Re}\,s$, $\mathrm{Re}\,\rho < 1$:

(73)

$$Z^{\mathrm{dis}}\left(\begin{matrix} s, & \nu \\ \rho, & \mu \end{matrix}\middle| h\right) + Z^{\mathrm{con}}\left(\begin{matrix} s, & \nu \\ \rho, & \mu \end{matrix}\middle| h\right)$$

$$= Z^{\mathrm{dis}}\left(\begin{matrix} \rho, & \nu \\ s, & \mu \end{matrix}\middle| h_0\right) + Z^{\mathrm{dis}}_{-1}\left(\begin{matrix} \rho, & \nu \\ s, & \mu \end{matrix}\middle| h_1\right) + Z^{\mathrm{con}}\left(\begin{matrix} \rho, & \nu \\ s, & \mu \end{matrix}\middle| h_0 + h_1\right)$$

$$+ Z^{\mathrm{cusp}}\left(\begin{matrix} \rho, & \nu \\ s, & \mu \end{matrix}\middle| h_0^*\right)$$

$$+ \frac{2(4\pi)^{2s-1}}{\zeta(2s)}\left\{ \frac{\zeta(2\nu)}{(4\pi)^{2\nu}} z(\rho + \nu; \mu, s)\hat{\varphi}(2\nu + 1 - 2s) \right.$$

$$\left. + (\text{the same with } \nu \mapsto 1 - \nu) \right\}$$

$$+ \frac{2(4\pi)^{2\rho-1}}{\zeta(2\rho)}\left\{ \frac{\zeta(2\mu)}{(4\pi)^{2\mu}} z(s + \mu; \nu, \rho)\hat{\varphi}(2\mu + 1 - 2\rho) \right.$$

$$\left. + (\text{the same with } \mu \mapsto 1 - \mu) \right\},$$

where h_0, h_1 are defined by the sequence of the integral transforms

(74)
$$h \mapsto \varphi \mapsto \begin{pmatrix} \Phi_0 \\ \Phi_1 \end{pmatrix} \mapsto \begin{pmatrix} h_0 \\ h_1 \end{pmatrix}$$

with

(75)
$$\varphi(x) = \int_{-\infty}^{\infty} R_0\left(x, \frac{1}{2} + ir\right) r \tanh(\pi r) h(r)\, dr,$$

(76) $\begin{pmatrix} \Phi_0 \\ \Phi_1 \end{pmatrix}(x)$

$$= x^{2s+2\rho-2} \iint_0^\infty \begin{pmatrix} R_0(\xi, \nu)R_0(\eta, \mu) + R_1(\xi, \nu)R_1(\eta, \mu) \\ R_0(\xi, \nu)R_1(\eta, \mu) + R_1(\xi, \nu)R_0(\eta, \mu) \end{pmatrix} \varphi\left(\frac{\xi\eta}{x}\right)$$

$$\cdot \xi^{1-2s}\eta^{1-2\rho}\, d\xi\, d\eta,$$

(77)
$$h_j(r) = \int_0^\infty R_j\left(x, \frac{1}{2} + ir\right)\Phi_j(x)\frac{dx}{x}, \qquad j = 0, 1.$$

Finally, $\hat{\varphi}$ denotes the Mellin transform of the function φ, and the sequence $h_0^* = \{h_0^{(2k-1)}\}_{k=1}^{\infty}$ is defined by the formal substitution $r = i(k - \frac{1}{2})$ in the formula for $h_0(r)$. This functional equation gives us an approach to the problem of the eighth power moment for the Riemann zeta-function.

For the special case

(78) $s = \mu = \tfrac{1}{2}, \qquad \rho = \nu = \tfrac{1}{2} + it, \quad t \in \mathbf{R},$

we choose the function $h(r)$ on the left side (73) so that the main contribution would be given from the integral over the continuous spectrum (for example, $h(r) = \exp(-\alpha r^2)$ with a large positive α). Then, after multiplying by the appropriate factor with modulo 1, we have on the left side the function that is near to $|\zeta(\tfrac{1}{2} + it)|^8$. At the same time for the special case (78) we have the product $\mathscr{H}_j^3(\tfrac{1}{2})\mathscr{H}_j(\tfrac{1}{2} + 2it)$ in the sum over a discrete spectrum and $|\zeta^6(\tfrac{1}{2} + ir)|\zeta(\tfrac{1}{2} + 2it + ir)\zeta(\tfrac{1}{2} + 2it - ir)$ in the integral over a continuous spectrum. Thus we can integrate over t in the exact form in the right-hand side. So the problem of the eighth moment will be reduced to the sixth power problem, and there exists a hope for progress in this difficult task.

REFERENCES

1. N. V. Kuznetsov, *The Petersson conjecture for cusp forms of weight zero and Linnik's conjecture*, Mat. Sb. **111** (1980), 334–383.

2. ____, *Convolution of the Fourier coefficients of the Eisenstein-Maass series*, Zap. Nauchn. Sem. Leningrad. Otdel. Mat. Inst. Steklov. (LOMI) **129** (1983), 43–84.

3. J.-M. Deshouillers and H. Iwaniex, *Kloosterman sums and Fourier coefficients of cusp forms*. I, Univ. de Bordeaux–Math. Inst. Polish Acad. Sci., 1982.

4. M. N. Huxley, *Introduction to Kloostermania*, Elementary and Analytic Theory of Numbers, Banach Center Publ., vol. 17, PWN, Warsaw, 1985, pp. 217–306.

ACADEMY OF SCIENCES OF THE USSR, KHABAROVSK CENTER, USSR

Proceedings of Symposia in Pure Mathematics
Volume **49** (1989), Part 2

The Existence of Maass Cusp Forms
and Kloosterman Sums

I. I. PIATETSKI-SHAPIRO

(Notes made by J. Cogdell)

1. Introduction.

1.1. Let $\mathcal{H} = \{z = x+iy \in \mathbf{C} | y > 0\}$ denote the Poincaré upper half-plane, equipped with the usual Poincaré metric. The associated Laplace-Beltrami operator is

$$\Delta = -y^2 \left(\frac{\partial^2}{\partial x^2} + \frac{\partial^2}{\partial y^2} \right).$$

Let $\Gamma \subset \mathrm{SL}_2(\mathbf{R})$ be a Fuchsian group of the first kind, that is, a discrete subgroup of $\mathrm{SL}_2(\mathbf{R})$ such that $\Gamma\backslash\mathcal{H}$ has finite volume. For our purposes, we will always assume that $\Gamma\backslash\mathcal{H}$ is not compact.

Let σ be a cusp of Γ. For each cusp σ, let N_σ be the unipotent radical of the Borel subgroup B_σ of $\mathrm{SL}_2(\mathbf{R})$ that stabilizes σ. If $\sigma = \infty$, then

$$N_\infty = \left\{ n(x) = \begin{pmatrix} 1 & x \\ 0 & 1 \end{pmatrix} \middle| x \in \mathbf{R} \right\}.$$

We will always assume ∞ is a cusp of Γ. Set $\Gamma_\sigma = \Gamma \cap N_\sigma$. A function $f \in L^2(\Gamma\backslash\mathcal{H})$ is called cuspidal, or a cusp form, if

$$(*) \qquad \int_{\Gamma_\sigma\backslash N_\sigma} f(nz)\, dn \equiv 0$$

for each cusp σ of Γ.

DEFINITION. A Maass cusp form for Γ is a cuspidal eigenfunction for Δ in $L^2(\Gamma\backslash\mathcal{H})$, i.e., $u \in L^2(\Gamma\backslash\mathcal{H})$ satisfying $(*)$ and such that $\Delta u = \lambda u$ for some $\lambda \in \mathbf{C}$.

If we consider the spectral decomposition of $L^2(\Gamma\backslash\mathcal{H})$ under Δ [6], then $L^2(\Gamma\backslash\mathcal{H})$ decomposes naturally into three spaces:

$$L^2(\Gamma\backslash\mathcal{H}) = L_0^2(\Gamma\backslash\mathcal{H}) \oplus L_r^2(\Gamma\backslash\mathcal{H}) \oplus L_{\mathrm{cont}}^2(\Gamma\backslash\mathcal{H}).$$

On $L_{\mathrm{cont}}^2(\Gamma\backslash\mathcal{H})$, Δ has a continuous spectrum spanned by the Eisenstein series associated to the cusps of Γ. The space $L_r^2(\Gamma\backslash\mathcal{H})$ is the finite-dimensional

1980 *Mathematics Subject Classification* (1985 *Revision*). Primary 11F42.

space spanned by the residues of the Eisenstein series, on which Δ has a discrete spectrum. $L_0^2(\Gamma \backslash \mathscr{H})$ is the space of all cusp forms on $\Gamma \backslash \mathscr{H}$. This space has a discrete spectrum with finite multiplicity with respect to Δ and is spanned by the Maass cusp forms. The space $L_{\text{cont}}^2(\Gamma \backslash \mathscr{H})$ is always nonempty and, as noted, can be constructed using the Eisenstein series. The space $L_r^2(\Gamma \backslash \mathscr{H})$ is also nonempty, since the constant function always occurs as the residue of an Eisenstein series. However, it is not known if there always exist Maass cusp forms, for there is no general method for producing cusp forms.

In his 1944 Göttingen lectures, Selberg made a conjecture about the existence of Maass cusp forms. For the full strength of this conjecture, see [3]. A weak form of his conjecture is

CONJECTURE (SELBERG). *For any Γ Fuchsian of the first kind, Γ has an infinite number of Maass cusp forms, or, equivalently, $L_0^2(\Gamma \backslash \mathscr{H})$ is infinite dimensional.*

For arithmetic Γ this conjecture is true and can be proved by many means. Here arithmetic means that it contains a congruence subgroup. Here a discrete subgroup Γ will be arithmetic iff $\exists$ a subgroup Γ_1 of Γ such that Γ_1 is conjugate to a congruence subgroup of $\mathrm{SL}_2(\mathbf{Z})$. For nonarithmetic Γ the situation is not so clear. Recent results of Colin de Verdiere [1], Phillips and Sarnak [9], and Deshouillers, Iwaniec, Phillips, and Sarnak [3], all based on the destruction of cusp forms under deformations, indicate that this conjecture may not be true. In fact, in [3] Deshouillers, Iwaniec, Phillips, and Sarnak conjecture that for certain generic Γ, Δ will have only a finite number of eigenvalues on $L_0^2(\Gamma \backslash \mathscr{H})$.

1.2. In this article we will describe our investigations centered around the Selberg conjecture. Our original idea was to use Kloosterman sums and the Kuznetsov trace formula as developed by Kuznetsov [7] and generalized by Proskurin [11]. We will describe the connections between Kloosterman sums and Maass cusp forms in §3. So far, this approach has not been successful. However, by completely elementary techniques, we were able to prove the following weaker version of Selberg's conjecture.

THEOREM 1. *If Γ is a Fuchsian group of the first kind, then Γ has a subgroup Γ' of finite index such that Γ' has an infinite number of Maass cusp forms, i.e., $\dim L_0^2(\Gamma' \backslash \mathscr{H}) = \infty$.*

Since giving our report on this theorem at the conference, we have discovered that this result is not new, but in fact was proven by Venkov [15]. However, our technique is different from Venkov's and readily generalizes to arbitrary semisimple groups, as we will see in §2.

2. The existence of Maass cusp forms.

2.1. Let us now prove Theorem 1. The first reduction is to Γ having finite-dimensional cuspidal representations in the following sense.

DEFINITION. A finite-dimensional representation (X, V_X) of Γ is called cuspidal if $\Gamma' = \text{Ker}(X)$ is of finite index in Γ and for every cusp σ of Γ we have

$$\sum_{\gamma \in \Gamma'_\sigma \backslash \Gamma_\sigma} X(\gamma) = 0,$$

where $\Gamma'_\sigma = \Gamma^1 \cap \Gamma_\sigma$.

PROPOSITION 1. *If Γ has a finite-dimensional cuspidal representation (X, V_X), then $\Gamma' = \text{Ker}(X)$ has an infinite number of Maass cusp forms.*

PROOF. Let M denote the quotient group $\Gamma' \backslash \Gamma$, so that X factors to a finite-dimensional representation of M.

Consider the space of L^2 vector-valued functions $L^2(\Gamma' \backslash \mathcal{H}, V_X)$ from $\Gamma' \backslash \mathcal{H}$ to V_X. There is a natural action of Γ on this space via left translation

$$(L(\gamma)F)(z) = F(\gamma^{-1}z), \qquad \gamma \in \Gamma,$$

which again factors through to an action of M. Let P_X denote the projection onto the X-isotypic component

$$(P_X F)(z) = \sum_{\gamma \in \Gamma/\gamma'} X(\gamma)F(\gamma^{-1}z).$$

The image of P_X consists of those $F \in L^2(\Gamma' \backslash \mathcal{H}, V_X)$ that satisfy $F(\gamma z) = X(\gamma)F(z)$. We claim that the components of such functions are cuspidal for Γ'. For indeed, if σ is a cusp of Γ' it is also a cusp for Γ. Then

$$\int_{\Gamma'_\sigma \backslash N_\sigma} F(nz)\,dn = \int_{\Gamma_\sigma \backslash N_\sigma} \sum_{\gamma \in \Gamma'_\sigma \backslash \Gamma_\sigma} F(\gamma nz)\,dn$$

$$= \int_{\Gamma_\sigma \backslash N_\sigma} \left(\sum_{\gamma \in \Gamma'_\sigma \backslash \Gamma_\sigma} X(\gamma) \right) F(nz)\,dz = 0$$

since X is a cuspidal representation of Γ. But $\int_{\Gamma'_\sigma \backslash N_\sigma} F(nz)\,dn = 0$ iff $\int_{\Gamma'_\sigma \backslash N_\sigma} f_i(nz)\,dn = 0$ for each component function $f_i(z)$ of $F(z)$. Hence these components are cuspidal for Γ'.

We must next show that the image of P_X is infinite dimensional. We have natural action of $M = \Gamma' \backslash \Gamma$ on $\Gamma' \backslash \mathcal{H}$ via $z \mapsto \Gamma' \gamma Z \pmod{\Gamma'}$ for $\Gamma' \gamma \in M$. Choose $z \in \Gamma' \backslash \mathcal{H}$ such that z has $|M|$ distinct images under M. Identify the orbit $M \cdot z$ with M. Then if we restrict the functions in $L^2(\Gamma' \backslash \mathcal{H}, V_X)$ to the orbit $M \cdot z$ we obtain all functions $f \colon M \to V_X$, which we denote $\mathscr{F}(M, V_X)$. This restriction, which we denote by Res_{Mz}, intertwines the action of M on $L^2(\Gamma' \backslash \mathcal{H}, V_X)$ via L with the left regular representation on $\mathscr{F}(M, V_X)$. As a representation of M, $\mathscr{F}(M, V_X)$ is isomorphic to $\dim V_X$ copies of the regular representation of M on $C[M]$. So we have

$$L^2(\Gamma' \backslash \mathcal{H}, V_X) \overset{\text{Res}_{M \cdot z}}{\to} (\dim V_X) \cdot C[M].$$

The projection P_X induces on $\mathbf{C}[M]$ the projection onto the X-isotypic component of the regular representation of M. The regular representation contains X with multiplicity $\dim V_X$. Hence

$$(\text{Im } P_X) \overset{\text{Res}_{M_z}}{\longrightarrow} (\dim V_X)^2 \cdot X$$

so that $\dim(\text{Im } P_X) \geq (\dim V_X)^3$. If we repeat this with N points $z_1, \ldots, z_N$ that are inequivalent mod Γ and such that all have orbits $M \cdot z_i$ under M of size $|M|$, then we obtain $\dim(\text{Im } P_X) \geq N(\dim V_X)^3$ for all $N > 0$. Hence $\text{Im } P_X$ is infinite dimensional.

Therefore $\Gamma' = \text{Ker}(X)$ has an infinite number of linearly independent Maass cusp forms. $\square$

2.2. We next introduce a notion of the weak arithmetic groups.

DEFINITION. A Fuchsian group of the first kind Γ is called weak arithmetic if there exists an algebraic number field k such that $\Gamma \subset \text{SL}_2(k)$ (up to a conjugation).

PROPOSITION 2. *If Γ is weak arithmetic, then Γ has a finite-dimensional cuspidal representation.*

PROOF. We assume k is an algebraic number field and ρ a homomorphism such that $\rho(\Gamma) \subset \text{SL}_2(k)$. Let θ be the ring of integers of k. Since Γ is Fuchsian of the first kind, it is finitely generated and among its generators we may take a primitive parabolic element $T_\sigma \in \Gamma$ for each Γ-equivalence class of cusps σ of Γ.

Choose a prime $y \subset \theta, y \nmid 2$, such that
(1) for each nonzero entry t_{ij} of each generator T of Γ we have $\text{ord } y(t_{ij}) = 0$, and
(2) y is a prime of degree 1, i.e., $\theta/y \simeq \mathbf{F}_p$, a prime field with p elements. If we then complete k at y, we have $\Gamma \subset \text{SL}_2(k_y)$ and by condition (i) we actually have $\Gamma \subset \text{SL}_2(\theta_y)$, where θ_y is the ring of integers of k_y.

Let $C(y)$ be the principal congruence subgroup of $\text{SL}_2(\theta_y)$,

$$C(y) = \{g \in \text{SL}_2(\theta_y) | g \equiv I \bmod y\},$$

and let $\Gamma(y) = \Gamma \cap C(y)$. Since y is a degree one prime, we have

$$1 \longrightarrow C(y) \longrightarrow \text{SL}_2(\theta_y) \overset{\pi}{\longrightarrow} \text{SL}_2(\mathbf{F}_p) \longrightarrow 1$$

$$\cup \qquad\qquad\qquad \cap$$

$$\Gamma(y) \longrightarrow \quad \Gamma.$$

We first claim that $\pi(\Gamma) \subset \text{SL}_2(\mathbf{F}_p)$. To see this note that since ∞ is a cusp of Γ then $\pi(\Gamma) \supset \bar{N} = \{\left(\begin{smallmatrix} 1 & x \\ 0 & 1 \end{smallmatrix}\right) \mid x \in \mathbf{F}_p\}$. By the structure of $\text{SL}_2(\mathbf{F}_p)$, any proper subgroup H with $\text{SL}_2(\mathbf{F}_p) \supsetneq H \supset \bar{N}$ must be contained in $\bar{B} = \{\left(\begin{smallmatrix} a & b \\ 0 & d \end{smallmatrix}\right)\} \subset \text{SL}_2(\mathbf{F}_p)$. If $\pi(\Gamma) \subset \bar{B}$ then for each generator T, $\pi(T) \subset \bar{B}$, which, by condition (i) on our choice of y, would imply that $\Gamma \subset B_\infty$. But then Γ could not have a finite volume quotient. Thus $\pi(\Gamma) = \text{SL}_2(\mathbf{F}_p)$.

It is well known [10] that $SL_2(\mathbf{F}_p)$ has cuspidal representations. Let (X, V_X) be one such, and view it as a representation of Γ via π. By the simplicity of $SL_2(\mathbf{F}_p)$, its kernel must be $\Gamma(y)$. X will be a cuspidal representation of Γ. For if σ is any cusp of Γ, then $\pi(\Gamma_\sigma)$ is not trivial; by (i) and (ii) it must be the full unipotent radical $\bar{U}$ of some Borel subgroup of $SL_2(\mathbf{F}_p)$. Thus

$$\sum_{\gamma \in \Gamma(y)\backslash\Gamma_\sigma} X(\gamma) = \sum_{\bar\gamma \in \bar U} X(\bar\gamma) = 0$$

since X is a cuspidal representation of $SL_2(\mathbf{F}_p)$. □

To complete the proof of Theorem 1 it suffices to note that every Γ can be deformed into a weak arithmetic group and hence will have finite-dimensional cuspidal representations.

2.3. The proof of Theorem 1 generalizes immediately to any semisimple group G. Let $H = G(\mathbf{R})/K(\mathbf{R})$ be the real symmetric space of G, and let $\Gamma \subset G(\mathbf{R})$ be a discrete group such that $\Gamma\backslash H$ has finite volume but is noncompact. Then we know the following facts.

(i) By Garland-Raghunathan [5], if $\operatorname{rank}_{\mathbf{R}} G = 1$ and $G(\mathbf{R})$ is not locally isomorphic to $SL_2(\mathbf{R})$, then Γ is finitely generated and weak arithmetic, i.e, there exists an algebraic number field $k \subset \mathbf{R}$ such that, up to conjugation in $G(\mathbf{R})$, $\Gamma \subset G(k)$.

(ii) If $\operatorname{rank}_{\mathbf{R}} G \geq 2$, then by Margulis [8], Γ is in fact arithmetic.

(iii) $G(\mathbf{F}_p)$ always has cuspidal representations. (See, for example, Deligne-Lusztig [2].)

Then the above proof gives

THEOREM 2. *If G is any semisimple group and $\Gamma \subset G(\mathbf{R})$ is discrete such that $\Gamma\backslash H$ is noncompact but has finite volume, then there is a subgroup $\Gamma' \subset \Gamma$ of finite index such that the space of cusp forms $L_0^2(\Gamma'\backslash H)$ is infinite dimensional.*

2.4. To conclude this section, let us point out that this method can never yield a proof of the full Selberg conjecture. The method here generalizes to the case of discrete subgroups $SL_2(K)$ where $K = F((t))$ is the field of formal power series in one variable over the finite field F_q and shows that for cofinite Γ there is always a subgroup Γ' of finite index for which there are unramified cusp forms. However, by the work of Drinfeld [4] we know that there are in fact certain arithmetic $\Gamma \subset SL_2(K)$ that have no cusp forms. So we cannot expect to press this method any further.*

3. Kloosterman sums and Maass cusp forms. Let us now come to the relation between Kloosterman sums and the existence of Maass cusp forms. We will work in a slightly more general situation. We consider discrete cofinite

---

* Recently I. Efrat has found many new examples $\Gamma \subset SL_2(K)$ without cusp forms. In this example volume of $\Gamma\backslash SL_2(K)$ is unbounded.

$\Gamma \subset \mathrm{SL}_2(K)$ for any local field K. In this context we will define Klooster-man sums, the associated Kloosterman-Selberg zeta function, and describe the connection with Maass cusp forms.

3.1. Before describing the Kloosterman sums we need some preliminary results. Let K be any local field, and let H denote the symmetric space of $\mathrm{SL}_2(K)$. If $K = \mathbf{R}$ then $H = \mathscr{H}$ as before, and if $K = \mathbf{C}$ then $\mathscr{H}$ is hyperbolic three space [12]. If K is a non-archimedean local field of residual degree q, then for H we take the homogeneous tree of degree $q + 1$ associated to $\mathrm{SL}_2(K)$ as in Serre [14]. We are interested in the existence of cusp forms in $L^2(\Gamma\backslash H)$ for discrete subgroups $\Gamma \subset \mathrm{SL}_2(K)$ such that $\Gamma\backslash H$ is noncompact but has finite volume. Over a non-archimedean field of characteristic O, the finiteness of the volume of $\Gamma\backslash H$ automatically implies that $\Gamma\backslash H$ is compact [14]. So we may restrict our attention to $K = \mathbf{R}$, $K = \mathbf{C}$, or $K = F_q((t))$, the field of formal power series over a finite field F_q with q elements.

Consider for a moment only the case $K = F_q((t))$. For basic structural facts we refer to Serre [14]. In this case, the role of the Laplacian Δ on H is played by the Hecke operator T given by

$$(Tf)(p) = \sum_{d(p,p')=1} f(p'),$$

i.e., "Δ=sum over nearest neighbors in H." The tree H has a natural boundary given by $\mathbf{P}^{\wedge}(K)$. The stabilizer of a boundary point $\sigma \in \mathbf{P}^{\wedge}(K)$ is a Borel subgroup $B_\sigma \subset \mathrm{SL}_2(K)$ with unipotent radical N_σ. If $\Gamma \subset \mathrm{SL}_2(K)$ is discrete with $\Gamma\backslash H$ of finite volume, then a boundary point $\sigma \in \mathbf{P}^{\wedge}(K)$ is called a cusp of Γ if $\Gamma_\sigma = \Gamma \cap N_\sigma$ is "large" in the sense that $\Gamma_\sigma\backslash N_\sigma$ is compact. A function $f \in L^2(\Gamma\backslash H)$ is called a cusp form if for every cusp σ of Γ we have

$$\int_{\Gamma_\sigma\backslash N_\sigma} f(np)\, dn \equiv 0.$$

A Maass cusp form is then a cuspidal eigenfunction of Δ ($= T$).

Return now to the general situation where K is a local field and Γ is discrete in $\mathrm{SL}_2(K)$ such that $\Gamma\backslash H$ has finite volume but is not compact. Let ∞ denote the cusp of Γ whose stabilizer in $\mathrm{SL}_2(K)$ is the standard Borel subgroup

$$B_\infty = \left\{ \begin{pmatrix} a & b \\ 0 & d \end{pmatrix} \right\} \subset \mathrm{SL}_2(K)$$

with unipotent radical

$$N_\infty = \left\{ n(x) = \begin{pmatrix} 1 & x \\ 0 & 1 \end{pmatrix} \,\middle|\, x \in K \right\}.$$

(We always assume ∞ is a cusp of Γ.) For any cusp σ of Γ, choose $g_\sigma \in \mathrm{SL}_2(K)$ such that $g_\sigma\infty = \sigma$. Define $\Delta_\sigma \subset N_\infty$ by

$$\Delta_\sigma = g_\sigma^{-1}\Gamma_\sigma g_\sigma.$$

Since $\Gamma_\sigma\backslash N_\sigma$ is compact, so will be $\Delta_\sigma\backslash N_\infty$.

Now fix two cusps σ, τ of Γ and set

$$\Omega_{\sigma,\tau}(\Gamma) = \left\{ c \in K^{\times} \mid N_{\infty} \begin{pmatrix} 0 & -c^{-1} \\ c & 0 \end{pmatrix} N_{\infty} \cap g_{\sigma}^{-1}\Gamma g_{\tau} \neq \varnothing \right\}.$$

Note that this is independent of the choice of g_{σ} and g_{τ}. For $c \in \Omega_{\sigma,\tau}(\Gamma)$ set

$$h_c = \begin{pmatrix} 0 & -c^{-1} \\ c & 0 \end{pmatrix} \quad \text{and} \quad \Gamma_{\sigma,\tau}(c) = N_{\infty} h_c N_{\infty} \cup g_{\sigma}^{-1}\Gamma g_{\tau}.$$

The group Δ_{σ} acts on $F_{\sigma,\tau}(c)$ by left multiplication and similarly Δ_{τ} acts on the right. Hence we may decompose this set into joint double cosets

$$\Gamma_{\sigma,\tau}(c) = \bigcup_i \Delta_{\sigma} \gamma_i \Delta_{\tau},$$

where each double coset representative γ_i can be written uniquely as $\gamma_i = n(x_i) h_c n(y_i)$ with $n(x_i)$ determined mod Δ_{σ} and $n(y_i)$ determined mod Δ_{τ}.

LEMMA 1. *If* $\gamma = n(x)h_c n(y) \in \Gamma_{\sigma,\tau}(c)$, *then* $n(x)$ *determines* $n(y)$ *uniquely* mod Δ_{τ}.

PROOF. Suppose $\gamma_i = n(x)h_c n(y_i) \in \Gamma_{\sigma,\tau}(c)$ for $i = 1, 2$. Then $\gamma_1^{-1}\gamma_2 = n(y_2 - y_1) \in g_{\tau}^{-1}\Gamma g_{\tau} \cap N_{\infty} = \Delta_{\tau}$. Therefore $n(y_2) \equiv n(y_1)$ mod Δ_{τ}. $\quad\square$

LEMMA 2. *For* $c \in \Omega_{\sigma,\tau}(\Gamma)$, $\Gamma_{\sigma,\tau}(c)$ *decomposes into a finite number of* $\Delta_{\sigma} - \Delta_{\tau}$ *double cosets.*

PROOF. Let $\gamma = n(x)h_c n(y) \in \Gamma_{\sigma,\tau}(c)$. Lemma 1 gives that $n(x)$ determines $n(y)$ mod Δ_{τ}. So to prove that there are only a finite number of double cosets occurring in $\Gamma_{\sigma,\tau}(c)$ it is enough to show that there are only a finite number of $n(x)$ that occur mod Δ_{σ}.

So suppose $\gamma_i = n(x_i)h_c n(y_i) \in \Gamma_{\sigma,\tau}(c)$ for $i = 1, 2$. Then

$$\gamma_1^{-1}\gamma_2 = \begin{pmatrix} * & * \\ c^{-2}(x_1 - x_2) & * \end{pmatrix} \in g_{\tau}^{-1}\Gamma g_{\tau}.$$

Since $\Delta_{\tau} \backslash N_{\infty}$ is compact, then standard arguments give that there exists a constant $d > 0$ such that if $\begin{pmatrix} a_1 & a_2 \\ a_3 & a_4 \end{pmatrix} \in g_{\tau}^{-1}\Gamma g_{\tau}$ with $a_3 \neq 0$ then $|a_3| \geq d$. So if $x_1 \neq x_2$, then there is a constant d such that $|x_1 - x_2| \geq d|c|^{-2}$. Therefore the set of $n(x)$ occurring in $\gamma \in \Gamma_{\sigma,\tau}(c)$ are discrete in N_{∞}. Since $\Delta_{\sigma} \backslash N_{\infty}$ is compact, there can only be a finite number of $n(x)$ mod Δ_{σ}. $\quad\square$

3.2. We are now ready to define the Kloosterman sums for Γ. Take two cusps σ, τ of Γ and two additive characters ψ_1, ψ_2 of K such that, when viewed as characters of N_{∞}, ψ_1 is trivial on Δ_{σ} and ψ_2 is trivial on Δ_{τ}.

DEFINITION. For $c \in \Omega_{\sigma,\tau}(\Gamma)$, the Kloosterman sum for Γ attached to ψ_1, ψ_2 as above is

$$Kl_{\sigma,\tau}(\psi_1, \psi_2; c) = \sum_{\substack{\gamma_i \in \Delta_{\sigma}\backslash\Gamma_{\sigma,\tau}(c)/\Delta_{\tau} \\ \gamma_i = n(x_i)h_c n(y_i)}} \psi_1(x_i)\psi_2(y_i).$$

By Lemma 2 we know that this is indeed a finite sum.

EXAMPLE. If we take $K = \mathbf{R}$, $\Gamma = \mathrm{SL}_2(\mathbf{Z})$, and $\sigma = \tau = \infty$, then $\Omega_{\sigma,\tau}(\Gamma) = \mathbf{Z}$ and $\Delta_\sigma = \Delta_\tau = \{n(x)|x \in \mathbf{Z}\}$. Fix $m, n \in \mathbf{Z}$ and let $\psi_1(x) = e^{2\pi m x i}$ and $\psi_2(y) = e^{2\pi i n y}$. Then

$$Kl_{\infty,\infty}(\psi_1, \psi_2; c) = S(m, n; c) = \sum_{\substack{u,v \bmod c \\ uv \equiv 1 \bmod c}} e^{2\pi i ((mu+nv)/c)},$$

which is the classical Kloosterman sum as in [7, 13]. (Note that in comparison with [13], we always take the trivial multiplier system.)

Similarly we obtain the Kloosterman sums of Proskurin [11] for $K = \mathbf{R}$ and arbitrary Γ and those of Sarnak [12] for $K = \mathbf{C}$.

The Kloosterman sums are related to the existence of Maass cusp forms through the Kloosterman-Selberg zeta function:

$$Z_{\sigma,\tau}(\psi_1, \psi_2; s) = \sum_{c \in \Omega_{\sigma,\tau}(\Gamma)} \frac{Kl_{\sigma,\tau}(\psi_1, \psi_2; c)}{|c|^{2s}}.$$

This series absolutely converges for $\mathrm{Re}(s) > 1$ and has a meromorphic continuation to the whole s-plane. The location of the poles of this function are related to the eigenvalues of the Laplacian Δ on $L^2(\Gamma \backslash H)$.

Before we give the location of the poles of $Z_{\sigma,\tau}(\psi_1, \psi_2; s)$, let us first briefly indicate how the relationship between $Z_{\sigma,\tau}(\psi_1, \psi_2; s)$ and the spectrum of Δ comes about. To each cusp σ of Γ and each character ψ_1 of $\Gamma_\sigma \backslash N_\sigma$ there is attached a special automorphic function in $L_2(\Gamma \backslash H)$, the Poincaré series $U_{\sigma,\psi_1}(z, s)$, depending on a complex parameter s. For $K = \mathbf{R}$ these were introduced by Selberg in [13]. If we compute the Fourier expansion of $U_{\sigma,\psi_1}(z, s)$ at the cusp τ of Γ, then for the character ψ_2 of $\Gamma_\tau \backslash N_\tau$ the ψ_2-Fourier coefficient will involve the zeta function $Z_{\sigma,\tau}(\psi_1, \psi_2; s)$. On the other hand, $U_{\sigma,\psi_1}(z, s)$ has a spectral decomposition in terms of the discrete and continuous eigenfunctions of Δ on $L^2(\Gamma \backslash H)$. Selberg obtained the relationship between the poles of $Z_{\sigma,\tau}(\psi_1, \psi_2; s)$ and the eigenvalues of the Laplacian by comparing the natural Fourier coefficients of $U_{\sigma,\psi_1}(z, s)$ with those computed from the spectral decomposition of $U_{\sigma,\psi_1}(z, s)$. Kuznetsov [7] recently has given another approach for $\Gamma = \mathrm{SL}_2(\mathbf{Z})$ by computing the inner products $\langle U_{\sigma,\psi_1}(\cdot, s), U_{\tau,\psi_2}(\cdot, s) \rangle$ of two Poincaré series in two different ways, one involving the Kloosterman sums and the other involving the spectral decompositions, to obtain the so-called "Kuznetsov trace formula" and from there deduced information on the poles of $Z_{\sigma,\tau}(\psi_1, \psi_2; s)$. (His method was generalized to arbitrary Fuchsian groups of the first kind by Proskurin [11].)

3.3. Let us now describe the location of the poles of the Kloosterman-Selberg zeta function $Z_{\sigma,\tau}(\psi_1, \psi_2; s)$ in the $K = \mathbf{R}$ following Selberg [13]. Let $\{u_j | j = 0, 1, \ldots\}$ be an orthonormal system of eigenfunctions for Δ on $L_\gamma^2(\Gamma \backslash \mathscr{H}) \oplus L_0^2(\Gamma \backslash \mathscr{H})$ with $\Delta u_j = \lambda_j u_j$. The eigenvalues for $u_j \in L_\tau^2(\Gamma \backslash \mathscr{H})$ all lie in the interval $[0, 1/4)$. The eigenvalues occurring in $L_0^2(\Gamma \backslash \mathscr{H})$ will lie on the half-line $(0, \infty)$. There will be at most a finite number of cuspidal eigenvalues lying in $(0, 1/4)$, called exceptional eigenvalues. The Selberg conjecture is then equivalent to there being an infinite number of cuspidal eigenvalues in the half line $[1/4, \infty)$.

For each eigenvalue λ_j set $K_j = \sqrt{\lambda_j - 1/4}$. If λ_j is an eigenvalue for $U_j \in L_\tau^2(\Gamma \backslash \mathscr{H})$ or an exceptional cuspidal eigenvalue, then K_j will be purely imaginary. If λ_j is a cuspidal eigenvalue occurring in $[1/4, \infty)$ then K_j will be real. The poles of $Z_{\sigma,\tau}(\psi_1, \psi_2; s)$ are then of the following types:

(i) Trivial poles occurring at the poles of $\Gamma(s)$, i.e., at $s = 0, -1, -2, \ldots$.

(ii) Poles coming from the cuspidal spectrum. These poles will be located at the points

$$s = \tfrac{1}{2} \pm iK_j - l, \qquad l = 0, 1, \ldots,$$

for $u_j \in L_0^2(\Gamma \backslash \mathscr{H})$. u_j will contribute a series of poles to $Z_{\sigma,\tau}(\psi_1, \psi_2; s)$ if it has a nonzero ψ_1-Fourier coefficient at the cusp σ and nonzero ψ_2-Fourier coefficient at the cusp τ. Note that each exceptional eigenfunction contributes an infinite sequence of poles that begin on the real line, while eigenfunctions u_j with $\lambda_j \leq 1/4$ will contribute an infinite sequence of poles that begin on the line $\mathrm{Re}(s) = 1/2$.

(iii) Poles coming from the residual spectrum, also located at the points

$$s = \tfrac{1}{2} \pm iK_j - l, \qquad l = 0, 1, \ldots,$$

for $u_j \in L_\tau^2(\Gamma \backslash \mathscr{H})$ with the same conditions as above. Note that for the residual spectrum the poles all lie on the real axis.

(iv) Contributions from the continuous spectrum. These poles all lie to the left of the line $\mathrm{Re}(s) = 1/2$. In the case of $\Gamma = \mathrm{SL}_2(\mathbf{Z})$, Kuznetsov has determined that the continuous spectrum contributes poles at the zeros of $\zeta(2s + 2l)\zeta(2 - 2s - 2l)$ for $l = 0, 1, \ldots$, which occur in the half-plane $\mathrm{Re}(s) < 1/2$ [7].

From this list we see that the only poles of $Z_{\sigma,\tau}(\psi_1, \psi_2; s)$ on the line $\mathrm{Re}(s) = 1/2$ come from the cuspidal spectrum. The Selberg conjecture is now equivalent to the zeta functions $Z_{\sigma,\tau}(\psi_1, \psi_2; s)$ having an infinite number of distinct poles on the line $\mathrm{Re}(s) = 1/2$ as the cusps σ, τ and the characters ψ_1, ψ_2 vary.

3.4. Now take $K = F_q((t))$ to be the field of formal power series in the indeterminant t over a field with q elements. We will restrict our attention to arithmetic Γ. So let Y be a smooth curve over a finite field F, and let $k = F(Y)$ denote its field of functions. Let v_0 be a point of Y such that the completion k_{v_0} of k at v_0 is K. Let A be the affine ring of the k consisting of functions that are regular outside of v_0. Then we take $\Gamma = \mathrm{SL}_2(A)$, which is discrete in $\mathrm{SL}_2(K)$ and has $\Gamma \backslash H$ of finite volume.

In this case we have established the analogous relation between the poles of $Z_{\sigma,\tau}(\psi_1, \psi_2; s)$ and the spectrum of Δ on $L^2(\Gamma \backslash H)$. The cuspidal spectrum will only give poles of $Z_{\sigma,\tau}(\psi_1, \psi_2; s)$ which lie on the line $\mathrm{Re}(s) = 1/2$ (thanks to Drinfeld's proof of the Ramanujan-Petersson conjecture for function fields [4]). The continuous spectrum will now produce poles related to the zeros of the zeta function of k and will still lie to the left of $\mathrm{Re}(s) = 1/2$.

In the function field case, the analogue of the Selberg conjecture is false. Drinfeld has shown that if the curve Y has genus 0 then there are no cusp forms in $L^2(\Gamma\backslash H)$, whereas if the genus of Y is greater than 0 and the residue field Fq is large enough, then $L^2(\Gamma\backslash H)$ will always have cusp forms.

ACKNOWLEDGMENT. I am grateful to J. Bernstein for very inspiring discussions. My warmest thanks to J. Cogdell, who made notes of this talk.

REFERENCES

1. Y. Colin de Verdiere, *Pseudo-Laplacians*. I, II, Ann. Inst. Fourier (Grenoble) **32** (1983), 275–268; **33** (1983), 87–113.

2. P. Deligne and G. Lusztig, *Representations of reductive groups over finite fields*, Ann. of Math. (2) **103** (1976), 103–161.

3. J. -M. Deshouillers, H. Iwaniec, R. S. Phillips, and P. Sarnack, *Maass cusp forms*, Proc. Nat. Acad. Sci. U. S. A. **82** (1985), 3533–3534.

4. V. G. Drinfeld, *Langlands conjecture for GL(2) over function fields*, Proc. Internat. Congr. Math. (Helsinki, 1978), Vol. 2, Acad. Sci. Fenn., Helsinki, 1980, pp. 565–574.

5. H. Garland and M. S. Raghunathan, *Fundamental domains for lattices in rank one semisimple Lie groups*, Ann. of Math. (2) **92** (1970), 279–326.

6. T. Kubota, *Elementary theory of Eisenstein series*, Kondasha, Tokyo, and Wiley, New York, 1973.

7. N. V. Kuznetsov, *Petersson's conjecture for cusp forms of weight zero and Linnik's conjecture; Sums of Kloosterman sums*, Math. USSR Sb. **39** (1981), 299–342.

8. G. A. Margulis, *Arithmeticity of the irreducible lattices in the semi-simple groups of rank greater than 1*, Invent. Math. **76** (1984), 93–120.

9. R. S. Phillips and P. Sarnak, *On cusp forms for co-finite subgroups of* PSL(2, **R**), Invent. Math. **80** (1985), 339–364.

10. I. I. Piatetski-Shapiro, *Complex representation of GL(2, K) for finite fields K*, Contemp. Math., no. 16, Amer. Math. Soc., Providence, R. I., 1983.

11. N. V. Proskurin, *Summation formulas for general Kloosterman sums*, J. Soviet Math. **18** (1982), 925–950.

12. P. Sarnak, *The arithmetic and geometry of some hyperbolic three manifolds*, Acta Math. **151** (1983), 243–295.

13. A. Selberg, *On the estimation of Fourier coefficients of modular forms*, Number Theory, Proc. Sympos. Pure Math., vol. 8, Amer. Math. Soc., Providence, R. I., 1965.

14. J. -P. Serre, *Trees*, Springer-Verlag, 1980.

15. A. B. Venkov, *Spectral theory of automorphic functions*, Proc. Steklov Inst. Math. **4** (1982), 163.

YALE AND TEL-AVIV UNIVERSITY

Combinatorics

Proceedings of Symposia in Pure Mathematics
Volume **49** (1989), Part 2

On the Complex Selberg Integral

K. AOMOTO

We consider the following integral over the complex plane:

$$(1) \quad F = \int \prod_{1 \le j < k \le N} |z_j - z_k|^{2\lambda_{j,k}} \left(\frac{i}{2}\right)^n \cdot dz_{m+1} \wedge d\bar{z}_{m+1} \wedge \cdots \wedge dz_N \wedge d\bar{z}_N,$$

for $0 \le m < N$, $(z_1, \ldots, z_m) \in \mathbf{C}^m$ and $\lambda_{j,k} \in \mathbf{R}$. The domain of integration is the product of n copies of $\mathbf{C}$ for $n = N - m$. Let Φ be a multiplicative function $\prod_{1 \le j < k \le N} (z_j - z_k)^{\lambda_{j,k}}$ defined by the logarithmic form $\omega = \sum_{1 \le j < k \le N} \lambda_{j,k} d \log(z_j - z_k)$ on the affine space $\mathbf{C}^N$. We denote by $V_{m,N}$ the space of points $(z_{m+1}, \ldots, z_N) \in \mathbf{C}^n$ such that $z_j \ne z_k$ for $1 \le j \le m$, $m + 1 \le k \le N$, and $m + 1 \le j, k \le N$, $j \ne k$. The n-dimensional twisted de Rham cohomology $H^n(V_{m,N}, \nabla_\omega)$ on $V_{m,N}$ is isomorphic to $\mathbf{C}^l$ for $l = (N - 2)(N - 3) \cdots (N - n - 1)$ and spanned by logarithmic forms

$$(2) \qquad\qquad d \log(i_1, j_1) \wedge \cdots \wedge d \log(i_n, j_n)$$

such that $(i_1, j_1), \ldots, (i_n, j_n)$ are linearly independent, where (i, j) denotes $x_i - x_j$, provided $\lambda_{j,k}$ are generic (see [A1]).

Holonomic linear differential and difference structures have been investigated in [A1]. The differential equations are expressed in classical Yang-Baxter equations, which are infinitesimal versions of pure braid transformations (see [K]).

We assume further that $\lambda_{j,k} = \lambda_j$ for $1 \le j \le m$ and $m + 1 \le k \le N$ and that $\lambda_{j,k} = \lambda$ for $m + 1 \le j < k \le N$. Since ω is symmetric in the arguments $z_{m+1}, \ldots, z_N$, the cohomology $H^n(V_{m,N}, \nabla_\omega)$ can be decomposed into irreducible $\mathfrak{S}_n$-modules. In particular the trivial part $\{H^n(V_{m,N}, \nabla_\omega)\}^{\mathfrak{S}_n}$ has dimension $(N - 2) \cdots (N - n - 1)/n!$.

When $m = 2$, it is equal to 1. This case is well known as the Selberg integral. It is expressed as products of Γ-functions (see [As] and [A2]).

The cohomology associated with the above integral has a product structure of $H^n(V_{m,N}, \nabla_\omega)$ and its complex conjugate:

$$(3) \qquad H^{2n}(V_{m,N} \times \overline{V}_{m,N}, \nabla_{\omega + \bar{\omega}}) \simeq H^n(V_{m,N}, \nabla_\omega) \otimes \overline{H^n(V_{m,N}, \nabla_\omega)}.$$

1980 *Mathematics Subject Classification* (1985 *Revision*). Primary 58F07.

Suppose that $m = 2$ and

(4) $\lambda_{1,k} = \lambda_1,\quad \lambda_{2,k} = \lambda_2$ and $\lambda_{j,k} = \lambda$ for $1 \leq j < k \leq N$.

Then (1) becomes very simple and can be expressed as a product of Γ-functions (see [A3] and also [D1]):

(5)
$$\prod_{j=1}^{n} \frac{\sin\left(\lambda_1 + \frac{j-1}{2}\lambda\right)\sin\left(\lambda_2 + \frac{j-1}{2}\lambda\right)\sin\frac{j\lambda}{2}\pi}{n!\sin\left(\lambda_1 + \lambda_2 + \frac{n+j-2}{2}\lambda\right)\sin\frac{\lambda}{2}\pi}$$
$$\cdot\left\{\prod_{j=1}^{n} \frac{\Gamma(\lambda_1 + 1 + \frac{j-1}{2}\lambda)\Gamma(\lambda_2 + 1 + \frac{j-1}{2}\lambda)\Gamma\left(1 + \frac{j\lambda}{2}\right)}{\Gamma\left(\lambda_1 + \lambda_2 + 2 + \frac{n+j-2}{2}\lambda\right)\Gamma\left(1 + \frac{\lambda}{2}\right)}\right\}^2.$$

The essential difference from the square of the Selberg integrals is a factor of monodromy, which is naturally given by the local system associated with the multiplicative function $\Phi(z) \cdot \overline{\Phi(z)}$ on $V_{m,N} \times \overline{V}_{m,N}$. It is known that this appears in the 2-dimensional conformal field theory (see [D1–D3]).

Under the condition (4) it seems interesting to study correlation functions as functions of n, too. We can show as in [A4] that the integral over $\mathbf{C}^n$, for $t_1, \ldots, t_m \in \mathbf{C}$,

(6)
$$\int \prod_{\substack{1 \leq j \leq n \\ 1 \leq \alpha \leq m}} |z_j - t_\alpha|^2 \prod |z_j|^{\lambda_1}|z_j - 1|^{2\lambda_2}$$
$$\cdot \prod_{1 \leq j < k \leq n} |z_j - z_k|^{2\lambda}\left(\frac{i}{2}\right)^n \cdot dz_1 \wedge d\overline{z}_1 \wedge \cdots \wedge dz_n \wedge d\overline{z}_n$$

satisfies a linear difference equation in n over the coefficients of polynomials. But this is rather complicated to write down in general form.

Prof. I. M. Gelfand and their collaborators have developed the theory of hypergeometric functions using the notion of "matroid." This seems to give a new insight to study the cohomology $H^n(V_{m,N}, \nabla_\omega)$ and related integrals.

REFERENCES

[A1] K. Aomoto, *Gauss-Manin connection of integral of difference products*, J. Math. Soc. Japan **39** (1987), 191–207.

[A2] ——, *Jacobi polynomials associated with Selberg integrals*, Siam. J. Math. Anal. **18** (1987), 545–549.

[A3] ——, *The complex Selberg integral*, Quart. J. Math. **38** (1987), 1–15.

[A4] ——, *Correlation functions of the Selberg integral*, Ramanujan Revisited, Acad. Press, 1988, 591–605.

[As] ——, *Some basic hypergeometric extensions of integrals of Selberg and Andrews*, SIAM J. Math. Anal. **11** (1980), 938–951.

[D1] V. Dotsenko and V. Fateev, *Conformal algebra and multipoint correlation functions in 2D statistical models*, Nuclear Phys. B **240** (1984), 312–348.

[D2] ——, *Four-point correlation functions and the operator algebra in 2D conformal invariant theories with central charge $C \leq 1$*, Nuclear Phys. B **251** (1985), 691–734.

[G1] I. M. Gelfand, *General theory of hypergeometric functions*, Dokl. Acad. Nauk SSSR **288** (1986), 573–577.

[G2] I. M. Gelfand and A. V. Zelevinskii, *Algebraic and combinatorial aspects of the general theory of hypergeometric functions*, Functional Anal. Appl. **20** (1986), 183–197.

[K] T. Kohno, *Série de Poincaré-Koszul associée aux groupes de tresse pures*, Invent. Math. **82** (1985), 57–75.

NAGOYA UNIVERSITY, JAPAN

Proceedings of Symposia in Pure Mathematics
Volume **49** (1989), Part 2

Mock Theta Functions

GEORGE E. ANDREWS

1. Introduction. The mock theta functions are the subject of Ramanujan's last letter to Hardy dated January, 1920. The mathematical portions of this letter have been reproduced in [25, pp. 127–131] (see also [24, pp. 354–355] and [29, pp. 56–61]), and we repeat them here at the beginning to lay the groundwork for this survey.

"If we consider a ϑ-function in the transformed Eulerian form, e.g.,

$$(A) \quad 1 + \frac{q}{(1-q)^2} + \frac{q^4}{(1-q)^2(1-q^2)^2} + \frac{q^9}{(1-q)^2(1-q^2)^2(1-q^3)^2} + \cdots,$$

$$(B) \quad 1 + \frac{q}{1-q} + \frac{q^4}{(1-q)(1-q^2)} + \frac{q^9}{(1-q)(1-q^2)(1-q^3)} + \cdots,$$

and determine the nature of the singularities at the points

$$q = 1, q^2 = 1, q^3 = 1, q^4 = 1, q^5 = 1, \ldots,$$

we know how beautifully the asymptotic form of the function can be expressed in a very neat and closed exponential form. For instance, when $q = e^{-t}$ and $t \to 0$,

$$(A) = \sqrt{\left(\frac{t}{2\pi}\right)} \exp\left(\frac{\pi^2}{6t} - \frac{t}{24}\right)^{*} + o(1)^{\dagger},$$

$$(B) = \sqrt{\left(\frac{2}{5-\sqrt{5}}\right)} \exp\left(\frac{\pi^2}{15t} - \frac{t}{60}\right)^{*} + o(1),$$

and similar results at other singularities.

"If we take a number of functions like (A) and (B), it is only in a limited number of cases the terms close as above; but in the majority of cases they

---

1980 *Mathematics Subject Classification* (*1985 Revision*). Primary 05A30, 11P68, and 33A35.

*It is not necessary that there should be only one term like this. There may be many terms but *the number of terms must be finite.*

†Also $o(1)$ may turn out to be $O(1)$. That is all. For instance, when $q \to 1$, the function $\{(1-q)(1-q^2)(1-q^3)\cdots\}^{-120}$ is equivalent to the sum of five terms like (*) together with $O(1)$ instead of $o(1)$.

never close as above. For instance, when $q = e^{-t}$ and $t \to 0$,

(C) $1 + \dfrac{q}{(1 - q^2)} + \dfrac{q^3}{(1 - q^2)(1 - q^2)^2} + \dfrac{q^6}{(1 - q^2)(1 - q^2)^2(1 - q^3)^2} + \cdots$

$$= \sqrt{\left(\dfrac{t}{2\pi\sqrt{5}}\right)} \exp\left[\dfrac{\pi^2}{5t}^{**} + a_1^{\dagger\dagger} t + a_2 t^2 + \cdots + O(a_k t^k)\right],$$

where $a_1 = 1/8\sqrt{5}$, and so on. The function (C) is a simple example of a function behaving in an unclosed form at the singularities.

"Now a very interesting question arises. Is the converse of the statements concerning the forms (A) and (B) true? That is to say: Suppose there is a function in the Eulerian form and suppose that all or an infinity of points are exponential singularities, and also suppose that at these points the asymptotic form of the function closes as neatly as in the cases of (A) and (B). The questions is: Is the function taken the sum of two functions one of which is an ordinary ϑ-function and the other a (trivial) function which is $O(1)$ at all the ponts $e^{2m\pi i/n}$? The answer is *it is not necessarily so.* When it is not so, I call the function a Mock ϑ-function. I have not proved rigorously that *it is not necessarily so.* But I have constructed a number of examples in which it is inconceivable to construct a ϑ-function to cut out the singularities of the original function. Also I have shown that if *it is necessarily so* then it leads to the following assertion—viz. it is possible to construct two power series in x, namely $\sum a_n x^n$ and $\sum b_n x^n$, both of which have *essential singularities* on the unit circle, are convergent when $|x| < 1$, and tend to *finite limits* at *every point* $x = e^{2r\pi i/s}$, and that at the same time the limit of $\sum a_n x^n$ at the point $x = e^{2r\pi i/s}$ is equal to the limit of $\sum b_n x^n$ at the point $x = e^{-2r\pi i/s}$.

"This assertion seems to be untrue. [H. Cohen, B. Gordon, and D. Hickerson have each pointed out to me that Ramanujan is incorrect; indeed Cohen's function $\varphi(q)$ (see (6.2) below) provides a counterexample with $\varphi(x^{24})$ and $-\varphi(x^{24})$.] Anyhow, we shall go to the examples and see how far our assertions are true.

"I have proved that, if

$$f(q) = 1 + \dfrac{q}{(1 + q^2)} + \dfrac{q^4}{(1 + q)^2(1 + q^2)^2} + \cdots,$$

then

$$f(q) + (1 - q)(1 - q^3)(1 - q^5) \cdots (1 - 2q + 2q^4 - 2q^4 + \cdots) = O(1)$$

at all the points $q = -1$, $q^3 = -1$, $q^5 = -1$, $q^7 = -1, \cdots$; and at the same time

$$f(q) - (1 - q)(1 - q^3)(1 - q^5) \cdots (1 - 2q + 2q^4 - 2q^9 \cdots) = O(1)$$

**The coefficient $1/t$ (sic) in the index of e happens to be $\pi^2/5$ in this particular case. It may be some other transcendental numbers in other cases.

$\dagger\dagger$The coefficients of $t, t^2, \ldots$ happen to be $1/8\sqrt{5}, \ldots$ in this case. In other cases they may turn out to be some other algebraic numbers.

at all the points $q^2 = -1, q^4 = -1, q^6 = -1, \ldots$. Also, obviously, $f(q) = O(1)$ at all the points $q = 1, q^3 = 1, q^5 = 1, \ldots$. And so $f(q)$ is a Mock ϑ-function.

"When $q = -e^{-t}$ and $t \to 0$,

$$f(q) + \sqrt{\left(\frac{\pi}{t}\right)} \exp\left(\frac{\pi^2}{24t} - \frac{t}{24}\right) \to 4.$$

"The coefficient of q^n in $f(q)$ is

$$(-1)^{n-1} \frac{\exp\left\{\pi\sqrt{\left(\frac{1}{6}n - \frac{1}{144}\right)}\right\}}{2\sqrt{\left(n - \frac{1}{24}\right)}} + O\left(\frac{\exp\left\{\frac{1}{2}\pi\sqrt{\left(\frac{1}{6}n - \frac{1}{144}\right)}\right\}}{\sqrt{\left(n - \frac{1}{24}\right)}}\right).$$

It is inconceivable that a single ϑ-function could be found to cut out the singularities of $f(q)$.

Mock ϑ-functions.

$$\phi(q) = 1 + \frac{q}{1+q^2} + \frac{q^4}{(1+q^2)(1+q^4)} + \cdots,$$

$$\psi(q) = \frac{q}{1-q} + \frac{q^4}{(1-q)(1-q^3)} + \frac{q^9}{(1-q)(1-q^3)(1-q^5)} + \cdots,$$

$$\chi(q) = 1 + \frac{q}{1-q+q^2} + \frac{q^4}{(1-q+q^2)(1-q^2+q^4)} + \cdots.$$

These are related to $f(q)$ as shown below.

$$2\phi(-q) - f(q) = f(q) + 4\psi(-q) = \frac{1 - 2q + 2q^4 - 2q^9 + \cdots}{(1+q)(1+q^2)(1+q^3)\cdots},$$

$$4\chi(q) - f(q) = 3\frac{(1 - 2q^3 + 2q^{12} - \cdots)^2}{(1-q)(1-q^2)(1-q^3)\cdots}.$$

These are of the 3rd order.

Mock ϑ-functions (of 5th order).

$$f(q) = 1 + \frac{q}{1+q} + \frac{q^4}{(1+q)(1+q^2)} + \cdots,$$

$$\phi(q) = 1 + q(1+q) + q^4(1+q)(1+q^3) + q^9(1+q)(1+q^3)(1+q^5) + \cdots,$$

$$\psi(q) = q + q^3(1+q) + q^6(1+q)(1+q^2) + q^{10}(1+q)(1+q^2)(1+q^3) + \cdots,$$

$$\chi(q) = 1 + \frac{q}{1-q^2} + \frac{q^2}{(1-q^3)(1-q^4)} + \frac{q^3}{(1-q^4)(1-q^5)(1-q^6)} + \cdots$$

$$= 1 + \frac{q}{1-q} + \frac{q^3}{(1-q^2)(1-q^3)} + \frac{q^5}{(1-q^3)(1-q^4)(1-q^5)} + \cdots,$$

$$F(q) = 1 + \frac{q^2}{1-q} + \frac{q^8}{(1-q)(1-q^3)} + \cdots,$$

$$\phi(-q) + \chi(q) = 2F(q),$$

$$f(-q) + 2F(q^2) - 2 = \phi(-q^2) + \psi(-q)$$
$$= 2\phi(-q^2) - f(q) = \frac{1 - 2q + 2q^4 - 2q^9 + \cdots}{(1-q)(1-q^4)(1-q^6)(1-q^9)\cdots},$$

$$\psi(q) - F(q^2) + 1 = q\frac{1 + q^2 + q^6 + q^{12} + \cdots}{(1-q^8)(1-q^{12})(1-q^{28})\cdots}.$$

Mock ϑ-functions (of 5th order).

$$f(q) = 1 + \frac{q^2}{1+q} + \frac{q^6}{(1+q)(1+q^2)} + \frac{q^{12}}{(1+q)(1+q^2)(1+q^3)} + \cdots,$$
$$\phi(q) = q + q^4(1+q) + q^9(1+q)(1+q^3) + \cdots,$$
$$\psi(q) = 1 + q(1+q) + q^3(1+q)(1+q^2) + q^6(1+q)(1+q^2)(1+q^3) + \cdots,$$
$$\chi(q) = \frac{1}{1-q} + \frac{q}{(1-q^2)(1-q^3)} + \frac{q^2}{(1-q^3)(1-q^4)(1-q^5)} + \cdots,$$
$$F(q) = \frac{1}{1-q} + \frac{q^4}{(1-q)(1-q^3)} + \frac{q^{12}}{(1-q)(1-q^3)(1-q^5)} + \cdots$$

have got similar relations as above.

Mock ϑ-functions (of 7th order).

$$1 + \frac{q}{1-q^2} + \frac{q^4}{(1-q^3)(1-q^4)} + \frac{q^9}{(1-q^4)(1-q^5)(1-q^6)} + \cdots,$$
$$\frac{q}{1-q} + \frac{q^4}{(1-q^2)(1-q^3)} + \frac{q^9}{(1-q^3)(1-q^4)(1-q^5)} + \cdots,$$
$$\frac{1}{1-q} + \frac{q^2}{(1-q^2)(1-q^3)} + \frac{q^6}{(1-q^3)(1-q^4)(1-q^5)} + \cdots.$$

These are not related to each other."

In this survey we shall try to make clear what has happened to mock theta functions since 1920 including an account of D. R. Hickerson's truly inspiring solution of the Mock Theta Conjectures [21].

2. The Watson-Selberg era. G. N. Watson wrote the first papers to elucidate the mock theta functions [29, 31]. The first of these is Watson's Presidential Address to the London Mathematical Society in 1935. He entitled it "The Final Problem: An Account of the Mock Theta Functions." He explained the title as follows: "I doubt whether a more suitable title could be found for it than the title used by John H. Watson, M. D., for what he imagined to be his final memoir on Sherlock Holmes."

In these two papers, Watson proves most of the assertions found in the letter of Ramanujan. The first paper considers only the third-order functions.

It provides three new mock theta functions not mentioned in the letter:

$$(2.1) \qquad \omega(q) = \sum_{n=0}^{\infty} \frac{q^{2n(n+1)}}{(1-q)^2(1-q^3)^2 \cdots (1-q^{2n+1})^2},$$

$$(2.2) \qquad \nu(q) = \sum_{n=0}^{\infty} \frac{q^{n(n+1)}}{(1+q)(1+q^3)\cdots(1+q^{2n+1})},$$

$$(2.3) \qquad \rho(q) = \sum_{n=0}^{\infty} \frac{q^{2n(n+1)}}{(1+q+q^2)(1+q^3+q^6)\cdots(1+q^{2n+1}+q^{4n+2})}.$$

The bulk of the paper is devoted to the modular transformations of these functions. For example if we let $q = e^{-\alpha}$, $\alpha\beta = \pi^2$, and $q_1 = e^{-\beta}$, then

$$(2.4)$$
$$q^{-1/24}f(q) = 2\left(\frac{2\pi}{\alpha}\right)^{1/2} q_1^{4/3}\omega(q_1^2) + 4\left(\frac{3\alpha}{2\pi}\right)^{1/2}\int_0^{\infty} e^{-3\alpha x^2/2}\frac{\sinh \alpha x}{\sinh(3\alpha x/2)}dx.$$

Indeed [29] is quite a thorough account of the third-order mock theta functions and provides a prototype for the general treatment of the subject. In [31], Watson moves on to the two families of fifth-order mock theta functions. He manages to prove all of Ramanujan's assertions about these functions; however he is unable to find any results like (2.4). Consequently he is unable to establish that, in fact, these functions are indeed new functions not included under Ramanujan's ϑ-function umbrella described in the last few paragraphs of his letter.

Watson's methods were generalized in [1, 3, 4, 5] to prove many extensions of Ramanujan's identities. In the "Lost" Notebook, we find a number of these extensions as well as (2.1), (2.2), and (2.3) and clear indications of how to do (2.4).

The seventh-order functions were mostly neglected by Watson perhaps because Ramanujan makes no positive assertions about them. Watson does briefly and cryptically mention them [29, p. 80], and they clearly are the motivation for his short paper on the dilogarithm [30]. However A. Selberg [28] provides a full account of the behavior of the seventh-order functions near the unit circle. This requires a very adroit comparison of the seventh-order functions with q-series that Selberg [27] had found earlier related to the modulus 7.

3. Asymptotics. Watson chooses not to treat Ramanujan's assertion that: "The coefficient of q^n in $f(q)$ is

$$(-1)^{n-1}\frac{\exp\left\{\pi\sqrt{(\frac{1}{6}n - \frac{1}{144})}\right\}}{2\sqrt{n - \frac{1}{24}}} + O\left(\frac{\exp\left\{\frac{\pi}{2}\sqrt{(\frac{1}{6}n - \frac{1}{144})}\right\}}{\sqrt{n - \frac{1}{24}}}\right)."$$

Watson [29, p. 62] states: "I have not troubled to verify this approximation; it is presumably derivable from the transformation formulae in the manner in which Hardy and Ramanujan [19] obtained the corresponding formula for $p(n)$, the number of partitions of n."

Indeed this is the case, and two of H. Rademacher's students (Dragonette [18] and Andrews [2]) carried out the full Hardy-Ramanujan-Rademacher expansion for these coefficients. The result they obtained was the following:

In the series $f(q) = \sum_{n \geq 0} A(n) q^n$,

$$A(n) = \sum_{0 < k \leq n^{1/2}} \frac{\lambda(k) \exp\left\{ \frac{\pi}{k} \left(\frac{1}{6} n - \frac{1}{144} \right)^{1/2} \right\}}{\sqrt{k \left(n - \frac{1}{24} \right)}} + E(n).$$

Dragonette [18] showed that $\lambda(1) = (-1)^{n-1}/2$, $\lambda(k)$ is a finite exponential sum, and $E(n) = O(n^{1/2} \log n)$. Andrews [2] showed that

$$\lambda(k) = \begin{cases} \frac{1}{2}(-1)^{(k+1)/2} A_{2k}(n), & k \text{ odd}, \\ \frac{1}{2}(-1)^{k/2} A_{2k}\left(n - \frac{k}{2} \right), & k \text{ even}, \end{cases}$$

where $A_k(n)$ is the exponential sum appearing in the Hardy-Ramanujan formula for $p(n)$, and $E(n) = O(n^\varepsilon)$. Numerical computations by Dragonette [18] suggest that $E(n) \to 0$ as $n \to \infty$. Indeed $E(100) = .206$ and $E(200) = -.153$.

4. q-series and indefinite quadratic forms. Within the last few years, significant discoveries have been made that greatly extend our knowledge of the mock theta functions. However these discoveries are really just a beginning.

The basis of these discoveries lies in the method of Bailey Chains [7, p. 278; 12, Chapter 3], which relies on the following result [12, pp. 25–26].

BAILEY'S LEMMA. *If for $n \geq 0$*

$$(4.1) \qquad \beta_n = \sum_{r=0}^{n} \frac{\alpha_r}{(q;q)_{n-r}(aq;q)_{n+r}},$$

then

$$(4.2) \qquad \beta_n' = \sum_{r=0}^{n} \frac{\alpha_r'}{(q;q)_{n-r}(aq;q)_{n+r}},$$

where

$$(4.3) \qquad \alpha_r' = \frac{(\rho_1;q)_r(\rho_2;q)_r(aq/\rho_1\rho_2)^r \alpha_r}{(aq/\rho_1;q)_r(aq/\rho_2;q)_r}$$

and

$$(4.4) \qquad \beta_r' = \sum_{j=0}^{n} \frac{(\rho_1;q)_j(\rho_2;q)_j(aq/\rho_1\rho_2;q)_{n-j}(aq/\rho_1\rho_2)^j \beta_j}{(q;q)_{n-j}(aq/\rho_1;q)_n(aq/\rho_2;q)_n}$$

where $(A; q)_n = (1 - A)(1 - Aq) \cdots (1 - Aq^{n-1})$.

In practice the α_n and β_n are sequences of rational functions in a, q, and other parameters. The power of Bailey's Lemma is that it allows the construction of infinitely many pairs of sequences $(\alpha_n^{(i)}, \beta_n^{(i)})$ merely by iteration because (4.2) is precisely (4.1) with (α_n, β_n) replaced by (α_n', β_n'). Such pairs are called Bailey Pairs, and the sequence of such pairs is called a Bailey Chain.

For our purposes here we note the much simpler instance of Bailey's Lemma when $\rho_1, \rho_2, n \to \infty$ in (4.2), (4.3), and (4.4). Thus if (4.1) holds, then

$$(4.5) \qquad \sum_{j=0}^{\infty} a^j q^{j^2} \beta_j = \frac{1}{(aq; q)_\infty} \sum_{r=0}^{\infty} a^r q^{r^2} \alpha_r.$$

Now Watson [29, p. 64] proved that

$$(4.6) \qquad f(q) \equiv \sum_{n=0}^{\infty} \frac{q^{n^2}}{(-q; q)_n^2} = \frac{1}{(q; q)_\infty} \left(1 + 4 \sum_{n=1}^{\infty} \frac{(-1)^n q^{n(3n+1)/2}}{1 + q^n} \right),$$

and it is not too difficult to show that his identity (which indeed Ramanujan recorded and generalized in his Lost Notebook [25, p. 202, first equation]) is an instance of (4.5) for the Bailey Pair

$$(4.7) \qquad \alpha_n = \begin{cases} 1, & n = 0, \\ \dfrac{4(-1)^n q^{n(n+1)/2}}{1 + q^n}, & n > 0, \end{cases}$$

$$(4.8) \qquad \beta_n = \frac{1}{(-q; q)_n^2}.$$

Watson [31, p. 274] (see also Andrews [10, pp. 113–114]) expressed his doubts about finding anything comparable to (4.6) for the fifth-order mock theta functions. Indeed it was only after IBM's symbolic algebra package SCRATCHPAD was employed in a significant way that comparable results were found for most of the other mock theta functions [10, §3]. For example (using Watson's notation for the fifth-order functions)

$$(4.9) \qquad f_0(q) \equiv \sum_{n=0}^{\infty} \frac{q^{n^2}}{(-q; q)_n}$$

$$= \frac{1}{(q; q)_\infty} \sum_{n \geq 0 \, |j| \leq n} \sum (-1)^j q^{n(5n+1)/2 - j^2} (1 - q^{4n+2}).$$

Similarly for the seventh-order mock theta functions

$$(4.10) \quad \mathscr{F}_1(q) \equiv \sum_{n=1}^{\infty} \frac{q^{n^2}}{(q^n; q)_n}$$

$$= \frac{1}{(q; q)_\infty} \sum_{n=1}^{\infty} (-1)^{n-1} q^{n(3n-1)/2} (1 - q^{2n}) \sum_{j=0}^{n-1} q^{j(n-1-j)},$$

which is a succinct restatement of [10, p. 132, (7.23)] (note that the minimal exponent on q in the nth term is $\approx 7n^2/4$).

Series similar to those on the right-hand sides of (4.9) and (4.10) have arisen previously in the work of Hecke [20], Rogers [26], and Kac-Peterson [23]. The important point is the appearance of an indefinite quadratic form in the exponent of q in the sum.

As predicted in [10, p. 114] such identities have very important applications in subsequent work. In the next section we illustrate perhaps the most striking example by considering a related function also due to Ramanujan but not in his mock theta function list.

5. Partitions with distinct parts. Here we shall consider

$$(5.1) \qquad \sigma(q) = 1 + \sum_{n=1}^{\infty} \frac{q^{n(n+1)/2}}{(1+q)(1+q^2)\cdots(1+q^n)} = \sum_{n=0}^{\infty} S(n)q^n$$

$$\equiv 1 + q - q^2 + 2q^3 + \cdots + 4q^{45} + \cdots + 6q^{1609} + \cdots.$$

This function appears in three identities stated in Ramanujan's Lost Notebook [25, p. 14]. These identities were proved in [9], and in [11] two conjectures were posed for $S(n)$. (The conjectures were proved in [15].)

CONJECTURE 1. $\limsup |S(n)| = +\infty$.

CONJECTURE 2. $S(n) = 0$ for infinitely many n.

Now $S(n)$ has a very simple interpretation in terms of partitions. The rank of a partition is defined as the largest part minus the number of parts. Let $\Delta_i(n)$ denote the number of partitions of n into distinct parts with rank $\equiv i \pmod 2$. Then it is easily shown [11] that

$$(5.2) \qquad\qquad S(n) = \Delta_0(n) - \Delta_1(n).$$

Thus since 3 has two partitions into distinct parts, 3 and $2 + 1$, and since each has even rank, we see that $S(3) = 2 - 0 = 2$.

By application of Bailey's Lemma [15, pp. 392–397], it was shown that

$$(5.3) \qquad \sigma(q) = \sum_{n=0}^{\infty} \sum_{|j|\leq n} (-1)^{n+j} q^{n(3n+1)/2-j^2}(1-q^{2n+1}),$$

a result closely resembling (4.9) and (4.10). From (5.3) it is possible to deduce the following identity [15, p. 392]:

$$(5.4) \qquad \sigma(q) = \sum_{n=0}^{\infty} S(n)q^n = \sum_{n=0}^{\infty} T(24n+1)q^n,$$

where $T(n)$ is an arithmetic function defined as follows: For (positive or negative) integers $m \equiv 1 \pmod{24}$, consider Pell's equation

$$(5.5) \qquad\qquad u^2 - 6v^2 = m.$$

Note that if (u, v) is a solution of this equation, then $u \equiv \pm 1 \pmod 6$ and v is even. We call two solutions (u, v) and (u', v') equivalent if

$$(5.6) \qquad\qquad u' + v'\sqrt{6} = \pm(5 + 2\sqrt{6})^r(u + v\sqrt{6})$$

for some integer r. By induction on $|r|$, it is easy to show that if (u, v) and (u', v') are equivalent, then $u + 3v \equiv \pm(u' + 3v') \pmod{12}$. Let $T(m)$ be the excess of the number of inequivalent solutions of (5.5) with $u + 3v \equiv \pm 1 \pmod{12}$ over the number of them with $u + 3v \equiv \pm 5 \pmod{12}$.

Once $T(m)$ is known, it is then an application of the arithmetic of $Q(\sqrt{6})$ to determine $T(m)$ fully [15, p. 401].

THEOREM 5.1. *Let $m \neq 1$ be an integer $\equiv 1 \pmod 6$. Suppose we write $m = p_1^{e_1} p_2^{e_2} \cdots p_r^{e_r}$, where each p_i is either a prime $\equiv 1 \pmod 6$ or the negative of a prime $\equiv 5 \pmod 6$. Then $T(m) = T(p_1^{e_1}) T(p_2^{e_2}) \cdots T(p_r^{e_r})$, where*

$$(5.7) \quad T(p^e) = \begin{cases} 0 & \text{if } p \not\equiv 1 \pmod{24} \text{ and } e \text{ is odd,} \\ 1 & \text{if } p \equiv 13 \text{ or } 19 \pmod{24} \text{ and } e \text{ is even,} \\ (-1)^{e/2} & \text{if } p \equiv 7 \pmod{24} \text{ and } e \text{ is even,} \\ e + 1 & \text{if } p \equiv 1 \pmod{24} \text{ and } T(p) = 2, \\ (-1)^e(e+1) & \text{if } p \equiv 1 \pmod{24} \text{ and } T(p) = -2. \end{cases}$$

In particular, $T(m) = 0$ if and only if there is some i for which $p_i \not\equiv 1 \pmod{24}$ and e_i is odd.

From this result and (5.4) one may quickly deduce [15, p. 401]

THEOREM 5.2. *$S(n)$ is almost always 0; that is, the set of n for which $S(n) \neq 0$ has density 0. On the other hand, $S(n)$ takes on every integer value infinitely often.*

This result overwhelmingly proves the two conjectures mentioned earlier.

H. Cohen, B. Gordon, and D. Hickerson have each pointed out that $\sigma(q)$ is not a mock theta function according to Ramanujan's description in §1. This is because [9, p. 157, (1.6)]

$$(5.8) \qquad \sigma(q) = 1 + \sum_{n=0}^{\infty} (-1)^n q^{n+1} (q; q)_n;$$

consequently $\sigma(q)$ has a finite limit as $q \to e^{2\pi i m/n}$ radially. Thus $\sigma(q)$ differs from the trivial theta function 0 by a function (namely itself) which is $O(1)$ at all points $e^{2\pi i m/n}$.

6. Cohen's extensions. H. Cohen [17] has extended the results of §6 using algebraic number theory in a very substantial way. Besides the function $\sigma(q)$, the function

$$(6.1) \qquad \sigma^*(q) = 2 \sum_{n \geq 1} \frac{(-1)^n q^{n^2}}{(1-q)(1-q^3) \cdots (1-q^{2n-1})}$$

was treated similarly in [15, pp. 404–405]. Cohen considers

$$(6.2) \qquad \varphi(q) = q^{1/24} \sigma(q) + q^{-1/24} \sigma^*(q) = \sum_{\substack{n \in \mathbf{Z} \\ n \equiv 1 \pmod{24}}} T(n) q^{|n|/24}.$$

He then restates several results of [15] in the following:

THEOREM 6.1. *For an ideal* $a = (\alpha) \subset \mathbb{Z}[\sqrt{6}]$ *coprime to* 6, *where* $\alpha = x + y\sqrt{6}$, *define* χ_1 *by*

(6.3)
$$\chi_1(a) = \begin{cases} i^{yx^{-1}}\left(\dfrac{12}{x}\right) & \text{if } y \text{ is even,} \\[2mm] i^{yx^{-1}+1}\left(\dfrac{12}{x}\right) & \text{if } y \text{ is odd.} \end{cases}$$

Then χ_1 *is a well-defined character of order* 2 *and conductor* $4(3 + \sqrt{6})$ *on ideals of* $\mathbb{Z}[\sqrt{6}]$, *and furthermore, setting as usual* $\chi_1(a) = 0$ *if a is not coprime to* 6, *we have the identity*

$$\varphi(q) = \sum_{a \subset \mathbb{Z}[\sqrt{6}]} \chi_1(a) q^{Na/24}.$$

This result is then embedded elegantly in the algebraic number theory related to the following diagram of number fields:

$$K = \mathbb{Q}(\sqrt{2}, \sqrt{3 + \sqrt{3}})$$
$$|$$
$$B = \mathbb{Q}(\sqrt{2}, \sqrt{3})$$
$$|$$
$$k_1 = \mathbb{Q}(\sqrt{6}) \quad k_2 = \mathbb{Q}(\sqrt{2}) \quad k_3 = \mathbb{Q}(\sqrt{3})$$
$$|$$
$$\mathbb{Q}$$

It is noted [17, p. 410] that $K/\mathbb{Q}$ is a Galois extension, that $\mathrm{Gal}(K/k_2) \cong \mathbb{Z}/4\mathbb{Z}$, that $\mathrm{Gal}(K/k_1) \cong \mathrm{Gal}(K/k_3) \cong (\mathbb{Z}/2\mathbb{Z}) \times (\mathbb{Z}/2\mathbb{Z})$, and consequently that $G = \mathrm{Gal}(K/\mathbb{Q}) \cong D_4$, the dihedral group with 8 elements.

It is now possible to go well beyond Theorem 6.1. Indeed [17, pp. 410–411], "the character χ_1 corresponds to a degree 1 representation of $\mathrm{Gal}(K/k_1)$. By induction to G one sees immediately that one obtains the unique irreducible representation ρ of degree 2 of G. Furthermore, Artin L-functions being preserved by induction, we have

(6.4)
$$L(\rho, s) = L(\chi_1, s) = \sum_{a \subset \mathbb{Z}[\sqrt{6}]} \chi_1(a)(Na)^{-s}.$$

Now ρ, being unique, is also induced by any one of the two characters χ_1, χ_2' of order 4 of $\mathrm{Gal}(K/k_2)$ and by two of the three characters of order 2 of $\mathrm{Gal}(K/k_3)$, say χ_3, χ_3'. Hence we have

(6.5)
$$L(\rho, s) = L(\chi_1, s) = L(\chi_2, s) = L(\chi_2', s) = L(\chi_3, s) = L(\chi_3', s).\text{"}$$

This powerful observation allows the deduction of 2 new combinatorial formulas each for $\sigma(q)$ and $\sigma^*(q)$. E.g.,

(6.6)
$$\sigma^*(q) = \sum_{\substack{|j| \geq |6n+1|/8 \\ j,n \in \mathbb{Z}}} (-1)^j q^{3j^2 - n(3n+1)/2},$$

(6.7)
$$\sigma^*(q) = \sum_{\substack{|j| \geq |6n+1|/6 \\ j,n \in \mathbb{Z}}} (-1)^{n(n+1)/2 + j} q^{2j^2 - n(3n+1)/2}.$$

Beyond the combinatorics, the related L-functions possess intriguing properties.

For any $j = 1, 2, 3$, set

(6.8) $$\Lambda(s) = (1152)^{s/2}\pi^{-s}\Gamma(s/2)^2 L(\chi_j, s).$$

Then Λ can be analytically continued to an entire function of order 1 on $\mathbb{C}$ satisfying the functional equation

(6.9) $$\Lambda(1 - s) = -\Lambda(s).$$

Cohen points out the importance of the factor $\Gamma(s/2)^2$ in (6.8). He notes that theta functions attached to positive definite binary quadratic forms are holomorphic modular forms of weight 1 on some congruence subgroup of the modular group due primarily to the fact that $\Gamma(s)$ itself is the Γ-factor of the associated L-function. However in Theorem 6.1, $\varphi(q)$ is a theta function attached to the indefinite form $x - 6y^2$. In the case of indefinite forms, the Γ-factor is $\Gamma(s/2)\Gamma((s+1)/2)$, $\Gamma(s/2)^2$, or $\Gamma((s+1)/2)^2$ if the infinity type of the character χ is respectively $+-$ (or $-+$), $++$, or $--$. Note that by the duplication formula

$$\Gamma(s/2)\Gamma((s+1)/2) = \sqrt{\pi}2^{1-s}\Gamma(s),$$

so that case 1 is essentially the same situation as the positive definite case. This fact serves to explain those identities found by Hecke connecting theta-type series with indefinite quadratic forms to classical modular forms.

While we have briefly summarized Cohen's contributions to the single example of $\varphi(q)$, it is clear from his paper that the methods apply to many similarly related algebraic number fields, and Cohen describes such examples.

7. **Hickerson's proof of the Mock Theta Conjectures.** In Ramanujan's "Lost" Notebook [25, pp. 18–20], we find ten important identities for the ten fifth-order mock theta functions. Each of these identities relates a specific fifth-order mock theta function to either

(7.1) $$\Phi(q) = -1 + \sum_{n=0}^{\infty} \frac{q^{5n^2}}{(q; q^5)_{n+1}(q^4; q^5)_n},$$

or

(7.2) $$\Psi(q) = -1 + \sum_{n=0}^{\infty} \frac{q^{5n^2}}{(q^2; q^5)_{n+1}(q^3; q^5)_n}.$$

For example [25, p. 19, fifth equation]

(7.3) $$f_0(q) \equiv \sum_{n=0}^{\infty} \frac{q^{n^2}}{(-q; q)_n} = \frac{(q^5; q^5)_\infty(q^5; q^{10})_\infty}{(q; q^5)_\infty(q^4; q^5)_\infty} - 2\Phi(q^2).$$

In [16], F. Garvan and I show that these ten identities split into two sets of five each and that in each class the five are either true or false together.

Furthermore one identity in each class has an especially simple formulation in terms of partitions. To state these conjectures we require the function $N(b, 5, n)$, the number of partitions of n with rank $\equiv b$ (mod 5).

FIRST MOCK THETA CONJECTURE. $N(1, 5, 5n) - N(0, 5, 5n)$ equals the number of partitions of n with unique smallest part and no parts exceeding the double of the smallest part.

SECOND MOCK THETA CONJECTURE. $2N(2, 5, 5n + 3) - N(1, 5, 5n + 3) - N(0, 5, 5n + 3) - 1$ equals the number of partitions of n with unique smallest part and all other parts at most one larger than the double of the smallest part.

In [13], the constant term method (see [12, Chapter 4] for background on constant term problems) was first applied to the fifth-order mock theta functions with the following wish [13, p. 48] . "It was our initial hope that by exhibiting the fifth-order mock theta functions as constant terms we could make some progress on the Mock Theta Conjectures described in [16]. So far the Mock Theta Conjectures remain unresolved."

Recently, D. R. Hickerson [21] proved much more explicit and powerful constant term identities than those in [13]. From his new method and discoveries he was able to prove the Mock Theta Conjectures.

Hickerson's proof rests on two different dissections of the function

$$(7.4) \qquad B(z) = \frac{z^2(-z, -q/z, q; q)_\infty (z, q^3/z, q^3; q^3)_\infty}{(z, q^2/z; q^2)_\infty},$$

where

$$(A_1, A_2, \ldots, A_r; q)_\infty = (A_1; q)_\infty (A_2; q)_\infty \cdots (A_r; q)_\infty.$$

Greatly extending the sorts of expansions considered in §4, he shows that

$$(7.5)$$

$$
\begin{aligned}
B(z) = q\, f_0(q) &\left[\sum_{\lambda=-\infty}^{\infty} (-1)^\lambda z^{5\lambda+1} q^{15\lambda^2-9\lambda} + \sum_{\lambda=-\infty}^{\infty} (-1)^\lambda z^{5\lambda+4} q^{15\lambda^2+9\lambda} \right] \\
+ f_1(q) &\left[\sum_{\lambda=-\infty}^{\infty} (-1)^\lambda z^{5\lambda+2} q^{15\lambda^2-3\lambda} + \sum_{\lambda=-\infty}^{\infty} (-1)^\lambda z^{5\lambda+3} q^{15\lambda^2+3\lambda} \right] \\
+ 2 \sum_{r=-\infty}^{\infty} &\frac{(-1)^r q^{15r^2+15r+3} z^{5r+5}}{1 - q^{6r+2}z} + 2 \sum_{r=-\infty}^{\infty} \frac{(-1)^r q^{15r^2+15r+3} z^{-5r}}{1 - q^{6r+2}z^{-1}},
\end{aligned}
$$

where

$$(7.6) \qquad f_1(q) = \sum_{n=0}^{\infty} \frac{q^{n^2+n}}{(-q; q)_n}$$

is a second fifth-order mock theta function.

From the form of (7.5) it is clear that one wants

$$(7.7) \qquad B(z) = \sum_{i=0}^{4} z^i B_i(z^5).$$

In particular, it is a straightforward deduction from (7.5) to see that $f_0(q)$ will be directly involved in the constant term for $B_1(z^5)$ that $f_1(z)$ will arise similarly in $B_2(z^5)$. For example, Hickerson derives

$$B_1(z^5) = q f_0(q) \sum_{\lambda=-\infty}^{\infty} (-1)^\lambda q^{15\lambda^2-9\lambda} z^{5\lambda}$$

(7.8)
$$+ 2 \sum_{r=-\infty}^{\infty} \frac{(-1)^r q^{15r^2+21r+5} z^{5r+5}}{1 - q^{30r+10} z^5}$$

$$+ 2 \sum_{r=-\infty}^{\infty} \frac{(-1)^r q^{15r^2+39r+11} z^{-5r-5}}{1 - q^{30r+10} z^{-5}}.$$

On the other hand, Hickerson uses (7.4) to find a pure theta function expansion for $B_1(z^5)$:

(7.9)

$$B_1(z^5) = \frac{q(q^5, q^5, q^{10}; q^{10})_\infty (q^2, q^3, q^5; q^5)_\infty \sum_{\lambda=-\infty}^{\infty} (-1)^\lambda q^{15\lambda^2-9\lambda} z^{5\lambda}}{(q;q)_\infty}$$

$$- \frac{2q^3 (q^{10}; q^{10})_\infty^2 \sum_{\lambda=-\infty}^{\infty} (-1)^\lambda q^{5\lambda^2-3\lambda} z^{5\lambda} \sum_{\mu=-\infty}^{\infty} (-1)^\mu q^{15\mu^2-15\mu} Z^{5\mu}}{(q^2, q^8, q^{10}; q^{10})_\infty \sum_{\nu=-\infty}^{\infty} (-1)^\nu q^{5\nu^2-5\nu} z^{5\nu}}.$$

In order to pick out the constant term in comparing (7.8) and (7.9), substantial work remains. Indeed Hickerson accomplishes this with a partial fractions type decomposition of the second term in (7.9). It is only then that he is able to read off (7.3), a result equivalent to the first Mock Theta Conjecture. Similar treatment of $B_2(z^5)$ yields the second conjecture as well.

In a second paper [22], Hickerson applies these methods to the seventh-order mock theta functions. Again his approach is totally successful, and he derives analogs of (7.3) for each of the seventh-order mock theta functions.

As was pointed out in [16, §5], the proof of these conjectures and their seventh-order counterparts establishes formulae that will clearly yield the behavior of the fifth- and seventh-order mock theta functions near the unit circle. Furthermore the asymptotic behavior of the resulting Mordell integrals (see (2.4) as an example, also [6]) should clearly establish that these functions just like the third-order functions are truly mock theta functions [29, p. 78, footnote] in the sense described by Watson. Dean Hickerson has noted a discrepancy between Watson's assertion and Ramanujan's original definition. He notes that for a function to be a mock theta function it must be of the form (θ-function)+$O(1)$ at each root of unity, but there must not be a single theta function that works for all roots of unity. Watson only proves that the third-order functions are not equal to θ-functions; that is, the $O(1)$ terms cannot be identically zero.

8. Combinatorics. The mock theta functions are closely allied with generating functions for certain polynomials that have arisen in the study of

partitions [9, 14]. This relationship provides possible combinatorial applications for whatever we learn subsequently about the mock theta functions. In
order to present this in a self-contained manner, we shall restrict ourselves
to three examples:

$$(8.1) \qquad M\theta_3(q,t) \equiv \sum_{n=0}^{\infty} \frac{t^{2n}q^{n^2}}{(t;q)_{n+1}(tq)_n}$$

$$= 1 + \sum_{N=1}^{\infty} t^N \sum_{m=0}^{N-1} q^m \begin{bmatrix} N-1 \\ m \end{bmatrix};$$

$$(8.2) \qquad M\theta_5(q,t) \equiv \sum_{n=0}^{\infty} \frac{t^{2n}q^{n^2}}{(t;q)_{n+1}}$$

$$= \sum_{N=0}^{\infty} t^N \sum_{m=-N}^{N} (-1)^\lambda q^{\lambda(5\lambda+1)/2} \begin{bmatrix} N \\ \left[\frac{N-5\lambda}{2}\right] \end{bmatrix},$$

where

$$(8.3) \qquad \begin{bmatrix} A \\ B \end{bmatrix} = \frac{(1-q^A)(1-q^{A-1})\cdots(1-q^{A-B+1})}{(1-q^B)(1-q^{B-1})\cdots(1-q)},$$

and

$$(8.4) \qquad [x] = \text{the largest integer not exceeding } x;$$

$$(8.5) \qquad M\theta_7(q,t) \equiv \sum_{n=0}^{\infty} \frac{t^{2n}q^{n^2}}{(t;q)_{n+1}(t^2q;q^2)_n}$$

$$= \sum_{N=0}^{\infty} t^N \sum_{m=-N}^{N} (-1)^m q^{m(7m+1)} \begin{bmatrix} N \\ \left[\frac{N-7m}{2}\right] \end{bmatrix}.$$

Each of the polynomials appearing in (8.1), (8.2), and (8.5) as coefficients
of t^N has an interpretation as a generating function for a certain class of
partitions.

In particular, $\sum_{m\geq 0} q^m \begin{bmatrix} N-1 \\ m \end{bmatrix}$ is the generating function for all partitions
wherein the largest part plus the number of parts is at most N. The polyno-
mial in (8.2) is the generating function for partitions with largest part $\leq N$
and difference at least 2 between parts. The polynomial in (8.5) is subject to
a somewhat more complicated interpretation [8, p. 14, (5.17)].

We remark that for $i = 3, 5, 7$

$$(8.6) \qquad \lim_{t\to 1}(1-t)M\theta_i(q,t) = \begin{cases} \left(\sum_{\lambda=-\infty}^{\infty}(-1)^\lambda q^{\lambda(3\lambda-1)/2}\right)^{-1}, & i = 3, \\ \dfrac{\sum_{\lambda=-\infty}^{\infty}(-1)^\lambda q^{\lambda(5\lambda-1)/2}}{\prod_{n=1}^{\infty}(1-q^n)}, & i = 5, \\ \dfrac{\sum_{\lambda=-\infty}^{\infty}(-1)^\lambda q^{\lambda(7\lambda-1)}}{\prod_{n=1}^{\infty}(1-q^n)}, & i = 7, \end{cases}$$

while

$$(8.7) \qquad 2M\theta_i(q,-1) = \begin{cases} f(q) \text{ in } (4.6), & i = 3, \\ f_0(q) \text{ in } (4.9), & i = 5, \\ \text{the first of the 7th-order} & i = 7. \\ \text{mock } \vartheta\text{-functions given in §1,} \end{cases}$$

Thus we see a close tie among mock theta functions in (8.7), classical modular functions in (8.6), and certain polynomial generating functions exemplified by (8.1), (8.2), and (8.5). Most of the mock theta functions can be placed in this sort of three-way relationship [8].

We do not know what more general relationships there are between these combinatorial observations and the work in §§4–7; however the fact that most of the mock theta functions arise as specializations of polynomial generating functions suggests that the study of such relationships may be fruitful.

9. Conclusion. I wish to thank H. Cohen, B. Gordon, and D. Hickerson for helpful conversations and letters that greatly assisted me in the preparation of this paper.

REFERENCES

1. R. P. Agarwal, *Certain basic hypergeometric identities associated with mock theta functions*, Quart J. Math. Oxford Ser. (2) **20** (1969), 121–128.

2. G. E. Andrews, *On the theorems of Watson and Dragonette for Ramanujan's mock theta functions*, Amer. J. Math. **88** (1966), 454–490.

3. ____, *On basic hypergeometric series, mock theta functions, and partitions. I*, Quart. J. Math. Oxford Ser. (2) **17** (1966), 64–80.

4. ____, *On basic hypergeometric series, mock theta functions, and partitions. II*, Quart J. Math. Oxford Ser. (2) **17** (1966), 132–143.

5. ____, *q-Identities of Auluck, Carlitz and Rogers*, Duke Math. J. **33** (1966), 575–582.

6. ____, *The Mordell integrals and Ramanujan's "lost" notebook*, Lecture Notes in Math., vol. 899, Springer-Verlag, 1981, pp. 10–48.

7. ____, *Multiple series Rogers-Ramanujan type identities*, Pacific J. Math. **114** (1984), 267–283.

8. ____, *Combinatorics and Ramanujan's "lost" notebook*, Surveys in Combinatorics 1985, I. Anderson, editor, London Math. Soc. Lecture Note Series, No. 103, Cambridge Univ. Press, 1985, pp. 1–23.

9. ____, *Ramanujan's "lost" notebook. V: Euler's partition identity*, Adv. in Math. **61** (1986), 156–164.

10. ____, *The fifth and seventh order mock theta functions*, Trans. Amer. Math. Soc. **293** (1986), 113–134.

11. ____, *Questions and conjectures in partition theory*, Amer. Math. Monthly **93** (1986), 708–711.

12. ____, *q-Series: Their development and application in analysis, number theory, combinatorics, physics and computer algebra*, CBMS Regional Conf. Ser. in Math., no. 66, Amer. Math. Soc., Providence, R.I., 1986.

13. ____, *Ramanujan's fifth order mock theta functions as constant terms*, Ramanujan Revisited, G. E. Andrews et al., editors, Academic Press, 1988, pp. 47–56.

14. G. E. Andrews, R. J. Baxter, D. Bressoud, W. H. Burge, P. J. Forrester, and G. Viennot, *Partitions with prescribed hook differences*, European J. Combin. **8** (1987), 341–350.

15. G. E. Andrews, F. J. Dyson, and D. R. Hickerson, *Partitions and indefinite quadratic forms*, Invent. Math. **91** (1988), 391–407.

16. G. E. Andrews and F. G. Garvan, *Ramanujan's "lost" notebook*. VI: *The mock theta conjectures*, Adv. in Math. (to appear).

17. H. Cohen, *q-Identities for Maass waveforms*, Invent. Math. **91** (1988), 409–422.

18. L. Dragonette, *Some asymptotic formulae for the mock theta series of Ramanujan*, Trans. Amer. Math. Soc. **72** (1952), 474–500.

19. G. H. Hardy and S. Ramanujan, *Asymptotic formulae in combinatory analysis*, Proc. London Math. Soc. (2) **17** (1918), 75–115; also in [**23**, pp. 276–309].

20. E. Hecke, *Über das Verhalten von* $\sum e^{2i\pi\tau|m^2 - 2n^2|/8}$ *und ähnlichen Funktionen bei Modulsubstitutionen*, Math. Werke, no. 25, Vandenhoeck und Ruprecht, Göttingen, 1970, pp. 487–498.

21. D. R. Hickerson, *A proof of the mock theta conjectures*, Invent. Math. (to appear).

22. ____, *On the seventh order mock theta functions*, Invent. Math. (to appear).

23. V. G. Kac and D. H. Peterson, *Affine Lie algebras and Hecke modular forms*, Bull. Amer. Math. Soc. (N.S.) **3** (1980), 1057–1061.

24. S. Ramanujan, *Collected papers*, Cambridge Univ. Press, 1927; reprinted by Chelsea, New York, 1962.

25. ____, *The Lost Notebook and other unpublished papers*, Narosa, New Delhi, 1987.

26. L. J. Rogers, *On the expansion of some infinite products*, Proc. London Math. Soc. **24** (1893), 337–352.

27. A Selberg, *Über einige arithmetische Identitäten*, Avhl. Norske Vid. **8** (1936), 23 pp.

28. ____, *Über die Mock-Thetafunktionen siebenter Ordnung*, Arch. Math. og Naturvidenskab **41** (1938), 3–15.

29. G. N. Watson, *The final problem: an account of the mock theta functions*, J. London Math. Soc. **11** (1936), 55–80.

30. ____, *A note on Spence's logarithmic transcendant*, Quart. J. Math. Oxford Ser. **8** (1937), 39–42.

31. ____, *The mock theta functions* (2), Proc. London Math. Soc. (2) **42** (1937), 274–304.

THE PENNSYLVANIA STATE UNIVERSITY

Proceedings of Symposia in Pure Mathematics
Volume **49** (1989), Part 2

Orthogonal Polynomials and Theta Functions

RICHARD ASKEY

Abstract. A few of the connections between orthogonal polynomials and theta functions are detailed. These include some sets of orthogonal polynomials whose weight function involves a theta function, and the original derivation of the Rogers-Ramanujan identities, which used orthogonal polynomials without Rogers being aware of most of the orthogonality relations.

1. Introduction. Orthogonal polynomials that are understood well enough to be useful come in two types. The most familiar are those orthogonal with respect to a positive measure on a subset of the real line.

$$(1.1) \qquad \int_{-\infty}^{\infty} p_n(x)p_m(x)\,d\alpha(x) = \delta_{m,n}/h_n, \qquad m,n = 0,1,\ldots,$$

where

$$(1.2) \qquad p_n(x) = \sum_{k=0}^{n} a_{k,n}x^k, \qquad a_{n,n} \neq 0.$$

The second class of orthogonal polynomials are those orthogonal on the unit circle with respect to a positive measure.

$$(1.3) \qquad \frac{1}{2\pi}\int_{-\pi}^{\pi} \varphi_n(e^{i\theta})\overline{\varphi_m(e^{i\theta})}\,d\alpha(\theta) = \delta_{m,n}/j_n$$

with

$$(1.4) \qquad \varphi_n(z) = \sum_{k=0}^{n} c_{k,n}z^k.$$

In the general theory it is useful to take $h_n = j_n = 1$, but for specific sets of polynomials it is often more convenient to use a different normalization.

A theta function will be

$$(1.5) \qquad \sum_{-\infty}^{\infty} q^{n^2}x^n$$

1980 *Mathematics Subject Classification* (1985 Revision). Primary 33A65.
Supported in part by National Science Foundation grant DMS-8701439.

or any function which can be formed by various specializations of this series.
Here $|q| < 1$ for convergence.

The study of specific instances of orthogonal polynomials and theta functions goes back to the eighteenth century, but the serious development of both subjects starts in the nineteenth century. For theta functions there are two very important formulas that were discovered in the early part of the nineteenth century. One determines the zeros.

$$(1.6) \qquad \sum_{-\infty}^{\infty} q^{n^2} x^n = \prod_{n=0}^{\infty} (1 - q^{2n+2})(1 + xq^{2n+1})(1 + x^{-1}q^{2n+1}).$$

The second is the modular transformation

$$(1.7) \qquad \sum_{-\infty}^{\infty} e^{-\pi t n^2 - 2\pi t z n} = \frac{e^{\pi t z^2}}{\sqrt{t}} \sum_{-\infty}^{\infty} e^{-\pi n^2 / t + 2\pi i n z}.$$

A proof of a more general result than (1.6) will be given below. There are a number of proofs of (1.7). One standard version uses the Poisson summation formula. A variant of this due to Pólya just uses the binomial theorem and sieves using roots of unity. See [34].

There is a more general class of series that needs to be introduced. These are basic hypergeometric series. A series $\sum c_n$ is called a hypergeometric series if the term ratio is a rational function of n. If

$$(1.8) \qquad \frac{c_{n+1}}{c_n} = \frac{(n + a_1) \cdots (n + a_p)x}{(n + b_1) \cdots (n + b_q)(n + 1)}$$

and $c_0 = 1$, then

$$(1.9) \qquad {}_pF_q \left(\begin{matrix} a_1, \ldots, a_p \\ b_1, \ldots, b_q \end{matrix}; x \right) := \sum_{n=0}^{\infty} \frac{(a_1)_n \cdots (a_p)_n}{(b_1)_n \cdots (b_q)_n} \cdot \frac{x^n}{n!}$$

with

$$(1.10) \qquad (a)_n = \Gamma(n + a)/\Gamma(a) = a(a + 1) \cdots (a + n - 1).$$

A basic hypergeometric series has term ratio a rational function of q^n for a fixed q. Usually $|q| < 1$, and this will be assumed for the rest of this paper. If

$$(1.11) \qquad \frac{c_{n+1}}{c_n} = \frac{(1 - a_1 q^n) \cdots (1 - a_{p+1} q^n)(-1)^r q^{rn} x}{(1 - b_1 q^n) \cdots (1 - b_p q^n)(1 - q^{n+1})}, \qquad c_0 = 1,$$

then

$$(1.12)$$
$$\quad {}_{p+1}\varphi_{p+r} \left(\begin{matrix} a_1, \ldots, a_{p+1} \\ b_1, \ldots, b_{p+r} \end{matrix}; q, x \right)$$
$$:= \sum_{-\infty}^{\infty} \frac{(a_1; q)_n \cdots (a_{p+1}; q)_n (-1)^{rn} q^{r\binom{n}{2}} x^n}{(b_1; q)_n \cdots (b_{p+r}; q)_n (q; q)_n}.$$

The theta series (1.5) is a bilateral series, so it is not contained in these basic hypergeometric series. To extend the definition to contain both basic hypergeometric series of the form (1.12) and theta series, consider

$$(1.13) \qquad \frac{c_{n+1}}{c_n} = \frac{(1-a,q^n)\cdots(1-a_pq^n)(-1)^r q^{rn} x}{(1-b,q^n)\cdots(1-b_pq^n)}, \qquad n = 0, \pm 1, \ldots.$$

Define

$$(1.14) \qquad {}_p\psi_{p+r}\left(\begin{matrix} a_1, \ldots, a_p \\ b_1, \ldots, b_{p+r} \end{matrix} ; q, x\right)$$
$$:= \sum_{-\infty}^{\infty} \frac{(a_1; q)_n \cdots (a_p; q)_n}{(b_1; q)_n \cdots (b_{p+r}; q)_n} (-1)^{rn} q^{r\binom{n}{2}} x^n.$$

The q analogue of the shifted factorial $(a)_n$ is $(a; q)_n$, where

$$(1.15) \qquad (a; q)_\infty = \prod_{n=0}^{\infty} (1 - aq^n)$$

and

$$(1.16) \qquad (a; q)_n = (a; q)_\infty / (aq^n; q)_\infty.$$

Observe that

$$\lim_{q \to 1} \frac{(q^a; q)_n}{(1-q)^n} = (a)_n.$$

The theta function (1.5) can be written as

$${}_0\psi_1\left(\frac{-}{0}; q^2, -xq\right) = \sum_{-\infty}^{\infty} (-1)^n q^{2\binom{n}{2}}(-xq)^n = \sum_{-\infty}^{\infty} q^{n^2} x^n.$$

The role of the factor $(1-q^{n+1})^{-1}$ in (1.11) is to get $(q; q)_n^{-1}$, and this vanishes for $n = -1, -2, \ldots$ from (1.16). One of the most important examples of a basic hypergeometric series is called the q-binomial series. Its sum is

$$(1.17) \qquad \frac{(ax; q)_\infty}{(x; q)_\infty} = \sum_{n=0}^{\infty} \frac{(a; q)_n}{(q; q)_n} x^n.$$

This is called the q-binomial theorem.

Ramanujan found a result that extends both the q-binomial theorem and the triple product (1.6). It is

$$(1.18) \qquad {}_1\psi_1\left(\frac{a}{b}; q, x\right) = \sum_{-\infty}^{\infty} \frac{(a; q)_n}{(b; q)_n} x^n$$
$$= \frac{(ax; q)_\infty (q/ax; q)_\infty (q; q)_\infty (b/a; q)_\infty}{(x; q)_\infty (b/ax; q)_\infty (b; q)_\infty (q/a; q)_\infty}$$

when $|b/a| < |x| < 1$.

There is a very attractive proof of (1.18) that starts like the classical proof of the triple product formula (1.6). Set

$$(1.19) \qquad f(x) = \frac{(ax; q)_\infty (q/ax; q)_\infty}{(x; q)_\infty (b/ax; q)_\infty} = \sum_{-\infty}^{\infty} c_n x^n.$$

Then

$$\frac{f(qx)}{f(x)} = \frac{(1-x)(1-1/ax)}{(1-ax)(1-b/aqx)} = \frac{(1-x)q}{(b-aqx)},$$

so

$$(1-x)q \sum_{-\infty}^{\infty} c_n x^n = (b-aqx) \sum_{-\infty}^{\infty} c_n q^n x^n.$$

Equating coefficients of x^{n+1} gives

(1.20) $$\frac{c_{n+1}}{c_n} = \frac{(1-aq^n)}{(1-bq^n)}.$$

Then

(1.21) $$c_n = [(a;q)_n/(b;q)_n]c_0$$

so

(1.22) $$\frac{(ax;q)_\infty (q/ax;q)_\infty}{(x;q)_\infty (b/ax;q)_\infty} = c_0 \sum_{-\infty}^{\infty} \frac{(a;q)_n}{(b;q)_n} x^n.$$

Multiply by $(1-x)$ and let $x \to 1$. The result is

(1.23) $$\frac{(a;q)_\infty (q/a;q)_\infty}{(q;q)_\infty (b/a;q)_\infty} = c_0 \frac{(a;q)_\infty}{(b;q)_\infty}$$

by Abel's lemma. Using c_0 in (1.22) gives (1.18). Observe that $b = q$ gives the q-binomial theorem, and $b = 0$, $x \to x/a$ followed by $a \to \infty$ gives the triple product. This proof was first given in Adiga et al. [2].

Notice the role played by the theta product in the numerator in the product in (1.19). A pair of infinite products will be called a theta product if the product of the variables is the base. In (1.19) this is $(ax)(q/ax) = q$. It is an interesting exercise to carry out the above argument on the reciprocal of the product in (1.19). The argument seems to work, but the resulting identity is clearly false. Details of this are given in [9], but I recommend doing the calculations yourself. If you can get the wrong result, and then find the error, you have learned enough to be able to do many of the proofs in this area.

A paper with a similar title was written by Carlitz [19]. He has a few observations that are not made here.

2. Rogers and the Rogers-Ramanujan identities. Some papers are written before they should be, and so are not appreciated. One example is [38]. Buried in this paper are the identities

(2.1) $$\frac{1}{(q;q^5)_\infty (q^4;q^5)_\infty} = \sum_{n=0}^{\infty} \frac{q^{n^2}}{(q;q)_n}$$

and

(2.2) $$\frac{1}{(q^2;q^5)_\infty (q^3;q^5)_\infty} = \sum_{n=0}^{\infty} \frac{q^{n(n+1)}}{(q;q)_n}.$$

These two identities become famous twenty years later because Ramanujan rediscovered them, used them to prove some incredible results like

$$(2.3) \qquad \cfrac{1}{1 + \cfrac{e^{-2\pi}}{1 + \cfrac{e^{-4\pi}}{1 + \cdots}}} = e^{2\pi/5} \left\{ \left[\frac{5 + \sqrt{5}}{2} \right]^{1/2} - \frac{\sqrt{5} + 1}{2} \right\}$$

yet neither he, nor Hardy, nor anyone Hardy wrote to could prove them. For details on (2.3) and other results of a similar type see Ramanathan [35, 36, 37].

When one has an identity like (2.1), one wants to know a number of things. First, is it true? MacMahon expanded both sides in a power series in q up to q^{89} and the coefficients agreed. This was enough to convince him that the result was true. To quote him: "there is practically no reason to doubt its truth; but it has not yet been established." [32, §276]. Thus it seemed reasonable to believe (2.1) and (2.2) as Ramanujan did, and to try to use them. In addition to Ramanujan's use of them through the continued fraction

$$\cfrac{1}{1 + \cfrac{q}{1 + \cfrac{q^2}{1 + \cdots}}} = \frac{(q; q^5)_\infty (q^4; q^5)_\infty}{(q^2; q^5)_\infty (q^3; q^5)_\infty},$$

which he proved under the assumption that (2.1) and (2.2) are true, MacMahon found very important combinatorial interpretations of these two series. For (2.1), the left-hand side is the generating function for the partitions of n with all parts congruent to 1 or 4 (mod 5). The right-hand side is the generating function for partitions of n where each part differs from any other part by at least one. For example, for 9, the partitions are

$$
\begin{array}{ll}
9 & 9 \\
6 + 1 + 1 + 1 & 8 + 1 \\
4 + 4 + 1 & 7 + 2 \\
4 + 1 + 1 + 1 + 1 + 1 & 6 + 3 \\
1 + 1 + \cdots + 1 & 5 + 3 + 1
\end{array}
$$

and there are five of each. There are other combinatorial interpretations of the sum side, but the one above has been the most important. See [3] for another combinatorial interpretation.

Once the importance of (2.1) was realized, it became important to find a proof—any proof. However one also wants specific types of proofs that answer different questions. First, one wants a combinatorial proof. Schur [42] found one that was almost what one wanted. For years George Andrews offered $50 for a purely combinatorial proof. The reason for this offer was not only that he wanted a proof that did not use infinite processes, but he felt that a combinatorial proof would have to introduce some new ideas that could be used elsewhere. This turned out to be the case. The first purely combinatorial proof was found by Garsia and Milne [25], and they discovered a very important involution principle as part of their proof.

One wants a proof that is simple enough to give to students in an undergraduate number theory course. The proof given by Bressoud [18] is such a proof. One wants a proof that starts with one side and generates the other side. Here there is still a lot of work to be done. For example,

$$
(2.4) \quad \frac{1}{(q;q^5)_\infty (q^4;q^5)_\infty} = \frac{(q^5;q^5)_\infty}{(q;q^5)_\infty (q^4;q^5)_\infty (q^5;q^5)_\infty}
$$
$$
= \frac{(q^2;q^5)_\infty (q^3;q^5)_\infty (q^5;q^5)_\infty}{(q;q)_\infty}.
$$

Since

$$
(2.5) \qquad (q;q)_\infty = (q;q^3)_\infty (q^2;q^3)_\infty (q^3;q^3)_\infty,
$$

both the right-hand sides of (2.4) are ratios of theta functions, and so are modular forms. Can one give a proof of (2.1) using the modular machine? So far no one has, and since I think such a proof would introduce an important new idea, I will pay $100 for a modular form proof. There has been inflation that raises $50 to $100, but this problem is also probably harder than the one Garsia and Milne solved.

Starting from the sum side of (2.1) and (2.2) there is a well-motivated proof in [23]. Also see [44].

One wants a proof that leads to other identities. There are a number of ways to extend (2.1). One was discovered as a combinatorial result by Gordon [26], and as an analytic identity by Andrews [4]. The series on one side are now multiple series. There is another proof that leads to the "right" one-dimensional sum identity that extends both (2.1) and (2.2). This proof was the way Rogers first proved (2.1) and (2.2). He did not set out to prove these identities; they came out as consequences of a more general identity.

One reason [38] was ignored is that the important results in it were way ahead of their time. Not only did Rogers get to (2.1) almost twenty years before Ramanujan did, he found

$$
(2.6) \qquad (q;q)_\infty^2 = \sum_{m=-\infty}^{\infty} \sum_{n \geq 2|m|} (-1)^{m+n} q^{\binom{n+1}{2} - 3m^2/2 + m/2}
$$

more than thirty years before Hecke did [28], and he found

$$
(2.7) \qquad \frac{1}{(q;q^7)_\infty (q^3;q^7)_\infty (q^4;q^7)_\infty (q^6;q^7)_\infty} = \sum_{n=0}^{\infty} \frac{q^{2n^2+2n}(q;q^2)_n}{(q^2;q^2)_{2n}}
$$

more than forty years before Selberg [43]. Not bad for a paper that no one seems to have read seriously before Ramanujan found it in 1916. However even after Ramanujan expressed admiration for this paper almost no one else seems to have read it until the last ten years. One exception was F. Dyson.

The real reason for the lack of knowledge of what is in this paper seems to be that Rogers did not understand his work very well, and so the paper is very hard to read and understand. However a year later Rogers wrote a sequel which is easy to read, and yet almost no one read it until ten years ago. Here is Rogers's derivation of (2.1) and (2.2) as it should have been written after he understood more as illustrated by [39].

From the q-binomial theorem, Rogers defined a set of polynomials in the following way,

$$
\begin{aligned}
(2.8) \quad \frac{(\beta r e^{i\theta}; q)_\infty (\beta r e^{-i\theta}; q)_\infty}{(r e^{i\theta}; q)_\infty (r e^{-i\theta}; q)_\infty} &= \sum_{n=0}^{\infty} \frac{(\beta; q)_n}{(q; q)_n} r^n e^{in\theta} \sum_{k=0}^{\infty} \frac{(\beta; q)_k}{(q; q)_k} r^k e^{-ik\theta} \\
&= \sum_{n=0}^{\infty} r^n \sum_{k=0}^{n} \frac{(\beta; q)_{n-k}(\beta; q)_k}{(q; q)_{n-k}(q; q)_k} e^{i(n-2k)\theta} = \sum_{n=0}^{\infty} C_n(\cos\theta; \beta|q) r^n,
\end{aligned}
$$

where

$$
(2.9) \quad C_n(\cos\theta; \beta|q) = \sum_{k=0}^{n} \frac{(\beta; q)_{n-k}(\beta; q)_k}{(q; q)_{n-k}(q; q)_k} \cos(n - 2k)\theta.
$$

In [39] Rogers showed that

$$
(2.10) \quad C_n(x; \gamma|q) = \sum_{k=0}^{\lfloor n/2 \rfloor} \beta^k \frac{(\gamma \beta^{-1}; q)_k (\gamma; q)_{n-k}}{(q; q)_k (\beta q; q)_{n-k}} \frac{(1 - \beta q^{n-2k})}{(1 - \beta)} C_{n-2k}(x; \beta|q).
$$

He proved this by induction using the recurrence relation

$$
\begin{aligned}
(2.11) \quad 2n(1 - \beta q^n) C_n(x; \beta|q) & \\
&= (1 - q^{n+1}) C_{n+1}(x; \beta|q) + (1 - \beta^2 q^{n-1}) C_{n-1}(x; \beta|q), \\
C_0(x; \beta|q) = 1, \quad & C_1(x; \beta|q) = 2x(1 - \beta)/(1 - q).
\end{aligned}
$$

It is easy to see that (2.11) and the generating function (2.8) are equivalent. Let

$$
(2.12) \quad D_n(\cos\theta|q) = \lim_{\beta \to \infty} \beta^{-n} C_n(\cos\theta; \beta|q).
$$

Then

$$
\begin{aligned}
(2.13) \quad (r e^{i\theta}; q)_\infty (r e^{-i\theta}; q)_\infty &= \sum_{n=0}^{\infty} D_n(\cos\theta|q) r^n \\
&= \sum_{n=0}^{\infty} \frac{q^{\binom{n}{2}} (-r)^n (1 - \beta q^n)}{(\beta q; q)_n (1 - \beta)} \sum_{k=0}^{\infty} \frac{q^{k^2 + nk - k} r^{2k}}{(\beta q^{n+1}; q)_k (q; q)_k} C_n(x; \beta|q)
\end{aligned}
$$

follows from (2.10) and (2.12). When $r = -q^{1/2}$,

$$
(q; q)_\infty (-q^{1/2} e^{i\theta}; q)_\infty (-q^{1/2} e^{-i\theta}; q)_\infty = \sum_{k=0}^{\infty} q^{n^2/2} e^{in\theta}
$$

$$
= 1 + 2 \sum_{m=1}^{\infty} q^{n^2/2} \cos n\theta
$$

by (1.6). But

(2.14)
$$\cos n\theta = \sum_{k=0}^{\lfloor n/2 \rfloor} \frac{\beta^k (\beta^{-1}; q)_k (q; q)_{n-k} (1 - q^n)(1 - \beta q^{n-2k})}{(q; q)_k (\beta q; q)_{n-k} (1 - q^{n-k}) 2 (1 - \beta)} C_{n-2k}(x; \beta|q),$$

$n \geq 1$, follows from (2.10) and

(2.15)
$$\lim_{\beta \to 1} \frac{(1 - q^n) C_n(\cos\theta; \beta|q)}{2(1 - \beta)} = T_n(\cos\theta)$$
$$= \cos n\theta, \qquad n = 1, 2, \ldots,$$

which is easy to prove. Then

$$(q; q)_\infty (-q^{1/2} e^{i\theta}; q)_\infty (-q^{1/2} e^{-i\theta}; q)_\infty$$

$$= 1 + 2 \sum_{n=1}^{\infty} q^{n^2/n} \cos n\theta$$

(2.16)
$$= \sum_{n=0}^{\infty} q^{n^2/2} \frac{(q; q)_n (1 - \beta q^n)}{(\beta q; q)_n (1 - \beta)} C_n(x; \beta|q)$$

$$\cdot \sum_{k=0}^{\infty} \frac{\beta^k (\beta^{-1}; q)_k (q^{n+1}; q)_k (1 - q^{n+2k})}{(q; q)_k (\beta q^{n+1}; q)_k (1 - q^{n+k})} q^{2k^2 + 2nk}.$$

Equating coefficients of $C_n(x; \beta|q)$ in (2.16) and (2.13) with $r = -q^{1/2}$ gives

$$\sum_{k=0}^{\infty} \frac{q^{k^2} q^{nk}}{(\beta q^{n+1}; q)_k (q; q)_k}$$

$$= \frac{1}{(q^{n+1}; q)_\infty} \sum_{k=0}^{\infty} \frac{\beta^k (\beta^{-1}; q)_k (q^{n+1}; q)_k (1 - q^{n+2k}) q^{2k^2 + 2nk}}{(q; q)_k (\beta q^{n+1}; q)_k (1 - q^{n+k})}.$$

Now let $q^n = a$ to get

(2.17)
$$\sum_{k=0}^{\infty} \frac{q^{k^2} a^k}{(a\beta q; q)_k (q; q)_k}$$

$$= \frac{1}{(aq; q)_\infty} \sum_{k=0}^{\infty} \frac{a^{2k} \beta^k (\beta^{-1}; q)_k (a; q)_k (1 - aq^{2k}) q^{2k^2}}{(q; q)_k (a\beta q; q)_k (1 - a)}.$$

Originally Rogers gave this argument when $\beta \to 0$. The identity is then

(2.18)
$$\sum_{k=0}^{\infty} \frac{q^{k^2} a^k}{(q; q)_k} = \frac{1}{(aq; q)_\infty} \sum_{k=0}^{\infty} \frac{(-1)^k q^{(5k^2 - k)/2} a^{2k} (a; q)_k (1 - aq^{2k})}{(q; q)_k (1 - a)}.$$

Finally $a = q$ gives

$$\sum_{k=0}^{\infty} \frac{q^{k^2 + k}}{(q; q)_k} = \frac{1}{(q; q)_\infty} \sum_{k=0}^{\infty} (-1)^k q^{(5k^2 + 3k)/2} (1 - q^{2k+1})$$

$$= \frac{\sum_{-\infty}^{\infty} (-1)^k q^{(5k^2 + 3k)/2}}{(q; q)_\infty} = \frac{(q^5; q^5)_\infty (q^4; q^5)_\infty (q; q^5)_\infty}{(q; q)_\infty}$$

$$= \frac{1}{(q^2; q^5)_\infty (q^3; q^5)_\infty}.$$

This is (2.2). Similarly $a \to 1$ gives (2.1). Bressoud [17] gave the above argument, but it is really due to Rogers as he says.

The only thing missing in the above proof seems to be orthogonal polynomials. They are there in all the polynomials used to derive the Rogers-Ramanujan identities. If $\{p_n(x)\}_{n=0}^{\infty}$ is a set of polynomials with $p_n(x)$ of exact degree n, and if $\{p_n(x)\}$ are orthogonal with respect to a positive measure on the real line,

$$(2.19) \qquad \int_{-\infty}^{\infty} p_n(x)p_m(x)\,d\alpha(x) = \delta_{m,n}/h_n, \qquad m, n = 0, 1, \ldots,$$

then

$$(2.20) \qquad xp_n(x) = A_n p_{n+1}(x) + B_n p_n(x) + C_n p_{n-1}(x)$$

with A_n, B_n, C_n real and $A_n C_{n+1} > 0$, $n = 0, 1, \ldots$. Conversely, if $\{p_n(x)\}$ satisfies (2.20) with the added restrictions in the next line, and if $p_0(x) = 1$, $p_{-1}(x) = 0$, then there is a positive measure $d\alpha(x)$ so that (2.19) holds, and

$$(2.21) \qquad h_n = \frac{A_0 \cdots A_{n-1}}{C_1 \cdots C_n} h_0.$$

See the comments to [48] for references.

The recurrence relation for $C_n(x; \beta|q)$ is (2.11). The orthogonality relation is

$$(2.22) \qquad \begin{aligned} &\frac{1}{2\pi} \int_{-1}^{1} C_n(x; \beta|q) C_m(x; \beta|q) w_\beta(x)\,dx \\ &= \delta_{m,n} \frac{(1-\beta)(\beta^2; q)_n}{(1 - \beta q^n)(q; q)_n} A(\beta) \end{aligned}$$

with

$$w_\beta(x) = \prod_{k=0}^{\infty} \left[\frac{1 - 2(2x^2 - 1)q^k + q^{2k}}{1 - 2(2x^2 - 1)\beta q^k + \beta^2 q^{2k}} \right] (1 - x^2)^{-1/2}$$

and

$$A(\beta) = \frac{(\beta; q)_\infty (\beta q; q)_\infty}{(\beta^2; q)_\infty (q; q)_\infty}$$

when $-1 < \beta, q < 1$.

See [11, 12, 13, 15] for various derivations of this orthogonality relation. In [12] the orthogonality (2.22) was used to derive (2.10). A second derivation of (2.10) using a divided difference operator is given in [10].

To return to the question of the different types of proofs of (2.1), the above proof can be extended to solve a natural question—how can one remove the quadratic powers in (2.18). First, why would one want to remove the quadratic powers? There are two reasons. A more general identity has more flexibility, and not only does this give more chances for interesting special cases to be discovered, the added flexibility often leads to identities that are easier to prove. For example, Ramanujan had discovered (2.1) and (2.2) but was unable to prove them until he found [38]. Then he knew that (2.18) could be used to obtain these identities. Once he knew this, it was easy to him to find a proof of (2.18). He used a difference equation, and without the extra parameter a there is no difference equation.

Once a new set of orthogonal polynomials is found, it is natural to look for a more general set of orthogonal polynomials. The natural extension of these polynomials of Rogers was found by extending a new set of orthogonal polynomials, but not those of Rogers. It was only later that the connection with Rogers's polynomials was discovered. J. Wilson [52] found a set of orthogonal polynomials with four free parameters that can be given as hypergeometric series. There are two cases, one when the orthogonality relation is discrete and the other when the weight function has nontrivial absolutely continuous part. The discrete case is closely related to the $6 - j$ symbols, or Racah coefficients, that arise in quantum angular momentum theory. The absolutely continuous measure is nicer looking (but not as useful yet), so it will be the one that is stated. It is also the one that was extended to a class of polynomials that contains those of Rogers.

Assume $a, b, c, d > 0$ and set

$$(2.23) \qquad W_n(x^2) = {}_4F_3 \left(\begin{matrix} -n, n + a + b + c + d - 1, a + ix, a - ix \\ a + b, a + c, a + d \end{matrix} ; 1 \right).$$

Then for $m \neq n$

$$(2.24)$$

$$\int_0^\infty W_n(x^2) W_m(x^2) \left| \frac{\Gamma(a + ix)\Gamma(b + ix)\Gamma(c + ix)\Gamma(d + ix)}{\Gamma(2ix)} \right|^2 dx = 0.$$

A basic hypergeometric extension was given in [15]. Set

$$(2.25) \qquad \frac{a^n W_n(x; a, b, c, d|q)}{(ab; q)_n (ac; q)_n (ad; q)_n} = {}_4\varphi_3 \left(\begin{matrix} q^{-n}, q^{n-1} abcd, ae^{i\theta} \\ ab, ac, ad \end{matrix} ; q, q \right).$$

If $-1 < a, b, c, d, q < 1$, then

$$(2.26)$$

$$\frac{1}{2\pi} \int_0^\pi W_n(\cos \theta) W_m(\cos \theta) \left| \frac{(e^{2i\theta}; q)_\infty}{(ae^{i\theta}; q)_\infty (be^{i\theta}; q)_\infty (ce^{i\theta}; q)(de^{i\theta}; q)_\infty} \right|^2 d\theta$$

$$= \begin{cases} 0, & m \neq n, \\ 1/h_n, & m = n, \end{cases}$$

$$(2.27) \qquad \frac{h_n}{(q^{n+1}; q)_\infty} = \frac{(abq^n; q)_\infty (acq^n; q)_\infty (adq^n; q)_\infty (bcq^n; q)_\infty (bdq^n; q)_\infty (cdq^n; q)_\infty}{(abcdq^{n-1}; q)_n (abcdq^{2n}; q)_\infty}.$$

See [15]. There are a number of properties of $W_n(x)$ that are useful. One is the symmetry in (a, b, c, d). The symmetry in (b, c, d) is obvious. The symmetry

$$(2.28) \qquad W_n(x; a, b, c, d|q) = W_n(x; b, a, c, d|q)$$

is not obvious, and is useful. It was first found by Watson. A second very useful fact is the solution of the connection coefficient problem when one parameter is changed

$$(2.29) \qquad W_n(x;\alpha,b,c,d|q) = \sum_{k=0}^{n} c_{k,n} W_k(x;a,b,c,d|q),$$

$$c_{k,n} = \frac{a^{n-k} q^{nk-k(k-1)/2}(-1)^k (q^{-n};q)_k (\alpha/a;q)_{n-k} (\alpha bcd q^{n-1};q)_k}{(q;q)_k (abcd q^{k-1};q)_k (abcd q^{2k};q)_{n-k}}$$
$$\cdot (bcq^k;q)_{n-k} (bdq^k;q)_{n-k} (cdq^k;q)_{n-k}.$$

See [15, (6.4)]. When $a = e^{i\theta}$ the left-hand side is the general balanced $_4\varphi_3$ and the right-hand side is a single sum. When this is translated into basic hypergeometric form, the result is equivalent to Watson's extension of Whipple's formula:

$$(2.30)$$
$$_4\varphi_3 \left(\begin{matrix} q^{-n}, aq/bc, d, e \\ deq^{-n}/a, aq/b, aq/c \end{matrix} ; q, q \right)$$
$$= \frac{(aq/d)_n (aq/e;q)_n}{(aq;q)_n (aq/de;q)_n}$$
$$\cdot {}_8\varphi_7 \left(\begin{matrix} a, qa^{1/2}, -qa^{1/2}, b, c, d, e, q^{-n} \\ a^{1/2}, -a^{1/2}, aq/b, aq/c, aq/d, aq/e, aq^{n+1} \end{matrix} ; q, \frac{a^2 q^{n+2}}{bcde} \right).$$

As Watson observed [50], letting $b,c,d,e,n \to \infty$ in (2.30) gives (2.18). Observe that the quadratic powers have been replaced by free parameters, one for each $q^{k^2/2}$ in (2.18). Since there are five of these factors, one expects five free parameters, and the sixth free parameter in (2.30) is there because there are two Rogers-Ramanujan identities.

In [27] Hardy wrote about the various proofs of the Rogers-Ramanujan identities that were known in the late 1930s. He mentioned seven proofs, the original one of Rogers [38], one he said was a simplified version of this proof, which Rogers also found [40], two proofs due to Schur [42], two much simpler proofs due to Rogers and Ramanujan separately [41], and Watson's proof [50], which Hardy wrote was based on quite different ideas. Schur's proofs are different than the others, but the other five are really all variants of the same proof, which is to fit the Rogers-Ramanujan identities into well-poised basic hypergeometric series. See Andrews [5] for an extension of Watson's transformation formula that contains the Gordon-Andrews version of the Rogers-Ramanujan result to higher moduli.

There are many instances of orthogonal polynomials whose weight function is a theta function or an elliptic function, which is the quotient to two theta functions, or even the product of such functions times an elementary function. One only needs to observe that

$$(2.31) \qquad (e^{2i\theta};q)_\infty = (e^{i\theta};q)_\infty (q^{1/2} e^{i\theta};q)_\infty (-e^{i\theta};q)_\infty (-q^{1/2} e^{i\theta};q)_\infty$$

and that taking a,b,c,d to be $\pm q^k$, $\pm q^{k+1/2}$, or 0 gives such weight functions. The first explicit example was discovered by Szegö [47]. It is the case $b = 1$, $c = -1$, $d = -q^{1/2}$, $a = 0$.

The polynomials $C_n(x; \beta|q)$ of Rogers fit into the $_4\varphi_3$ polynomials by taking $a = \beta^{1/2}$, $b = aq^{1/2}$, $c = -a$, $d = -b$. The weight functions then agree, and one can compare coefficients of x^n to show that

$$(2.32) \quad C_n(x; \beta|q) = \frac{(\beta;q)_n(\beta^2;q)_n}{(q;q)_n(\beta^2;q)_{2n}} W_n(x; \beta^{1/2}, (\beta q)^{1/2}, -\beta^{1/2}, -(\beta q)^{1/2}|q).$$

3. The q-Hermite polynomials of Rogers and Szegö's polynomials. The special case $\beta = 0$ of the polynomials of Rogers played a central role in the derivation he gave of (2.1) and (2.2). Their weight function is

$$(1 - x^2)^{-1/2} \prod_{k=0}^{\infty} [1 - 2(2x^2 - 1)q^k + q^{2k}] dx = (e^{2i\theta};q)_\infty (e^{-2i\theta};q)_\infty \, d\theta.$$

As a function of θ this is only one factor off a theta function. These polynomials of Rogers are q-extensions of Hermite polynomials.

Ultraspherical polynomials $C_n^\lambda(x)$ are defined by the recurrence relation

$$(3.1) \quad 2x(n + \lambda)C_n^\lambda(x) = (n + 1)C_{n+1}^\lambda(x) + (n + 2\lambda - 1)C_{n-1}^\lambda(x),$$
$$C_0^\lambda(x) = 1, \qquad C_1^\lambda(x) = 2\lambda x.$$

They are orthogonal on $[-1, 1]$ with respect to $(1 - x^2)^{\lambda-1/2}$ and

$$\lim_{q \to 1} C_n(x; q^\lambda|q) = C_n^\lambda(x).$$

Since

$$\lim_{\lambda \to \infty} (1 - x^2/\lambda)^{\lambda-1/2} = e^{-x^2},$$

the ultraspherical polynomials converge to the polynomials orthogonal with respect to e^{-x^2} on $(-\infty, \infty)$. These polynomials are called Hermite polynomials. There are a number of different notations for them, but the standard books by Szegö [49] and Erdélyi et al. [24] use

$$(3.2) \quad H_n(x)/n! = \lim_{\lambda \to \infty} \lambda^{-n/2} C_n^\lambda(x\lambda^{-1/2}).$$

From this point of view the polynomials $C_n(x; 0|q)$ are q-analogues of Hermite polynomials and the integral

$$(3.3) \quad \int_{-1}^{1} \prod_{k=0}^{\infty} [1 - 2(2x^2 - 1)q^k + q^{2k}](1 - x^2)^{-1/2} dx = \frac{1}{(q;q)_\infty}$$

is an extension of the normal integral

$$(3.4) \quad \int_{-\infty}^{\infty} e^{-x^2} dx = \sqrt{\pi}.$$

There are a number of other ways in which the theta function is related to the normal integral. The most elementary is

$$\lim_{\delta \to 0} \sqrt{\delta} \sum_{-\infty}^{\infty} e^{-\delta n^2} = \int_{-\infty}^{\infty} e^{-x^2} dx.$$

or more generally

$$(3.5) \qquad \lim_{\delta \to 0} \sqrt{\delta} \sum_{-\infty}^{\infty} e^{-\delta n^2 - 2a\delta^{1/2}n} = \int_{-\infty}^{\infty} e^{-x^2 - 2ax} \, dx = e^{a^2} \sqrt{\pi}.$$

The integral on the right is just a shifted normal integral, and the series on the left is a theta series. The modular transformation (1.7) gives an error estimate in (3.5).

The $\beta = 0$ polynomials of Rogers are often renormalized, so that

$$(3.6) \qquad \frac{H_n(x|q)}{(q;q)_n} = C_n(x;0|q).$$

Then

$$(3.7) \qquad \sum_{n=0}^{\infty} \frac{H_n(x|q)}{(q;q)_n} r^n = \frac{1}{(re^{i\theta};q)_\infty (re^{-i\theta};q)_\infty}, \qquad x = \cos \theta.$$

This generating function for the continuous q-Hermite polynomials can be used to evaluate

$$(3.8) \qquad \int_0^\pi \left| \frac{(e^{2i\theta};q)_\infty}{(ae^{i\theta};q)_\infty (be^{i\theta};q)_\infty (ce^{i\theta};q)_\infty (de^{i\theta};q)_\infty} \right|^2 d\theta.$$

For this integral is

(3.9)

$$\sum_{j,k,l,m \geq 0} \frac{a^{j+k+l+m}}{(q;q)_j (q;q)_k (q;q)_l (q;q)_m}$$

$$\cdot \int_0^\pi H_j(\cos \theta|q) H_k(\cos \theta|q) H_l(\cos \theta|q) H_m(\cos \theta|q) |(e^{2i\theta};q)_\infty|^2 d\theta.$$

Rogers [39] showed that

$$(3.10) \qquad C_n(x;\beta|q) C_m(x;\beta|q) = \sum_{k=0}^{\min(m,n)} a(k,m,n) C_{m+n-2k}(x;\beta|q)$$

with

$$(3.11) \quad a(k,m,n) = \frac{(q;q)_{m+n-2k}(\beta;q)_{m-k}(\beta;q)_{n-k}(\beta;q)_k(\beta^2;q)_{m+n-k}(1 - \beta q^{m+n-2k})}{(\beta^2;q)_{m+n-2k}(q;q)_{m-k}(q;q)_{n-k}(q;q)_k(\beta q;q)_{m+n-k}(1 - \beta)}.$$

For $\beta = 0$ this is

$$(3.12) \qquad \frac{H_m(x|q)}{(q;q)_m} \frac{H_n(x|q)}{(q;q)_n} = \sum_{k=0}^{\min(m,n)} \frac{H_{m+n-2k}(x|q)}{(q;q)_{m-k}(q;q)_{n-k}(q;q)_k}.$$

Rogers proved this in [38]. Rogers also showed that

$$(3.13) \qquad \sum_{n=0}^{\infty} \frac{H_n(\cos \theta|q) H_n(\cos \varphi|q)}{(q;q)_n} r^n = \frac{(r^2;q)_\infty}{|(re^{i(\theta+\varphi)};q)_\infty (re^{i(\theta-\varphi)};q)_\infty|^2}.$$

This is a q-extension of Mehler's formula for Hermite polynomials, or in other words, it is the Poisson kernel for q-Hermite series. In (3.9), use (3.12) twice, then the orthogonality of $H_n(x|q)$. Two of the resulting sums can be evaluated by the q-binomial theorem, and the remaining triple sum can be put in the form of (3.13), and so summed. The result is (2.26) and (2.27) when $m = n = 0$. See [30] for the details, and [31] for a related combinatorial evaluation of this integral.

While one can let $q \to 1$ in the final result, it is still an open problem to evaluate

$$(3.14) \qquad \int_0^\infty \left| \frac{\Gamma(a + ix)\Gamma(b + ix)\Gamma(c + ix)\Gamma(d + ix)}{\Gamma(2ix)} \right|^2 dx$$

by either of these methods. Wilson [52] evaluated (3.14) directly, but the extra parameter q in (3.8) has so far been necessary for an argument like that discovered by Ismail and Stanton.

The polynomials

$$H_n(\cos\theta|q) = \sum_{k=0}^n \begin{bmatrix} n \\ k \end{bmatrix}_q \cos(n - 2k)\theta = e^{in\theta} \sum_{k=0}^n \begin{bmatrix} n \\ k \end{bmatrix}_q e^{-2ik\theta}$$

have a variation that was studied by Szegö [47]. Set

$$(3.15) \qquad \varphi_n(z) = \sum_{k=0}^n \begin{bmatrix} n \\ k \end{bmatrix}_q q^{-k/2} z^k.$$

Then

$$\int_0^{2\pi} \varphi_n(e^{i\theta}) e^{-im\theta} \sum_{j=-\infty}^\infty q^{j^2/2} e^{ij\theta} \, d\theta$$

$$= \sum_{k=0}^n \begin{bmatrix} n \\ k \end{bmatrix}_q q^{-k/2} \sum_{j=-\infty}^\infty q^{j^2/2} \int_0^{2\pi} e^{i(k+j-m)\theta} \, d\theta$$

$$= 2\pi \sum_{k=0}^n \frac{(q;q)_n}{(q;q)_k(q;q)_{n-k}} q^{-k/2} q^{(m-k)^2/2}$$

$$= 2\pi q^{m^2/2} \sum_{k=0}^n \frac{(q^{-n};q)_k}{(q;q)_k} q^{(n-m)k} = 2\pi q^{m^2/2}(q^{-m};q)_n.$$

Thus
(3.16)

$$\int_0^{2\pi} \varphi_n(e^{i\theta})\overline{\varphi_m(e^{i\theta})}(-q^{1/2}e^{i\theta};q)_\infty(-q^{1/2}e^{-i\theta};q)_\infty \, d\theta = 0, \qquad m \neq n.$$

In a footnote Szegö [47] remarked that
(3.17)

$$\int_{-1}^1 B_n(x)B_m(x) \prod_{k=1}^\infty (1 - 2xq^{k-1/2} + q^{2k-1})(1 - x^2)^{-1/2} \, dx = 0, \qquad m \neq n,$$

when

$$(3.18) \qquad B_n(\cos\theta) = \begin{bmatrix} 2n \\ n \end{bmatrix} + \sum_{k=1}^n \begin{bmatrix} 2n \\ n+k \end{bmatrix} (q^{k/2} + q^{-k/2}) \cos k\theta.$$

His argument was not given, but he had discovered the following fact that gives (3.17) and (3.18) from (3.15) and (3.16).

If $\psi_n(z)$ is a polynomial of degree n in z and $w(\cos\theta)$ is a nonnegative function with

(3.19)
$$\int_0^{2\pi} \psi_n(e^{i\theta})\overline{\psi_m(e^{i\theta})}w(\cos\theta)\,d\theta = 0, \qquad m \neq n,$$

then

(3.20)
$$p_n(x) = z^n\psi_{2n}(z^{-1}) + z^{-n}\psi_{2n}(z)$$

with $x = (z + z^{-1})/2$ satisfies

(3.21)
$$\int_{-1}^1 p_n(x)p_m(x)w(x)(1-x^2)^{-1/2}\,dx = 0, \qquad m \neq n.$$

See [49, Theorem 11.5].

I had not read this footnote carefully, so have misrepresented it on other occasions. To obtain (3.17) and (3.18) replace θ by $\theta + \pi$ in (3.16). Then (3.20) becomes (3.18) and (3.21) is (3.17) when $\psi = \varphi$. There is a second way to show that (3.18) satisfies (3.17). In (2.29) take $a = q^{1/2}$, $b = 1$, $c = -1$, $d = -q^{1/2}$, and $\alpha = 0$. A simple calculation gives (3.18).

There is a second formula like (3.18) that represents a set of orthogonal polynomials. The case $\alpha = 0$, $a = 1$, $b = q^{1/2}$, $c = -1$, $d = -q^{1/2}$ gives the representation

(3.22)
$$C_n(\cos\theta|q) = \begin{bmatrix} 2n \\ n \end{bmatrix} + 2\sum_{k=1}^n \begin{bmatrix} 2n \\ n+k \end{bmatrix} \frac{(q^{1/2};q)_k}{(q;q)_k}\cos k\theta$$

for the polynomials that satisfy
(3.23)
$$\int_{-1}^1 C_n(x|q)C_m(x|q)\prod_{k=0}^\infty (1 - 2xq^k + q^{2x})(1-x^2)^{-1/2}\,dt = 0, \qquad m \neq n.$$

In (3.18) and (3.22) the q-binomial coefficient is defined by

(3.24)
$$\begin{bmatrix} n \\ k \end{bmatrix} = \begin{bmatrix} n \\ k \end{bmatrix}_q = \frac{(q;q)_n}{(q;q)_k(q;q)_{n-k}}.$$

4. The Stieltjes-Wigert polynomials. Stieltjes found other connections between orthogonal polynomials and theta functions. The first instance [46, pp. 507–508 in reprint] was only done by him for one value of the parameter q, and extended to the general case by Wigert [51]. Here are these polynomials and Wigert's orthogonality.

Let

(4.1)
$$q = \exp(-1/(2k^2)),$$

(4.2)
$$w(x) = \frac{k}{\sqrt{\pi}}\exp(-k^2(\log x)^2),$$

and

$$
\begin{aligned}
p_n(x) &= \sum_{k=0}^{n} \frac{(q;q)_n}{(q;q)_k(q;q)_{n-k}} q^{k(k+1)}(-1)^k q^{-k/2} x^k \\
&= \sum_{k=0}^{n} \frac{(q^{-n};q)_k}{(q;q)_k} q^{k^2/2}(q^{n+1}x)^k
\end{aligned}
$$

(4.3)

Then

$$
\int_0^\infty p_n(x)p_m(x)w(x)\,dx = 0, \qquad m \neq n.
$$

(4.4)

Now you can see the orthogonal polynomials, but the reader must be wondering where the theta functions are hidden. To uncover them we give yet another q-extension of the beta integral. Euler's integral can be transformed to

$$
\int_0^\infty \frac{t^{\alpha-1}}{(1+t)^{\alpha+\beta}}\,dt = \frac{\Gamma(\alpha)\Gamma(\beta)}{\Gamma(\alpha+\beta)}.
$$

(4.5)

This was extended by Ramanujan to

$$
\int_0^\infty \frac{t^{\alpha-1}(-tq^{\alpha+\beta};q)_\infty\,dt}{(-t;q)_\infty} = \frac{\Gamma(\alpha)\Gamma(1-\alpha)\Gamma_q(\beta)}{\Gamma_q(1-\alpha)\Gamma_q(\alpha+\beta)},
$$

(4.6)

where

$$
\Gamma_q(\alpha) = \frac{(q;q)_\infty(1-q)^{1-\alpha}}{(q^\alpha;q)_\infty}.
$$

(4.7)

One thing lacking in (4.6) is the symmetry in α and β that occurs in (4.5). There is an extension of (4.6) that restores the symmetry. Consider

$$
f(a,b) = f_\gamma(a,b) = \int_0^\infty \frac{t^{\gamma-1}(-at;q)_\infty(-qb/t;q)_\infty}{(-t;q)_\infty(-q/t;q)_\infty}\,dt.
$$

(4.8)

Then

$$
\begin{aligned}
f(a,b) &= \int_0^\infty t^{\gamma-1}(1+at)\frac{(-aqt;q)_\infty(-qb/t;q)_\infty}{(-t;q)_\infty(-q/t;q)_\infty}\,dt \\
&= f(aq,b) + a\int_0^\infty t^\gamma \frac{(-aqt;q)_\infty(-qb/t;q)_\infty}{(-t;q)_\infty(-q/t;q)_\infty}\,dt \\
&= f(aq,b) + \frac{a}{q^{\gamma+1}}\int_0^\infty t^\gamma \frac{(-at;q)_\infty(-q^2b/t;q)_\infty}{(-t/q;q)_\infty(-q^2/t;q)_\infty}\,dt \\
&= f(aq,b) + \frac{a}{q^\gamma}\int_0^\infty t^{\gamma-1}\frac{(-at;q)_\infty(-q^2b/t;q)_\infty}{(-t;q)_\infty(-q/t;q)_\infty}\,dt \\
&= f(aq,b) + aq^{-\gamma}f(a,bq).
\end{aligned}
$$

The same argument with a factor of $(-qb/t;q)_\infty$ being removed leads to

$$
f(a,b) = f(a,bq) + bq^\gamma f(aq,b).
$$

These give

(4.9)
$$f(a, b) = \frac{(1 - ab)}{(1 - bq^\gamma)} f(a, bq)$$
$$= \frac{(ab; q)_\infty}{(bq^\gamma; q)_\infty} f(a, 0).$$

But $f(a, 1)$ is just Ramanujan's integral, so $f(a, b)$ can be evaluated from

$$f(a, b) = \frac{(ab; q)_\infty (q^\gamma; q)_\infty}{(bq^\gamma; q)_\infty (a; q)_\infty} f(a, 1).$$

When $a = q^{\beta+\gamma}$, $b = q^{\alpha-\gamma}$, then

(4.10)
$$\int_0^\infty t^{\gamma-1} \frac{(-tq^{\beta+\gamma}; q)_\infty (-q^{\alpha+q-\gamma}; q)_\infty}{(-t; q)_\infty (-q/t; q)_\infty} dt$$
$$= \frac{\Gamma(\gamma)\Gamma(1-\gamma)}{\Gamma_q(\gamma)\Gamma_q(1-\gamma)} \frac{\Gamma_q(\alpha)\Gamma_q(\beta)}{\Gamma_q(\alpha+\beta)}.$$

Ramanujan's integral can be evaluated by the same argument. See [7]. Formula (4.10) was first derived in [14] as a corollary of (4.6). The reason for giving the alternative proof here is to see the role played by the theta product. We can rewrite (4.10) as

(4.11)
$$\int_0^\infty \frac{t^{\gamma-1}(-at; q)_\infty(-bq/t; q)_\infty}{(-t; q)_\infty(-q/t; q)_\infty} dt = \frac{\pi(q^\gamma; q)_\infty(q^{1-\gamma}; q)_\infty(ab; q)_\infty}{\sin \pi\gamma (aq^{-\gamma}; q)_\infty(bq^\gamma; q)_\infty(q; q)_\infty}.$$

Then $a = b = 0$ gives

(4.12)
$$\int_0^\infty \frac{t^{\gamma-1}}{(-t; q)_\infty(-q/t; q)_\infty} dt = \frac{\pi(q^\gamma; q)_\infty(q^{1-\gamma}; q)_\infty}{\sin \pi\gamma (q; q)_\infty} = c(\gamma).$$

The moments of this measure are $c(\gamma + n)$, and

(4.13)
$$c(\gamma + n) = q^{-\gamma n - \binom{n}{2}} c(\gamma).$$

Using (4.2) we see that

(4.14)
$$d(n) = \int_0^\infty x^n w(x) \, dx = e^{n^2/4k^2 + n/2k^2 + 1/4k^2} = q^{-n^2/2 + n - 1}$$

so

(4.15)
$$\frac{d(n)}{d(0)} = q^{-n^2/2 + n}.$$

Then $\gamma = -3/2$ gives the same moments. Thus

(4.16)
$$\int_0^\infty \frac{p_n(x)p_m(x)x^{-5/2}}{(-x; q)_\infty(-q/x; q)_\infty} dx = 0, \qquad m \neq n.$$

Now the theta function is made explicit.

It is a natural question to ask how these polynomials fit into the scheme of classical orthogonal polynomials. The Laguerre polynomials are

(4.17)
$$L_n^\alpha(x) = \frac{(\alpha + 1)_n}{n!} {}_1F_1\left(\begin{matrix} -n \\ \alpha+1 \end{matrix}; x\right)$$

and their orthogonality is

(4.18) $$\int_0^\infty L_n^\alpha(x)L_m^\alpha(x)x^\alpha e^{-x}\,dx = 0, \qquad m \neq n.$$

Hermite polynomials arise in two ways from Laguerre polynomials.

(4.19)
$$H_{2n}(x) = (-1)^n 2^{2n} n! L_n^{-1/2}(x^2),$$
$$H_{2n+1}(x) = (-1)^n 2^{2n+1} n! x L_n^{1/2}(x^2)$$

is the well-known way. The other way comes from shifting

$$x^\alpha e^{-x} = \exp(-x + \alpha \log x)$$

to put the maximum at $x = 0$, and then rescaling and letting $\alpha \to \infty$. The result is

(4.20) $$\lim_{\alpha \to \infty} (2/\alpha)^{n/2} L_n^\alpha(\alpha + (2\alpha)^{1/2}x) = (-1)^n H_n(x)/n!.$$

It is the second limit that leads to the Stieltjes-Wigert polynomials. See [8]. Since this moment problem is indeterminate, there are many other positive measures that can be used. See [21] for others.

The q-Hermite polynomials given above are not the only q-extensions of Hermite polynomials. Another is given in [6] as a limit of the discrete q-ultraspherical polynomials, or the symmetric case of the big q-Jacobi polynomials. Here the weight function is not a theta function, but is a different extension of the exponential function.

5. A different view. There is a minor but attractive use of the modular transformation (1.7). For Jacobi polynomials there is an important fractional integral connection between Jacobi polynomials whose eigenvalues are the same. One version is

(5.1)
$$(1-x)^{\alpha+\mu}\frac{P_n^{(\alpha+\mu,\beta-\mu)}(x)}{P_n^{(\alpha+\mu,\beta-\mu)}(1)}$$
$$= \frac{\Gamma(\alpha+\mu+1)}{\Gamma(\alpha+1)\Gamma(\mu)}$$
$$\cdot \int_x^1 (1-y)^\alpha (y-x)^{\mu-1}\frac{P_n^{(\alpha,\beta)}(y)}{P_n^{(\alpha,\beta)}(1)}\,dy, \qquad \alpha > 1, \ \mu > 0,$$

where

(5.2)
$$P_n^{(\alpha,\beta)}(x) = \frac{(\alpha+1)_n}{n!}\,{}_2F_1\left(\begin{matrix}-n, n+\alpha+\beta+1 \\ \alpha+1\end{matrix}; \frac{1-x}{2}\right)$$
$$= \frac{(1-x)^{-\alpha}(1+x)^{-\beta}(-1)^n}{2^n n!}\frac{d^n}{dx^n}[(1-x)^{n+\alpha}(1+x)^{n+\beta}].$$

The eigenvalue of the standard equation satisfied by $P_n^{(\alpha,\beta)}(x)$ is $-n(n+\alpha+\beta+1)$.

Observe that this is constant for fixed n when $\alpha + \beta$ is fixed.

$$\int_{-1}^{1} P_n^{(\alpha,\beta)}(x)P_m^{(\alpha,\beta)}(x)(1-x)^\alpha(1+x)^\beta \, dx = 0, \qquad m \neq n.$$

There are two surprising and important facts in (5.1). One is the nonnegativity of the kernel. The second is the vanishing of the kernel for $-1 < y < x$.

There are a number of q-extensions of Jacobi polynomials. For the polynomials $W_n(x; a, b, c, d|q)$ there are some special cases that look very similar to Jacobi polynomials. One is

(5.3)

$$P_n^{(\alpha,\beta)}(x|q)$$

$$= \frac{(q^{\alpha+1}; q)_n}{(q; q)_n} {}_4\varphi_3 \left(\begin{matrix} q^{-n}, q^{n+\alpha+\beta+1}, q^{(\alpha/2+1/4)}e^{i\theta}, q^{(\alpha/2+1/4)}e^{-i\theta} \\ q^{\alpha+1}, -q^{(\alpha+\beta+1)/2}, -q^{(\alpha+\beta+2)/2} \end{matrix} ; q, q \right),$$

where $x = \cos\theta$ as usual. Nassrallah and Rahman [33] showed that

(5.4)

$$\frac{1}{2\pi} \int_{-1}^{1} w(y; a, b, \mu e^{i\theta}, \mu e^{-i\theta}) W_n(y; a, b, c, d) \, dy$$

$$= \frac{(ab\mu^2 q^n; q)_\infty \mu^n W_n(x; a\mu, b\mu, c\mu^{-1}, d\mu^{-1}|q)}{(q; q)_\infty (\mu^2; q)_\infty (abq^n; q)_\infty |(a\mu e^{i\theta}; q)_\infty (b\mu e^{i\theta}; q)_\infty|^2}.$$

Here

$$w(\cos\varphi; a, b, c, d) = \frac{(e^{2i\varphi}; q)_\infty (e^{-2i\varphi}; q)_\infty (\sin\varphi)^{-1}}{(ae^{i\varphi}; q)_\infty (be^{i\varphi}; q)_\infty (ce^{i\varphi}; q)_\infty (de^{i\varphi}; q)_\infty (ae^{-i\varphi}; q)_\infty}$$

$$\cdot \frac{1}{(be^{-i\varphi}; q)_\infty (ce^{-i\varphi}; q)_\infty (de^{-i\varphi}; q)_\infty}.$$

The choice $a = q^{(\alpha+1/2)/2}$, $b = aq^{1/2}$, $c = -q^{(\beta+1/2)/2}$, $d = cq^{1/2}$ gives the q-version of Jacobi polynomials above. When α, β are nonnegative and μ is real and less than one in absolute value, the kernel is nonnegative. However the vanishing no longer holds, since $w(y; a, b, \mu e^{i\theta}, \mu e^{i\theta})$ is positive when $-1 < y < 1$ and the above conditions hold, i.e., $a = q^{(\alpha+1/2)/2}$, $b = -q^{(\beta+1/2)/2}$ with $\alpha, \beta > 0$ and $-1 < \mu < 1$. It is a natural question to see how the special case of (5.4) for a, b, c, d given above reduces to (5.1) when $q \to 1$. The key to proving this easily is the modular transformation (1.7). See [33] for details. The kernel in (5.4) is real analytic in y for $-1 < y < 1$, and the limit when $q \to 1$ is an analytic function of y for $x < y < 1$ and $-1 < y < x$, but they are different analytic functions.

6. Other instances. There are other connections between theta and elliptic functions and orthogonal polynomials or solutions to three-term recurrences. Probably the earliest connection between three-term recurrences and hyperelliptic curves is in Abel [1]. This was pointed out by Gregory Chudnovsky [22], and he mentioned many later authors who considered this connection.

There is a second class of examples that Stieltjes found [45; 46, pp. 549–555 in reprint]. These polynomials were studied later by Carlitz [20]. Laguerre polynomials are the polynomials orthogonal on $[0, \infty)$ whose recurrence relation has linear coefficients with the outside terms having the same rate of growth. In this case the recurrence is

$$(6.1) \quad -xL_n^\alpha(x) = (n+1)L_{n+1}^\alpha(x) - (2n+\alpha+1)L_n^\alpha(x) + (n+\alpha)L_{n-1}^\alpha(x).$$

Note that $(n+1)$ and $(n+\alpha)$ have the same rate of growth as $n \to \infty$. The associated Laguerre polynomials are the polynomials formed from (6.1) when n in the coefficients is replaced by $n + \mu$. The orthogonality relation can be found explicitly when the weight function has no discrete mass points. See [16], and also see [29] for a second derivation of this orthogonality and an orthogonality relation for a variant of the associated Laguerre polynomials.

When the outside coefficients are linear in n but have a different rate of growth, the polynomials are called Meixner polynomials.

$$(6.2) \qquad\qquad M_n(x; \beta, c) = {}_2F_1\left({-n, -x \atop \beta}; 1 - \frac{1}{c}\right).$$

Their recurrence relation is
(6.3)
$$-xM_n(x) = \frac{c(n+\beta)}{1-c}M_{n+1}(x) - \left[\frac{c(n+\beta)+n}{1-c}\right]M_n(x) + \frac{n}{1-c}M_{n-1}(x).$$

When $0 < c < 1$ and $\beta > 0$ the orthogonality is

$$(6.4) \qquad\qquad \sum_{x=0}^{\infty} M_n(x)M_k(x)\frac{(\beta)_x}{x!}c^x = 0, \qquad k \neq n.$$

Quadratic growth of the same rate comes from the continuous dual Hahn polynomials

$$(6.5) \qquad\qquad R_n(x^2) = {}_3F_2\left({-n, a+ix, a-ix \atop a+b, a+c}; 1\right).$$

When $a, b, c > 0$ the orthogonality is
(6.6)
$$\int_{-\infty}^{\infty} R_n(x^2)R_m(x^2)\left|\frac{\Gamma(a+ix)\Gamma(b+ix)\Gamma(c+ix)}{\Gamma(2ix)}\right|^2 dx = 0, \qquad m \neq n.$$

See Wilson [52]. The recurrence relation is
(6.7)
$$-(x^2 + a^2)R_n(x^2) = (n+a+b)(n+a+c)R_{n+1}(x^2)$$
$$- [(n+a+b)(n+a+c) + n(n+b+c-1)]R_n(x^2)$$
$$+ n(n+b+c-1)R_{n-1}(x^2).$$

One can ask if there is an extension similar to the Meixner polynomials. Stieltjes [45, 46] found four examples. One is

$$(6.8) \qquad \begin{aligned} -xp_n(x) &= k^2(2n+1)(2n+2)p_{n+1}(x) \\ &\quad - (2n+1)^2(k^2+1)p_n(x) + (2n)(2n+1)p_{n-1}(x), \end{aligned}$$

which is almost of the form of the Meixner polynomials. The orthogonality is discrete with the mass points located on a quadratic sequence $x_j = a(2j+1)^2$, $j = 0, 1, 2, \ldots$, for an appropriate constant a. See Carlitz [20]. This natural extension of Meixner polynomials leads to the Lamé differential equation, so it is surprising that anything can be done explicitly.

These examples of Stieltjes were mentioned in one of the talks David and Gregory Chudnovsky gave at this meeting. The Chudnovskys should be contacted by anyone interested in arithmetic applications of a number of the polynomials mentioned in this paper.

BIBLIOGRAPHY

1. N. H. Abel, *Sur l'intégration de la formule différentièlle* $\rho \, dx / \sqrt{R}$, *R et* ρ *étant des fonctions entières*, J. Reine Angew. Math. **1** (1826); reprinted, *Oeuvres*. Vol. 1, Nouvelle édition, Christiania, 1881, pp. 104–144.

2. C. Adiga, B. C. Berndt, S. Bhargava, and G. N. Watson, *Chapter 16 of Ramanujan's second notebook: Theta functions and q-series*, Mem. Amer. Math. Soc. No. 315 (1985).

3. G. E. Andrews, *On the Alder polynomials and a new generalization of the Rogers-Ramanujan identities*, Trans. Amer. Math. Soc. **204** (1975), 40–64.

4. ____, *An analytic generalization of the Rogers-Ramanujan identities for odd moduli*, Proc. Nat. Acad. Sci. U.S.A. **71** (1974), 4082–4085.

5. ____, *Problems and prospects for basic hypergeometric functions*, Theory and Application of Special Functions, R. Askey, editor, Academic Press, 1975, pp. 191–224.

6. G. E. Andrews and R. Askey, *Classical orthogonal polynomials*, Polynômes Orthogonaux et Applications, C. Brezinski et al., editors, Lecture Notes in Math., vol. 1171, Springer-Verlag, 1985, pp. 36–62.

7. R. Askey, *Ramanujan's extensions of the gamma and beta functions*, Amer. Math. Monthly **87** (1980), 346–358.

8. ____, *Limits of some q-Laguerre polynomials*, J. Approx. Theory **46** (1986), 213–216.

9. ____, *Ramanujan's* $_1\psi_1$ *and formal Laurent series*, Indian J. Math. **29** (1987), 101–105.

10. ____, *Divided difference operators and classical orthogonal polynomials*, Rocky Mountain J. Math. **19** (1989), 1–5.

11. R. Askey and M. Ismail, *The Rogers q-ultraspherical polynomials*, Approximation Theory. III, E. W. Cheney, editor, Academic Press, 1980, pp. 175–182.

12. ____, *A generalization of ultraspherical polynomials*, Studies in Pure Mathematics, P. Erdös, editor, Birkäuser, 1983, pp. 55–78.

13. ____, *Recurrence relations, continued fractions, and orthogonal polynomials*, Mem. Amer. Math. Soc. No. 300 (1984).

14. R. Askey and R. Roy, *More q-beta integrals*, Rocky Mountain J. Math. **16** (1986), 365–372.

15. R. Askey and J. Wilson, *Some basic hypergeometric orthogonal polynomials that generalize Jacobi polynomials*, Mem. Amer. Math. Soc. No. 319 (1985).

16. R. Askey and J. Wimp, *Associated Laguerre and Hermite polynomials*, Proc. Roy. Soc. Edinburgh **96A** (1984), 15–37.

17. D. M. Bressoud, *On partitions, orthogonal polynomials and the expansion of certain infinite products*, Proc. London Math. Soc. **42** (1981), 478–500.

18. ____, *An easy proof of the Rogers-Ramanujan identities*, J. Number Theory **16** (1983), 235–241.

19. L. Carlitz, *Note on orthogonal polynomials related to theta functions*, Publ. Math. Debrecen **5** (1958), 222–228.

20. ____, *Some orthogonal polynomials related to elliptic functions*, Duke Math. J. **27** (1960), 443–460.

21. T. S. Chihara, *A characterization and a class of distribution functions for the Stieltjes-Wigert polynomials*, Canad. Math. Bull. **13** (1970), 529–532.

22. G. V. Chudnovsky, *An explicit solution of classical and quantum field theory models and parallel arithmetic problems: A unified approach*, informal notes for invited address at A.M.S. meeting, Jan. 7–10, 1981, unpublished.

23. J. M. Dobbie, *A simple proof of the Rogers-Ramanujan identities*, Quart. J. Math. Oxford Ser. (2) **13** (1962), 31–34.

24. A. Erdélyi et al., *Higher transcendental functions*, 2, McGraw-Hill, New York, 1953; reprinted, Krieger, Melbourne, Florida, 1981.

25. A. Garsia and S. Milne, *A Rogers-Ramanujan bijection*, J. Combin. Theory Ser. A **31** (1981), 289–339.

26. B. Gordon, *A combinatorial generalization of the Rogers-Ramanujan identities*, Amer. J. Math. **83** (1961), 393–399.

27. G. H. Hardy, *Ramanujan*, Cambridge Univ. Press, 1940; reprinted, Chelsea, New York, 1959.

28. E. Hecke, *Über einem Zusammenhang zwischen elliptischen Modulfunktionen und indefiniten quadratischen Formen*, Nach. K. Gesellschaft Wissen, zu Göttingen. Math. Physik. Kl. 1925, 35–44; reprinted in *Math. Werke*, Vandenhoeck and Ruprecht, Göttingen, 1959, pp. 418–427.

29. M. Ismail, J. Letessier, and G. Valent, *Linear birth and death models and associated Laguerre and Meixner polynomials* (to appear).

30. M. Ismail and D. Stanton, *On the Askey-Wilson and Rogers polynomials*, Canad. J. Math. (to appear).

31. M. Ismail, D. Stanton, and G. Viennot, *The combinatorics of q-Hermite polynomials and the Askey-Wilson integral*, European J. Combin. **8** (1987), 379–392.

32. P. A. MacMahon, *Combinatory analysis*. Vol. 2, Cambridge Univ. Press, London, 1916; reprinted, Chelsea, New York, 1960.

33. B. Nassrallah and M. Rahman, *Projection formulas, a reproducing kernel and a generating function for q-Wilson polynomials*, SIAM J. Math. Anal. **16** (1985), 186–197.

34. G. Pólya, *Elementarer Beweis einer Thetaformel*, Sitz. Berich. Preuss. Akad. Wissen., Phys.-Math. Kl. (1927), 158–161; reprinted, *Collected Papers*. Vol. 1, M.I.T. Press, Cambridge, Mass., 1974, pp. 303–306.

35. R. G. Ramanathan, *On Ramanujan's continued fraction*, Acta Arith. **43** (1983), 93–110.

36. K. G. Ramanathan, *On the Rogers-Ramanujan continued fraction*, Proc. Indian Acad. Sci. Math. Sci. **93**, Nos. 2& 3 (1984), 67–77.

37. ____, *Ramanujan's continued fraction*, Indian J. Pure Appl. Math. **16** (1985), 695–724.

38. L. J. Rogers, *Second memoir on the expansion of certain infinite products*, Proc. London Math. Soc. (1) **25** (1894), 318–343.

39. ____, *Third memoir on the expansion of certain infinite products*, Proc. London Math. Soc. (1) **26** (1895), 15–32.

40. ____, *On two theorems of combinatory analysis and some allied identities*, Proc. London Math. Soc. (2) **16** (1917), 315–336.

41. L. J. Rogers and S. Ramanujan, *Proof of certain identities in combinatory analysis (with a prefatory note by G. H. Hardy)*, Proc. Cambridge Philos. Soc. **19** (1919), 211–216.

42. I. J. Schur, *Ein Beitrag zur additiven Zahlentheorie und zur Theories der Kettenbrüche*, Sitz. Preuss. Akad. Wissen. 1917, Phys.-Math. Kl. 302–321; reprinted, *Gesammelte Abhandlungen*. Vol. II, Springer-Verlag, 1973, pp. 117–136.

43. A. Selberg, *Über einige arithmetische Identitäten*, Abh. Norske. Vidensk. Akad. Oslo Mat. Naturvidensk. (1936) Kl. 8.

44. G. W. Starcher, *On identities arising from solutions of q-difference equations and some interpretations in number theory*, Amer. J. Math. **53** (1931), 801–816.

45. T. J. Stieltjes, *Sur la réduction en fraction continue d'une série procédant suivant les puissances descendantes d'une variable*, Ann. Fac. Sci. Toulouse 3 (1889), H.1–17; reprinted, *Oeuvres complètes*. Vol. II, Groningen, 1918, pp. 184–200.

46. ____, *Recherches sur les fractions continues*, Ann. Fac. Sci. Toulouse **8** (1894), J. 1–122, **9** (1895), A 1–47; reprinted, *Oeuvres complètes*. Vol. II, Groningen, 1918, pp. 402–566.

47. G. Szegö, *Ein Beitrag zur Theorie der Thetafunktionen*, Sitz. Preuss. Akad. Wiss., Phys.-Math. Klasse, 1926, 242–252; reprinted, *Collected Papers*. Vol. 1, Birkhäuser, 1982, pp. 795–805.

48. ____, *An outline of the history of orthogonal polynomials*, Proc. Conf. Orthogonal Expansions and Their Continuous Analogues, D. Haimo, editor, Southern Ill. Univ. Press, Carbondale, Ill., 1968, pp. 3–11; reprinted, *Collected Papers*. Vol. 3, Birkhäuser, 1982, pp. 857–865. Comments are on pages 866–869.

49. ____, *Orthogonal polynomials*, 4th ed., Amer. Math. Soc. Colloq. Publ., vol. 23, Amer. Math. Soc., Providence, R.I., 1975.

50. G. N. Watson, *A new proof of the Rogers-Ramanujan identities*, J. London Math. Soc. **4** (1929), 4–9.

51. S. Wigert, *Sur les polynomes orthogonaux et l'approximation des fonctions continues*, Ark. Mat., Astron. Fysik **17** (1923), no. 18, 15 pp.

52. J. Wilson, *Some hypergeometric orthogonal polynomials*, SIAM J. Math. Anal. **11** (1980), 690–701.

UNIVERSITY OF WISCONSIN, MADISON

Proceedings of Symposia in Pure Mathematics
Volume **49** (1989), Part 2

The Multidimensional $_1\Psi_1$ Sum and Macdonald Identities for $A_l^{(1)}$

S. C. MILNE

Abstract. We survey the summation theorems and transformation properties for a new class of multiple q-series, which were inspired by certain aspects of mathematical physics and the unitary groups U(n). Our theory provides a rich generalization of the main summation theorems and transformations for classical one-variable q-series. As an application we obtain an elegant common U(n) generalization of Ramanujan's $_1\Psi_1$ summation, the $_6\Psi_6$ summation theorem and the Macdonald identities for $A_l^{(1)}$.

Introduction. Classical basic hypergeometric functions of one-variable (q-series) have many significant applications in several areas of pure and applied mathematics. More familiar applications include classical analysis, combinatorics, and additive number theory. Recently, q-series have enriched new developments in physics, Lie algebras, transcendental number theory, and statistics. Extensive references to and accounts of the general theory and applications of basic hypergeometric functions can be found in books by G. E. Andrews [**3, 4, 5**], W. N. Bailey [**35**], R. J. Baxter [**38**], L. J. Slater [**86**], and in the papers of C. Adiga, B. C. Berndt, S. Bhargava, and G. N. Watson [**1**], G. E. Andrews [**6–26**], G. E. Andrews and R. Askey [**27**], G. E. Andrews, R. J. Baxter, and P. J. Forrester [**29**], R. Askey and M. Ismail [**33**], R. Askey and J. A. Wilson [**34**], R. J. Baxter [**36, 37**], W. Hahn [**49–52**], M. E. H. Ismail [**56**], V. G. Kac and D. H. Peterson [**61**], J. Lepowsky [**63**], I. G. Macdonald [**67–69**], and D. Zeilberger and D. M. Bressoud [**92**].

The symmetries, transformation properties, and summation theorems for basic hypergeometric series are responsible for many of the above applications. Several higher-dimensional generalizations of some of these results for one-variable q-series are contained in the theory of (basic) Appell and Lauricella functions of several variables, which have been studied in [**3, 5, 9, 10, 30, 35, 70, 86**]. Generally, none of these multiple series has as much depth,

1980 *Mathematics Subject Classification* (1985 *Revision*). Primary 33A35, 05A19; Secondary 17B65, 05A15.

richness, or contact with other major areas of mathematics and physics as does classical (one-variable) basic hypergeometric series. Formally, there is simply too much freedom in the number of ways one can move from a single sum to multidimensional analogs.

Recently, modern mathematical physics has provided a new natural multivariable generalization of ordinary classical hypergeometric series that has subsequently led to multiple basic hypergeometric series that appear to have the required depth and rich interaction with other parts of mathematics and physics. Furthermore, extensions in [81] of these new multiple hypergeometric series contain (basic) Appell and Lauricella functions as a special limiting case.

The three mathematical physicists L. C. Biedenharn, W. J. Holman III, and J. D. Louck showed in [54, 55] how the classical work on ordinary hypergeometric series is intimately related to the irreducible representations of the compact group SU(2). Similarly, they also initiated the study of the generalized multiple hypergeometric series $W_m^{(n)}$ and $F^{(n)}$, which arise in the theory of Wigner coefficients for SU(n). This work was done in the context of the quantum theory of angular momentum [39, 40] and the special unitary groups SU(n).

By utilizing explicit expressions for the matrix elements of multiplicity-free Wigner and Racah coefficients in U(n) both L. C. Biedenharn, W. J. Holman III, J. D. Louck [55] and W. J. Holman III [54] established generalizations of known one-variable hypergeometric summation theorems for $W_m^{(n)}$ and $F^{(n)}$, respectively. Essentially, the same matrix element is computed in two different ways. The methods rely upon the representation theory of U(n). Recently, R. A. Gustafson [45] significantly extended this program by proving a generalized Biedenharn-Elliott identity for multiplicity-free U(n) Racah coefficients. As a consequence of an explicit evaluation of a special case of this identity, he then derived an elegant generalization of Whipple's [89, 90] classical transformation of a very well poised $_7F_6(1)$ into a balanced $_4F_3(1)$, to the multiple hypergeometric series $W_m^{(n)}$ and $F^{(n-1)}$, respectively. Suitable special cases and limits of this U(n) generalization of Whipple's transformation yield multivariable generalizations of most classical summation theorems, including W. J. Holman III's [54] U(n) generalization of the summation theorems of Gauss of Pfaff-Saalschutz.

In several recent papers [72–83] we have introduced and studied natural q-analogs $[W]_m^{(n)}$ and $[F]^{(n)}$ of $W_m^{(n)}$ and $F^{(n)}$, respectively. The series $[W]_m^{(n)}$ satisfy a general q-difference equation [73] that generalizes our corresponding difference equation for $W_m^{(n)}$ from [71]. In addition [77], we have discovered basic hypergeometric series $[H]^{(n)}$ very well poised in U(n), which provides an elegant, explicit, multivariable generalization of the one-variable classical, very well poised basic hypergeometric series responsible for the Rogers-Ramanujan-Schur identities [3, 5].

Recently [78, 79], we found direct, elementary proofs of a q-analog of W. J. Holman III's [54] $U(n)$ generalization of the summation theorems of Gauss and Pfaff-Saalschutz. The $U(n)$ Gauss summation theorem [78] for $[F]^{(n)}$ series was obtained by iterating a suitable multivariable generalization of the Gauss reduction formula for a $_2F_1(1)$. The $U(n)$ Pfaff-Saalschutz summation theorem [79] for $[F]^{(n)}$ series is a consequence of q-$U(n)$ Gauss and a new general q-difference equation for "balanced" basic hypergeometric series $[F]^{(n)}$ in $U(n)$. By utilizing q-$U(n)$ Pfaff-Saalschutz and our q-difference equations for $[W]_m^{(n)}$ and $[F]^{(n)}$, we found in [80] a direct, elementary proof of a q-analog of R. A. Gustafson's $U(n)$ Whipple's transformation. Our q-difference equation approach to the theory of the multiple q-series $[W]_m^{(n)}, [F]^{(n)}$, and $[H]^{(n)}$ is complementary to the Biedenharn-Gustafson-Holman III-Louck representation-theoretic treatment of the ordinary multiple series $W_m^{(n)}$ and $F^{(n)}$. The paper [48] strongly suggests that $[W]_m^{(n)}, [F]^{(n)}$, and $[H]^{(n)}$ can also be studied from the viewpoint of representation theory.

Very recently [81] we extended our earlier work in [79] and provided two new multivariable generalizations of Jackson's classical q-analog of the balanced $_3F_2$ or Pfaff-Saalschutz summation theorem. These results apply to a new class of "balanced" multiple basic hypergeometric series analogous to the series $[F]^{(n)}$ in $U(n)$ and yield the Gauss summation theorem [78] for $[F]^{(n)}$ series in $U(n)$ as a special limiting case.

Subsequently, in [82] we obtained a $U(n)$ generalization of G. E. Andrews's [10] infinite family of extensions of Watson's q-analog of Whipple's transformation as well as his corresponding [13] multiple series generalization of Jackson's classical q-analog of the balanced $_3F_2$ summation theorem. Our work here involves multiple sums that are themselves iterations of other multiple sums. One of the simpler special cases of these results yields a new, different, and more symmetrical multiple series generalization of Watson's q-analog of Whipple's transformation than the one appearing in [80]. The $q = 1$ case of this new q-$U(n)$ Whipple's transformation is also much more symmetrical than the corresponding result in [45].

In a different direction, several of the results in [79, 81, 82] are utilized in [83] to derive multiple series extensions of the little q-Jacobi orthogonal polynomials of Andrews and Askey [27] and Andrews [13]. The paper [83] also contains an elegant solution of the general connection coefficient problem for these new multiple series little q-Jacobi polynomials.

In [75] basic properties of $[F]^{(n)}$ led to a $U(n)$ multiple series refinement of the q-binomial theorem [3, 35, 86], and a direct, elementary derivation of the Macdonald identities [43, 62, 64, 67, 91] for $A_l^{(1)}$. Subsequently, M. E. H. Ismail's proof from [52] was extended in [76] to give a $U(n)$ multiple series generalization of Ramanujan's $_1\Psi_1$ summation [7, 8, 28, 31, 51, 59] directly from the $U(n)$ q-binomial theorem. As a consequence, we obtained a new generalization of the Macdonald identities for $A_l^{(1)}$, with extra free parameters. Very recently, R. A. Gustafson [46] put together an induction argument based upon the noterminating $U(n)_6\Phi_5$ summation theorem [77] and certain multiple contour integrals to prove a $U(n)$ generalization of the $_6\Psi_6$ summation theorem [10].

In this paper we describe how the results from [44–47, 71–80] provide an elegant common $U(n)$ generalization of Ramanujan's $_1\Psi_1$ summation, the $_6\Psi_6$ summation theorem, and the Macdonald identities for $A_l^{(1)}$. We then survey the summation theorems and transformation properties for the multiple q-series $[W]_m^{(n)}$, $[F]^{(n)}$, and $[H]^{(n)}$.

The contents of this paper are organized as follows.

In §2 we review the definitions of classical one-variable basic hypergeometric series and bilateral basic series. We state W. N. Bailey's $_6\Psi_6$ summation theorem and recall how both Jacobi's triple product identity [3, 5] and Ramanujan's $_1\Psi_1$ summation [7, 8, 28, 31, 51, 56, 59] are consequences of the q-binomial theorem. These applications of the q-binomial theorem motivated much of our work in [75, 76].

The multiple q-series $[W]_m^{(n)}$ and $[F]^{(n)}$ are introduced in §3. We describe here how the Macdonald identities for $A_l^{(1)}$ and our $U(n)$ multiple series generalization of Ramanujan's $_1\Psi_1$ summation are both consequences of the $U(n)$ q-binomial theorem [75].

§4 gives R. A. Gustafson's [46] $U(n)$ generalization of the $_6\Psi_6$ summation theorem and his extension of the $U(n)$ $_1\Psi_1$ summation theorem in [76]. We include the hypergeometric series $[H]^{(n)}$ very well poised in $U(n)$ as part of this section.

Our q-analogs [78, 79] of W. J. Holman III's [54] $U(n)$ generalizations of the summation theorems of Gauss and Pfaff-Saalschutz are presented in §5.

§6 contains a discussion based upon [80] of our q-analogs of R. A. Gustafson's $U(n)$ Whipple's transformation and $U(n)$ generalization of J. Dougall's summation theorem.

Our most recent extension of much of the work in [79, 80] appears in [81–83]. We are currently utilizing the results in [81–83] to derive a $U(n)$ generalization of Bailey's lemma [20], Bailey's transform [20, 86], the Bailey lattice [2, 4], and q-Lagrange inversion [44, 87]. This latest work will appear elsewhere.

2. Background information. We begin with the classical q-binomial theorem [3, 35, 86] given by

THEOREM 2.1 (q-binomial theorem). *If* $|q| < 1$, $|t| < 1$, *then*

$$(2.2) \qquad \sum_{n=0}^{\infty} \frac{(a)_n}{(q)_n} \cdot t^n = \frac{(at)_\infty}{(t)_\infty},$$

where $(A)_n$ *and* $(A)_\infty$ *are defined by*

$$(2.3a) \qquad (A)_n = (1 - A)(1 - Aq) \cdots (1 - Aq^{n-1}),$$

and

(2.3b)
$$(A)_\infty = \lim_{n\to\infty} (A)_n = \prod_{r=0}^{\infty}(1 - Aq^r).$$

Note that we may define $(A)_n$ for all real numbers n by

(2.4)
$$(A)_n = (A)_\infty/(Aq^n)_\infty.$$

In particular,

(2.5a)
$$(A)_0 = 1,$$

and

(2.5b) $(A)_{-n} = (1 - A/q^n)^{-1} \cdots (1 - A/q)^{-1} = (-A)^{-n}q^{n(n+1)/2} \cdot (q/A)_n^{-1}.$

Theorem 2.1 is originally due to Cauchy [**42**, p. 40] in 1843. To prove Theorem 2.1, just note that both sides of (2.2) satisfy the q-difference equation and initial condition

(2.6a)
$$(1 - t)F(t) = (1 - at)F(tq),$$

(2.6b)
$$F(0) = 1.$$

Now, Cauchy and Gauss independently utilized Theorem 2.1 to give a simple, elegant proof of the fundamental Jacobi triple product identity [**60**] given by

THEOREM 2.7 (JACOBI). *If* $x \neq 0$, $|q| < 1$, *then*

(2.8)
$$\sum_{n=-\infty}^{+\infty} x^n q^{(n^2+n)/2} = \prod_{n=0}^{\infty}(1 + xq^{n+1})(1 + x^{-1}q^n)(1 - q^{n+1})$$
$$= (-xq)_\infty(-x\Gamma^{-1})_\infty(q)_\infty.$$

They first set

(2.9)
$$a = q^{-2N} \quad \text{and} \quad t = -xq^{1+N}$$

in (2.2) and rewrite the resulting identity in the form

(2.10a)
$$\sum_{n=-N}^{N} x^n q^{(n^2+n)/2} \cdot \{(q)_{2N}/((q)_{N-n}(q)_{N+n})\}$$

(2.10b)
$$= \prod_{n=1}^{N}(1 + xq^n)(1 + x^{-1}q^{n-1}).$$

Letting $N \to \infty$ in (2.10) immediately yields (2.8).

Important applications of Theorem 2.7 to the theories of elliptic functions and partitions are described in the books of Andrews [**3**] and Hardy and Wright [**53**].

Observing that

(2.11)
$$\frac{1}{(q)_n} = \frac{(q^{n+1})_\infty}{(q)_\infty}$$

vanishes when $n = -1, -2, \ldots$, one sees immediately that Theorem 2.1 is the $b = q$ case of

THEOREM 2.12 (Ramanujan's $_1\Psi_1$ summation). *Let*

(2.13a) $0 < |q| < 1$

and

(2.13b) $|b/a| < |t| < 1.$

Then

(2.14) $$\sum_{n=-\infty}^{+\infty} \frac{(a)_n}{(b)_n} \cdot t^n = \left\{ \frac{(at)_\infty (q/(at))_\infty}{(t)_\infty (b/(at))_\infty} \right\} \cdot \left\{ \frac{(b/a)_\infty (q)_\infty}{(q/a)_\infty (b)_\infty} \right\}.$$

It is not difficult to see that replacing a, q, b, and t by $(-1/c), q, 0$, and (qxc), respectively, in Theorem 2.12, and letting $c \to 0$ yields Theorem 2.7. Thus, Ramanujan's $_1\Psi_1$ summation contains both the q-binomial theorem and the Jacobi triple product identity as special cases. In [31] Theorem 2.12 is viewed as an extension of the beta function given as an integral on $[0, \infty)$. Applications to orthogonal polynomials defined by basic hypergeometric series are discussed in [28, 31].

In [7, 8, 28, 31, 51, 59] clever rearrangements of series or q-difference equations are utilized to prove Theorem 2.12. The most elegant proof appears in [56]. First, observe that both sides of (2.14) are analytic in $z = b$ for $|z|$ sufficiently small. The q-binomial theorem immediately implies that both sides of (2.14) agree when $z = q^m$, for $m = 0, 1, 2, \ldots$. Both sides of (2.14) are now equal for general b (subject to (2.13)) since 0 is an interior point of the domain of analyticity. We generalized this proof to higher dimensions in [76].

Motivated by Theorems 2.1, 2.7, and 2.12 we now recall the definitions of classical basic and bilateral basic hypergeometric series. We first have

DEFINITION 2.15 (classical basic hypergeometric series). We let

$$_m\Phi_n \left[\begin{array}{cc} a, & (\alpha) \\ & (\beta) \end{array} \bigg| q; x \right]$$

denote the basic hypergeometric series

$$_m\Phi_n \left[\begin{array}{cc} a, & (\alpha) \\ & (\beta) \end{array} \bigg| q; x \right] \equiv {}_m\Phi_n \left[\begin{array}{cc} a, & \alpha_1, \alpha_2, \ldots, \alpha_{m-1} \\ & \beta_1, \beta_2, \ldots, \beta_n \end{array} \bigg| q; x \right]$$

(2.16a) $$= \sum_{l=0}^{\infty} \frac{(a)_l (\alpha_1)_l (\alpha_2)_l \cdots (\alpha_{m-1})_l}{(\beta_1)_l (\beta_2)_l \cdots (\beta_n)_l} \cdot \frac{x^l}{(q)_l},$$

where

(2.16b) $(A)_l \equiv (A; q)_l = (1 - A)(1 - Aq) \cdots (1 - Aq^{l-1}),$

(2.16c) $|x| < 1, \quad |q| < 1, \quad \beta_i \neq q^{-N},$

for any nonnegative integer N. The basic hypergeometric series in (2.16a) is well poised provided that

(2.16d) $m = n + 1 \quad \text{and} \quad aq = \alpha_1 \beta_1 = \cdots = \alpha_n \beta_n,$

and very well poised if, in addition,

(2.16e) $\alpha_1 = q\sqrt{a}, \quad \alpha_2 = -q\sqrt{a}, \quad \beta_1 = \sqrt{a}, \quad \beta_2 = -\sqrt{a}.$

Finally, the series (2.16a) is balanced if

(2.16f) $(\beta_1\beta_2\cdots\beta_n) = (a\alpha_1\alpha_2\cdots\alpha_{m-1})q.$

The a and α_i are numerator parameters and the β_i are denominator parameters.

Note that if (2.16a) is very well poised we have

(2.17) $\dfrac{(q\sqrt{a})_l(-q\sqrt{a})_l}{(\sqrt{a})_l(-\sqrt{a})_l} = \dfrac{(1-aq^{2l})}{(1-a)}.$

Bilateral basic series are determined by

DEFINITION 2.18 (classical bilateral basic series). Let $n \leq m$, $|q| < 1$, and $b_i \neq q^{-N}$, $a_i \neq q^{(N+1)}$, for all nonnegative integers N. Then,

$$_m\Psi_n\begin{bmatrix} a_1,\ldots,a_m; q; z \\ b_1,\ldots,b_n \end{bmatrix}$$

(2.19)
$$= \sum_{l=-\infty}^{\infty} \frac{(a_1)_l(a_2)_l\cdots(a_m)_l}{(b_1)_l(b_2)_l\cdots(b_n)_l}\cdot z^l,$$

where

(2.20a) $|z| < 1, \quad \text{if } n < m,$

and

(2.20b) $|(b_1\cdots b_n)/(a_1\cdots a_m)| < |z| < 1, \quad \text{if } n = m.$

We will sometimes use the product notation

$$\prod[(a);(b)] \equiv \prod \begin{bmatrix} a_1,\ldots,a_m; \\ b_1,\ldots,b_n; \end{bmatrix} q$$

(2.21)
$$= \frac{(a_1)_\infty(a_2)_\infty\cdots(a_m)_\infty}{(b_1)_\infty(b_2)_\infty\cdots(b_n)_\infty}.$$

One of the most fundamental results in the theory of basic hypergeometric series is Watson's [88] q-analog of Whipple's [89, 90] classical transformation of a very well poised $_7F_6(1)$ into a balanced $_4F_3(1)$. This result leads to q-analogs of all the main summation theorems for classical ordinary hypergeometric series. Watson's transformation is given by

THEOREM 2.22 (WATSON).

(2.23a) $_8\Phi_7\begin{bmatrix} a, q\sqrt{a}, -q\sqrt{a}, b, c, d, e, q^{-n} \\ \sqrt{a}, -\sqrt{a}, \dfrac{aq}{b}, \dfrac{aq}{c}, \dfrac{aq}{d}, \dfrac{aq}{e}, aq^{n+1} \end{bmatrix} \dfrac{a^2q^{n+2}}{bcde}$

(2.23b) $= \left\{ (aq)_n((aq)/(de))_n(aq/d)_n^{-1}(aq/e)_n^{-1} \right\}$

(2.23c) $\cdot {}_4\Phi_3\begin{bmatrix} ((aq)/(bc)), d, e, q^{-n} \\ ((de)/(aq^n)), (aq/b), (aq/c) \end{bmatrix} q.$

We conclude this section with Bailey's $_6\Psi_6$ summation theorem in

THEOREM 2.24 (BAILEY).

(2.25a)
$$_6\Psi_6\left[\begin{array}{c} q\sqrt{a},-q\sqrt{a},b,c,d,e, \\ \sqrt{a},-\sqrt{a},\frac{aq}{b},\frac{aq}{c},\frac{aq}{d},\frac{aq}{e}; \end{array} ;q;\; \frac{a^2q}{bcde}\right]$$

(2.25b)
$$= \prod\left[\begin{array}{c} aq,\frac{aq}{bc},\frac{aq}{bd},\frac{aq}{be},\frac{aq}{cd},\frac{aq}{ce},\frac{aq}{de},q,\frac{q}{a}; \\ \frac{q}{b},\frac{q}{c},\frac{q}{d},\frac{q}{e},\frac{aq}{b},\frac{aq}{c},\frac{aq}{d},\frac{aq}{e},\frac{a^2q}{bcde}; \end{array} q\right].$$

The identity (2.25) is probably the most general summation identity known for bilateral basic hypergeometric series. Andrews [9, §3] deduces many important diverse results in number theory from (2.25). Other q-series identities that follow from the $_6\Psi_6$ summation are given by Andrews [26, 29] and Slater [85].

3. U(n) generalization of Ramanujan's $_1\Psi_1$ summation. Just as Theorems 2.7 and 2.12 are consequences of Theorem 2.1, their higher-dimensional generalizations in this section follow from

THEOREM 3.1 (U(n) refinement of the q-binomial theorem). *If* $0 < |q| < 1$, $|t| < 1$, *then*

(3.2a)
$$\sum_{y_1,\ldots,y_n\geq 0}\left(\prod_{1\leq i<r\leq n}(1-(z_i/z_r)\cdot q^{y_i-y_r})/(1-(z_i/z_r))\right)$$

(3.2b)
$$\cdot\left(\prod_{1\leq i,r\leq n}(a_i(z_r/z_i))_{y_r}\right)\cdot\left(\prod_{1\leq i,r\leq n}(q(z_r/z_i))_{y_r}\right)^{-1}$$

(3.2c)
$$\cdot q^{(y_2+2y_3+\cdots+(n-1)y_n)}\cdot t^{(y_1+\cdots+y_n)}$$

(3.2d)
$$= (a_1a_2\cdots a_nt)_\infty/(t)_\infty,$$

where

(3.3a)
$$(z_i/z_r)\neq q^p, \quad \text{if } 1\leq i<r\leq n \text{ and } p\in\mathbf{Z};$$

and

(3.3b)
$$(qz_r/z_i)\neq q^{-p}, \quad \text{if } 1\leq r\neq i\leq n,\ p\in\mathbf{Z},\ \text{and } p\geq -1.$$

Clearly, the $n = 1$ case of Theorem 3.1 is Theorem 2.1. Theorem 3.1 was derived in [75] by making use of the basic hypergeometric series $[F]^{(n)}$ in U(n), introduced in [73]. We will discuss the proof of Theorem 3.1 at the end of this section.

Our higher-dimensional generalization of Theorem 2.12 is given by

THEOREM 3.4 (U(n) generalization of Ramanujan's $_1\Psi_1$ summation). *Let*

(3.5a)
$$0 < |q| < 1$$

and

(3.5b)
$$|(b^nq^{(1-n)})/(a_1a_2\cdots a_n)| < |t| < 1.$$

Then

$$
(3.6a) \quad \sum_{y_1=-\infty}^{+\infty} \cdots \sum_{y_n=-\infty}^{+\infty} \left(\prod_{1 \le i < r \le n} (1 - (z_i/z_r) \cdot q^{y_i - y_r})/(1 - (z_i/z_r)) \right)
$$

$$
(3.6b) \quad \cdot \left(\prod_{1 \le i, r \le n} (a_i(z_r/z_i))_{y_r} \right) \cdot \left(\prod_{1 \le i, r \le n} (b(z_r/z_i))_{y_r} \right)^{-1}
$$

$$
(3.6c) \quad \cdot q^{(y_2 + 2y_3 + \cdots + (n-1)y_n)} \cdot t^{(y_1 + \cdots + y_n)}
$$

$$
(3.6d) \quad = \left\{ \frac{(a_1 a_2 \cdots a_n t)_\infty (q(a_1 a_2 \cdots a_n t)^{-1})_\infty}{(t)_\infty (q^{(1-n)} b^n (a_1 a_2 \cdots a_n t)^{-1})_\infty} \right\}
$$

$$
(3.6e) \quad \cdot \left\{ \prod_{1 \le i, r \le n} \left[\frac{((b/a_i)(z_i/z_r))_\infty (q(z_r/z_i))_\infty}{((q/a_i)(z_i/z_r))_\infty (b(z_r/z_i))_\infty} \right] \right\},
$$

where

$$
(3.7a) \quad (z_i/z_r) \ne q^p, \quad \text{if } 1 \le i < r \le n \text{ and } p \in \mathbf{Z};
$$
$$
(3.7b) \quad (q z_r/z_i) \ne q^{-p}, \quad \text{if } 1 \le r \ne i \le n, \ p \in \mathbf{Z}, \text{ and } p \ge -1;
$$
$$
(3.7c) \quad (z_i/z_r) \ne a_i q^{-p}, \quad \text{if } 1 \le i, r \le n, \ p \in \mathbf{Z}, \text{ and } p \ge 1;
$$
$$
(3.7d) \quad t \ne q^{-p}, \quad \text{if } p \in \mathbf{Z} \text{ and } p \ge 0;
$$
$$
(3.7e) \quad (a_1 a_2 \cdots a_n t) \ne q^p, \quad p \in \mathbf{Z};
$$

and

$$
(3.8a) \quad b \ne q^{-p} \cdot (z_i/z_r), q^{-p}(a_i z_r) z_i,
$$

or

$$
(3.8b) \quad (a_1 a_2 \cdots a_n t)^{1/n} \cdot q^{(1 - 1/n)} \cdot q^{-(p/n)},
$$

with

$$
(3.8c) \quad 1 \le i, r \le n, \quad p \in \mathbf{Z}, \quad \text{and} \quad p \ge 0.
$$

It is immediate that the $n = 1$ case of Theorem 3.4 is Theorem 2.12, and that the $b = q$ case of Theorem 3.4 is Theorem 3.1.

Theorem 3.4 was first proven in [76]. Both sides of (3.6) are analytic in $z = b$ for $|z|$ sufficiently small. Theorem 3.1 implies that both sides of (3.6) agree when $z = q^m$, for $m = 0, 1, 2, \ldots$. Thus, both sides of (3.6) are now equal for general b, subject to (3.8).

From Theorem 2.12 it is not hard to see that the products in (3.6d–e) can be written as

(3.9a)
$$\left\{ \sum_{M=-\infty}^{+\infty} \frac{(a_1 a_2 \cdots a_n)_M}{(q^{(1-n)} \cdot b^n)_M} \cdot t^M \right\}$$

(3.9b)
$$\cdot \left\{ \frac{(q/(a_1 a_2 \cdots a_n))_\infty (q^{(1-n)} \cdot b^n)_\infty}{(q^{(1-n)} \cdot b^n/(a_1 a_2 \cdots a_n))_\infty (q)_\infty} \right\}$$

(3.9c)
$$\cdot \left\{ \prod_{1 \le i, r \le n} \left[\frac{((b/a_i)(z_i/z_r))_\infty (q(z_r/z_i))_\infty}{((q/a_i)(z_i/z_r))_\infty (b(z_r/z_i))_\infty} \right] \right\}.$$

It is now immediate that equating coefficients of t^M in the Laurent series (3.6a–c) and (3.9a–c) yields

THEOREM 3.10. *Let* $M \in \mathbf{Z}$, $|q| < 1$, *and assume that* (3.5b), (3.7), *and* (3.8) *hold. We then have*

(3.11a)
$$\sum_{\substack{y_1 + \cdots + y_n = M \\ y_i \in \mathbf{Z}}} \left(\prod_{1 \le i < r \le n} (1 - (z_i/z_r) \cdot q^{y_i - y_r}) \right) \cdot \left(\prod_{1 \le i, r \le n} (a_i(z_r/z_i))_{y_r} \right)$$

(3.11b)
$$\cdot \left(\prod_{1 \le i, r \le n} (b(z_r/z_i))_{y_r} \right)^{-1} \cdot q^{(y_2 + 2y_3 + \cdots + (n-1)y_n)}$$

(3.12a)
$$= \left\{ \left[\frac{(a_1 a_2 \cdots a_n)_M}{(q^{(1-n)} \cdot b^n)_M} \right] \cdot \left(\prod_{1 \le i < r \le n} (1 - (z_i/z_r)) \right) \right\}$$

(3.12b)
$$\cdot \left\{ \frac{(q/(a_1 a_2 \cdots a_n))_\infty (q^{(1-n)} \cdot b^n)_\infty}{(q^{(1-n)} \cdot b^n/(a_1 a_2 \cdots a_n))_\infty (q)_\infty} \right\}$$

(3.12c)
$$\cdot \left\{ \prod_{1 \le i, r \le n} \left[\frac{((b/a_i)(z_i/z_r))_\infty (q(z_r/z_i))_\infty}{((q/a_i)(z_i/z_r))_\infty (b(z_r/z_i))_\infty} \right] \right\}.$$

The product formula for a Vandermonde determinant and an interchange of summation transforms the $M = 0$ case of Theorem 3.10 into

THEOREM 3.13 ($_1\Psi_1$ generalization of the Macdonald identities for $A_l^{(1)}$).
Let $|q| < 1$. Assume that (3.5b), (3.7), and (3.8) hold. We then have

$$(3.14a) \quad \sum_{\sigma \in S_n} \varepsilon(\sigma) \prod_{r=1}^{n} z_{\sigma(r)}^{(r-\sigma(r))} \sum_{\substack{y_1+\cdots+y_n=0 \\ y_i \in \mathbf{Z}}} q^{[(y_{\sigma(2)}+2y_{\sigma(3)}+\cdots+(n-1)y_{\sigma(n)})]}$$

$$(3.14b) \quad \cdot \left(\prod_{1 \le i,r \le n} (a_i(z_r/z_i))_{y_r} \right) \cdot \left(\prod_{1 \le i,r \le n} (b(z_r/z_i))_{y_r} \right)^{-1}$$

$$(3.15a) \quad = \left\{ \frac{(q/(a_1a_2\cdots a_n))_\infty (q^{(1-n)} \cdot b^n)_\infty}{(q^{(1-n)} \cdot b^n/(a_1a_2\cdots a_n))_\infty (q)_\infty} \right\}$$

$$(3.15b) \quad \cdot \left\{ \left(\prod_{1 \le i < r \le n} (1-(z_i/z_r)) \right) \right.$$

$$\left. \cdot \left(\prod_{1 \le i,r \le n} \left[\frac{((b/a_i)(z_i/z_r))_\infty (q(z_r/z_i))_\infty}{((q/a_i)(z_i/z_r))_\infty (b(z_r/z_i))_\infty} \right] \right) \right\},$$

where $\varepsilon(\sigma)$ is the sign of the permutation σ, and $\mathbf{Z}$ is the set of all integers.

Replacing b and a_i by 0 and $(-1/c)$, respectively in Theorem 3.13, simplifying, and then letting $c \to 0$ yields (4.3) of [75] given by

$$(3.16a) \quad \left\{ (q)_\infty^{n-1} \cdot \left(\prod_{1 \le i < r \le n} (z_i/z_r)_\infty (q z_r/z_i)_\infty \right) \right\}$$

$$(3.16b) \quad = \sum_{\sigma \varepsilon S_n} \varepsilon(\sigma) \prod_{r=1}^{n} z_{\sigma(r)}^{(r-\sigma(r))} \sum_{\substack{y_1+\cdots+y_n=0 \\ y_i \in \mathbf{Z}}} \left(\prod_{r=1}^{n} z_{\sigma(r)}^{(ny_{\sigma(r)})} \right)$$

$$(3.16c) \quad \cdot \{ q^{[n/2(y_{\sigma(1)}^2+\cdots+y_{\sigma(n)}^2)+(y_{\sigma(2)}+2y_{\sigma(3)}+\cdots+(n-1)y_{\sigma(n)})]} \}.$$

In [75] it is proven that (3.16) is equivalent to the Macdonald identities for $A_l^{(1)}$. Thus, Theorem 3.13 may be viewed as a generalization of the Macdonald identities for $A_l^{(1)}$ with the extra parameters $a_1, \ldots, a_n$, and b.

Recall that Theorem 2.7 is a limiting case of Theorem 2.12. A similar situation holds in higher dimensions. Consider the specialization

$$(3.17a) \qquad\qquad q \to q,$$
$$(3.17b) \qquad\qquad b \to 0,$$
$$(3.17c) \qquad a_i \to (-1/c), \qquad 1 \le i \le n,$$
$$(3.17d) \qquad\qquad t \to (-1)^{(n-1)} q x c^n.$$

In [76] we showed that substituting (3.17) into both sides of (3.6), simplifying, and then letting $c \to 0$ gives

THEOREM 3.18 (U(n) generalization of the Jacobi triple product identity). *If $x \neq 0$, $|q| < 1$, and $\{z_1, \ldots, z_n\}$ are indeterminants such that*

(3.19) $z_i/z_r \neq q^p$,   *if $1 \leq i < r \leq n$ and p is any integer,*

then we have

(3.20a)
$$\sum_{y_1=-\infty}^{+\infty} \cdots \sum_{y_n=-\infty}^{+\infty} \left(\prod_{1 \leq i < r \leq n} (1 - (z_i/z_r) \cdot q^{y_i - y_r}) \right)$$

(3.20b) $\cdot \left(\prod_{i=1}^{n} z_i^{(n y_i - (y_1 + \cdots + y_n))} \right) \cdot (-1)^{(n-1)(y_1 + \cdots + y_n)}$

(3.20c) $\cdot q^{[n(\binom{y_1}{2} + \cdots + \binom{y_n}{2}) + (y_1 + 2y_2 + \cdots + n y_n)]} \cdot x^{(y_1 + \cdots + y_n)}$

(3.20d) $= [(-xq)_\infty (-x^{-1})_\infty (q)_\infty] \cdot \left[(q)_\infty^{n-1} \left(\prod_{1 \leq i < r \leq n} (z_i/z_r)_\infty (q z_r/z_i)_\infty \right) \right].$

Theorem 3.18 was first derived in [75] from Theorem 3.1 in the same way that Theorem 2.7 follows from Theorem 2.1. The substitution of

(3.21a) $a_i \to q^{-2N}$,   if $1 \leq i \leq n$,
(3.21b) $t \to -xq^{(1+nN)}$,

into both sides of (3.2) leads to a multiple series generalization of (2.10), which yields (3.20) in the limit as $N \to \infty$.

Now, (3.20) also contains the Macdonald identities for $A_l^{(1)}$. We use the same type of argument that led from (3.6) to (3.16). It is immediate from Theorem 2.7 that (3.20d) can be written as

(3.22) $\left[(q)_\infty^{(n-1)} \left(\prod_{1 \leq i < r \leq n} (z_i/z_r)_\infty (q z_r/z_i)_\infty \right) \right] \cdot \sum_{M=-\infty}^{+\infty} x^M q^{\binom{1+M}{2}}.$

Equating coefficients of x^M in the Laurent series (3.20a–c) and (3.22), considering the case $M = 0$, and utilizing the product formula for a Vandermonde determinant and an interchange of summation gives the identity in (3.16). But (3.16) is equivalent to the Macdonald identities for $A_l^{(1)}$.

We close this section by describing our proof of Theorem 3.1. To this end, we recall the definitions of the multiple q-series $[W]_m^{(n)}$ and $[F]^{(n)}$ from [73].

DEFINITION 3.23 (basic hypergeometric series well poised in SU(n)). We define

(3.24a)

$$[W]_m^{(n)} \left(\begin{array}{c|c|c|c} (A_{rs}) & (a_{rs}) & (b_{rs}) & (x_i) \\ 1 \le r < s \le n & \begin{array}{c} 1 \le r \le n \\ 1 \le s \le k \end{array} & \begin{array}{c} 1 \le r \le n \\ 1 \le s \le j \end{array} & 1 \le i \le n \end{array} \right)$$

$$\equiv (1-q)^{(j-k-1)m}(q)_m \sum_{\substack{y_1+\cdots+y_n=m \\ y_i \ge 0}} \left(\prod_{1 \le i < r \le n} (1 - A_{ir} \cdot q^{y_i - y_r})/(1 - A_{ir}) \right)$$

(3.24b)

$$\cdot \left(\prod_{i=1}^{k} \prod_{r=1}^{n} (a_{ri})_{y_r} \right) \cdot \left(\prod_{i=1}^{j} \prod_{r=1}^{n} (b_{ri})_{y_r} \right)^{-1} \cdot \left(\prod_{i=1}^{n} x_i^{y_i} \right)$$

to be well poised in SU(n) if m is a nonnegative integer, $|q| < 1$, $(A)_r$ is given by (2.3a), $(b_{ri})_{y_r} \ne 0$, $A_{ir} \ne 1$, $n \ge 2$, and

(3.25a) $j \ge n$,

(3.25b) $A_{ir}/A_{is} = A_{sr}$,   for $s < r$,

(3.25c) $a_{ir}/a_{sr} = A_{is}$,   for $i < s$,

(3.25d) $b_{ir}/b_{sr} = A_{is}$,   for $i < s$,

(3.25e) $b_{ii} = q$, $1 \le i \le n$.

We will call the a's numerator parameters and the b's denominator parameters. We denote the series in (3.24) by $[W]_m^{(n)}((A)|(a)|(b)|(x))$, or just $[W]_m^{(n)}$.

The related series $[F]^{(n)}((A))|(a)|(b)|(x))$ are determined by

DEFINITION 3.26 (basic hypergeometric series in U(n)).

(3.27a)

$$[F]^{(n)} \left(\begin{array}{c|c|c|c} (A_{rs}) & (a_{rs}) & (b_{rs}) & (x_i) \\ 1 \le r < s \le n & \begin{array}{c} 1 \le r \le n \\ 1 \le s \le k \end{array} & \begin{array}{c} 1 \le r \le n \\ 1 \le s \le j \end{array} & 1 \le i \le n \end{array} \right)$$

(3.27b)

$$\equiv \sum_{m=0}^{\infty} \frac{(1-q)^m}{(q)_m}$$

$$\cdot [W]_m^{(n)} \left(\begin{array}{c|c|c|c} (A_{rs}) & (a_{rs}) & (b_{rs}) & (x_i) \\ 1 \le r < s \le n & \begin{array}{c} 1 \le r \le n \\ 1 \le s \le k \end{array} & \begin{array}{c} 1 \le r \le n \\ 1 \le s \le j \end{array} & 1 \le i \le n \end{array} \right)$$

$$\equiv \sum_{y_1,\ldots,y_n \ge 0} (1-q)^{(j-k)(y_1+\cdots+y_n)} \cdot \left(\prod_{1 \le i < r \le n} (1 - A_{ir} \cdot q^{y_i - y_r})/(1 - A_{ir}) \right)$$

(3.27c)

$$\cdot \left(\prod_{i=1}^{k} \prod_{r=1}^{n} (a_{ri})_{y_r} \right) \cdot \left(\prod_{i=1}^{j} \prod_{r=1}^{n} (b_{ri})_{y_r} \right)^{-1} \cdot \left(\prod_{i=1}^{n} x_i^{y_i} \right),$$

where the arrays (A), (a), and (b) satisfy the conditions in (3.25). We continue to call the a's numerator parameters and the b's denominator parameters. The series in (3.27) is balanced provided that $k \geq n$ and

$$(3.28) \quad \left(\prod_{s=n+1}^{j} (b_{is}) \right) = [(a_{11}a_{22} \cdots a_{nn})q] \cdot \left(\prod_{s=n+1}^{k} (a_{is}) \right), \quad \text{for } 1 \leq i \leq n.$$

REMARK 3.29. The conditions in (3.25) reduce the number of free parameters in (3.27). Just set

$$(3.30a) \qquad\qquad A_{rs} = (z_r/z_s), \quad \text{for } 1 \leq r < s \leq n,$$

and then note that the relations in (3.25) transform the arrays (a_{rs}) and (b_{rs}) into

$$(3.30b) \qquad\qquad b_{rs} = q(z_r/z_s), \quad \text{if } 1 \leq r,s \leq n,$$
$$(3.30c) \quad b_{r,n+s} = (b_{n,n+s})(z_r/z_n), \quad \text{if } 1 \leq r \leq n \text{ and } 1 \leq s \leq j - n,$$
$$(3.30d) \qquad\qquad a_{rs} = (a_{ss})(z_r/z_s), \quad \text{if } 1 \leq r,s \leq n,$$
$$(3.30e) \quad a_{r,n+s} = (a_{n,n+s})(z_r/z_n), \quad \text{if } 1 \leq r \leq n \text{ and } 1 \leq s \leq k - n.$$

We have the free parameters $\{z_1, \ldots, z_n\}$, $\{b_{n,n+s} | 1 \leq s \leq j - n\}$, $\{a_{ss} | 1 \leq s \leq n\}$, and $\{a_{n,n+s} | 1 \leq s \leq k - n\}$ subject to the condition that each term in (3.27) is always defined.

Theorem 3.1 is an immediate consequence of (3.27b–c), Remark 3.29, the

$$(3.31) \qquad\qquad a = a_{11}a_{22} \cdots a_{nn} \equiv a_1 a_2 \cdots a_n$$

case of Theorem 2.1, and the summation theorem given by

THEOREM 3.32. *Let* $[W]_m^{(n)}((A)|(a)|(b)|(x))$ *be defined by Definition 3.23. We then have*

$$(3.33a) \quad [W]_m^{(n)} \left(\begin{array}{c|c|c|c} (A_{rs}) & (a_{rs}) & (b_{rs}) & q^{(i-1)} \\ 1 \leq r < s \leq n & 1 \leq r,s \leq n & 1 \leq r,s \leq n & 1 \leq i \leq n \end{array} \right)$$

$$(3.33b) \qquad = \frac{(a_{11}a_{22} \cdots a_{nn})_m}{(1 - q)^m},$$

where $(A)_r$ *is defined by* (2.3a).

We derived Theorem 3.32 in [75] by induction on m, the general q-difference equation for $[W]_m^{(n)}$ given by Theorem 1.31 of [73], and the fundamental

LEMMA 3.34. *Let* $\{x_1, \ldots, x_n\}$ *and* $\{y_1, \ldots, y_n\}$ *be indeterminants with the* y_i *distinct. We then have*

$$(3.35) \qquad 1 - x_1 x_2 \cdots x_n = \sum_{p=1}^{n} (1 - x_p) \cdot \prod_{\substack{i=1 \\ i \neq p}}^{n} \left(\frac{y_p - x_i y_i}{y_p - y_i} \right).$$

Lemma 3.34 was first proven in §2 of [73]. This proof relied upon elementary properties of symmetric functions and an important summation theorem (see Theorem 1.20 of [73]) due to Louck and Biedenharn [65, 66], which appears frequently in dealing with the explicit matrix elements that arise in the unitary groups. Louck and Biedenharn [65, 66] gave a rather complicated analytic proof of their result. In [47] we utilized the determinantal definition of Schur functions [68] and the Laplace expansion formula for the determinant of an $n \times n$ matrix to give a simple direct proof of an elegant identity, involving Schur functions, which contains the result of Louck and Biedenharn as a special case.

In §§7 and 8 of [78] we presented additional elementary proofs of Lemma 3.34, which were based on a partial fraction expansion, the Lagrange interpolation formula, or an induction argument using symmetry and recursion, respectively.

Several recent applications of Lemma 3.34 are of independent interest. Lemma 3.34 is responsible for the q-difference equations in [73], which support the analytical results in [74–80]. Lemma 3.34 is also central to the work in [48]. Furthermore, as described in this section (and explained in detail in [75]) the Macdonald identities for $A_l^{(1)}$ are a consequence of Lemma 3.34, Theorem 2.1, and the product formula for a Vandermonde determinant (classical denominator formula for $\mathrm{sl}(l+1,\mathbf{C})$).

4. A $\mathrm{U}(n)$ generalization of the $_6\Psi_6$ summation theorem. R. A. Gustafson's higher-dimensional generalization of Theorem 2.24 is given by

THEOREM 4.1 ($\mathrm{U}(n)$ generalization of the $_6\Psi_6$ summation theorem). *Let* $0 < |q| < 1$, $n \geq 2$, *and*

$$(4.2) \qquad |q^{(1-n)}(b_1 \cdots b_n)/(a_1 \cdots a_n)| < 1.$$

We then have

$$(4.3a) \qquad \sum_{\substack{y_1+\cdots+y_n=0 \\ y_1 \in \mathbf{Z}}} \left(\prod_{1 \leq i < r \leq n} (1 - (z_i/z_r)q^{y_i-y_r})/(1 - (z_i/z_r)) \right)$$

$$(4.3b) \qquad \cdot \left(\prod_{1 \leq i,r \leq n} ((a_i(z_r/z_i))_{y_r} \cdot (b_i(z_r/z_i))_{y_r}^{-1}) \right)$$

$$(4.3c) \qquad \cdot q^{(y_2+2y_3+\cdots+(n-1)y_n)}$$

$$(4.3d) \qquad = \left\{ \frac{(q/(a_1a_2\cdots a_n))_\infty (q^{(1-n)}(b_1\cdots b_n))_\infty}{(q^{(1-n)}(b_1\cdots b_n)/(a_1\cdots a_n))_\infty (q)_\infty} \right\}$$

$$(4.3e) \qquad \cdot \left\{ \prod_{1 \leq i,r \leq n} \left[\frac{((b_r/a_i)(z_i/z_r))_\infty (q(z_r/z_i))_\infty}{((q/a_i)(z_i/z_r))_\infty (b_i(z_r/z_i))_\infty} \right] \right\},$$

where

(4.4a) $z_i \neq z_r$,   *if* $1 \leq i < r \leq n$;

(4.4b) $(a_i z_r / z_i) \neq q^p$,   *if* $1 \leq i, r \leq n$, $p \in \mathbf{Z}$, *and* $p \geq 1$;

(4.4c) $(b_i z_r / z_i) \neq q^p$,   *if* $1 \leq i, r \leq n$, $p \in \mathbf{Z}$, *and* $p \leq 0$.

It is not difficult to see that the $n = 2$ case of Theorem 4.1 is equivalent to Theorem 2.24.

In [46] Gustafson proved Theorem 4.1 by an induction argument involving certain multiple contour integrals. The starting point for the induction was the U(n) nonterminating $_6\Phi_5$ summation theorem from [77] that appears as Theorem 4.13 below. This induction was on the number of summation indices that assume negative as well as positive integer values. The most difficult part of Gustafson's proof was finding the correct integrand that allowed this type of induction to work.

By an argument similar to the one in [76] that led from Theorem 3.4 to Theorem 3.10 and 3.13, Gustafson [46] utilized Theorems 2.12 and 4.1 to derive

THEOREM 4.5 (distinct b_i case of U(n) generalization of Ramanujan's $_1\Psi_1$ summation). *Let* $n \geq 1$,

(4.6a) $0 < |q| < 1$

and

(4.6b) $|q^{(1-n)}(b_1 \cdots b_n)/(a_1 \cdots a_n)| < |t| < 1$.

Then

$$(4.7a) \quad \sum_{y_1=-\infty}^{+\infty} \cdots \sum_{y_n=-\infty}^{+\infty} \left(\prod_{1 \leq i < r \leq n} (1 - (z_i/z_r) \cdot q^{y_i - y_r})/(1 - (z_i/z_r)) \right)$$

$$(4.7b) \quad \cdot \left(\prod_{1 \leq i, r \leq n} ((a_i(z_r/z_i))_{y_r} (b_i(z_r/z_i))_{y_r}^{-1}) \right)$$

$$(4.7c) \quad \cdot q^{(y_2 + 2y_3 + \cdots + (n-1)y_n)} \cdot t^{(y_1 + \cdots + y_n)}$$

$$(4.7d) \quad = \left\{ \frac{(a_1 a_2 \cdots a_n t)_\infty (q(a_1 a_2 \cdots a_n t)^{-1})_\infty}{(t)_\infty (q^{(1-n)}(b_1 \cdots b_n)/(a_1 a_2 \cdots a_n t))_\infty} \right\}$$

$$(4.7e) \quad \cdot \left\{ \prod_{1 \leq i, r \leq n} \left[\frac{((b_r/a_i)(z_i/z_r))_\infty (q(z_r/z_i))_\infty}{((q/a_i)(z_i/z_r))_\infty (b_i(z_r/z_i))_\infty} \right] \right\},$$

where the conditions in (4.4) *hold.*

Clearly, the $n = 1$ case of Theorem 4.5 is the classical Ramanujan $_1\Psi_1$ summation in Theorem 2.12. It is also immediate that (4.7) is identical to (3.6) when the b_i's are all equal to b. Furthermore, this $b_i = b$ case of Theorem 4.1 is exactly the $M = 0$ case of Theorem 3.10, and is also equivalent to Theorem 3.13. In fact, the same argument that leads from Theorem 3.4 to Theorem 3.10 also gives Theorem 4.1 directly from Theorem 4.5. Consequently, Theorem 4.5 is a common generalization of Ramanujan's $_1\Psi_1$ summation, the $_6\Psi_6$ summation theorem, and the Macdonald identities for $A_l^{(1)}$.

In particular, the $n = 2$ case of Theorem 4.5 implies the $n = 2$ case of Theorem 4.1, which is equivalent to the classical $_6\Psi_6$ summation theorem in (2.25). That is, the classical $_6\Psi_6$ summation is much more natural when viewed as a 2-dimensional result—a simple consequence of the 2-dimensional $_1\Psi_1$ summation theorem. Note that the infinite products in (2.25b) are much more symmetrical when written as in the $n = 2$ case of (4.3d–e).

A very useful special limiting case of (2.14) is obtained by replacing t by (t/a), simplifying, and then letting $a \to \infty$. In Chapter 4 of [5] Andrews applied this identity, together with constant term arguments, to the original pair of Rogers-Ramanujan-Schur identities to derive Rogers's additional identities of the same type. Here, for future reference, we write down the higher-dimensional generalization of Andrews's identity. We have

COROLLARY 4.8. Let $0 < |q| < 1$, $n \geq 1$, and

(4.9) $$|b_1 \cdots b_n| < |q^{(n-1)}t|.$$

If in (4.7) we make the substitution

(4.10) $$t \to (t/(a_1 \cdots a_n)),$$

and then let each $a_i \to \infty$, we obtain the identity

(4.11a) $$\sum_{y_1=-\infty}^{+\infty} \cdots \sum_{y_n=-\infty}^{+\infty} \left(\prod_{1 \leq i < r \leq n} (1 - (z_i/z_r)q^{y_i-y_r})/(1 - (z_i/z_r)) \right)$$

(4.11b) $$\cdot \left(\prod_{1 \leq i,r \leq n} (b_i(z_r/z_i))_{y_r}^{-1} \right) \cdot \left(\prod_{i=1}^{n} z_i^{(ny_i - (y_1 + \cdots + y_n))} \right)$$

(4.11c) $$\cdot q^{n[\binom{y_1}{2} + \cdots + \binom{y_n}{2}]} \cdot q^{(y_2 + 2y_3 + \cdots + (n-1)y_n)}$$

(4.11d) $$\cdot (-1)^{n(y_1 + \cdots + y_n)} \cdot t^{(y_1 + \cdots + y_n)}$$

(4.11e) $$= \left\{ \frac{(t)_\infty (q/t)_\infty}{(q^{1-n} \cdot t^{-1} \cdot (b_1 \cdots b_n))_\infty} \right\}$$

(4.11f) $$\cdot \left\{ \prod_{1 \leq i,r \leq n} \frac{(q(z_r/z_i))_\infty}{(b_i(z_r/z_i))_\infty} \right\}.$$

REMARK 4.12. Setting each $b_i = 0$ in (4.11) gives a version of (3.20). If we then set $t = -qx$, we get exactly (3.20).

Since it was the starting point for deriving Theorem 4.1, we now state the multidimensional $_6\Phi_5$ summation theorem from [77].

THEOREM 4.13 (U(n) generalization of the nonterminating $_6\Phi_5$ summation theorem). *Let*

(4.14a) $$0 < |q| < 1$$

and

(4.14b) $$|q/(a_1 a_2 \cdots a_n c)| < 1.$$

Then

(4.15a) $$\left\{ \left(\frac{(q(a_1 \cdots a_n)^{-1})_\infty (q(a_n c)^{-1})_\infty}{(q(a_n)^{-1})_\infty (q(a_1 \cdots a_n c)^{-1})_\infty} \right) \right.$$

(4.15b) $$\left. \cdot \left(\prod_{i=1}^{n-1} \frac{((qz_i)/(z_n))_\infty ((qz_i)/(cz_n a_i))_\infty}{((qz_i)/(z_n a_i))_\infty ((qz_i)/(cz_n))_\infty} \right) \right\}$$

(4.16a) $$= \sum_{y_1,\ldots,y_{n-1} \geq 0} \left\{ \left[\left(\prod_{i=1}^{n-1} (z_i/z_n)_{(y_1+\cdots+y_{n-1})} \right) \cdot \left(\prod_{i=1}^{n-1} (q)_{y_i} \right)^{-1} \right] \right.$$

(4.16b) $$\cdot \left[\left(\prod_{i=1}^{n-1} (1 - (z_i/z_n) \cdot q^{[y_i+(y_1+\cdots+y_{n-1})]}) / (1 - z_i/z_n) \right) \right]$$

(4.16c) $$\cdot \left[(c)_{(y_1+\cdots+y_{n-1})} \cdot \left(\prod_{i=1}^{n-1} ((qz_i)/(cz_n))_{y_i} \right)^{-1} \right]$$

(4.16d) $$\cdot \left[\left(\prod_{1 \leq r < s \leq n-1} (1 - (z_r/z_s) \cdot q^{y_r - y_s}) / (1 - (z_r/z_s)) \right) \right]$$

(4.16e) $$\cdot \left[\left(\prod_{1 \leq r < s \leq n-1} ((qz_r)/(z_s))_{y_r} ((qz_s)/(z_r))_{y_s} \right)^{-1} \right]$$

(4.16f) $$\cdot \left[\left(\prod_{1 \leq r < s \leq n-1} (((a_s z_r)/(z_s))_{y_r} \cdot ((a_r z_s)/(z_r))_{y_s}) \right) \right]$$

(4.16g) $$\cdot \left[\left(\prod_{i=1}^{n-1} (a_i)_{y_i} \right) \cdot \left(\prod_{i=1}^{n-1} ((qz_i)/(z_n a_i))_{(y_1+\cdots+y_{n-1})} \right)^{-1} \right]$$

(4.16h) $$\cdot\left[\left(\prod_{i=1}^{n-1}((a_n z_i)/(z_n))_{y_i}\right)\cdot\left((q/a_n)_{(y_1+\cdots+y_{n-1})}\right)^{-1}\right]\Bigg\}$$

(4.16i) $$\cdot\{(a_1 a_2\cdots a_n c)^{-(y_1+\cdots+y_{n-1})}\cdot q^{(y_1+2y_2+\cdots+(n-1)y_{n-1})}\}$$

where $\{z_1,\ldots,z_n\}$ are indeterminants such that (3.29) holds, the corresponding arrays (A), (a), and (b) satisfy the conditions in (3.25), and we have

(4.17a) $(z_r/z_s) \neq q^p$, if $1 \leq r < s \leq n-1$,

(4.17b) $(z_i/z_n) \neq 1$, if $1 \leq i \leq n-1$,

(4.17c) $(qz_i)/(cz_n) \neq q^{-p}$, $p \geq 0$ and $1 \leq i \leq n-1$,

(4.17d) $(qz_i)/(z_n a_i) \neq q^{-p}$, $p \geq 0$ and $1 \leq i \leq n-1$,

(4.17e) $(a_1\cdots a_n c) \neq q^p$, $p \geq 1$,

(4.17f) $(a_1\cdots a_n c) \neq 0$,

(4.17g) $a_n \neq q^p$, $p \geq 1$,

with $p \in \mathbf{Z}$ restricted as shown.

There are two natural ways to terminate the multiple sum in (4.16). The first is to set

(4.18) $$c = q^{-N},$$

for N a nonnegative integer. The second is to make the substitution

(4.19) $$a_i = q^{-N}, \quad \text{for } 1 \leq i \leq n,$$

and each N_i a nonnegative integer.

Our original proof of Theorem 4.13 in [77] involved (4.18). In particular, we first showed that Theorem 3.32 is equivalent to the (4.18) case of the identity (4.15)–(4.16). Next, we observed that both (4.15) and (4.16) are analytic functions of $z = (1/c)$ in a disk of positive radius about the origin. But, by the (4.18) case of (4.15)–(4.16), these two analytic functions agree when $z = q^m$, $m = 0, 1, 2, \ldots$. Thus, the general identity (4.15)–(4.16) follows.

From the above outline of the proof of Theorem 4.13 it is immediate that the (4.19) case of (4.15)–(4.16) is a consequence of the (4.18) case. Nonetheless, the (4.19) case of (4.15)–(4.16) is very interesting and we are studying it further elsewhere.

Our proof of Theorem 4.13 helped motivate the multiple very well poised $[H]^{(n)}$ series determined by

DEFINITION 4.20 (basic hypergeometric series very well poised in $U(n)$).
We define

(4.21)

$$
[H]^{(n)}\left[\begin{array}{c|c|c|c}
(A_{rs}) & (a_{rs};k) & (b_{rs};n-1),(b_{in}/c),(b_{rs};j) & x_1 \\
 & & & \vdots \\
(A_{in}/c) & (ca_{ni};k) & (cb_{ni};n-1),(b_{nn}),(cb_{ni};j) & x_n
\end{array}\right]
$$

$$
\equiv \sum_{y_1,\ldots,y_{n-1}\geq 0}\left\{\left[\left(\prod_{i=1}^{n-1}(A_{in})_{(y_1+\cdots+y_{n-1})}\right)\cdot\left(\prod_{i=1}^{n-1}(q)_{y_i}\right)^{-1}\right]\right.
$$

$$
\cdot\left[\left(\prod_{i=1}^{n-1}(1-A_{in}\cdot q^{[y_i+(y_1+\cdots+y_{n-1})]})\right)/(1-A_{in})\right]
$$

(4.22a)
$$
\cdot\left[(c)_{(y_1+\cdots+y_{n-1})}\cdot\left(\prod_{i=1}^{n-1}((qA_{in})/c)_{y_i}\right)^{-1}\right]\Bigg\}
$$

(4.22b)
$$
\cdot\left\{\left(\prod_{1\leq i<l\leq n-1}(1-A_{il}q^{y_i-y_l})/(1-A_{il})\right)\right.
$$

$$
\left.\cdot\left(\prod_{1\leq r<s\leq n-1}(qA_{rs})_{y_r}(q/A_{rs})_{y_s}\right)^{-1}\right\}
$$

(4.22c)
$$
\cdot\left[\frac{\left(\prod_{i=1}^{k}\prod_{l=1}^{n-1}(a_{li})_{y_l}\right)\cdot\left(\prod_{i=n+1}^{j}(q/b_{ni})_{(y_1+\cdots+y_{n-1})}\right)}{\left(\prod_{i=1}^{k}(q/a_{ni})_{(y_1+\cdots+y_{n-1})}\right)\cdot\left(\prod_{i=n+1}^{j}\prod_{l=1}^{n-1}(b_{li})_{y_l}\right)}\right]
$$

(4.22d)
$$
\cdot\left\{(-1)^{(j+k)(y_1+\cdots+y_{n-1})}\cdot q^{\left[n(y_1+\cdots+y_{n-1})-(j-k)\binom{1+(y_1+\cdots+y_{n-1})}{2}\right]}\right\}
$$

(4.22e)
$$
\cdot\left[\frac{(b_{n,n+1}b_{n,n+2}\cdots b_{n,j})}{(c)(a_{n1}a_{n2}\cdots a_{nk})(A_{1n}A_{2n}\cdots A_{n-1,n})}\right]^{(y_1+\cdots+y_{n-1})}
$$

(4.22f)
$$
\cdot\left(\prod_{i=1}^{n-1}(x_i/x_n)^{y_i}\right)
$$

to be very well-poised in $U(n)$ if $|q|<1$, $(A)_l$ is given by (2.3a), $(b_{li})_{y_l}\neq 0$,
$A_{il}\neq 1$, $n\geq 2$, and the relations in (3.25) hold. We will call c and the a's
numerator parameters and the b's denominator parameters. We sometimes
denote the multiple series in (4.21) by the notation

(4.23) $[H]^{(n)}[(A)|(a),c|(b)|(x)]$.

The very well poised $_m\Phi_n$ series corresponds to a well-poised $_{m-2}\Phi_{n-2}$ series in the same way that

$$(4.24) \qquad [H]^{(n)}[(A)|(a), c|(b)|(x)]$$

is related to the $n \to (n-1)$ case of the multiple series

$$(4.25) \qquad [F]^{(n)}((A)|(a)|(b)|(x)).$$

When comparing (4.24) to (4.25) it is useful to view (4.25) as simply a multivariable generalization of the classical (general) basic hypergeometric series in (2.16).

REMARK 4.26. The multiple series (4.16) is obtained from (4.22) by first setting

$$(4.27) \qquad k = j = n,$$
$$(4.28) \qquad x_i = q^{i-1}, \qquad 1 \le i \le n,$$

and then replacing A_{rs} by

$$(4.29) \qquad A_{rs} = (z_r/z_s), \quad \text{for } 1 \le r < s \le n.$$

That is, (4.22a–c) becomes (4.16a–h), and (4.22d–f) is written in (4.16i).

The $[H]^{(n)}$ series appear again in §6.

REMARK 4.30. §6 deals with the $k = j$ case of Definition 4.20 in which

$$(4.31) \qquad a_{ii} = q^{-N_i}, \quad \text{for } 1 \le i \le n-1 \text{ and } N_i \text{ nonnegative integers.}$$

In order to see why these multiple series in (4.21–4.22) are a natural generalization of terminating classical (1-variable) very well poised basic hypergeometric series we combine the relations in (3.30) with the relabelling of parameters given by

$$(4.32a) \qquad z_i = z_i, \quad \text{for } 1 \le i \le n-1,$$
$$(4.32b) \qquad (z_n)^{-1} = a/z_{n-1},$$
$$(4.32c) \qquad c = b_{(j-n+1)},$$
$$(4.32d) \qquad b_{n,n+(j-n)-i+1} = (q/b_i), \quad \text{for } 1 \le i \le (j-n),$$
$$(4.32e) \qquad a_{n,n+(j-n)-i+1} = c_i(z_n/z_{n-1}) = c_i/a, \quad \text{for } 1 \le i \le (j-n+1).$$

Applying both (3.30) and (4.32) to the parameters in (4.22) leads to the relations

(4.33a) $\qquad\qquad A_{in} = a(z_i/z_{n-1}), \quad \text{for } 1 \le i \le n-1,$

(4.33b) $\qquad\qquad\qquad c = b_{(j-n+1)},$

(4.33c) $\quad (qA_{in}/c) = (qa/b_{(j-n+1)})(z_i/z_{n-1}), \quad \text{for } 1 \le i \le n-1,$

(4.33d) $\qquad\qquad A_{rs} = (z_r/z_s), \quad \text{for } 1 \le r < s \le n-1,$

(4.33e) $\qquad\qquad a_{rs} = q^{-N_s} \cdot (z_r/z_s), \quad \text{for } 1 \le r,s \le n-1,$

(4.33f)
$$a_{l,n+s} = c_{(j-n+1-s)}(z_l/z_{n-1}), \quad \text{for } 1 \le l \le n-1 \text{ and } 0 \le s \le j-n,$$

(4.33g) $\qquad (q/a_{ni}) = aq^{(1+N_i)} \cdot (z_i/z_{n-1}), \quad \text{for } 1 \le i \le n-1,$

(4.33h) $\qquad (q/a_{n,n+s}) = (aq/c_{(j-n+1-s)}), \quad \text{for } 0 \le s \le j-n,$

(4.33i) $\qquad (q/b_{n,n+s}) = b_{(j-n+1-s)}, \quad \text{for } 1 \le s \le j-n,$

(4.33j) $\quad b_{l,n+s} = (aq/b_{(j-n+1-s)})(z_l/z_{n-1}), \quad \text{for } 1 \le l \le n-1$
$$\text{and } 1 \le s \le j-n,$$

and

(4.34a)
$$\frac{(b_{n,n+1}b_{n,n+2}\cdots b_{n,j})}{(c)(a_{n1}a_{n2}\cdots a_{nj})(A_{1n}A_{2n}\cdots A_{n-1,n})}$$

(4.34b)
$$= q^{(N_1+\cdots+N_{n-1})} \cdot \frac{a^{(j-n+1)}q^{(j-n)}}{(b_1 c_1 b_2 c_2 \cdots b_{(j-n+1)}c_{(j-n+1)})}.$$

It is clear from (4.33–4.34) that the $k = j$ and (4.31) case of (4.21–4.22) depends only on the additional parameters

(4.35) $\qquad\qquad \{b_1, \ldots, b_{(j-n+1)}\}, \quad \{c_1, \ldots, c_{(j-n+1)}\}, \quad \{a\},$

and

(4.36) $\qquad\qquad\qquad \{z_1, \ldots, z_{n-1}\},$

and, of course, q. Note that the classical ($n = 2$) case only depends upon (4.35) and q. That is, the higher-dimensional case uses the "classical" parameters in (4.35) along with the "new" parameters in (4.36). It is now not hard to see from (4.33–4.34), and setting

(4.37) $\qquad\qquad\qquad (j - n + 1) \to k,$

that Definition 4.20 generalizes terminating classical very well poised basic hypergeometric series.

5. A q-analog of the Guass summation theorem for $[F]^{(n)}$ series. In this section we present our q-analogs from [78, 79] of W. J. Holman III's [54] U(n) generalization of the summation theorems of Gauss and Pfaff-Saalschutz.

We first state

THEOREM 5.1 (q-U(n) Gauss summation theorem for $[F]^{(n)}$).

(5.2a) $[F]^{(n)} \left(\begin{array}{c|cc} (z_r/z_s) & (a_s)(z_r/z_s), & b(z_i/z_n) \\ 1 \le r < s \le n & 1 \le r,s \le n & 1 \le i \le n \end{array} \right.$

$$\left. \begin{array}{cc|c} q(z_r/z_s), & c(z_i/z_n) & x_i \\ 1 \le r,s \le n & 1 \le i \le n & 1 \le i \le n \end{array} \right)$$

(5.2b) $$= \sum_{y_1,\ldots,y_n \ge 0} \left(\prod_{1 \le r < s \le n} (1 - (z_r/z_s) \cdot q^{y_r - y_s})/(1 - (z_r/z_s)) \right)$$

(5.2c) $$\cdot \left[\left(\prod_{1 \le r,s \le n} \frac{(a_s z_r/z_s)_{y_r}}{(q z_r/z_s)_{y_r}} \right) \cdot \left(\prod_{i=1}^{n} \frac{(b z_i/z_n)_{y_i}}{(c z_i/z_n)_{y_i}} \right) \right]$$

(5.2d) $$\cdot \left(\prod_{i=1}^{n} [q^{(i-1)} \cdot c/(a_1 a_2 \cdots a_n b)]^{y_i} \right)$$

(5.2e) $$= \frac{(c/b)_\infty}{(c/(a_1 a_2 \cdots a_n b))_\infty} \cdot \left(\prod_{i=1}^{n} \frac{((c z_i)/(z_n a_i))_\infty}{((c z_i)/z_n)_\infty} \right),$$

where $n \ge 1$, $0 \le |q| < 1$, and

(5.3a) $$x_i = q^{(i-1)} \cdot c/(a_1 a_2 \cdots a_n b), \quad \text{if } 1 \le i \le n;$$

(5.3b) $$|c| < |a_1 a_2 \cdots a_n b|,$$

(5.3c) $$(c z_i)/z_n \ne q^{-p}, \quad \text{if } p \ge 0 \text{ and } 1 \le i \le n,$$

(5.3d) $$(z_r/z_s) \ne q^p, \quad \text{if } 1 \le r < s \le n,$$

with p any integer.

We prove Theorem 5.1 in [78] by iterating a multivariable q-analog of the Gauss reduction formula (for a classical $_2F_1$) that is a direct consequence of the general contiguous relation for $[F]^{(n)}$ series in

THEOREM 5.4. *Let $[F]^{(n)}((A)|(a)|(b)|(x))$ be defined by Definition 3.26, and $_j\Phi_{j-1}$ by Definition 2.15. Also assume that $k = j \ge n$, $1 \le v \le n$, and $\{b_{v,n+1}, \ldots, b_{v,j}\}$ are distinct. We then have*

(5.5a) $$[1 - (b_{v,n+1} \cdots b_{v,j})(a_{11} \cdots a_{vn})^{-1}(a_{v,n+1} \cdots a_{v,j})^{-1}]$$

(5.5b) $$\cdot [F]^{(n)} \left(\begin{array}{c|c|c|c} (A_{rs}) & (a_{rs}) & (b_{rs}) & (x_i) \\ 1 \le r < s \le n & \begin{array}{c} 1 \le r \le n \\ 1 \le s \le j \end{array} & \begin{array}{c} 1 \le r \le n \\ 1 \le s \le j \end{array} & 1 \le i \le n \end{array} \right)$$

$$= \sum_{p=1}^{n} \left\{ \left[(q^{(1-p)} \cdot x_p) - (b_{v,n+1} \cdots b_{v,j})(a_{11} \cdots a_{nn})^{-1}(a_{v,n+1} \cdots a_{v,j})^{-1} \right] \right.$$

(5.5c)
$$\cdot \left[(-1)^{(p-1)} \cdot \left(\prod_{i=1}^{p-1} \frac{(A_{ip})}{(1 - A_{ip})} \right) \left(\prod_{i=p+1}^{n} (1 - A_{pi}) \right)^{-1} \right.$$

$$\left. \cdot \left(\prod_{i=1}^{j} (1 - a_{pi}) \right) \cdot \left(\prod_{i=n+1}^{j} (1 - b_{pi}) \right)^{-1} \right] \right\}$$

(5.5d)
$$\cdot {}_j\Phi_{j-1} \left[\left. \begin{matrix} qa_{p1}, \ldots, qa_{pj} \\ b_{p1}, \ldots, b_{p,p-1}, b_{p,p+1}, \ldots, b_{p,n}, qb_{p,n+1}, \ldots, qb_{p,j} \end{matrix} \right|_{x_p \cdot q^{(1-p)}} \right]$$

(5.5e)
$$\cdot [F]^{(n-1)} \left(\begin{matrix} (\overline{A}_{rs}) \\ 1 \le r < s \le n - 1 \end{matrix} \left| \begin{matrix} (\overline{a}_{rs}) \\ 1 \le r \le n - 1 \\ 1 \le s \le j \end{matrix} \right. \right.$$

$$\left. \left| \begin{matrix} (\overline{b}_{rs}) \\ 1 \le r \le n - 1 \\ 1 \le s \le j \end{matrix} \right| \begin{matrix} (\overline{x}_i) \\ 1 \le i \le n - 1 \end{matrix} \right)$$

(5.5f)
$$+ \sum_{p=n+1}^{j} \left\{ \left(\prod_{i=1}^{n} (1 - (b_{ip}/a_{ii}))/(1 - b_{ip}) \right) \right.$$

$$\left. \cdot \left(\prod_{i=n+1}^{j} (1 - (b_{vp}/a_{vi})) \right) \left(\prod_{\substack{i=n+1 \\ i \ne p}}^{j} (1 - (b_{vp}/b_{vi})) \right)^{-1} \right\}$$

(5.5g)
$$\cdot [F]^{(n)} \left(\begin{matrix} (A_{rs}) \\ 1 \le r < s \le n \end{matrix} \left| \begin{matrix} (a_{rs}) \\ 1 \le r \le n \\ 1 \le s \le j \end{matrix} \right. \right.$$

$$\left. \left| \begin{matrix} (b_{rs}), & (qb_{i,p}), & (b_{rs}) \\ 1 \le r \le n & 1 \le i \le n \ 1 \le r \le n \\ 1 \le s \le p - 1 & & p < s \le j \end{matrix} \right| \begin{matrix} (qx_i) \\ 1 \le i \le n \end{matrix} \right),$$

where $(\overline{A}_{rs}), (\overline{a}_{rs}), (\overline{b}_{rs}), (\overline{x}_i)$ *are given by*

$$(5.6a)$$
$$(5.6b)$$
$$(5.6c)$$
$$\overline{A}_{rs} = \begin{cases} A_{rs}, & \text{if } 1 \le r < s < p, \\ A_{r,s+1}, & \text{if } 1 \le r < p \le s < n, \\ A_{r+1,s+1}, & \text{if } p \le r < s < n; \end{cases}$$

$$(5.7a)$$
$$(5.7b)$$
$$\overline{a}_{rs} = \begin{cases} a_{rs}, & \text{if } 1 \le r < p \text{ and } 1 \le s \le j, \\ a_{r+1,s}, & \text{if } p \le r < n \text{ and } 1 \le s \le j; \end{cases}$$

$$(5.8a)$$
$$(5.8b)$$
$$(5.8c)$$
$$(5.8b)$$
$$(5.6c)$$
$$(5.8d)$$
$$\overline{b}_{rs} = \begin{cases} b_{rs}, & \text{if } 1 \le r, s < p, \\ b_{r,s+1}, & \text{if } 1 \le r < p \text{ and } p \le s < j, \\ b_{r+1,s}, & \text{if } p \le r < n \text{ and } 1 \le s < p, \\ b_{r+1,s+1}, & \text{if } p \le r < n \text{ and } p \le s < j, \\ b_{rp}, & \text{if } 1 \le r < p \text{ and } s = j, \\ b_{r+1,p}, & \text{if } p \le r < n \text{ and } s = j; \end{cases}$$

and the $\overline{x}_1, \ldots, \overline{x}_{n-1}$ *are given by*

$$(5.9a)$$
$$(5.9b)$$
$$\overline{x}_i = \begin{cases} qx_i, & \text{if } 1 \le i < p, \\ x_{i+1}, & \text{if } p \le i < n. \end{cases}$$

Furthermore, $\overline{A}_{rs}, \overline{a}_{rs},$ *and* $\overline{b}_{rs}$ *satisfy the well poised conditions in (3.25).*

Motivated by §§33 and 48 of [84] and §I.7 of [86] we derive Theorem 5.4 in [78] from certain q-differential operators in [74] and two special cases of Lemma 3.34.

The special value of x_i in (5.3a) was discovered by setting each coefficient in the $j = (n+1)$ case of (5.5c) equal to 0. It turns out that (5.5d) and (5.5e) are absolutely convergent for this choice of the x_i. Thus, each of the n terms in (5.5c–e) becomes 0, and (5.5a–b) and (5.5f-g) yield a two-term q-difference equation, which is our multivariable q-analog of the Gauss reduction formula. Iterating this relation an infinite number of times gives Theorem 5.1.

Two equivalent forms of the Vandermonde summation theorem for $[F]^{(n)}$ are consequences of Theorem 5.1. We write these results in terms of the original arrays (A), (a), and (b) in (3.25).

We first have

THEOREM 5.10 (terminating Gauss summation theorem for $[F]^{(n)}$ series).

$$(5.11a) \quad [F]^{(n)} \left(\begin{array}{c} (A_{rs}) \\ 1 \le r < s \le n \end{array} \middle| \begin{array}{c} (a_{rs}) \\ 1 \le r \le n \\ 1 \le s \le n+1 \end{array} \middle| \begin{array}{c} (b_{rs}) \\ 1 \le r \le n \\ 1 \le s \le n+1 \end{array} \middle| \begin{array}{c} x_i \\ 1 \le i \le n \end{array} \right)$$

$$(5.11b) \quad = (b_{n,n+1}/a_{n,n+1})_{(N_1 + \cdots + N_n)} \cdot \left(\prod_{i=1}^{n} (b_{i,n+1})_{N_i} \right)^{-1},$$

where $n \geq 1$, q is arbitrary, and

(5.12a) $a_{ii} = q^{-N_i}$,   *for* $1 \leq i \leq n$ *with* N_i *nonnegative integers,*

(5.12b) $x_i = q^{(i-1)} \cdot q^{(N_1 + \cdots + N_n)} \cdot (b_{n,n+1}/a_{n,n+1})$,   *for* $1 \leq i \leq n$,

(5.12c) $b_{li} \neq q^{-p}$,   *if* $p \geq -1$, $1 \leq l \neq i \leq n$,

(5.12d) $b_{l,n+1} \neq q^{-p}$,   *if* $p \geq 0$, $1 \leq l \leq n$,

(5.12e) $A_{il} \neq q^p$,   *if* $1 \leq i < l \leq n$,

(5.12f)   *with* p *any integer.*

It is clear from (2.4) that if a_{ii} is given by (5.12a) then (5.2e) becomes (5.11b). Furthermore, since (5.11a) is a finite sum (terminates) and (5.11b) are finite products, (5.11) is equivalent to a polynomial identity in q. Thus, (5.11) is valid whenever each term is defined and the restrictions $0 < |q| < 1$ and (5.3b) can be eliminated.

The $n = 1$ case of Theorem 5.10 is the same identity as equation (3.3.2.6) on p. 97 of [**86**], and is known as one of the two q-analogs of Vandermonde's summation theorem [**35, 84, 86**].

Motivated by the classical ($n = 1$) case [**32**] we next invert the base q, or reverse the order of summation, in Theorem 5.10 to obtain

THEOREM 5.13 (second q-analog of the Vandermonde summation theorem for $[F]^{(n)}$ series). *Let* $n \geq 1$, *and* q *be arbitrary. Also assume that* a_{ii} *is given by* (5.12a), *and that* (5.12c–f) *holds. We then have*

(5.14a)
$$[F]^{(n)} \left(\begin{array}{c|c|c|c} (A_{rs}) & (a_{rs}) & (b_{rs}) & q^i \\ 1 \leq r < s \leq n & \begin{array}{c} 1 \leq r \leq n \\ 1 \leq s \leq n+1 \end{array} & \begin{array}{c} 1 \leq r \leq n \\ 1 \leq s \leq n+1 \end{array} & 1 \leq i \leq n \end{array} \right)$$

(5.14b)
$$= \left\{ (b_{n,n+1}/a_{n,n+1})(N_1 + \cdots + N_n) \cdot \left(\prod_{i=1}^{n} (b_{i,n+1})_{N_i} \right)^{-1} \right\}$$

(5.14c)
$$\cdot \left\{ \left(\prod_{i=1}^{n} (a_{i,n+1})^{N_i} \right) \cdot q^{-\sigma_2(N_1,\ldots,N_n)} \right\},$$

where $\sigma_2(N_1, \ldots, N_n)$ *is the second elementary symmetric function of* $\{N_1, \ldots, N_n\}$.

The $n = 1$ case of Theorem 5.13 is equivalent to equation (3.3.2.7) on p. 97 of [**86**]. Slater views (3.3.2.7) as the "usual" q-analog of the Vandermonde summation and, despite the discussion on p. 88 of [**86**], regards (3.3.2.6) and (3.3.2.7) as distinct identities. However, as pointed out in [**32**], and generalized in [**78, 79**], (3.3.2.6) and (3.3.2.7) are in fact equivalent.

Inverting the base q in terminating $[F]^{(n)}$ series in which

(5.15) $a_{ii} = q^{-N_i}$,   for $1 \leq i \leq n$ and N_i nonnegative integers, transforms $[F]^{(n)}((A)|(a)|(b)|(x))$ into a multiple series of the same type, but with the base q replaced by q^{-1}. This transformation depends upon various special cases of

$$(5.16) \qquad (A;q)_l = (A^{-1};q^{-1})_l \cdot ((-1)^l A^l q^{l(l-1)/2}),$$

noting that

$$(5.17) \qquad (a_{ii})^{-1} = q^{N_i} = (q^{-1})^{-N_i}, \quad \text{for } 1 \leq i \leq n,$$

and observing that

$$(5.18) \qquad \begin{aligned} &(1 - A_{il} \cdot q^{(y_i - y_l)})/(1 - A_{il}) \\ &= q^{(y_i - y_l)} \cdot ((1 - (A_{il})^{-1} \cdot (q^{-1})^{(y_i - y_l)})/(1 - (A_{il})^{-1})), \end{aligned}$$

and

$$(5.19) \qquad \left(\prod_{1 \leq i < l \leq n} q^{(y_i - y_l)} \right) = \prod_{i=1}^{n} q^{(n - 2i + 1) y_i}.$$

In §4 of [78] we deduce Theorem 5.13 from 5.10 by inverting the base q in (5.14a), applying Theorem 5.10, and then simplifying the resulting products. In fact, these two theorems are equivalent under this transformation.

Theorems 5.13 and 5.10 are also equivalent under multivariable reversal of terminating $[F]^{(n)}$ series where

$$(5.20a) \qquad \qquad k \geq n$$

and

$$(5.20b) \qquad a_{ii} = q^{-N_i}, \quad \text{for } 1 \leq i \leq n \text{ and } N_i \text{ nonnegative integers.}$$

Reversal of $[F]^{(n)}$ series, subject to (5.20), consists of making the substitution

$$(5.21) \qquad \qquad y_l \rightarrow (N_l - y_l), \quad \text{for } 1 \leq l \leq n,$$

in $[F]^{(n)}((A)|(a)|(b)|(x))$ and then applying certain special cases of

$$(5.22) \qquad (a;q)_{N-m} = \frac{(a;q)_N \cdot q^{m(m+1)/2}}{(a^{-1}q^{1-N};q)_m \cdot (-a)^m \cdot q^{Nm}},$$

and some algebra, to obtain a new $[F]^{(n)}$ series that is equal to the original. Writing everything out explicitly gives the identity between $[F]^{(n)}$ series contained in Theorem 6.6 of [79]. The special case in which $j = k$ is Corollary 6.29 of [79] and is given by

LEMMA 5.23 ($k = j \geq n$ case of reversal of $[F]^{(n)}$ series). *Let* (5.20) *hold, $j = k$, and suppose that x_i are independent of $\{y_1, \ldots, y_n\}$ for $1 \leq i \leq n$. We*

then have
(5.24a)
$$[F]^{(n)} \left(\begin{array}{c|c|c|c} (A_{rs}) & (a_{rs}) & (b_{rs}) & (x_i) \\ 1 \le r < s \le n & \begin{array}{c} 1 \le r \le n \\ 1 \le s \le j \end{array} & \begin{array}{c} 1 \le r \le n \\ 1 \le s \le j \end{array} & 1 \le i \le n \end{array} \right)$$

(5.24b)
$$= \left\{ \left(\prod_{i=1}^{n} ((-1)^{N_i} \cdot q^{-iN_i} \cdot q^{-\binom{N_i}{2}} \cdot x_i^{N_i}) \right) \cdot q^{-\sigma_2(N_1,\ldots,N_n)} \right\}$$

(5.24c)
$$\cdot \left\{ \left(\prod_{i=n+1}^{j} \prod_{l=1}^{n} ((a_{li})_{N_l} \cdot (b_{li})_{N_l}^{-1}) \right) \right\}$$

(5.24d)
$$\cdot [F]^{(n)} \left(\begin{array}{c|c|c|c} (\overline{A}_{rs}) & (\overline{a}_{rs}) & (\overline{b}_{rs}) & (\overline{x}_i) \\ 1 \le r < s \le n & \begin{array}{c} 1 \le r \le n \\ 1 \le s \le j \end{array} & \begin{array}{c} 1 \le r \le n \\ 1 \le s \le j \end{array} & 1 \le i \le n \end{array} \right),$$

where $(\overline{A}_{rs})$, $(\overline{a}_{rs})$, $(\overline{b}_{rs})$, $(\overline{x}_i)$ *are given by*

(5.25a) $\qquad \overline{A}_{rs} = (A_{rs})^{-1} \cdot q^{(N_s - N_r)}, \quad \text{if } 1 \le r < s \le n,$

(5.25b) $\qquad \overline{a}_{rs} = ((b_{rs})^{-1} \cdot q^{1-N_r}), \quad \text{if } 1 \le r \le n \text{ and } 1 \le s \le j,$

(5.25c) $\qquad \overline{b}_{rs} = ((a_{rs})^{-1} \cdot q^{1-N_r}), \quad \text{if } 1 \le r \le n \text{ and } 1 \le s \le j,$

(5.25d) $\qquad \overline{x}_i = \left\{ \left(\prod_{l=1}^{j} (b_{il}/a_{il}) \right) \cdot q^{(2i-n-1)} \cdot (x_i)^{-1} \right\}, \quad \text{if } 1 \le i \le n.$

The identity (5.24) preserves the conditions in both (3.25) and (3.28).

As an application of Lemma 5.23 we obtain Theorem 5.13 from 5.10 in §6 of [79]. Similarly, Theorem 5.10 follows from Theorem 5.13.

One of our primary applications of Theorems 5.10 and 5.13 has been to derive a q-analog of Holman's [54] $U(n)$ generalization of the Pfaff-Saalschutz or balanced $_3F_2$ summation theorem [35, 84, 86] contained in

THEOREM 5.26 (q-analog of the $U(n)$ balanced $_3F_2$ summation theorem). *Assume* $n \ge 1$ *and that* N_i *is a nonnegative integer for* $1 \le i \le n$. *We then have*

(5.27a)
$$[F]^{(n)} \left(\begin{array}{c|c} (z_r/z_s) & \begin{array}{ccc} (q^{-N_s}(z_r/z_s)), & (a(z_i/z_n)), & (b(z_i/z_n)) \\ 1 \le r, s \le n & 1 \le i \le n & 1 \le i \le n \end{array} \\ 1 \le r < s \le n & \end{array} \right.$$
$$\left. \begin{array}{ccc|c} (q(z_r/z_s)), & (c(z_i/z_n)), & (d(z_i/z_n)) & (q^i) \\ 1 \le r, s \le n & 1 \le i \le n & 1 \le i \le n & 1 \le i \le n \end{array} \right)$$

(5.27b) $\quad = \{(c/a)_{(N_1 + \cdots + N_n)} \cdot (c/b)_{(N_1 + \cdots + N_n)}\}$

(5.27c)
$$\cdot \left\{ \prod_{i=1}^{n} ((c(z_i/z_n))_{N_i}^{-1} \cdot ((c/(ab))(z_n/z_i) q^{[(N_1 + \cdots + N_n) - N_i]})_{N_i}^{-1}) \right\},$$

where

(5.28) $cd = (ab)q^{1-(N_1+\cdots+Nh_n)}.$

Theorem 5.26 provides a multivariable generalization of Jackson's [57] classical q-analog of the balanced $_3F_2$ summation theorem.

We prove Theorem 5.26 in [79] by an induction argument based upon a new general q-difference equation for balanced $[F]^{(n)}$ series and both Theorems 5.10 and 5.13. Our proof of Theorem 5.26 amounts to inductively factoring (5.27a) into lower order $[F]^{(n)}$ series, each of which can be summed by either Theorem 5.10 or 5.13.

R. A. Gustafson's [45] elegant generalization of F. J. W. Whipple's [89, 90] classical transformation of a very well-poised $_7F_6(1)$ into a balanced $_4F_3(1)$ was partly responsible for our discovery of the new q-difference equations for $[F]^{(n)}$. Gustafson found a transformation between the ordinary $(q = 1)$ multiple hypergeometric series $W_m^{(n)}$ and $F^{(n-1)}$. This transformation allowed us to deduce difference equations for special $F^{(n-1)}$ series from the corresponding general difference equations for $W_m^{(n)}$ that appeared in [71, 73]. These crucial examples led us to general difference equations for $F^{(n)}$, which in turn made it much easier to discover our q-difference equations for $[F]^{(n)}$ in Theorems 1.46 and 1.49 of [79]. We made use of elementary series manipulations and standard partial fraction expansions to give a direct proof of these q-difference equations for $[F]^{(n)}$.

In §6 of [79] we showed that Theorem 5.26 is invariant under both inversion of the base q in (5.15–19) and multivariable reversal of terminating $[F]^{(n)}$ series in (5.20–25). These symmetry results show that Theorems 5.10, 5.13, and 5.26 are particularly natural generalizations of the classical one-variable theory.

6. A q-analog of a Whipple's transformation for $[F]^{(n)}$ series.

The q-difference equations in Theorems 1.46 and 1.49 of [79] not only lead to Theorem 5.26. They also enable us to prove

THEOREM 6.1 (special case of first form of q-U(n) Whipple). *Let* $[W]_m^{(n)}((A)|(a)|(b)|(x))$ *and* $[F]^{(n)}((A)|(a)|(b)|(x))$ *be defined by Definitions 3.23 and 3.26, respectively. Also assume that* $n \geq 2$, m *is a nonnegative integer, and*

(6.2) $a_{ii} = q^{-N_i},$ *for* $1 \leq i \leq n-1$ *and* N_i *nonnegative integers.*

We then have

(6.3a)

$$[W]_m^{(n)}\left(\begin{array}{c|c|c|c}(A_{rs}) & (a_{rs}) & (b_{rs}) & q^{i-1} \\ 1 \le r < s \le n & 1 \le r \le n & 1 \le r \le n & 1 \le i \le n \\ & 1 \le s \le n+1 & 1 \le s \le n+1 & \end{array}\right)$$

(6.3b)
$$= \left\{\left[\prod_{i=1}^{n-1}((b_{in})_{N_i}((a_{nn}a_{i,n+1})^{-1} \cdot q^{[1-m+(N_1+\cdots+N_{n-1})-N_i]})_{N_i})\right]\right.$$

(6.3c)
$$\cdot [(q^{1-m}/a_{nn})^{-1}_{(N_1+\cdots+N_{n-1})} \cdot (q^{1-m}/a_{n,n+1})^{-1}_{(N_1+\cdots+N_{n-1})}]$$

(6.3d)
$$\cdot [(a_{nn})_m (a_{n,n+1})_m (b_{n,n+1})_m^{-1} (a_{11} \cdots a_{n-1,n-1})^m \cdot (1-q)^{-m}]\right\}$$

$$\cdot [F]^{(n-1)}\left(\begin{array}{c|c}(A_{rs}) & (a_{rs}) \qquad , (b_{i,n+1}q^m) \\ 1 \le r < s \le n-1 & 1 \le r \le n-1 \; 1 \le i \le n-1 \\ & 1 \le s \le n+1 \end{array}\right.$$

(6.3e)
$$\left.\begin{array}{c|c}(b_{rs}) \qquad , ((a_{11}a_{22} \cdots a_{nn})(a_{i,n+1}q^m)) & q^i \\ 1 \le r \le n-1 \qquad 1 \le i \le n-1 & 1 \le i \le n-1 \\ 1 \le s \le n+1 \end{array}\right).$$

In [80] we utilized q-difference equations and induction on m to prove Theorem 6.1. The general q-difference equations for $[F]^{(n)}$ in Theorem 1.46 of [79] imply that both sides of (6.3) satisfy the same special case of the general q-difference equation for $[W]_m^{(n)}$ in [73]. Our proof of Theorem 6.1 is completed by uniqueness and the fact that the $m = 0$ case of (6.3) is equivalent to Theorem 5.26.

R. A. Gustafson's [45] transformation of $W_m^{(n)}$ into $F^{(n-1)}$ can be written as the $q = 1$ case of Theorem 6.1.

It turns out that Theorem 6.1 is a key step in deriving the higher-dimensional generalization of Theorem 2.22 given by

THEOREM 6.4 (general case of second form of q-U(n) Whipple). *Let* $[F]^{(n)}((A)|(a)|(b)|(x))$ *and* $[H]^{(n)}[(A)|(a), c|(b)|(x)]$ *be defined by Definitions 3.26 and 4.20, respectively. Also assume that $n \ge 2$ and that (6.2) holds. We then have*

(6.5a)
$$[H]^{(n)}\left[\begin{array}{c|c|c|c}(A_{rs}) & (a_{rs}; n+1) & (b_{rs}; n-1),(b_{in}/c),(b_{i,n+1}) & q^{i-1} \\ (A_{in}/c) & (ca_{ni}; n+1) & (cb_{ni}; n-1),(b_{nn}),(cb_{n,n+1}) & 1 \le i \le n \end{array}\right]$$

(6.5b)
$$= \left\{\left[\prod_{i=1}^{n-1}((b_{in})_{N_i} \cdot ((a_{nn}a_{i,n+1})^{-1} \cdot q^{[1+(N_1+\cdots+N_{n-1})-N_i]})_{N_i})\right]\right.$$

(6.5c)
$$\left.\cdot \left[(q/a_{nn})^{-1}_{(N_1+\cdots+N_{n-1})} \cdot (q/a_{n,n+1})^{-1}_{(N_1+\cdots+N_{n-1})}\right]\right\}$$

$$\cdot [F]^{(n-1)}\left(\begin{array}{c|c}(A_{rs}) & (a_{rs}) \qquad , (b_{i,n+1}/c) \\ 1 \le r < s \le n-1 & 1 \le r \le n-1 \; 1 \le i \le n-1 \\ & 1 \le s \le n+1 \end{array}\right.$$

$$\begin{vmatrix} (b_{rs}) & ,(b_{in}/c) & ,(b_{i,n+1}) & ,((a_{11}\cdots a_{nn})a_{i,n+1}) \\ 1\le r\le n-1 & 1\le i\le n-1 & 1\le i\le n-1 & 1\le i\le n-1 \\ 1\le s\le n-1 & & & \end{vmatrix}$$

(6.5d)
$$\left. \begin{vmatrix} q^i \\ 1\le i\le n-1 \end{vmatrix} \right).$$

The connection between Theorems 6.1 and 6.4 is provided by

LEMMA 6.6. *Let m be a nonnegative integer and*

(6.7)
$$c = q^{-m}.$$

We then have the identity

(6.8a)
$$[H]^{(n)}\left[\begin{array}{c|c|c|c} (A_{rs}) & (a_{rs};k) & (b_{rs};n-1),(q^m b_{in}),(b_{rs};j) & x_i \\ (q^m A_{in}) & (q^{-m}a_{ni};k) & (q^{-m}\cdot b_{ni};n-1),(b_{nn}),(q^{-m}b_{ni};j) & 1\le i\le n \end{array}\right]$$

$$= \left\{ \left(\prod_{i=1}^{k}(q^{-m}\cdot a_{ni})_m\right)^{-1}\cdot \left(\prod_{i=n+1}^{j}(q^{-m}\cdot b_{ni})_m\right)\cdot \left(\prod_{i=1}^{n-1}(qA_{in})_m\right)\right.$$

(6.8b)
$$\cdot ((-1)^{(n-1)}\cdot x_n)^{-m}\cdot (1-q)^{(k+1-j)m}$$

$$\left. \cdot (A_{1n}A_{2n}\cdots A_{n-1,n})^{-m}\cdot q^{-(n-1)\binom{m}{2}}\right\}$$

(6.8c)
$$\cdot [W]_m^{(n)}\left(\begin{array}{c|c} (A_{rs}) & (a_{rs};k) \\ (q^m A_{in}) & (q^{-m}a_{ni};k) \end{array}\right.$$

$$\left. \begin{array}{c|c} (b_{rs};n-1),(q^m b_{in}),(b_{rs};j) & x_1 \\ & \vdots \\ (q^{-m}\cdot b_{ni};n-1),(b_{nn}),(q^{-m}b_{ni},j) & x_n \end{array}\right),$$

where $[H]^{(n)}$ and $[W]_m^{(n)}$ are defined by Definitions (4.20) and (3.23), respectively.

Lemma 6.6 is proved in [77] by elementary series manipulations.

Now, it is not hard to see that both sides of (6.5) are analytic functions of

(6.9)
$$z = (1/c)$$

in a disk of positive radius about the origin. From Theorem 6.1 and Lemma 6.6 it follows that these two analytic functions agree when

(6.10)
$$z = (1/c) = q^m, \quad \text{for } m = 0, 1, 2, \ldots.$$

Thus, both sides of (6.5) are equal for general c if $0 < |q| < 1$. However, the condition $0 < |q| < 1$ can then be removed by analytic continuation, and we finally obtain Theorem 6.4. Theorem 6.4 is contained in [80].

This proof of Theorem 6.4 from Theorem 6.1 is the same argument that led from Theorem 3.32 to 4.13, from Theorem 3.1 to 3.4, and from Theorem 2.1 to 2.12.

It follows by relabelling parameters as in Remark 4.30 that Theorem 2.22 is the $n = 2$ case of Theorem 6.4.

We conclude this section with some applications from [80] of Theorem 6.4.

The terminating case of Theorem 4.13 in which (4.19) holds is a direct consequence of Theorem 6.4. By setting

$$(6.11) \qquad a_{i,n+1} = b_{i,n+1}, \quad \text{for } 1 \leq i \leq n,$$

in Theorem 6.4, using Theorem 5.26 to sum the resulting $[F]^{(n-1)}$ series in (6.5d), and simplifying, we obtain (4.15–16), subject to (4.19).

Theorem 6.4 is also responsible for a higher-dimensional generalization of Jackson's q-analog of Dougall's [58] $_7F_6$ summation theorem. If we assume that

$$(6.12) \qquad (b_{i,n+1}/c) = (a_{11}a_{22} \cdots a_{nn})a_{i,n+1}, \quad \text{for } 1 \leq i \leq n-1,$$

in (6.5), then it is not hard to see that Theorem 5.26 allows us to sum the series in (6.5d). Recalling the relations

$$(6.13a) \qquad a_{nn} = (a_{in}q/b_{in}), \quad \text{for } 1 \leq i \leq n,$$

$$(6.13b) \qquad (q/a_{nn}) = (b_{n-1,n}/a_{n-1,n}),$$

and

$$(6.13c) \qquad (q/a_{n,n+1}) = (b_{n-1,n}/a_{n-1,n+1}),$$

and simplifying the products resulting from (6.5b–d), we are immediately led to

THEOREM 6.14 (U(n) generalization of Jackson's q-analog of Dougall's theorem). *Let $n \geq 2$ and assume that (6.2) holds. Furthermore, suppose that*

$$(6.15) \qquad (b_{in}b_{i,n+1}) = c(a_{in}a_{i,n+1})(a_{11} \cdots a_{n-1,n-1})q, \quad \textit{for } 1 \leq i \leq n.$$

We then have

(6.16a)
$$[H]^{(n)}\left[\begin{array}{c|c|c|c} (A_{rs}) & (a_{rs};n+1) & (b_{rs};n-1),(b_{in}/c),(b_{i,n+1}) & q^{i-1} \\ (A_{in}/c) & (ca_{ni};n+1) & (cb_{ni};n-1),(b_{nn}),(cb_{n,n+1}) & 1\le i\le n \end{array}\right]$$

(6.16b)
$$= \left\{\prod_{i=1}^{n-1}((b_{in})_{N_i}(b_{in}/c)_{N_i}^{-1})\right\}$$

$$\cdot\left\{\left[\prod_{i=1}^{n-1}((b_{in}a_{ii})\cdot[(a_{in}a_{i,n+1})(a_{11}\cdots a_{n-1,n-1})]^{-1})_{N_i}\right]\right.$$

(6.16c)
$$\left.\cdot\left[\prod_{i=1}^{n-1}((b_{in}a_{ii})\cdot[(ca_{in}a_{i,n+1})(a_{11}\cdots a_{n-1,n-1})]^{-1})_{N_i}^{-1}\right]\right\}$$

$$\cdot\left\{\left[(b_{n-1,n}/(ca_{n-1,n}))_{(N_1+\cdots+N_{n-1})}\cdot(b_{n-1,n}/a_{n-1,n})_{(N_1+\cdots+N_{n-1})}^{-1}\right]\right.$$

(6.16d)
$$\left.\cdot\left[(b_{n-1,n}/(ca_{n-1,n+1}))_{(N_1+\cdots+N_{n-1})}\cdot(b_{n-1,n}/a_{n-1,n+1})_{(N_1+\cdots+N_{n-1})}^{-1}\right]\right\}.$$

The $n=2$ case of (6.16) is Jackson's q-analog of Dougall's theorem.

When $c=q^{-m}$, Lemma 6.6 and a fair amount of algebraic simplification transforms this case of Theorem 6.14 into

THEOREM 6.17 (U(n) generalization of Jackson's q-analog of Dougall's theorem for $[W]_m^{(n)}$ series). *Let $n\ge 2$, m a nonnegative integer, and suppose that (6.2) holds. Also, assume that*

(6.18) $(b_{in}b_{i,n+1})=(a_{in}a_{i,n+1})(a_{11}\cdots a_{n-1,n-1})q$, *for* $1\le i\le n-1$.

We then have

(6.19a)
$$[W]_m^{(n)}\left(\begin{array}{c|c|c|c} (A_{rs}) & (a_{rs}) & (b_{rs}) & q^{i-1} \\ 1\le r<s\le n & \begin{array}{c}1\le r\le n\\1\le s\le n+1\end{array} & \begin{array}{c}1\le r\le n\\1\le s\le n+1\end{array} & 1\le i\le n \end{array}\right)$$

(6.19b)
$$= \{(1-q)^{-m}\cdot[(b_{n-1,n+1}/a_{n-1,n+1})_m\cdot(b_{n-1,n+1}/a_{n-1,n})_m\cdot(b_{n,n+1})_m^{-1}]\}$$

(6.19c)
$$\cdot\left\{\prod_{i=1}^{n-1}((b_{i,n+1}/a_{ii})_m\cdot(b_{i,n+1})_m^{-1})\right\}.$$

It is also possible to obtain Theorem 6.17 directly from Theorem 6.1 and Theorem 5.26.

R. A. Gustafson's [45] generalization of Dougall's summation theorem for $W_m^{(n)}$ can be simplified to the $q=1$ case of Theorem 6.17.

Letting $q\to 1$ in Theorem 6.14 involves the ordinary series $H^{(n)}$ very well poised in U(n) which were introduced in [77].

Further applications of Theorem 6.4 appear in [80]. These include a U(n) generalization of the Rogers-Selberg identity [6], which is a major ingredient in several proofs of the Rogers-Ramanujan-Schur identities [3, 4, 5].

REFERENCES

1. C. Adiga, B. C. Berndt, S. Bhargava, and G. N. Watson, *Chapter 16 of Ramanujan's second notebook: theta-functions and q-series*, Mem. Amer. Math. Soc. No. 315 (1985).

2. A. K. Agarwal, G. E. Andrews, D. M. Bressoud, *The Bailey lattice* (to appear).

3. G. E. Andrews, *"The Theory of Partitions"*, Vol. 2, *"Encyclopedia of Mathematics and Its Applications"* (G.-C. Rota, Ed.), Addison-Wesley, Reading, Mass., 1976.

4. ____, *Partitions: yesterday and today*, New Zealand Math. Soc., Wellington, 1979.

5. ____, *q-Series: Their development and application in analysis, number theory, combinatorics, physics and computer algebra*, CBMS Regional Conf. Ser. in Math., no. 66, Conf. Board Math. Sci., Washington, D.C., 1986.

6. ____, *An analytic proof of the Rogers-Ramanujan-Gordon identities*, Amer. J. Math. **88** (1966), 844–846.

7. ____, *On Ramanujan's summation of $_1\Psi_1(a, b, z)$*, Proc. Amer. Math. Soc. **22** (1969), 552–553.

8. ____, *On a transformation of bilateral series with applications*, Proc. Amer. Math. Soc. **25** (1970), 554–558.

9. ____, *Applications of basic hypergeometric functions*, SIAM Rev. **16** (1974), 441–484.

10. ____, *Problems and prospects for basic hypergeometric functions*, Theory and Applications of Special Functions, R. Askey, editor, Academic Press, 1975, pp. 191–224.

11. ____, *Partitions, q-series and the Lusztig-Macdonald-Wall conjectures*, Invent. Math. **41** (1977), 91–102.

12. ____, *An introduction to Ramanujan's "lost" notebook*, Amer. Math. Monthly **86** (1979), 89–108.

13. ____, *Connection coefficient problems and partitions*, D. Ray-Chaudhuri, editor, Proc. Sympos. Pure Math., vol. 34, Amer. Math. Soc., Providence, R.I., 1979, pp. 1–24.

14. ____, *Ramanujan's "lost" notebook. I: partial theta functions*, Adv. in Math. **41** (1981), 137–172.

15. ____, *Ramanujan's "lost" notebook. II: θ-function expansions*, Adv. in Math. **41** (1981), 173–185.

16. ____, *Ramanujan's "lost" notebook. III: The Rogers-Ramanujan continued fraction*, Adv. in Math. **41** (1981), 186–208.

17. ____, *The Mordell integrals and Ramanujan's "lost" notebook*, Analytic Number Theory, M. I. Knopp, editor, Lectures Notes in Math., vol. 899, Springer-Verlag, 1981, pp. 10–48.

18. ____, *Generalized Frobenius partitions*, Mem. Amer. Math. Soc. No. 301 (1984).

19. ____, *Ramanujan and SCRATCHPAD*, Proc. of the 1984 MACSYMA Users' Conference, General Electric, Schenectady, New York, 1984, pp. 384–408.

20. ____, *Multiple series Rogers-Ramanujan type identities*, Pacific J. Math. **114** (1984), 267–283.

21. ____, *Hecke modular forms and the Kac-Peterson identities*, Trans. Amer. Math. Soc. **283** (1984), 451–458.

22. ____, *Ramanujan's "lost" notebook. IV: stacks and alternating parity in partitions*, Adv. in Math. **53** (1984), 55–74.

23. ____, *Combinatorics and Ramanujan's "lost" notebook*, London Math. Soc. Lecture Note Ser., no. 103, Cambridge Univ. Press, London, 1985, pp. 1–23.

24. ____, *The fifth and seventh order mock theta functions*, Trans. Amer. Math. Soc. **293** (1986), 113–134.

25. ____, *Ramanujan's "lost" notebook. V: Euler's partition identity*, Adv. in Math. **61** (1986), 156–164.

26. ____, *The Rogers-Ramanujan's identities without Jacobi's triple product*, Rocky Mountain J. Math. **17** (1987), 659–672.

27. G. E. Andrews and R. Askey, *Enumeration of partitions: the role of Eulerian series and q-orthogonal polynomials*, Higher Combinatorics, M. Aigner, editor, Reidel, Dordrecht, 1977, pp. 3–26.

28. G. E. Andrews and R. A. Askey, *A simple proof of Ramanujan's summation of the $_1\Psi_1$*, Aequationes Math. **18** (1978), 333–337.

29. G. E. Andrews, R. J. Baxter, and P. J. Forrester, *Eight-vertex SOS model and generalized Rogers-Ramanujan-type identities*, J. Statist. Phys. **35** (1984), 193–266.

30. P. Appéll and J. Kampé de Fériet, "*Fonctions hypergéometriques et hypersphériques; Polynomes d'Hermites*", Gauthier-Villars, Paris, 1926.

31. R. Askey, *Ramanujan's extension of the gamma and beta functions*, Amer. Math. Monthly **87** (1980), 346–359.

32. ____, *Book review in "Zentralblatt für Mathematik (Mathematics Abstracts)"*, **514** (1984), 161–163. Review #33001.

33. R. Askey and M. Ismail, *The Rogers q-ultraspherical polynomials*, Approximation Theory. III, E. W. Cheney, editor, Academic Press, 1980, pp. 175–182.

34. R. Askey and J. A. Wilson, *Some basic hypergeometric orthogonal polynomials that generalize Jacobi polynomials*, Mem. Amer. Math. Soc. No. 319 (1985).

35. W. N. Bailey, *Generalized hypergeometric series*, Cambridge Math. Tract, no. 32, Cambridge Univ. Press, Cambridge, 1935.

36. R. J. Baxter, *Hard hexagons: exact solution*, J. Phys. A **13** (1980), L61–L70.

37. ____, *Rogers-Ramanujan identities in the hard hexagon model*, J. Statist. Phys. **26** (1981), 427–452.

38. ____, *Exactly solved models in statistical mechanics*, Academic Press, 1982.

39. L. C. Biedenharn and J. D. Louck, *Angular momentum in quantum physics: Theory and applications*, Encyclopedia of Mathematics and Its Applications, Vol. 8, G.-C. Rota, editor, Addison-Wesley, Reading, Mass., 1981.

40. ____, *The Racah-Wigner algebra in quantum theory*, Encyclopedia of Mathematics and Its Applications, Vol. 9, G.-C. Rota, editor, Addison-Wesley, Reading, Mass., 1981.

41. D. M. Bressoud, *A matrix inverse*, Proc. Amer. Math. Soc. **88** (1983), 446–448.

42. A. Cauchy, *Oeuvres*, 1st Sér., Vol. 8, Gauthier-Villars, Paris, 1893, p. 45.

43. F. J. Dyson, *Missed opportunities*, Bull. Amer. Math. Soc. **78** (1972), 635–653.

44. I. Gessel and D. Stanton, *Applications of q-Lagrange inversion to basic hypergeometric series*, Trans. Amer. Math. Soc. **277** (1983), 173–201.

45. R. A. Gustafson, *A Whipple's transformation for hypergeometric series in $U(n)$ and multivariable hypergeometric orthogonal polynomials*, SIAM J. Math. Anal. **18** (1987), 495–530.

46. ____, *Multilateral summation theorems for ordinary and basic hypergeometric series in $U(n)$*, SIAM J. Math. Anal. (to appear).

47. R. A. Gustafson and S. C. Milne, *Schur functions, Good's identity, and hypergeometric series well poised in $SU(n)$*, Adv. in Math. **48** (1983), 177–188.

48. ____, *A q-analog of transposition symmetry for invariant G-functions*, J. Math. Anal. Appl. **114** (1986), 210–240.

49. W. Hahn, *Über orthogonal polynome, die q-differenzengleichungen genügen*, Math. Nachr. Berlin **2** (1949), 4–34.

50. ____, *Beitrage zur theorie der Heineschen reihen, die 24 integrale der hypergeometrischen q-differenzengleichung das q-analogen der Laplace transformation*, Math. Nachr. **2** (1949), 263–278.

51. ____, *Beitrage zur Theorie der Heineschen Reihen*, Math. Nachr. **2** (1949), 340–379.

52. ____, *Über die hoheren Heineschen Rechen und eine einheitliche Theorie der Sogennanten speziellen Funktionen*, Math. Nachr. **3** (1950), 257–294.

53. G. H. Hardy and E. M. Wright, *An introduction to the theory of numbers*, 5th ed., Oxford Univ. Press, London/New York, 1979.

54. W. J. Holman III, *Summation theorems for hypergeometric series in $U(n)$*, SIAM J. Math. Anal. **11** (1980), 523–532.

55. W. J. Holman III, L. C. Biedenharn, and J. D. Louck, *On hypergeometric series well-poised in $SU(n)$*, SIAM J. Math. Anal. **7** (1976), 529–541.

56. M. E. H. Ismail, *A simple proof of Ramanujan's $_1\Psi_1$ sum*, Proc. Amer. Math. Soc. **63** (1977), 185–186.

57. F. H. Jackson, *Transformations of q-series*, Messenger of Math. **39** (1910), 145–153.

58. ____, *Summation of q-hypergeometric series*, Messenger of Math. **47** (1917), 101–112.

59. M. Jackson, *On Lerch's transcendent and the basic bilateral hypergeometric series* $_2\Psi_2$, J. London Math. Soc. **25** (1950), 189–196.

60. C. G. J. Jacobi, *Fundamenta nova theoriese functionum ellipticarum* (1829), Regiom-noti, fratrum Bornträger (reprinted in Gesammelte Werke, Vol. 1, Reimer, Berlin, 1881, pp. 49–239).

61. V. G. Kac and D. H. Peterson, *Affine Lie algebras and Hecke modular forms*, Bull. Amer. Math. Soc. (N.S.) **3** (1980), 1057–1061.

62. V. G. Kac, *Infinite dimensional Lie algebras*, Prog. in Math., vol. 44, Birkhäuser, Boston, 1983.

63. J. Lepowsky, *Affine Lie algebras and combinatorial identities*, Lie Algebras and Related Topics, Rutgers Univ. Press, New Brunswick, N.J., 1981, pp. 130–156; Lecture Notes in Math., vol. 933, Springer-Verlag, 1982.

64. J. Lepowsky and S. Milne, *Lie algebraic approaches to classical partition identities*, Adv. in Math. **29** (1978), 15–59.

65. J. D. Louck, *Theory of angular momentum in N-dimensional space*, Los Alamos Scientific Laboratory Report LA-2451, 1960; and the cited contribution of E. D. Cashwell.

66. J. D. Louck and L. C. Biedenharn, *Canonical unit adjoint tensor operators in* $U(n)$, J. Math. Phys. **11** (1970), 2368–2414.

67. I. G. Macdonald, *Affine root systems and Dedekind's η-function*, Invent. Math. **15** (1972), 91–143.

68. ____, *Symmetric functions and Hall polynomials*, Oxford Univ. Press, London/New York, 1979.

69. ____, *Some conjectures for root systems*, SIAM J. Math. Anal. **13** (1982), 988–1007.

70. A. M. Mathai and R. K. Saxena, *Generalized hypergeometric functions with applications in statistics and physical sciences*, Lecture Notes in Math., vol. 348, Springer-Verlag, 1973.

71. S. C. Milne, *Hypergeometric series well-poised in* $SU(n)$ *and a generalization of Biedenharn's G-functions*, Adv. in Math. **36** (1980), 169–211.

72. ____, *A new symmetry related to* $SU(n)$ *for classical basic hypergeometric series*, Adv. in Math. **57** (1985), 71–90.

73. ____, *A q-analog of hypergeometric series well-poised in* $SU(n)$ *and invariant G-functions*, Adv. in Math. **58** (1985), 1–60.

74. ____, *A q-analog of the* $_5F_4(1)$ *summation theorem for hypergeometric series well-poised in* $SU(n)$, Adv. in Math. **57** (1985), 14–33.

75. ____, *An elementary proof of the Macdonald identities for* $A_l^{(1)}$, Adv. in Math. **57** (1985), 34–70.

76. ____, *A* $U(n)$ *generalization of Ramanujan's* $_1\Psi_1$ *summation*, J. Math. Anal. Appl. **118** (1986), 263–277.

77. ____, *Basic hypergeometric series very well-poised in* $U(n)$, J. Math. Anal. Appl. **122** (1987) 223–256.

78. ____, *A q-analog of the Gauss summation theorem for hypergeometric series in* $U(n)$, Adv. in Math **72** (1988), 59–131.

79. ____, *A q-analog of the balanced* $_3F_2$ *summation theorem for hypergeometric series in* $U(n)$, Adv. in Math., accepted.

80. ____, *A q-analog of a Whipple's transformation for hypergeometric series in* $U(n)$, Adv. in Math., accepted.

81. ____, *New families of balanced* $_3\Phi_2$ *summation theorems for* $U(n)$ *basic hypergeometric series* (in preparation).

82. ____, *Iterated multiple series expansions of basic hypergeometric series very-well poised in* $U(n)$ (in preparation).

83. ____, *An extension of little q-Jacobi polynomials for basic hypergeometric series in* $U(n)$ (in preparation).

84. E. D. Rainville, *Special functions*, Macmillan, New York, 1960.

85. L. J. Slater, *A new proof of Roger's transformation of infinite series*, Proc. London Math. Soc. (2) **53** (1951), 460–475.

86. ____, *Generalized hypergeometric functions*, Cambridge Univ. Press, London and New York, 1966.

87. D. Stanton, *Recent results for the q-Lagrange inversion formula* (to appear).

88. G. N. Watson, *A new proof of the Rogers-Ramanujan identities*, J. London Math. Soc. **4** (1929), 4–9.

89. F. J. W. Whipple, *On well-poised series, generalized hypergeometric series having parameters in pairs, each pair with the same sum*, Proc. London Math. Soc. (2) **24** (1924), 247–263.

90. ____, *Well-poised series and other generalized hypergeometric series*, Proc. London Math. Soc. (2) **25** (1926), 525–544.

91. L. Winquist, *An elementary proof of* $p(11m + 6) \equiv 0 \pmod{11}$, J. Combin. Theory **6** (1969), 56–59.

92. D. Zeilberger and D. M. Bressoud, *A proof of Andrews' q-Dyson conjecture*, Discrete Math. **54** (1985), 201–224.

UNIVERSITY OF KENTUCKY

List of Participants

List of Participants

Organizing committee: Enrico Arbarello, David Chudnovsky, Gregory Chudnovsky, Takahiro Kawai, Henry McKean, and co-chairmen Leon Ehrenpreis and Robert C. Gunning.

ABLOWITZ, Mark J.	Clarkson University, Potsdam, NY
ACCOLA, Robert D. M.	Brown University, Providence, RI
ADDINGTON, Susan L.	Harvard University, Cambridge, MA
ADLER, Mark	Brandeis University, Waltham, MA
ANDREWS, George E.	Pennsylvania State University, University Park, PA
AOKI, Kenichiro	Princeton University, Princeton, NJ
AOMOTO, Kazuhiko	Nagoya University, Nagoya, Japan
ARBARELLO, Enrico	University of Rome, Rome, Italy
ASKEY, Richard A.	University of Wisconsin, Madison, WI
AU-YANG, Helen	State University of New York, Stony Brook, NY
BEAUVILLE, Arnaud	University of Paris, le Pecq, France
BERNSTEIN, Joseph N.	Harvard University, Cambridge, MA
BLUHER, Antonia	Princeton University, Fine Hall, Princeton, NJ
BLUHER, Grigory	Princeton University, Fine Hall, Princeton, NJ
BOCHERER, Siegfreid	Math. Instut der Univ, Freiberg, West Germany
BORWEIN, Jonathan	Dalhousie University, Halifax, Nova Scotia, Canada
BORWEIN, Peter	Dalhousie University, Halifax, Nova Scotia, Canada
BREEN, Lawrence	Centre de Mathematiques, de l'Ecole Polytechnique, Palaiseau, France
BUONOCORE, Michael H.	University of California, Sacramento, CA
CANDILERA, Maurizio	Universita degli Studi, Padova, Italy
CHAI, Ching-Li	Princeton University, Princeton, NJ
CHARI, Vyjayanthi	Rutgers University, New Brunswick, NJ
CHUDNOVSKY, David	Columbia University, New York, NY
CHUDNOVSKY, Gregory	Columbia University, New York, NY
COGDEL, James W.	Oklahoma State University, Stillwater, OK
COHN, Harvey	City College, City University of New York, New York, NY
COLLINO, Alberto	Verzuolo (CN), Italy
CRISTANTE, Valentino	Universita degli Studi, Padova, Italy
CUKIERMAN, Fernando	University of California, Los Angeles, CA
D'HOKER, Eric	Princeton University, Princeton, NJ
DATE, Etsuro	Max Planck Institute, Bonn, West Germany
DEBARRE, Olivier	Université de Paris-SUD, Orsay, France
DUCROT, Francois	Universite D'Angers, Angers Cedex, France
DUKE, William	Rutgers University, New Brunswick, NJ
DUNCAN, Tyrone E.	The University of Kansas, Lawrence, KS
EARLE, Clifford J.	Cornell University, Ithaca, NY
EHRENPREIS, Leon	Temple University, Philadelphia, PA
ERCOLANI, Nicholas	Courant Institute, New York, NY

363

FARKAS, Hershel	Hebrew University, Jerusalem, Israel
FAY, John D.	Haverford College, Haverford, PA
FEINGOLD, Alex J.	State University of New York, Binghamton, NY
FLASCHKA, Hermann	University of Arizona, Tucson, AZ
FRENKEL, Igor	Yale University, New Haven, CT
FRIED, Michael D.	University of Florida, Gainsville, FL
GAETA, Federico	Universitas Complutense, Madrid, Spain
GARVAN, Francis	University of Minnesota, Minneapolis, MN
GEL'FAND, Israil	Moscow University, Moscow, USSR
GILBERT, Gerald	CalTech, Pasadena, CA
GOMEZ-MONT, Carlos	University of Mexico/UNAM, Mexico, D. F., Mexico
GRANT, David	University of Michigan, Ann Arbor, MI
GRUNBAUM, F. Alberto	University of California, Berkeley, CA
GUNNING, Robert C.	Princeton University, Princeton, NJ
GURARIE, David	Case Western Reserve University, Case Institute of Technology, Cleveland, OH
GUSTAFSON, Robert	Texas A & M, College Station, TX
HAINE, Luc	University of Arizona, Tuscon, AZ
HAMMOND, William F.	State University of New York, Albany, NY
HELMINCK, G. F.	University of Twente, AE Enschede, the Netherlands
HENDON, M. D.	Athens, GA
HOYT, William L.	Rutgers University, New Brunswick, NJ
HSIEH, June	Academia Sinica, Nankang, Taipei, Taiwan, Republic of China
IGUSA, Jun-Ichi	The Johns Hopkins University, Baltimore, MD
JABLOW, Eric	State University of New York, Stony Brook, NY
JIJTCHENKO, Alexei B.	Academy of Sciences USSR, Moscow, USSR
JIMBO, Michio	Kyoto University, Kyoto, Japan
JORGENSON, Jay. A	Stanford University, Stanford, CA
JUN, Sung-Tae	University of Rochester, Rochester, NY
KAC, Victor	Massachusetts Institute of Technology, Cambridge, MA
KANEV, Vassil	Bulgarian Academy of Sciences, Sofia, Bulgaria
KAREL, Martin L.	Wilmington, DE
KASHIWARA, Masaki	Kyoto University, Kyoto, Japan
KASTOR, David	University of Chicago, Chicago, IL
KATZ, Talbot Michael	Hunter College, New York, NY
KAWAI, Takahiro	Kyoto University, Kyoto, Japan
KAZHDAN, David	Harvard University, Cambridge, MA
KENT, Adrian	University of Chicago, Chicago, IL
KNOPP, Marvin	Temple University, Philadelphia, PA
KRUSEMEYER, Mark	Carleton College, Northfield, MN
KUZNETSOV, Nikolai V.	ul P. Komarouva, Khabarousk, USSR
KUNIBA, Atsuo	University of Tokyo, Tokyo, Japan
LAGARIAS, Jeffery	A T & T Bell Laboratories, Murray Hill, NJ
LANDESMAN, Peter	New York, NY
LATHAM, Geoff	University of California, Berkeley, CA
LEE, Ronnie	Yale University, New Haven, CT
LEHNER, Joseph	Jamesburg, NJ
LEWIS, Adrian	Dalhousie University, Halifax, Nova Scotia, Canada
LIM, Chjan Chin	University of Michigan, Ann Arbor, MI
LITTLE, John B.	College of the Holy Cross, Worcester, MA
LU, Shirong	Massachusetts Institute of Technology, Cambridge, MA
MARTENS, Henrick H.	Norges Tekniske Hogskole, Trondheim, Norway
MARTINEC, Emil	University of Chicago, Chicago, IL
MAYER, Alan L.	Brandeis University, Waltham, MA
McCOY, Barry M.	State University of New York, Stony Brook, NY
McDANIEL, Andrew	George Mason University, Fairfax, VA
McKEAN, Henry	New York University, New York, NY

MENEZES, Maria Lucia	Pontifica Universidade Catolica, Rio de Janeiro, Brazil
MILNE, Stephen C.	University of Kentucky, Lexington, KY
MIWA, Tetsuji	Kyoto University, Kyoto, Japan
MOLL, Victor H.	Tulane University, New Orleans, LA
MORENO, Carlos J.	City University of New York, New York, NY
MULASE, Motohico	University of California, Los Angeles, CA
MUNOZ-PORRAS, J. M.	Universidad de Salamanca, Salamanca, Spain
MURTY, V. Kumar	University of Toronto, Toronto, Ontario, Canada
NAMIKAWA, Yukihiko	Max Planck Institute, Bonn, West Germany
NARANJO del val, Juan C.	Universidad de Barcelona, Barcelona, Spain
NARASIMHAN, Mudumbai S.	Tata Institute for Fundamental Research, Bombay, India
NEMCHENOK, Jacob	Dartmouth College, Hanover, NH
NEVEU, Andre	USTL, France
NORMAN, Peter	University of Massachusetts, Amherst, MA
NOURI-MOGHADM, M. R.	King's College, University of London, London, England
OKADO, Masato	Kyoto University, Kyoto, Japan
PALMER, John	University of Arizona, Tuscon, AZ
PARKER, Phillip E.	Wichita State University, Wichita, KA
PERIWAL, Vipul	Princeton University, Princeton, NJ
PERK, Jacques H. H.	State University of New York, Stony Brook, NY
PERLINE, Ronald K.	Drexel University, Philadelphia, PA
PETRI, Monica	Massachusetts Institute of Technology, Cambridge, MA
PHONG, Duong	Columbia University, New York, NY
PIATETSKI-SHAPIRO, I. I.	Tel-Aviv University, Ramat Aviv, Israel
PILA, Jonathan	Stanford University, Stanford, CA
PIOVAN, Luis	Brandeis University, Waltham, MA
PIROLA, Gian Pietro	University di Pavia, Pavia, Italy
PIZER, Arnold K.	University of Rochester, Rochester, NY
POOR, Chris	Princeton University, Princeton, NJ
PRESSLEY, Andrew N.	Rutgers University, New Brunswick, NJ
PREVIATO, Emma	Imperial College, London, England
RAMANAN, Sundararaman	Tata Inst. of Fund. Research, Bombay, India
REACH, Michael	Berkeley, CA
RHODES, John A.	Harvard University, Cambridge, MA
RIBET, Kenneth	IMES, Bures sur Yvette, France
RIES, John F. X.	State University of New York, Binghamton, NY
RIGGS, Harold	University of Chicago, Chicago, IL
ROCHA-CARIDI, Alvany	Baruch College, City University of New York, New York, NY
ROHM, Ryan M.	California Institute of Technology, Pasadena, CA
ROSS, Shepley L., II	Bates College, Lewiston, ME
SACHS, Robert L.	Pennsylvania State University, University Park, PA
SALVATI-MANNI, R.	University de Roma, Rome, Italy
SATO, Mikio	Kyoto University, Kyoto, Japan
SCHILLING, Randolph J.	Louisiana State University, Baton Rouge, LA
SCHLICHENMAIER, Martin	Universität Mannheim, Mannheim, West Germany
SEILER, Wolfgang K.	Universität Mannheim, Mannheim, West Germany
SEKIGUCHI, Tsutomu	Chuo University, Tokyo, Japan
SHEMANSKE, Thomas R.	Dartmouth College, Hanover, NH
SHIOTA, Takahiro	Institute for Advanced Study, Princeton, NJ
SINCLAIR, Paul	Witchita State University, Witchita, KS
SINNOU, David	University of Paris VI, Paris, France
SIPE, Patricia L.	Smith College, Northampton, MA
SMITH, Roy	University of Georgia, Athens, GA
STANTON, Dennis	University of Minnesota, Minneapolis, MN
SULLIVAN, John M.	Princeton University, Princeton, NJ
TAI, Y.-S.	Haverford College, Haverford, PA
TAKASAKI, Kanehisa	Kyoto University, Kyoto, Japan